Arak M. Mathai and Hans J. Haubold
Linear Algebra
De Gruyter Textbook

Also of Interest

Probability and Statistics. A Course for Physicists and Engineers
Arak M. Mathai, Hans J. Haubold, 2017
ISBN 978-3-11-056253-8, e-ISBN (PDF) 978-3-11-056254-5,
e-ISBN (EPUB) 978-3-11-056260-6

Advanced Calculus. Differential Calculus and Stokes' Theorem
Pietro-Luciano Buono, 2016
ISBN 978-3-11-043821-5, e-ISBN (PDF) 978-3-11-043822-2,
e-ISBN (EPUB) 978-3-11-042911-4

Probability Theory and Statistical Applications.
A Profound Treatise for Self-Study
Peter Zörnig, 2016
ISBN 978-3-11-036319-7, e-ISBN (PDF) 978-3-11-040271-1,
e-ISBN (EPUB) 978-3-11-040283-4

Complex Analysis. A Functional Analytic Approach
Friedrich Haslinger, 2017
ISBN 978-3-11-041723-4, e-ISBN (PDF) 978-3-11-041724-1,
e-ISBN (EPUB) 978-3-11-042615-1

Functional Analysis. A Terse Introduction
Gerardo Chacón, Humberto Rafeiro, Juan Camilo Vallejo, 2016
ISBN 978-3-11-044191-8, e-ISBN (PDF) 978-3-11-044192-5,
e-ISBN (EPUB) 978-3-11-043364-7

Arak M. Mathai and Hans J. Haubold

Linear Algebra

A Course for Physicists and Engineers

DE GRUYTER

Mathematics Subject Classification 2010
15-01, 15A03, 15A04, 15A05, 15A09, 15A15, 15A16, 15A18, 15A21, 15A63

Authors

Prof. Dr Arak M. Mathai
McGill University
Department of Mathematics and Statistics
805 Sherbrooke St. West
Montreal, QC H3A 2K6
Canada
mathai@math.mcgill.ca

Prof. Dr Hans J. Haubold
United Nations Office for Outer Space Affairs
Vienna International Centre
P.O. Box 500
1400 Vienna
Austria
hans.haubold@gmail.com

ISBN 978-3-11-056235-4
e-ISBN (PDF) 978-3-11-056250-7
e-ISBN (EPUB) 978-3-11-056259-0

Library of Congress Cataloging-in-Publication Data
A CIP catalog record for this book has been applied for at the Library of Congress.

Bibliographic information published by the Deutsche Nationalbibliothek
The Deutsche Nationalbibliothek lists this publication in the Deutsche Nationalbibliografie;
detailed bibliographic data are available on the Internet at http://dnb.dnb.de.

© 2017 Arak M. Mathai, Hans J. Haubold, published by Walter de Gruyter GmbH, Berlin/Boston.
The book is published with open access at www.degruyter.com.
Typesetting: VTeX UAB, Lithuania
Printing and binding: CPI books GmbH, Leck
Cover image: Pasieka, Alfred / Science Photo Library
♾ Printed on acid-free paper
Printed in Germany

www.degruyter.com

MIX
Papier aus verantwor-
tungsvollen Quellen
FSC
www.fsc.org   FSC® C083411

Basic properties of vectors, matrices, determinants, eigenvalues and eigenvectors are discussed. Then, applications of matrices and determinants to various areas of statistical problems such as principal components analysis, model building, regression analysis, canonical correlation analysis, design of experiments etc. are examined. Applications of vector/matrix derivatives in the simplification of Taylor expansions of functions of many real scalar variables are considered. Jacobians of matrix transformations of real-valued scalar functions of matrix argument, maxima/minima problems, optimizations of linear forms, quadratic forms, bilinear forms with linear and quadratic constraints are examined. Matrix sequences and series, convergence of matrix series etc. and applications in physical sciences, chemical sciences, social sciences, input-analysis, linear programming problem, non-linear least squares and dynamic programming problems etc. are studied in this book.

Each topic is motivated by real-life situations and each concept is illustrated with examples and counter examples. The book is class-tested since 1999. It is written with the experience of teaching fifty years in various universities around the world. The first three Modules of the Centre for Mathematical and Statistical Sciences (CMSS)are combined to make this book. These Modules are used for intensive undergraduate mathematics training camps of CMSS. Each camp is a 10-day intensive training course with 40 hours of lectures and 40 hours of problem-solving sessions. Thirty such camps are already conducted by CMSS. Only high school level mathematics is assumed. The book is written as a self-study material. Each topic is brought from fundamentals to the senior undergraduate to graduate level. Usual doubts of the students on various topics are answered in the book.

Since 2004, the material in this book was made available to UN-affiliated Regional Centres for Space Science and Technology Education, located in India, China, Morocco, Nigeria, Jordan, Brazil, and Mexico (http://www.unoosa.org/oosa/en/ourwork/psa/regional-centres/index.html).

Since 1988 the material was taken into account for the development of education curricula in the fields of remote sensing and geographic information systems, satellite meteorology and global climate, satellite communications, space and atmospheric science, and global navigation satellite systems (http://www.unoosa.org/oosa/en/ourwork/psa/regional-centres/study_curricula.html).

As such the material was considered to be a prerequisite for applications, teaching, and research in space science and technology. It was also a prerequisite for the nine-months post-graduate courses in the five disciplines of space science and technology, offered by the Regional Centres on an annual basis to participants from all 194 Member States of the United Nations.

Since 1991, whenever suitable at the research level, the material in this book was utilized in lectures in a series of annual workshops and follow-up projects of the so-

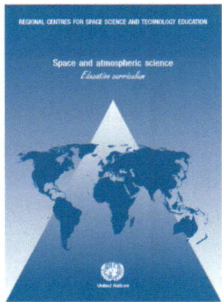

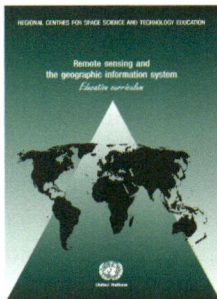
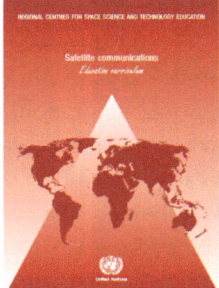

called Basic Space Science Initiative of the United Nations (http://www.unoosa.org/oosa/en/ourwork/psa/bssi/index.html).

As such the material was considered a prerequisite for teaching and research in astronomy and physics.

RIPPLE SIGHTING The cosmic dance of two black holes warped spacetime as the pair spiraled inward and merged, creating gravitational waves (illustration below). Advance Laser Interferometer Gravitational-Wave Observatory (LIGO) detected these ripples, produced by black holes eight and 14 times the mass of the sun, on December 26, 2015. Einstein's theory of general relativity was 100 years old in 2015. It has been very important in applications such as GPS (GNSS), and tremendously successful in understanding astrophysical systems like black holes. Gravitational waves, which are ripples in the fabric of space and time produced by violent events in the distant universe – for example, by the collision of two black holes or by the cores of supernova explosions – were predicted by Albert Einstein in 1916 as a consequence of his general theory of relativity. Gravitational waves are emitted by accelerating masses much in the same way electromagnetic waves are produced by accelerating charges, such as radio waves radiated by electrons accelerating in antennas. As they travel to Earth, these ripples in the space–time fabric carry information about their violent origins and about the nature of gravity that cannot be obtained by traditional astronomical observations using light. Gravitational waves have now been detected directly. Scien-

tists do, however, have great confidence that they exist because their influence on a binary pulsar system (two neutron stars orbiting each other) has been measured accurately and is in excellent agreement with the predictions. Directly detecting gravitational waves has confirmed Einstein's prediction in a new regime of extreme relativistic conditions, and open a promising new window into some of the most violent and cataclysmic events in the cosmos. The GNSS education curricula provides opportunities to teach navigation and do research in astrophysics (basic space science). The development of the education curricula (illustrated above) started in 1988 at UN Headquarters in New York, the specific GNSS curriculum emanated only in 1999 after the UNISPACE III Conference, held at and hosted by the United Nations at Vienna.

Usually students from other areas, other than mathematics, are intimidated by seeing theorems and proofs. Hence no such phrase as "theorem" is used in the book. Main results are called "results" and are written in bold so that the material will be user-friendly.

This book can be used as a textbook for a beginning undergraduate level course on vectors, matrices and determinants, and their applications, for students from all disciplines.

Preface

The basic material in this book originated from a course given by the first author at the University of Texas at El Paso in 1998–1999 academic year. Students from mathematics, engineering, biology, economics, physics and chemistry were in the class. The textbook assigned to the course did not satisfy the students from any of the disciplines, including mathematics. Hence Dr Mathai started developing a course from fundamentals, assuming no background, with lots of examples and counter examples taken from day to day life. All sections of the students enjoyed the course. Dr Mathai gave courses on calculus and linear algebra and for both of these courses he developed his own materials in close interaction with students. The El Paso experiment was initially for one semester only but, due to the popularity, extended to more semesters.

During 2000 to 2006 these notes were developed into CMSS Modules and based on these Modules, occasional courses were given for teachers and students at various levels in Kerala, India, as per requests from teachers. From 2007 onward CMSS became a Department of Science and Technology, Government of India centre for mathematical and statistical sciences. Modules in other areas were also developed during this period, and by 2014, ten Modules were developed.

As a Life Member of CMSS, the second author is an active participant of all programs at CMSS, including the undergraduate mathematics training camps, Ph.D training etc. and he is also a frequent visitor to CMSS to participate in and contribute to various activities.

Chapter 1 is devoted to all basic properties of vectors as ordered set of real numbers, Each definition is motivated by real-life examples. After introducing major properties of vectors with the real elements, vectors in the complex domain are considered and more rigorous definitions are introduced. Chapter 1 ends with Gram–Schmidt orthogonalization process.

Chapter 2 deals with matrices. Again, all definitions and properties are introduced from real-life situations. Roles of elementary matrices and elementary operations in solving linear equations, checking consistency of linear systems, checking linear dependence of vectors, evaluating rank of a matrix, canonical reductions of quadratic and bilinear forms, triangularizations and diagonalizations of matrices, computing inverses of matrices etc. are highlighted.

Chapter 3 deals with determinants. An axiomatic definition is introduced. Various types of expansions of determinants are given. Role of elementary matrices in evaluating determinants is highlighted. This chapter melts into Chapter 4 on eigenvalues and eigenvectors and their properties.

Chapters 5 and 6 are on applications of matrices and determinants to various disciplines. Applications to maxima/minima problems, constrained maxima/minima, optimization of linear, quadratic and bilinear forms, with linear and quadratic constraints are considered. For each optimization, at least one practical procedure such as principal components analysis, canonical correlation analysis, regression analysis etc. is illustrated. Some additional topics are also developed in Chapter 6. Matrix polynomials, matrix sequences and series, convergence, norms of matrices, singular value decomposition of matrices, simultaneous reduction of matrices to diagonal forms etc. are also discussed in Chapter 6.

A. M. Mathai
H. J. Haubold

14th March 2017

Acknowledgement

Several people have contributed directly or indirectly to make these Modules to the present levels. The financial support from the Department of Science and Technology, Government of India (DST), during the period 2007 to 2014 helped in printing and reprinting of the Modules. This helped in improving the quality of the material. Dr B.D. Acharya, then Dr A. K. Singh and then Dr P. K. Malhotra of the mathematical sciences division of DST, New Delhi, deserve special mention in providing research funds to CMSS. At the termination of DST support, Dr V. N. Rajasekharan Pillai, former Executive Vice-President of the Kerala State Council for Science, Technology and Environment (KSCSTE), a man with vision, took steps to support CMSS so that its activities of research, undergraduate mathematics training camps and Ph.D. training could continue uninterrupted. Dr T. Princy of CMSS was kind enough to reset all figures in the current book. She deserves special mention. Ms Sini Devassy, the office manager of CMSS deserves special mention. Former Ph.D. graduates from CMSS, Dr Seema S. Nair, Dr Nicy Sebastian, Dr Dhannya P. Joseph, Dr Dilip Kumar, Dr P. Prajitha, Dr T. Princy, Dr Naiju M. Thomas, Dr Anita Thomas, Dr Shanoja S. Pai, Dr Ginu Varghese, Dr Sona P. Jose and the many other graduate and undergraduate students helped in the developments of the Modules in many ways. They all deserve special thanks.

Contents

List of Symbols

$(\cdot), [\cdot]$	vector/matrix notation (Section 1.1, p. 1)
$\frac{\partial}{\partial X}$	partial differential operator (Section 1.1, p. 4)
O	null vector/matrix (Section 1.1, p. 7)
A'	transpose of A (Section 1.1, p. 7)
$\|(\cdot)\|$	length of $(\cdot)$ (Section 1.1, p. 8)
$U.V$	dot product of U and V (Section 1.1, p. 10)
J	vector of unities (Section 1.1, p. 11)
$\vec{U}$	vector as arrowhead (Section 1.2, p. 15)
$\alpha(i) + (j)$	α times i-th row added to j-th row (Section 1.3, p. 32)
$\dim(S)$	dimension of the vector subspace (Section 1.3, p. 47)
$M_X(T)$	moment generating function (Section 1.4, p. 55)
O	null matrix (Section 2.1, p. 61)
I	identity matrix (Section 2.1, p. 63)
$E(\cdot)$	expected value of $(\cdot)$ (Section 2.2, p. 91)
$\otimes$	Kronecker product (Section 2.7, p. 175)
$|A|$	determinant of A (Section 3.1, p. 181)
$\vec{a} \times \vec{b}$	cross product (Section 3.3.1, p. 223)
J	Jacobian (Section 3.3.3, p. 229)
$A > O, A \geq O$	positive definite, positive semi-definite (Definition 3.3.6, p. 247)
$A < O, A \leq O$	negative definite, negative semi-definite (Definition 3.3.6, p. 247)
$\sqrt{-1}$	complex number (Section 4.3.1, p. 280)
$\mathrm{Ker}(A)$	kernel of the matrix A (Problem 4.4.27, p. 323)

1 Vectors

1.0 Introduction

We start with vectors as ordered sets in order to introduce various aspects of these objects called vectors and the different properties enjoyed by them. After having discussed the basic ideas, a formal definition, as objects satisfying some general conditions, will be introduced later on. Several examples from various disciplines will be introduced to indicate the relevance of the concepts in various areas of study. As the students may be familiar, a collection of well-defined objects is called a *set*. For example $\{2, \alpha, B\}$ is a set of 3 objects, the objects being a number 2, a Greek letter α and the capital letter B. Sets are usually denoted by curly brackets {list of objects}. Each object in the set is called an *element* of the set. Let the above set be denoted by S, then $S = \{2, \alpha, B\}$. Then 2 is an element of S. It is usually written as $2 \in S$ (2 in S or 2 is an element of S). Thus we have

$$S = \{2, \alpha, B\}, \quad 2 \in S, \ \alpha \in S, \ B \in S, \ 7 \notin S, \ -\gamma \notin S \tag{1.0.1}$$

where $\notin$ indicates "not in". That is, 7 is not in S and $-\gamma$ (gamma) is not an element of S.

For a set, the order in which the elements are written is unimportant. We could have represented S equivalently as follows:

$$S = \{2, \alpha, B\} = \{2, B, \alpha\} = \{\alpha, 2, B\}$$
$$= \{\alpha, B, 2\} = \{B, 2, \alpha\} = \{B, \alpha, 2\} \tag{1.0.2}$$

because all of these sets contain the same objects and hence they represent the same set. Now, we consider ordered sets. In (1.0.2) there are 6 ordered arrangements of the 3 elements. Each permutation (rearrangement) of the objects gives a different ordered set. With a set of n distinct objects we can have a total of $n! = (1)(2) \ldots (n)$ ordered sets.

1.1 Vectors as ordered sets

For the time being we will define a vector as an ordered set of objects. More rigorous definitions will be given later on in our discussions. Vectors or these ordered sets will be denoted by ordinary brackets (*ordered list of elements*) or by square brackets [*ordered list of elements*]. For example, if the ordered sequences are taken from (1.0.2) then we have six vectors. If these are denoted by $V_1, V_2, \ldots, V_6$ respectively, then we have

$$V_1 = (2, \alpha, B), \quad V_2 = (2, B, \alpha), \quad V_3 = (\alpha, 2, B),$$
$$V_4 = (\alpha, B, 2), \quad V_5 = (B, 2, \alpha), \quad V_6 = (B, \alpha, 2).$$

We could have also represented these by square brackets, that is,

$$V_1 = [2, a, B], \quad \ldots, \quad V_6 = [B, a, 2]. \tag{1.1.1}$$

As a convention, we will use either all ordinary brackets (·) or all square brackets [·] when we discuss a given collection of vectors. The two notations will not be mixed up in the same collection. We could have also written the ordered sequences as columns, rather than as rows. For example,

$$U_1 = \begin{bmatrix} 2 \\ a \\ B \end{bmatrix}, \quad \ldots, \quad U_6 = \begin{bmatrix} B \\ a \\ 2 \end{bmatrix} \quad \text{or} \quad U_1 = \begin{pmatrix} 2 \\ a \\ B \end{pmatrix}, \quad \ldots, \quad U_6 = \begin{pmatrix} B \\ a \\ 2 \end{pmatrix} \tag{1.1.2}$$

also represent the same collection or ordered sets or vectors. In (1.1.1) they are written as row vectors whereas in (1.1.2) they are written as column vectors.

Definition 1.1.1 (An n-vector). It is an ordered set of n objects written either as a row (a row n-vector) or as a column (a column n-vector).

Example 1.1.1 (Stock market gains). A person has invested in 4 different stocks. Taking the January 1, 1998 as the base the person is watching the gain/loss, from this base value, at the end of each week.

	Stock 1	Stock 2	Stock 3	Stock 4
Week 1	100	150	−50	50
Week 2	50	−50	70	−50
Week 3	−150	−100	−20	0

The performance vector at the end of week 1 is then $(100, 150, -50, 50)$, a negative number indicating the loss and a positive number denoting a gain. The performance vector of stock 1 over the three weeks is $\begin{bmatrix} 100 \\ 50 \\ -150 \end{bmatrix}$. Observe that we could have also written weeks as columns and stocks as rows instead of the above format. Note also that for each element the position where it appears is relevant, in other words, the elements above are ordered.

Example 1.1.2 (Consumption profile). Suppose the following are the data on the food consumption of a family in a certain week, where q denotes quantity (in kilograms) and p denotes price per unit (per kilogram).

	Beef	Pork	Chicken	Vegetables	cereals
q	10	15	20	10	5
p	\$2.00	\$1.50	\$0.50	\$1.00	\$3.45

The vector of quantities consumed is $[10, 15, 20, 10, 5]$ and the price vector is $[2.00, 1.50, 0.50, 1.00, 3.45]$.

Example 1.1.3 (Discrete statistical distributions). If a discrete random variable takes values $x_1, x_2, \ldots, x_n$ with probabilities $p_1, \ldots, p_n$ respectively where $p_i > 0$, $i = 1, \ldots, n$, $p_1 + \cdots + p_n = 1$ then this distribution can be represented as follows:

$$
\begin{array}{lcccc}
x\text{-values} & x_1 & x_2 & \ldots & x_n \\
\text{probabilities} & p_1 & p_2 & \ldots & p_n
\end{array}
$$

As an example, if x takes the values $0, 1, -1$, (such as a gambler gains nothing, gains one dollar, loses one dollar) with probabilities $\frac{1}{2}, \frac{1}{4}, \frac{1}{4}$ respectively then the distribution can be written as

$$
\begin{array}{lccc}
x\text{-values} & 0 & 1 & -1 \\
\text{probabilities} & \frac{1}{2} & \frac{1}{4} & \frac{1}{4}
\end{array}
$$

Here the observation vector is $(0, 1, -1)$ and the corresponding probability vector is $(\frac{1}{2}, \frac{1}{4}, \frac{1}{4})$. Note that when writing the elements of a vector, the elements may be separated by sufficient spaces, or by commas if there is possibility of confusion. Any vector $(p_1, \ldots, p_n)$ such that $p_i > 0$, $i = 1, \ldots, n$, $p_1 + \cdots + p_n = 1$ is called a *discrete probability distribution*.

Example 1.1.4 (Transition probability vector). Suppose at El Paso, Texas, there are only two possibilities for a September day. It can be either sunny and hot or cloudy and hot. Let these be denoted by S (sunny) and C (cloudy). A sunny day can be followed by either a sunny day or a cloudy day and similarly a cloudy day can follow either a sunny or a cloudy day. Suppose that the chances (transition probabilities) are the following:

$$
\begin{array}{ccc}
 & S & C \\
S & 0.95 & 0.05 \\
C & 0.90 & 0.10
\end{array}
$$

Then for a sunny day the transition probability vector is $(0.95, 0.05)$ to be followed by a sunny and a cloudy day respectively. For a cloudy day the corresponding vector is $(0.90, 0.10)$.

Example 1.1.5 (Error vector). Suppose that an automatic machine is filling 5 kg bag of potatoes. The machine is not allowed to cut or chop to make the weight exactly 5 kg. Naturally, if one such bag is taken then the actual weight can be less than or greater than or equal to 5 kg. Let ϵ denote the error = observed weight minus the expected weight(5 kg). [One could have defined "error" as expected value minus the observed value]. Suppose 4 such bags are selected and weighed. Suppose the observation vector, denoted by X, is

$$
X = (5.01, 5.10, 4.98, 4.92).
$$

Then the error vector, denoted by ϵ, is

$$\epsilon = (0.01, 0.10, -0.02, -0.08)$$
$$= (5.01 - 5.00, 5.10 - 5.00, 4.98 - 5.00, 4.92 - 5.00).$$

Note that we could have written both X and ϵ as column vectors as well.

Example 1.1.6 (Position vector). Suppose a person walks on a straight path (horizontal) for 4 miles and then along a perpendicular path to the left for another 6 miles. If these distances are denoted by x and y respectively then her position vector is, taking the starting points as the origin,

$$(x, y) = (4, 6).$$

Example 1.1.7 (Vector of partial derivatives). Consider $f(x_1, \ldots, x_n)$, a scalar function of n real variables $x_1, \ldots, x_n$. As an example,

$$f(x_1, x_2, x_3) = 3x_1^2 + x_2^2 + x_3^2 - 2x_1 x_2 + 5x_1 x_3 - 2x_1 + 7.$$

Here $n = 3$ and there are 3 variables in f. Consider the partial derivative operators $\frac{\partial}{\partial x_1}, \frac{\partial}{\partial x_2}, \frac{\partial}{\partial x_3}$, that is, $\frac{\partial}{\partial x_1}$ operating on f means to differentiate f with respect to x_1 treating x_2 and x_3 as constants. For example, $\frac{\partial}{\partial x_1}$ operating on the above f gives

$$\frac{\partial f}{\partial x_1} = 6x_1 - 2x_2 + 5x_3 - 2.$$

Consider the partial differential operator

$$\frac{\partial}{\partial X} = \left(\frac{\partial}{\partial x_1}, \ldots, \frac{\partial}{\partial x_n} \right).$$

Then $\frac{\partial}{\partial X}$ operating on f gives the vector

$$\frac{\partial f}{\partial X} = \left(\frac{\partial f}{\partial x_1}, \ldots, \frac{\partial f}{\partial x_n} \right).$$

For the above example,

$$\frac{\partial f}{\partial X} = \left(\frac{\partial f}{\partial x_1}, \frac{\partial f}{\partial x_2}, \frac{\partial f}{\partial x_3} \right)$$
$$= (6x_1 - 2x_2 + 5x_3 - 2, 2x_2 - 2x_1, 2x_3 + 5x_1).$$

Example 1.1.8 (Students' grades). Suppose that Miss Gomez, a first year student at UTEP, is taking 5 courses, Calculus I (course 1), Linear Algebra (course 2),…, (course 5). Suppose that each course requires 2 class tests, a set of assignments to be submitted and a final exam. Suppose that Miss Gomez' performance profile is the following (all grades in percentages):

	course 1	course 2	course 3	course 4	course 5
test 1	80	85	80	90	95
test 2	85	85	85	95	100
assignments	100	100	100	100	100
final exam	90	95	90	92	95

Then for example, her performance profiles on courses 1 and 4 are

$$\begin{bmatrix} 80 \\ 85 \\ 100 \\ 90 \end{bmatrix} \quad \text{and} \quad \begin{bmatrix} 90 \\ 95 \\ 100 \\ 92 \end{bmatrix}$$

respectively. Her performances on all courses is the vector $(80, 85, 80, 90, 95)$ for test 1.

Example 1.1.9 (Fertility data). Fertility of women is often measured in terms of the number of children produced. Suppose that the following data represent the average number of children in a particular State according to age and racial groups:

	group 1	group 2	group 3	group 4
≤ 16	1	0.8	1.5	0.5
16 to ≤ 18	1	1	0.8	0.9
18 to ≤ 35	4	2	3	2
35 to ≤ 50	1	0	2	0
> 50	0	0	1	0

The first row vector in the above table represents the performance of girls 16 years or younger over the 4 racial groups. Column 2 represents the performance of racial group 2 over the age groups, and so on.

Example 1.1.10 (Geometric probability law). Suppose that a person is playing a game of chance in a casino. Suppose that the chance of winning at each trial is 0.2 and that of losing is 0.8. Suppose that the trials are independent of each other. Then the person can win at the first trial, or lose at the first trial and win at the second trial, or lose at the first two trials and win at the third trial, and so on. Then the chance of winning at the x-th trial, $x = 1, 2, 3, \ldots$ is given by the vector

$$[0.2, (0.8)(0.2), (0.8)^2(0.2), (0.8)^3(0.2), \ldots].$$

It is an n-vector with $n = +\infty$. Note that the number of ordered objects, representing a vector, could be finite or infinitely many (countable, that is one can draw a one-to-one correspondence to the natural numbers $1, 2, 3, \ldots$).

In Example 1.1.1 suppose that the gains/loses were in US dollars and suppose that the investor was a Canadian and she would like to convert the first week's gain/loss into Canadian dollar equivalent. Suppose that the exchange rate is US\$ 1=CA\$ 1.60.

Then the first week's performance is available by multiplying each element in the vector by 1.6. That is,

$$1.6(100, 150, -50, 50) = ((1.6)(100), (1.6)(150), (1.6)(-50), (1.6)(50))$$
$$= (160, 240, -80, 80).$$

Another example of this type is that someone has a measurement vector in feet and that is to be converted into inches, then each element is multiplied by 12 (one foot = 12 inches), and so on.

Definition 1.1.2 (Scalar multiplication of a vector). Let c be a scalar, a 1-vector, and $U = (u_1, \ldots, u_n)$ an n-vector. Then the scalar multiple of U, namely cU, is defined as

$$cU = (cu_1, \ldots, cu_n). \tag{1.1.3}$$

As a convention the scalar quantity c is written on the left of U and not on the right, that is, not as Uc but as cU. As numerical illustrations we have

$$-3\begin{bmatrix} 1 \\ -1 \\ 2 \end{bmatrix} = \begin{bmatrix} -3 \\ 3 \\ -6 \end{bmatrix}; \quad 0\begin{pmatrix} 1 \\ -1 \\ 2 \end{pmatrix} = \begin{pmatrix} 0 \\ 0 \\ 0 \end{pmatrix}; \quad \frac{1}{2}\begin{bmatrix} 1 \\ -1 \\ 2 \end{bmatrix} = \begin{bmatrix} \frac{1}{2} \\ -\frac{1}{2} \\ 1 \end{bmatrix};$$
$$4(2, -1) = (8, -4).$$

In Example 1.1.1 if the total (combined) gain/loss at the end of the second week is needed then the combined performance vector is given by

$$(100 + 50, 150 - 50, -50 + 70, 50 - 50) = (150, 100, 20, 0).$$

If the combined performance of the first three weeks is required then it is the above vector added to the third week's vector, that is,

$$(150, 100, 20, 0) + (-150, -100, -20, 0) = (0, 0, 0, 0).$$

Definition 1.1.3 (Addition of vectors). Let $a = (a_1, \ldots, a_n)$ and $b = (b_1, \ldots, b_n)$ be two n-vectors. Then the sum is defined as

$$a + b = (a_1 + b_1, \ldots, a_n + b_n), \tag{1.1.4}$$

that is, the vector obtained by adding the corresponding elements.

Note that vector addition is defined only for vectors of the same category and order. Either both are row vectors of the same order or both are column vectors of the same order. In other words, if U is an n-vector and V is an m-vector then $U + V$ is not defined unless $m = n$, and further, both are either row vectors or column vectors.

Definition 1.1.4 (A null vector). A vector with all its elements zeros is called a null vector and it is usually denoted by a big O.

In Example 1.1.1 the combined performance of the first 3 weeks is a null vector. In other words, after the first 3 weeks the performance is back to the base level. From the above definitions the following properties are evident. If U, V, W are three n-vectors (either all row vectors or all column vectors) and if a, b, c are scalars then

$$U + V = V + U; \quad U + (V + W) = (U + V) + W$$
$$U - V = U + (-1)V; \quad U + O = O + U = U; \quad U - U = O;$$
$$a[bU + cV] = abU + acV = b(aU) + c(aV). \tag{1.1.5}$$

Some numerical illustrations are the following:

$$2\begin{bmatrix} 1 \\ 0 \\ -1 \end{bmatrix} - 3\begin{bmatrix} 0 \\ 1 \\ -2 \end{bmatrix} + \begin{bmatrix} 0 \\ 0 \\ 0 \end{bmatrix} = \begin{bmatrix} 2 \\ 0 \\ -2 \end{bmatrix} + \begin{bmatrix} 0 \\ -3 \\ 6 \end{bmatrix} + \begin{bmatrix} 0 \\ 0 \\ 0 \end{bmatrix}$$
$$= \begin{bmatrix} 2+0+0 \\ 0-3+0 \\ -2+6+0 \end{bmatrix} = \begin{bmatrix} 2 \\ -3 \\ 4 \end{bmatrix};$$
$$(1,-7) + 6(0,-1) + (0,0) = (1,-7) + (0,-6) + (0,0)$$
$$= (1+0+0, -7-6+0) = (1,-13);$$
$$(1,1,2) - (1,1,2) = (1,1,2) + (-1,-1,-2)$$
$$= (1-1, 1-1, 2-2) = (0,0,0).$$

Definition 1.1.5 (Transpose of a vector). [Standard notations: $U' = $ transpose of U, $U^T = $ transpose of U.] If U is a row n-vector then U' is the same written as a column and vice versa.

Some numerical illustrations are the following, where "$\Rightarrow$" means "implies":

$$U = \begin{bmatrix} -3 \\ 0 \\ 1 \end{bmatrix} \Rightarrow U' = [-3, 0, 1]$$

$$V = [1, 5, -1] \Rightarrow V' = \begin{bmatrix} 1 \\ 5 \\ -1 \end{bmatrix} = V^T.$$

Note that in the above illustration $U + V$ is not defined but $U + V'$ is defined. Similarly $U' + V$ is defined but $U' + V'$ is not defined. Also observe that if z is a 1-vector (a scalar quantity) then $z' = z$, that is, the transpose is itself.

In Example 1.1.6 the position vector is $(x, y) = (4, 6)$. Then the distance of this position from the starting point is obtained from Pythagoras' rule as,

$$\sqrt{x^2 + y^2} = \sqrt{4^2 + 6^2} = \sqrt{52}.$$

This then is the straight distance from the starting point $(0, 0)$ to the final position $(4, 6)$. We will formally define the length of a vector as follows, the idea will be clearer when we consider the geometry of vectors later on:

Definition 1.1.6 (Length of a vector). Let U be a real n-vector (either a column vector or a row vector). If the elements of U are $u_1, \ldots, u_n$ then the length of U, denoted by $\|U\|$, is defined as

$$\|U\| = \sqrt{u_1^2 + \cdots + u_n^2}, \tag{1.1.6}$$

when the elements are real numbers. When the elements are not real then the length will be redefined later on. Some numerical illustrations are the following:

$$U = \begin{bmatrix} 1 \\ -1 \\ 0 \end{bmatrix} \Rightarrow \|U\| = \sqrt{(1)^2 + (-1)^2 + (0)^2} = \sqrt{2};$$

$$V = (1, 1, -2) \Rightarrow \|V\| = \sqrt{(1)^2 + (1)^2 + (-2)^2} = \sqrt{6};$$

$$O = \begin{bmatrix} 0 \\ 0 \\ 0 \end{bmatrix} \Rightarrow \|O\| = 0; \quad e_1 = (1, 0, 0, 0) \Rightarrow \|e_1\| = 1;$$

$$Z = \left(\frac{1}{\sqrt{2}}, -\frac{1}{\sqrt{2}} \right) \Rightarrow \|Z\| = 1.$$

Note that the "length", by definition, is a non-negative quantity. It is either zero or a positive quantity and it cannot be negative. For a null vector the length is zero. The length of a vector is zero iff (if and only if) the vector is a null vector.

Definition 1.1.7 (A unit vector). A vector whose length is unity is called a unit vector.

Some numerical illustrations are the following:

$$e_4 = (0, 0, 0, 1) \Rightarrow \|e_4\| = 1.$$

But $U = (1, -2, 1) \Rightarrow \|U\| = \sqrt{6}$, U is not a unit vector whereas

$$V = \frac{1}{\|U\|} U = \frac{1}{\sqrt{6}} (1, -2, 1) = \left(\frac{1}{\sqrt{6}}, -\frac{2}{\sqrt{6}}, \frac{1}{\sqrt{6}} \right) \Rightarrow \|V\| = 1,$$

that is, V is a unit vector. Observe the following: A null vector is not a unit vector. If the length of any vector is non-zero (the only vector with length zero is the null vector) then taking a scalar multiple, where the scalar is the reciprocal of the length, a unit vector can be created out of the given non-null vector. In general, if $U = (u_1, \ldots, u_n)$, where $u_1, \ldots, u_n$ are real, then

$$\|U\| = \|U'\| = \sqrt{u_1^2 + \cdots + u_n^2}$$

and

$$V = \frac{1}{\|U\|} U \;\Rightarrow\; \|V\| = 1 \tag{1.1.7}$$

when $\|U\| \neq 0$.

From the definition of length itself the following properties are obvious. If U and V are n-vectors of the same type and if a, b, c are scalars, then

$$\|cU\| = |c|\,\|U\|; \quad \|cU + cV\| = |c|\,\|U + V\|$$
$$\|U + V\| \leq \|U\| + \|V\|;$$
$$\|aU + bV\| \leq |a|\,\|U\| + |b|\,\|V\| \tag{1.1.8}$$

where, for example, $|c|$ means the absolute value of c, that is, the magnitude of c, ignoring the sign. For example,

$$\|-2(1, -1, 1)\| = |-2|\,\sqrt{(1)^2 + (-1)^2 + (1)^2} = 2\sqrt{3};$$
$$\|2(1, -1, 1)\| = 2\sqrt{3};$$

$$U = \begin{bmatrix} 1 \\ -1 \\ 1 \end{bmatrix}, \quad V = \begin{bmatrix} 1 \\ 2 \\ -3 \end{bmatrix} \;\Rightarrow\; U + V = \begin{bmatrix} 2 \\ 1 \\ -2 \end{bmatrix};$$

$$\|U\| = \sqrt{(1)^2 + (-1)^2 + (1)^2} = \sqrt{3};$$

$$\|U + V\| = \sqrt{(2)^2 + (1)^2 + (-2)^2} = \sqrt{9} = 3 < \|U\| + \|V\| = \sqrt{3} + \sqrt{14}.$$

Now, we will look at another concept. In Example 1.1.2 the family's total expense of the week on those food items is available by multiplying the quantities with unit prices and then adding up. That is, if the quantity vector is denoted by Q and the per unit price vector is denoted by P then

$$Q = (10, 15, 20, 10, 5)$$

and

$$P = (2.00, 1.50, 0.50, 1.00, 3.45).$$

Thus the total expense of that family for that week on these 5 items is obtained by multiplying and adding the corresponding elements in P and Q. That is,

$$(10)(2.00) + (15)(1.50) + (20)(0.50) + (10)(1.00) + (5)(3.45) = \$79.75.$$

It is a scalar quantity (1-vector) and not a 5-vector, even though the vectors Q and P are 5-vectors. For computing quantities such as the one above we define a concept called the *dot product or the inner product* between two vectors.

Definition 1.1.8 (Dot product or inner product). Let U and V be two real n-vectors (either both row vectors or both column vectors or one row vector and the other column vector). Then the dot product between U and V, denoted by $U.V$ is defined as

$$U.V = u_1 v_1 + \cdots + u_n v_n$$

that is, the corresponding elements are multiplied and added, where $u_1, \ldots, u_n$ and $v_1, \ldots, v_n$ are the elements (real) in U and V respectively. (Vectors in the complex field will be considered in a later chapter.)

Some numerical illustrations are the following: In the above example, the family's consumption for the week is $Q.P = P.Q = 79.75$.

$$U_1 = \begin{pmatrix} 0 \\ 1 \\ 2 \end{pmatrix}, \quad U_2 = \begin{pmatrix} 1 \\ -1 \\ 1 \end{pmatrix} \Rightarrow$$

$$U_1.U_2 = (0)(1) + (1)(-1) + (2)(1) = 1.$$
$$V_1 = (3, 1, -1, 5), \quad V_2 = (-1, 0, 0, 1) \Rightarrow$$
$$V_1.V_2 = (3)(-1) + (1)(0) + (-1)(0) + (5)(1) = 2.$$

From the definition itself the following properties are evident:

$$U.O = 0, \quad aU.V = (aU).V = U.(aV)$$

where a is a scalar.

$$U.V = V.U, \quad (aU).(bV) = ab(U.V)$$

where a and b are scalars.

$$U.(V + W) = U.V + U.W = (W + V).U. \tag{1.1.9}$$

The notation with a dot, $U.V$, is an awkward one. But unfortunately this is a widely used notation. A proper notation in terms of transposes and matrix multiplication will be introduced later. Also, further properties of dot products will be considered later, after looking at the geometry of vectors as ordered sets.

Exercises 1.1

1.1.1. Are the following defined? Whichever is defined compute the answers.

(a) $\begin{bmatrix} 0 \\ -1 \\ 1 \end{bmatrix} + \begin{bmatrix} 2 \\ 3 \end{bmatrix}$;   (b) $\begin{bmatrix} 1 \\ 0 \\ 1 \end{bmatrix} - 3 \begin{bmatrix} 2 \\ 0 \\ 0 \end{bmatrix}$;

(c) $(3, -1, 4) - (2, 1)$;   (d) $5(1, 0) - 3(-2, -1)$.

1.1.2. Compute the lengths of the following vectors. Normalize the vectors (create a vector with unit length from the given vector) if possible:

(a) $(0,0,0)$;   (b) $(1,1,-1)$;

(c) $\begin{bmatrix} 2 \\ -1 \\ 1 \end{bmatrix}$;   (d) $\begin{bmatrix} 5 \\ 0 \\ -1 \end{bmatrix}$;   (e) $3\begin{bmatrix} 1 \\ -1 \\ 1 \end{bmatrix}$.

1.1.3. Convert the stock market performance vectors in Example 1.1.1 to the following: First week's performance into pound sterling (1 $ = 0.5 pounds sterling); the second week's performance into Italian lira (1 $ = 2 000 lira).

1.1.4. In Example 1.1.3 compute the expected value of the random variable. [The expected value of a discrete random variable is denoted as $E(x)$ and defined as $E(x) = x_1 p_1 + \cdots + x_n p_n$ if x takes the values $x_1, \ldots, x_n$ with probabilities $p_1, \ldots, p_n$ respectively.] If it is a game of chance where the person wins $0, $1, $(-1) (loses a dollar) with probabilities $\frac{1}{2}, \frac{1}{4}, \frac{1}{4}$ respectively how much money can the person expect to win in a given trial of the game?

1.1.5. In Example 1.1.3 if the expected value is denoted by $\mu = X.P$ (μ the Greek letter mu), where $X = (x_1, \ldots, x_n)$ and $P = (p_1, \ldots, p_n)$ then the variance of the random variable is defined as the dot product between $((x_1 - \mu)^2, \ldots, (x_n - \mu)^2)$ and P. Compute the variance of the random variable in Example 1.1.3. [Variance is the square of a measure of scatter or spread in the random variable.]

1.1.6. In Example 1.1.5 compute the sum of squares of the errors [Hint: If ϵ is the error vector then the sum of squares of the errors is available by taking the dot product $\epsilon.\epsilon$.]

1.1.7. In Example 1.1.8 suppose that for each course the distribution of the final grade is the following: 20 points each for each test, 10 points for assignments and 50 points for the final exam. Compute the vector of final grades of the student for the 5 courses by using the various vectors and using scalar multiplications and sums.

1.1.8. From the chance vector in Example 1.1.10 compute the chance of ever winning (sum of the elements) and the expected number of trials for the first win, $E(x)$ (note that x takes the values $1, 2, \ldots$ with the corresponding probabilities).

1.1.9. Consider an n-vector of unities denoted by $J = (1, 1, \ldots, 1)$. If $X = (x_1, \ldots, x_n)$ is any n-vector then compute (a) $X.J$; (b) $\frac{1}{n}X.J$.

1.1.10. For the quantities in Exercise 1.1.9 establish the following:

(a) $(X - \bar{\mu}).J = 0$ where $\bar{\mu} = \left(\frac{1}{n}X.J, \ldots, \frac{1}{n}X.J \right)$.

[This holds whatever be the values of $x_1, \ldots, x_n$. Verify by taking some numerical values.]

(b) $\quad (X - \bar{\mu}).(X - \bar{\mu}) = X.X - n\left(\dfrac{1}{n}X.J\right)^2$

$$= X.X - \dfrac{1}{n}(X.J)(X.J)$$

whatever be the values of $x_1, \ldots, x_n$.

(c) Show that the statement in (a) above is equivalent to the statement $\sum_{i=1}^{n}(x_i - \bar{x}) = 0$ where $\bar{x} = \sum_{i}^{n} \dfrac{x_i}{n}$ with $\sum$ denoting a sum.

(d) Show that the statement in (b) is equivalent to the statements

$$\sum_{i=1}^{n}(x_i - \bar{x})^2 = \sum_{i=1}^{n} x_i^2 - n\bar{x}^2$$

$$= \sum_{i=1}^{n} x_i^2 - \dfrac{1}{n}\left(\sum_{i=1}^{n} x_i\right)^2 .$$

A note on $\sum$ notation. This is a convenient notation to denote a sum.

$$\sum_{i=1}^{n} a_i = a_1 + \cdots + a_n,$$

that is, i is replaced by $1, 2, \ldots, n$ and the elements are added up.

$$\sum_{i=1}^{n} 4 = 4 + 4 + \cdots + 4 = 4n;$$

$$\sum_{i=1}^{n} a_i b_i = a_1 b_1 + \cdots + a_n b_n = a.b$$

where $a = (a_1, \ldots, a_n)$ and $b = (b_1, \ldots, b_n)$.

$$\sum_{i=1}^{n} a_i^2 = a_1^2 + \cdots + a_n^2 = a.a;$$

$$\sum_{i=1}^{n}(5a_i) = 5a_1 + \cdots + 5a_n = 5(a_1 + \cdots + a_n) = 5\sum_{i=1}^{n} a_i;$$

$$\sum_{i=1}^{n}\sum_{j=1}^{m} a_i b_j = \sum_{i=1}^{n} a_i \left[\sum_{j=1}^{m} b_j\right]$$

$$= \sum_{i=1}^{n} a_i(b_1 + \cdots + b_m) = (a_1 + \cdots + a_n)(b_1 + \cdots + b_m)$$

$$= \sum_{j=1}^{m}\sum_{i=1}^{n} a_i b_j;$$

$$\sum_{i=1}^{m}\sum_{j=1}^{n} a_{ij} = a_{11} + \cdots + a_{1n}$$

$$+ a_{21} + \cdots + a_{2n}$$

$$\vdots$$

$$+ a_{m1} + \cdots + a_{mn} = \sum_{j=1}^{n} \sum_{i=1}^{m} a_{ij};$$

$$\sum_{i=1}^{n} (a_i + 3b_i) = \sum_{i=1}^{n} a_i + 3 \sum_{i=1}^{n} b_i;$$

$$\bar{x} = \sum_{j=1}^{n} \frac{x_j}{n} = \frac{1}{n}(x_1 + \cdots + x_n) = \frac{1}{n} X.J$$

where

$$X = (x_1, \dots, x_n) \quad \text{and} \quad J = (1, 1, \dots, 1);$$

$$\sum_{i=1}^{n} (x_i - \bar{x}) = \sum_{i=1}^{n} x_i - \sum_{i=1}^{n} \bar{x}$$

$$= \sum_{i=1}^{n} x_i - n\bar{x} = \sum_{i=1}^{n} x_i - n \left(\sum_{i=1}^{n} \frac{x_i}{n} \right)$$

$$= \sum_{i=1}^{n} x_i - \sum_{i=1}^{n} x_i = 0$$

whatever be $x_1, \dots, x_n$.

1.1.11. When searching for maxima/minima of a scalar function f of many real scalar variables the critical points (the points where one may find a maximum or a minimum or a saddle point) are available by operating with $\frac{\partial}{\partial X}$, equating to a null vector and then solving the resulting equations. For the function

$$f(x_1, x_2) = 3x_1^2 + x_2^2 - 2x_1 + x_2 + 5$$

evaluate the following: (a) the operator $\frac{\partial}{\partial X}$, (b) $\frac{\partial f}{\partial X}$, (c) $\frac{\partial f}{\partial X} = O$, (d) the critical points.

1.1.12. For the following vectors U, V, W compute the dot products $U.V$, $U.W$, $V.W$ where

$$U = (1, 1, 1), \quad V = (1, -2, 1), \quad W = (1, 0, -1).$$

1.1.13. If V_1, V_2, V_3 are n vectors, either $n \times 1$ column vectors or $1 \times n$ row vectors and if $\|V_j\|$ denotes the length of the vector V_j then show that the following results hold in general:
(i) $\|V_1 - V_2\| > 0$ and $\|V_1 - V_2\| = 0$ iff $V_1 = V_2$;
(ii) $\|cV_1\| = |c| \|V_1\|$ where c is a scalar;
(iii) $\|V_1 - V_2\| + \|V_2 - V_3\| \geq \|V_1 - V_3\|$.

1.1.14. Verify (i), (ii), (iii) of Exercise 1.1.13 for

$$V_1 = (1, 0, -1), \quad V_2 = (0, 0, 2), \quad V_3 = (2, 1, -1).$$

1.1.15. Let $U = (1, -1, 1, -1)$. Construct three non-null vectors V_1, V_2, V_3 such that $U.V_1 = 0$, $U.V_2 = 0$, $U.V_3 = 0$, $V_1.V_2 = 0$, $V_1.V_3 = 0$, $V_2.V_3 = 0$.

1.2 Geometry of vectors

From the position vector in Example 1.1.6 it is evident that $(x, y) = (4, 6)$ can be denoted as a point in a 2-space (plane) with a rectangular coordinate system. In general, since an n-vector of real numbers is an ordered set of real numbers it can be represented as a point in a Euclidean n-space.

1.2.1 Geometry of scalar multiplication

If the position $(4, 6)$, which could also be written as $\binom{x}{y} = \binom{4}{6}$, is marked in a 2-space then we have the following Figure 1.2.1. One can also think of this as an arrowhead starting at $(0, 0)$ and going to $(4, 6)$. In this representation the vector has a length and a direction. In general, if U is an arrowhead from the origin $(0, 0, \dots, 0)$ in n-space to the point $U = (u_1, \dots, u_n)$ then $-U$ will represent an arrowhead with the same length but going in the opposite direction. Then cU will be an arrowhead in the same direction with length $c\|U\|$ if $c > 0$ and in the opposite direction with length $|c|\,\|U\|$ if $c < 0$, where $|c|$ denotes the absolute value or the magnitude of c, and it is the origin itself if $c = 0$. In physics, chemistry and engineering areas it is customary to denote a vector with an arrow on top such as $\vec{U}$, meaning the vector $\vec{U}$.

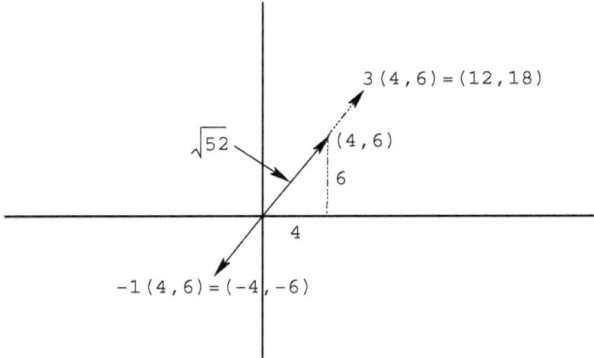

Figure 1.2.1: Geometry of vectors.

1.2.2 Geometry of addition of vectors

Scalar multiplication is interpreted geometrically as above. Then, what will be the geometrical interpretation for a sum of two vectors? For simplicity, let us consider a 2-space. If $\vec{U} = \binom{u_1}{u_2}$ and $\vec{V} = \binom{v_1}{v_2}$ then algebraically

$$\vec{U} + \vec{V} = \begin{pmatrix} u_1 + v_1 \\ u_2 + v_2 \end{pmatrix}$$

which is the arrowhead representing the diagonal of the parallelogram as shown in Figure 1.2.2. From the geometry of vectors one can notice that a vector, as an ordered set of real numbers, possesses two properties basically, namely, a length and a direction. Hence we can give a coordinate-free definition as an arrowhead with a length and a direction.

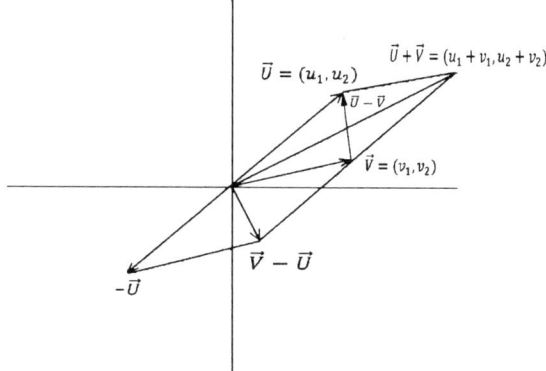

Figure 1.2.2: Sum of two vectors.

1.2.3 A coordinate-free definition of vectors

Definition 1.2.1 (A coordinate-free definition for a vector). It is defined as an arrowhead with a given length and a given direction.

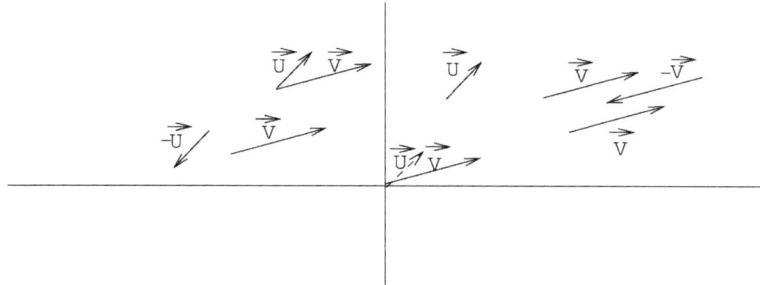

Figure 1.2.3: Coordinate-free definition of vectors.

In this definition, observe that all arrowheads with the same length and same direction are taken to be one and the same vector as shown in Figure 1.2.3. We can move an arrowhead parallel to itself. All such arrowheads obtained by such displacements are taken as one and the same vector. If one has a coordinate system then move the vector parallel to itself so that the tail-end (the other end to the arrow tip) coincides with the origin of the coordinate system. Thus the position vectors are also included in this general definition. In a coordinate-free definition one can construct $\vec{U} + \vec{V}$ and $\vec{U} - \vec{V}$ as follows: Move $\vec{U}$ or $\vec{V}$ parallel to itself until the tail-ends coincide. Complete the parallelogram. The leading diagonal gives $\vec{U} + \vec{V}$ and the diagonal going from the head of $\vec{U}$ to the head of $\vec{V}$ gives $\vec{V} - \vec{U}$ and the one the other way around is $-(\vec{V} - \vec{U}) = \vec{U} - \vec{V}$.

1.2.4 Geometry of dot products

Consider a Euclidean 2-space and represent the vectors $\vec{U} = (u_1, u_2)$ and $\vec{V} = (v_1, v_2)$ as points in a rectangular coordinate system. Let the angles, the vectors $\vec{U}$ and $\vec{V}$ make with the x-axis be denoted by θ_1 and θ_2 respectively. Let

$$\theta = \theta_1 - \theta_2.$$

Then

$$\cos\theta_1 = \frac{u_1}{\sqrt{u_1^2 + u_2^2}}, \quad \cos\theta_2 = \frac{v_1}{\sqrt{v_1^2 + v_2^2}},$$

$$\sin\theta_1 = \frac{u_2}{\sqrt{u_1^2 + u_2^2}}, \quad \sin\theta_2 = \frac{v_2}{\sqrt{v_1^2 + v_2^2}}.$$

But

$$\cos\theta = \cos(\theta_1 - \theta_2) = \cos\theta_1 \cos\theta_2 + \sin\theta_1 \sin\theta_2$$

$$= \frac{u_1 v_1 + u_2 v_2}{\sqrt{u_1^2 + u_2^2}\sqrt{v_1^2 + v_2^2}} = \frac{\vec{U}.\vec{V}}{\|\vec{U}\|\,\|\vec{V}\|} \tag{1.2.1}$$

whenever $\|\vec{U}\| \neq 0$ and $\|\vec{V}\| \neq 0$. Thus

$$\vec{U}.\vec{V} = \|\vec{U}\|\,\|\vec{V}\|\cos\theta, \quad \|\vec{U}\| \neq 0, \ \|\vec{V}\| \neq 0. \tag{1.2.2}$$

The dot product is the product of the lengths multiplied by the cosine of the angle between the vectors. This result remains the same whatever be the space. That is, it holds in 2-space, 3-space, 4-space and so on. The Figure 1.2.4 shows the situation when $0 \le \theta_1 \le \pi/2$, $0 \le \theta_2 \le \pi/2$, $\theta_1 > \theta_2$. The student may verify the result for all possible cases of θ_1 and θ_2, as an exercise. From (1.2.1) we can obtain an interesting result. Since $\cos\theta$, in absolute value, is less than or equal to 1 we have a result known as Cauchy–Schwartz inequality:

$$|\cos\theta| = \left|\frac{\vec{U}.\vec{V}}{\|\vec{U}\|\,\|\vec{V}\|}\right| \le 1 \implies |\vec{U}.\vec{V}| \le \|\vec{U}\|\,\|\vec{V}\|.$$

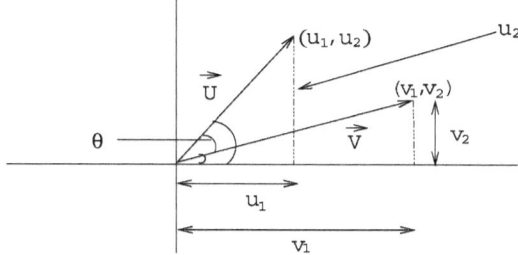

Figure 1.2.4: Geometry of the dot product.

1.2.5 Cauchy–Schwartz inequality

$$|\vec{U}.\vec{V}| \leq \|\vec{U}\| \, \|\vec{V}\|.$$

In other words, if $\vec{U} = (u_1, \ldots, u_n)$ and $\vec{V} = (v_1, \ldots, v_n)$ then for real $u_1, \ldots, u_n$ and $v_1, \ldots, v_n$,

$$|u_1 v_1 + \cdots + u_n v_n| \leq \sqrt{u_1^2 + \cdots + u_n^2} \, \sqrt{v_1^2 + \cdots + v_n^2}. \tag{1.2.3}$$

When the angle θ between the vectors $\vec{U}$ and $\vec{V}$ is zero or $2n\pi$, $n = 0, 1, \ldots$ then $\cos\theta = 1$ which means that the two vectors are scalar multiples of each other. Thus we have an interesting result:

(i) When equality in the Cauchy–Schwartz inequality holds the two vectors are scalar multiples of each other, that is, $\vec{U} = c\vec{V}$ where c is a scalar quantity.

When $\theta = \pi/2$ then $\cos\theta = 0$ which means $\vec{U}.\vec{V} = 0$. When the angle between the vectors $\vec{U}$ and $\vec{V}$ is $\pi/2$, we may say that the vectors are orthogonal to each other, then the dot product is zero. Orthogonality will be taken up later.

Example 1.2.1. A girl is standing in a park and looking at a bird sitting on a tree. Taking one corner of the park as the origin and the rectangular border roads as the (x, y)-axes the positions of the girl and the tree are $(1, 2)$ and $(10, 15)$ respectively, all measurements in feet. The girl is 5 feet tall to her eye level and the bird's position from the ground is 20 feet up. Compute the following items: (a) The vector from the girl's eyes to the bird and its length; (b) The vector from the foot of the tree to the girl's feet and its length; (c) When the girl is looking at the bird the angle this path makes with the horizontal direction; (d) The angle this path makes with the vertical direction.

Solution 1.2.1. The positions of the girl's eyes and the bird are respectively $\vec{U} = (1, 2, 5)$ and $\vec{V} = (10, 15, 20)$.

(a) The vector from the girl's eyes to the bird is then

$$\vec{V} - \vec{U} = (10 - 1, 15 - 2, 20 - 5) = (9, 13, 15)$$

and its length is then

$$\|\vec{V} - \vec{U}\| = \sqrt{(9)^2 + (13)^2 + (15)^2} = \sqrt{475}.$$

(b) The foot of the tree is $\vec{V}_1 = (10, 15, 0)$ and the position of the girl's feet is $\vec{U}_1 = (1, 2, 0)$. The vector from the foot of the tree to the girl's feet is then

$$\vec{U}_1 - \vec{V}_1 = (1, 2, 0) - (10, 15, 0) = (-9, -13, 0)$$

and its length is

$$\|\vec{U}_1 - \vec{V}_1\| = \sqrt{(-9)^2 + (-13)^2 + (0)^2} = \sqrt{250}.$$

(c) From the girl's eyes the vector in the horizontal direction to the tree is

$$\vec{V}_2 - \vec{U}_2 = (10, 15, 5) - (1, 2, 5) = (10 - 1, 15 - 2, 5 - 5) = (9, 13, 0)$$

and its length is

$$\|\vec{V}_2 - \vec{U}_2\| = \sqrt{(9)^2 + (13)^2 + (0)^2} = \sqrt{250}.$$

Let θ be the angle between the vectors $\vec{V} - \vec{U}$ and $\vec{V}_2 - \vec{U}_2$. Then

$$\cos\theta = \frac{(\vec{V} - \vec{U}).(\vec{V}_2 - \vec{U}_2)}{\|\vec{V} - \vec{U}\| \|\vec{V}_2 - \vec{U}_2\|}$$
$$= \frac{(9, 13, 15).(9, 13, 0)}{\sqrt{475}\sqrt{250}} = \frac{\sqrt{250}}{\sqrt{475}} = \sqrt{\frac{10}{19}}.$$

Then the angle θ is given by

$$\theta = \cos^{-1}\sqrt{\frac{10}{19}}.$$

(d) The angle in the vertical direction is

$$\frac{\pi}{2} - \theta = \frac{\pi}{2} - \cos^{-1}\sqrt{\frac{10}{19}}.$$

1.2.6 Orthogonal and orthonormal vectors

Definition 1.2.2 (Orthogonal vectors). Two real vectors $\vec{U}$ and $\vec{V}$ are said to be orthogonal to each other if the angle between them is $\frac{\pi}{2} = 90°$ or equivalently, if $\cos\theta = 0$ or equivalently, if $\vec{U}.\vec{V} = 0$.

It follows, trivially, that every vector is orthogonal to a null vector since the dot product is zero.

Definition 1.2.3 (Orthonormal system of vectors). A system of real vectors $\vec{U}_1, \dots, \vec{U}_k$ is said to be an orthonormal system if $\vec{U}_i.\vec{U}_j = 0$ for all i and j, $i \neq j$ (all different vectors are orthogonal to each other or they form an *orthogonal system*) and in addition, $\|\vec{U}_j\| = 1$, $j = 1, 2, \dots, k$ (all vectors have unit length).

As an illustrative example, consider the vectors

$$\vec{U}_1 = (1,1,1), \quad \vec{U}_2 = (1,0,-1), \quad \vec{U}_3 = (1,-2,1).$$

Then

$$\vec{U}_1.\vec{U}_2 = (1)(1) + (1)(0) + (1)(-1) = 0;$$
$$\vec{U}_1.\vec{U}_3 = (1)(1) + (1)(-2) + (1)(1) = 0;$$
$$\vec{U}_2.\vec{U}_3 = (1)(1) + (0)(-2) + (-1)(1) = 0.$$

Thus $\vec{U}_1, \vec{U}_2, \vec{U}_3$ form an orthogonal system. Let us normalize the vectors in order to create an orthonormal system. Let us compute the lengths

$$\|\vec{U}_1\| = \sqrt{(1)^2 + (1)^2 + (1)^2} = \sqrt{3}, \quad \|\vec{U}_2\| = \sqrt{2}, \quad \|\vec{U}_3\| = \sqrt{6}.$$

Consider the vectors

$$\vec{V}_1 = \frac{1}{\|\vec{U}_1\|}\vec{U}_1 = \frac{1}{\sqrt{3}}(1,1,1) = \left(\frac{1}{\sqrt{3}}, \frac{1}{\sqrt{3}}, \frac{1}{\sqrt{3}}\right)$$
$$\vec{V}_2 = \frac{1}{\|\vec{U}_2\|}\vec{U}_2 = \left(\frac{1}{\sqrt{2}}(1,0,-1)\right)$$
$$\vec{V}_3 = \frac{1}{\|\vec{U}_3\|}\vec{U}_3 = \left(\frac{1}{\sqrt{6}}(1,-2,1)\right).$$

Then $\vec{V}_1, \vec{V}_2, \vec{V}_3$ form an orthonormal system.

As another example, consider the vectors,

$$e_1 = (1,0,\dots,0), \quad e_2 = (0,1,0,\dots,0), \quad \dots, \quad e_n = (0,\dots,0,1).$$

Then evidently

$$e_i.e_j = 0, \quad i \neq j, \quad \|e_i\| = 1, \quad i,j = 1,\dots,n.$$

Hence $e_1, \dots, e_n$ is an orthonormal system.

Definition 1.2.4 (Basic unit vectors). The above vectors $e_1, \ldots, e_n$ are called the basic unit vectors in n-space. [One could have written them as column vectors as well.]

Engineers often use the notation

$$\vec{i} = (1,0), \quad \vec{j} = (0,1) \quad \text{or} \quad \vec{i} = \begin{pmatrix} 1 \\ 0 \end{pmatrix}, \quad \vec{j} = \begin{pmatrix} 0 \\ 1 \end{pmatrix} \tag{1.2.4}$$

to denote the basic unit vectors in 2-space and

$$\vec{i} = (1,0,0), \quad \vec{j} = (0,1,0), \quad \vec{k} = (0,0,1) \quad \text{or}$$

$$\vec{i} = \begin{bmatrix} 1 \\ 0 \\ 0 \end{bmatrix}, \quad \vec{j} = \begin{bmatrix} 0 \\ 1 \\ 0 \end{bmatrix}, \quad \vec{k} = \begin{bmatrix} 0 \\ 0 \\ 1 \end{bmatrix} \tag{1.2.5}$$

to denote the basic unit vectors in 3-space. One interesting property is the following:

(ii) Any n-vector can be written as a linear combination of the basic unit vectors $e_1, \ldots, e_n$.

For example, consider a general 2-vector $\vec{U} = (a, b)$. Then

$$a\vec{i} + b\vec{j} = a(1,0) + b(0,1) = (a,0) + (0,b) = (a,b) = \vec{U}. \tag{1.2.6}$$

If $\vec{V} = (a,b,c)$ is a general 3-vector then

$$a\vec{i} + b\vec{j} + c\vec{k} = a(1,0,0) + b(0,1,0) + c(0,0,1)$$
$$= (a,0,0) + (0,b,0) + (0,0,c) = (a,b,c) = \vec{V}. \tag{1.2.7}$$

Note that the same notation $\vec{i}$ and $\vec{j}$ are used for the unit vectors in 2-space as well as in 3-space. There is no room for confusion since we will not be mixing 2-vectors and 3-vectors at any stage when these are used. In general, we can state a general result. Let $\vec{U}$ be an n-vector with the elements $(u_1, \ldots, u_n)$ then

$$\vec{U} = u_1 e_1 + \cdots + u_n e_n. \tag{1.2.8}$$

[Either all row vectors or all column vectors.]

The geometry of the above result can be illustrated as follows: We take a 2-space for convenience.

The vector $\vec{i}$ is in the horizontal direction with unit length. Then $a\vec{i}$ will be of length $|a|$ and in the same direction if $a > 0$ and in the opposite direction if $a < 0$. Similarly $\vec{j}$ is a unit vector in the vertical direction and $b\vec{j}$ is of length $|b|$ and in the same direction

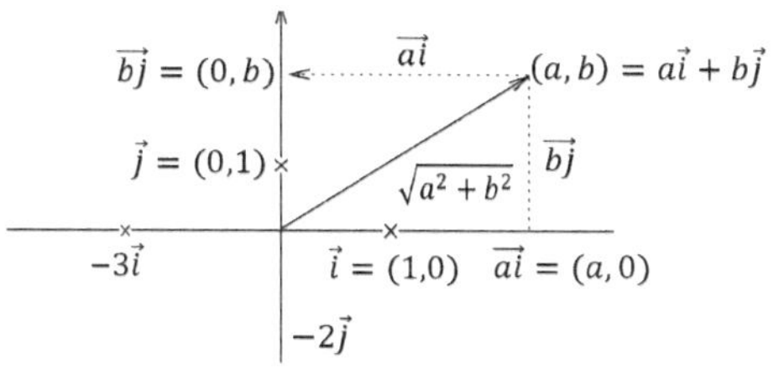

Figure 1.2.5: Geometry of linear combinations.

if $b > 0$ and in the opposite direction if $b < 0$ as shown in Figure 1.2.5. Then the point (a, b), as an arrowhead, is $a\vec{i} + b\vec{j}$. If the angle the vector

$$\vec{U} = (a, b) = a\vec{i} + b\vec{j}$$

makes with the x-axis is θ then

$$\cos\theta = \frac{(a\vec{i} + b\vec{j}).(a\vec{i})}{\|a\vec{i} + b\vec{j}\| \, \|a\vec{i}\|} = \frac{(a)(a) + (b)(0)}{\sqrt{a^2 + b^2}\sqrt{a^2}}$$

$$= \frac{a}{\sqrt{a^2 + b^2}} \tag{1.2.9}$$

and

$$\sin\theta = \sqrt{1 - \cos^2\theta} = \frac{b}{\sqrt{a^2 + b^2}}. \tag{1.2.10}$$

Observe that (1.2.9) and (1.2.10) are consistent with the notions in ordinary trigonometrical calculations as well.

1.2.7 Projections

If $\vec{U} = (a, b)$ then the projection of $\vec{U}$ in the horizontal direction is

$$a = \sqrt{a^2 + b^2} \cos\theta = \|\vec{U}\| \cos\theta$$

which is the shadow on the x-axis if light beams come parallel to the y-axis and hit the vector (arrowhead), and the projection in the vertical direction is

$$b = \sqrt{a^2 + b^2} \sin\theta = \|\vec{U}\| \sin\theta$$

which is the shadow on the y-axis if light beams come parallel to the x-axis and hit the vector. These results hold in n-space also. Consider a plane on which the vector $\vec{V}$ in n-space lies. Consider a horizontal and a vertical direction in this plane with the tail-end of the vector at the origin and let θ be the angle $\vec{V}$ makes with the horizontal direction. Then

$$\|\vec{V}\| \cos\theta = \text{projection of } \vec{V} \text{ in the horizontal direction} \tag{1.2.11}$$

and

$$\|\vec{V}\|\sin\theta = \text{projection of } \vec{V} \text{ in the vertical direction.} \qquad (1.2.12)$$

In practical terms one can explain the horizontal and vertical components of a vector as follows: Suppose that a particle is sitting at the position $(0,0)$. A wind with a speed of $5\cos 45° = \frac{5}{\sqrt{2}}$ units is blowing in the horizontal direction and a wind with a speed of $5\sin 45° = \frac{5}{\sqrt{2}}$ units is blowing in the vertical direction. Then the particle will move at $45°$ angle to the x-axis and move at a speed of 5 units.

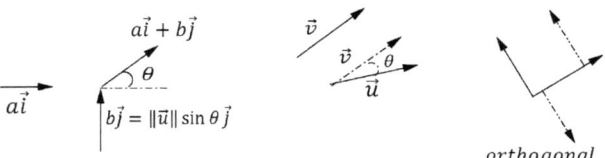

Figure 1.2.6: Movement of a particle.

Consider two arbitrary vectors $\vec{U}$ and $\vec{V}$ (coordinate-free definitions). What is the projection of $\vec{V}$ in the direction of $\vec{U}$? We can move $\vec{V}$ parallel to itself so that the tail-end of $\vec{V}$ coincides with the tail-end of $\vec{U}$. Consider the plane where these two vectors lie and let θ be the angle this displaced $\vec{V}$ makes with $\vec{U}$. Then the projection of $\vec{V}$ onto $\vec{U}$ is $\|\vec{V}\|\cos\theta$ as shown in Figure 1.2.6 (b). But

$$\cos\theta = \frac{\vec{U}.\vec{V}}{\|\vec{U}\|\,\|\vec{V}\|} \quad\Rightarrow$$

$$\|\vec{V}\|\cos\theta = \frac{\vec{U}.\vec{V}}{\|\vec{U}\|} = \text{projection of } \vec{V} \text{ onto } \vec{U}. \qquad (1.2.13)$$

If $\vec{U}$ is a unit vector then $\|\vec{U}\| = 1$ and then the projection of $\vec{V}$ in the direction of $\vec{U}$ is the dot product between $\vec{U}$ and $\vec{V}$.

Definition 1.2.5 (Projection vector of $\vec{V}$ in the direction of a unit vector $\vec{U}$). A vector in the direction of $\vec{U}$ with a length equal to $\|\vec{V}\|\cos\theta$, the projection of $\vec{V}$ onto $\vec{U}$, is called the projection vector of $\vec{V}$ in the direction of $\vec{U}$.

Then the projection vector $\vec{V}$ in the direction of $\vec{U}$ is given by

$$(\vec{U}.\vec{V})\vec{U} \quad \text{if } \vec{U} \text{ is a unit vector}$$

and

$$(\vec{U}.\vec{V})\frac{\vec{U}}{\|\vec{U}\|^2} \quad \text{if } \vec{U} \text{ is any non-null vector.} \qquad (1.2.14)$$

Example 1.2.2. Evaluate the projection vector $\vec{V}$ in the direction of $\vec{U}$ if

(a) $\vec{V} = 2\vec{i} + \vec{j} - \vec{k}, \quad \vec{U} = \frac{1}{\sqrt{3}}(\vec{i} + \vec{j} + \vec{k});$

(b) $\vec{V} = \vec{i} - \vec{j} + \vec{k}, \quad \vec{U} = 2\vec{i} + \vec{j} + \vec{k};$

(c) $\vec{V} = \vec{i} + \vec{j} + \vec{k}, \quad \vec{U} = \vec{i} - \vec{k}.$

Solution 1.2.2. (a) Here $\vec{U}$ is a unit vector and hence the required vector is

$$(\vec{U}.\vec{V})\vec{U} = \left[(2,1,-1).\left(\frac{1}{\sqrt{3}}, \frac{1}{\sqrt{3}}, \frac{1}{\sqrt{3}} \right) \right]\left(\frac{\vec{i}}{\sqrt{3}} + \frac{\vec{j}}{\sqrt{3}} + \frac{\vec{k}}{\sqrt{3}} \right)$$

$$= \frac{2}{\sqrt{3}}\left(\frac{\vec{i}}{\sqrt{3}} + \frac{\vec{j}}{\sqrt{3}} + \frac{\vec{k}}{\sqrt{3}} \right) = \frac{2}{3}(\vec{i} + \vec{j} + \vec{k}).$$

(b) Here $\vec{U}$ is not a unit vector. Let us create a unit vector in the direction of $\vec{U}$, namely

$$\vec{U}_1 = \frac{\vec{U}}{\|\vec{U}\|} = \frac{1}{\sqrt{6}}(2\vec{i} + \vec{j} + \vec{k}).$$

Now apply the formula on $\vec{V}$ and $\vec{U}_1$. The required vector is the following:

$$(\vec{V}.\vec{U}_1)\vec{U}_1 = \left[(1,-1,1).\left(\frac{2}{\sqrt{6}}, \frac{1}{\sqrt{6}}, \frac{1}{\sqrt{6}} \right) \right]\left(\frac{2}{\sqrt{6}}\vec{i} + \frac{1}{\sqrt{6}}\vec{j} + \frac{1}{\sqrt{6}}\vec{k} \right)$$

$$= \frac{2}{\sqrt{6}}\left(\frac{2}{\sqrt{6}}\vec{i} + \frac{1}{\sqrt{6}}\vec{j} + \frac{1}{\sqrt{6}}\vec{k} \right)$$

$$= \frac{2}{6}(2\vec{i} + \vec{j} + \vec{k}).$$

(c) Here $\vec{V}.\vec{U} = (1,1,1).(1,0,-1) = 0$. Hence the projection vector is the null vector.

Definition 1.2.6 (Velocity vector). In the language of engineers and physicists, the velocity is a vector with a certain direction and magnitude (length of the vector) and speed is the magnitude of the velocity vector.

For example, if $\vec{V} = (a,b)$ is the velocity vector as in Figure 1.2.6 then the direction of the vector is shown by the arrowhead there and the speed in this case is $\sqrt{a^2 + b^2} = \|\vec{V}\|$. If the velocity vector of a wind is $\vec{V} = 2\vec{i} + \vec{j} + \vec{k}$ in a 3-space then its speed is $\|\vec{V}\| = \sqrt{(2)^2 + (1)^2 + (-1)^2} = \sqrt{6}$.

Example 1.2.3. A plane is flying straight East horizontally at a speed of 200 km/hour and another plane is flying horizontally North-East at a speed of 600 km/hour. Draw the velocity vectors for both the planes.

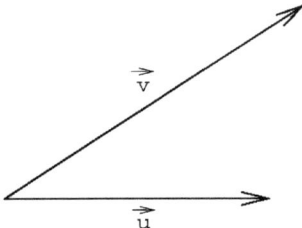

Figure 1.2.7: Velocity vectors.

Solution 1.2.3. If the velocity vectors for the two planes are denoted by $\vec{U}$ and $\vec{V}$ respectively, as shown in Figure 1.2.7 then the given information is that

$$\|\vec{U}\| = 200 \quad \text{and} \quad \|\vec{V}\| = 600.$$

If the direction of $\vec{U}$ is taken as the x-axis on the plane where the two vectors lie (displaced if necessary so that the tail-ends meet at $(0,0)$) then on this plane

$$\vec{U} = 200\vec{i} \quad \text{and} \quad \vec{V} = (600\cos 45°)\vec{i} + (600\sin 45°)\vec{j}$$

$$= \frac{600}{\sqrt{2}}\vec{i} + \frac{600}{\sqrt{2}}\vec{j}.$$

Example 1.2.4. A sail boat is steered to move straight East. There is a wind with a velocity in the North-East direction and with a speed of 50 km/hour. What is the speed of the boat if (a) the only force acting on the boat is the wind, (b) in addition to the wind the sail boat has a motor which is set for a speed of 20 km/hour.

Solution 1.2.4. (a) The only component here is the component of the wind velocity vector in the direction of the boat which is $\|\vec{V}\|\cos\theta$ if $\vec{V}$ is the velocity vector and θ is the angle $\vec{V}$ makes with the East direction (East direction is taken as the x-axis direction). We are given $\|\vec{V}\| = 50$ and $\theta = 45°$. Then the speed of the boat is $\|\vec{V}\|\cos\theta = \frac{50}{\sqrt{2}}$ and the velocity is $\vec{U} = \frac{50}{\sqrt{2}}\vec{i}$.

(b) In this case the above component plus the speed set by the engine are there. Then the combined speed is $\frac{50}{\sqrt{2}} + 20$ and the velocity vector is

$$\vec{U} = \left(\frac{50}{\sqrt{2}} + 20\right)\vec{i}.$$

1.2.8 Work done

When a force of magnitude F is applied on an object and the object is moved in the same direction of the force for a distance d then we say that the work done is Fd (F multiplied by d). For example if the force vector has the magnitude 20 units and the distance moved in the same direction of the force is 10 units then the work done is 200 units (force, distance and work are measured in different units such as force in *new-*

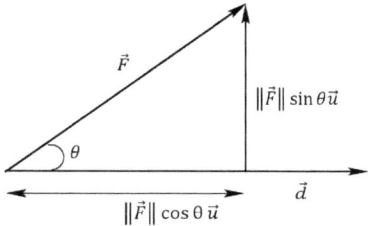

Figure 1.2.8: Work done.

tons, distance in kilometers and work in *joules*). Suppose that the force vector is in a certain direction and the distance moved is in another direction then what will be the work done? Let $\vec{F}$ be the force vector and $\vec{d}$ the displacement vector as shown in Figure 1.2.8.

Let the force vector $\vec{F}$ make an angle θ with the displacement vector $\vec{d}$. Then the projection of $\vec{F}$ in the direction of $\vec{d}$ is $\|\vec{F}\|\cos\theta$ [and the projection vector is $\|\vec{F}\|(\cos\theta)\vec{U}$ where $\vec{U}$ is a unit vector in the direction of $\vec{d}$. The component vector of $\vec{F}$ in the perpendicular direction to $\vec{d}$ is

$$\|\vec{F}\|\sin\theta = \vec{F} - \|\vec{F}\|(\cos\theta)\vec{U}.$$

This is not required in our computations]. Then the work done, denoted by w, is

$$w = \|\vec{F}\|\cos\theta\|\vec{d}\|$$

$$= \|\vec{F}\|\frac{(\vec{F}.\vec{d})}{\|\vec{F}\|\,\|\vec{d}\|}\|\vec{d}\| = \vec{F}.\vec{d}. \tag{1.2.15}$$

Example 1.2.5. The ground force $\vec{F} = 5\vec{i} + 2\vec{j}$ of a wind moved a stone in the direction of the displacement $\vec{d} = \vec{i} + 3\vec{j}$. What is the work done by this wind in moving the stone?

Solution 1.2.5. According to (1.2.15) the work done is

$$w = \vec{F}.\vec{d} = (5,2).(1,3) = (5)(1) + (2)(6) = 11.$$

Example 1.2.6. Consider a triangle ABC with the angles denoted by A, B, C and the lengths of the sides opposite to these angles by a, b, c, as shown in Figure 1.2.9. Then show that

$$a^2 = b^2 + c^2 - 2bc\cos A.$$

Solution 1.2.6. Consider the vectors $\vec{AB}$ and $\vec{AC}$, starting from A and going to B and C respectively.

Then the vector $\vec{BC} = \vec{AC} - \vec{AB}$. Therefore

$$\|\vec{BC}\|^2 = \|\vec{AC} - \vec{AB}\|^2$$

$$= \|\vec{AC}\|^2 + \|\vec{AB}\|^2 - 2\|\vec{AC}\|\,\|\vec{AB}\|\cos A.$$

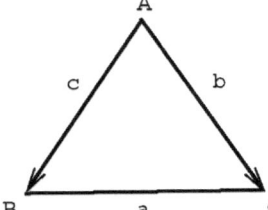

Figure 1.2.9: A triangle.

That is, $a^2 = b^2 + c^2 - 2bc \cos A$. Here we have used the fact that the square of the length is the dot product with itself:

$$a^2 = \|\vec{AC} - \vec{AB}\|^2$$
$$= (\vec{AC} - \vec{AB}).(\vec{AC} - \vec{AB})$$
$$= (\vec{AC}.\vec{AC}) - (\vec{AC}).(\vec{AB}) - (\vec{AB}.\vec{AC}) + (\vec{AB}).(\vec{AB})$$
$$= \|\vec{AC}\|^2 + \|\vec{AB}\|^2 - 2(\vec{AC}.\vec{AB})$$
$$= \|\vec{AC}\|^2 + \|\vec{AB}\|^2 - 2\|\vec{AC}\| \|\vec{AB}\| \cos A$$
$$= b^2 + c^2 - 2bc \cos A.$$

Exercises 1.2

1.2.1. Give geometric representation to the following vectors:

(a) $\vec{U} = 2\vec{i} - 3\vec{j}$,   (b) $2\vec{U}$,   (c) $-2\vec{U}$,

(d) $\vec{V} = \vec{i} + \vec{j}$,   (e) $\vec{U} + \vec{V}$,

(f) $\vec{U} - \vec{V}$,   (g) $\vec{V} - 2\vec{U}$,   (h) $2\vec{U} + 3\vec{V}$.

1.2.2. Compute the angle between the following vectors:

(a) $\vec{U} = \vec{i} + \vec{j} - \vec{k}$, $\vec{V} = 2\vec{i} - \vec{j} + 3\vec{k}$;

(b) $\vec{U} = \vec{i} + \vec{j} + \vec{k}$, $\vec{V} = \vec{i} - 2\vec{j} + \vec{k}$;

(c) $\vec{U} = (1, -1, 2, 3, 5, -1)$, $\vec{V} = (2, 0, 0, -1, 1, 2)$.

1.2.3. Verify Cauchy–Schwartz inequality for $\vec{U}$ and $\vec{V}$ in the three cases in Exercise 1.2.2.

1.2.4. Normalize the following vector $\vec{U}$, then construct two vectors which are orthogonal among themselves as well as both are orthonormal to $\vec{U}$, where $\vec{U} = (1, 1, 1, 1)$.

1.2.5. Given the two vectors $\vec{U}_1 = (1, 1, 1, 1)$ and $\vec{U}_2 = (1, 2, -1, 1)$ construct two vectors $\vec{V}_1$ and $\vec{V}_2$ such that $\vec{V}_1$ is the normalized $\vec{U}_1$, $\vec{V}_2$ is a normalized vector orthogonal to $\vec{V}_1$ and both $\vec{V}_1$ and $\vec{V}_2$ are linear functions of $\vec{U}_1$ and $\vec{U}_2$.

1.2.6. Let $P = (x_0, y_0, z_0)$ a fixed point in 3-space, $Q = (x, y, z)$ an arbitrary point in 3-space. Construct the vector going from P to Q. Derive the equation to the plane where the vector $\vec{PQ}$ lies on the plane as well as another vector $\tilde{N} = (a, b, c)$ is normal to this plane (Normal to a plane means orthogonal to every vector lying on the plane).

1.2.7. If $x - y + z = 7$ is a plane, (i) is the point $(1, 1, 1)$ on this plane? (ii) construct a normal to this plane with length 5, (iii) construct a plane parallel to the given plane and passing through the point $(1, 1, 2)$, (iv) construct a plane orthogonal to the given plane and passing through the point $(1, -1, 4)$.

1.2.8. Derive the equation to the plane passing through the points

$$(1, 1, -1), \quad (2, 1, 2), \quad (2, 1, 0).$$

1.2.9. Find the area of the parallelogram formed by the vectors (by completing it as in Figure 1.2.2 on the plane determined by the two vectors),

$$\vec{U} = 2\vec{i} + \vec{j} - \vec{k} \quad \text{and} \quad \vec{V} = \vec{i} - \vec{j} + 3\vec{k}.$$

1.2.10. Find the work done by the force $\vec{F} = 2\vec{i} - \vec{j} + 3\vec{k}$ for the displacement $\vec{d} = 3\vec{i} + \vec{j} - \vec{k}$.

1.2.11. A boat is trying to cross a river at a speed of 20 miles/hour straight across. The river flow downstream is 10 miles/hour. Evaluate the eventual direction and speed of the boat.

1.2.12. In Exercise 1.2.11 if the river flow speed is the same what should be the direction and speed of the boat so that it can travel straight across the river?

1.2.13. Evaluate the area of the triangle whose vertices are $(1, 0, 1)$, $(2, 1, 5)$, $(1, -1, 2)$ by using vector method.

1.2.14. Find the angle between the planes (angle between the normals to the planes)

$$x + y - z = 7 \quad \text{and} \quad 2x + y - 3z = 5.$$

1.2.15. In some engineering problems of signal processing a concept called *convolution of two vectors* is defined. Let $X = (x_1, \dots, x_n)$ and $Y = (y_1, \dots, y_n)$ be two row vectors of the same order. Then the convolution, denoted by $X * Y$, is defined as follows: It is again a $1 \times n$ vector where the i-th element in $X * Y$ is given by

$$x_1 y_i + x_2 y_{i-1} + \cdots + x_i y_1$$
$$+ x_{i+1} y_n + x_{i+2} y_{n-1} + \cdots + x_n y_{i+1}.$$

For example, for $n = 2$

$$X * Y = (x_1, x_2) * (y_1, y_2) = (x_1 y_1 + x_2 y_2, x_1 y_2 + x_2 y_1)$$

(a) Write down the explicit expression for $(x_1, x_2, x_3) * (y_1, y_2, y_3)$.

(b) Show that the operator $*$ is commutative as well as associative for a general n.
(c) Evaluate $(1, 0, -1, 2) * (3, 4, 5, -2)$.

1.2.16. Find the angle between the planes

$$x - y + z = 2 \quad \text{and} \quad 2x + 3y - 4z = 8.$$

1.2.17. Evaluate the area of the triangle whose vertices are $(1, 1, 1)$, $(2, 5, 3)$, $(1, -1, -1)$.

1.2.18. Evaluate the area of the parallelogram determined by the vectors $U = (1, -1, 2, 5)$ and $V = (1, 1, -1, -1)$.

1.3 Linear dependence and linear independence of vectors

Consider the vectors $U_1 = (1, 0, -1)$ and $U_2 = (1, 1, 1)$. For arbitrary scalars a_1 and a_2 let us try to solve the equation

$$a_1 U_1 + a_2 U_2 = O \tag{1.3.1}$$

to see whether there exist nonzero a_1 and a_2 such that (1.3.1) is satisfied.

$$a_1 U_1 + a_2 U_2 = O \Rightarrow$$
$$a_1(1, 0, -1) + a_2(1, 1, 1) = O = (0, 0, 0).$$

That is,

$$(a_1 + a_2, a_2, -a_1 + a_2) = (0, 0, 0) \Rightarrow$$
$$a_1 + a_2 = 0, \quad a_2 = 0, \quad -a_1 + a_2 = 0.$$

The only values of a_1 and a_2 satisfying the three equations $a_1 + a_2 = 0$, $a_2 = 0$ and $-a_1 + a_2 = 0$ are $a_1 = 0$ and $a_2 = 0$. This means that the only solution for a_1 and a_2 in (1.3.1) is $a_1 = 0$ and $a_2 = 0$. Observe that $a_1 = 0$, $a_2 = 0$ is always a solution to the equation (1.3.1). But here we have seen that $a_1 = 0$, $a_2 = 0$ is the only solution. Now, let us look at another situation. Consider the vectors

$$U_1 = (1, 1, 1), \quad U_2 = (1, -1, 2), \quad U_3 = (2, 0, 3).$$

Solve the equation

$$a_1 U_1 + a_2 U_2 + a_3 U_3 = O \tag{1.3.2}$$

for a_1, a_2, a_3. Then

$$a_1 U_1 + a_2 U_2 + a_3 U_3 = O \Rightarrow$$
$$a_1(1, 1, 1) + a_2(1, -1, 2) + a_3(2, 0, 3) = (0, 0, 0).$$

That is,

$$(a_1 + a_2 + 2a_3, a_1 - a_2, a_1 + 2a_2 + 3a_3) = (0, 0, 0).$$

This means,

$$a_1 + a_2 + 2a_3 = 0, \qquad \text{(i)}$$
$$a_1 - a_2 = 0, \qquad \text{(ii)}$$
$$a_1 + 2a_2 + 3a_3 = 0. \qquad \text{(iii)}$$

From (ii), $a_1 = a_2$; substituting in (i), $a_1 = a_2 = -a_3$; substituting in (iii) the equation is satisfied. Then there are infinitely many non-zero a_1, a_2, a_3 for which (1.3.2) is satisfied. For example, $a_1 = 1 = a_2, a_3 = -1$ will satisfy (1.3.2). In the above considerations we have two systems of vectors. In one system the only possibility for the coefficient vector is the null vector which means that no vector can be written as a linear function of the other vectors. In the other case the coefficient vector is not null which means that at least one of the vectors there can be written as a linear combination of others.

Definition 1.3.1 (Linear independence). Let $U_1, U_2, \dots, U_k$ be k given non-null n-vectors, where k is finite. Consider the equation

$$a_1 U_1 + a_2 U_2 + \cdots + a_k U_k = O \qquad (1.3.3)$$

where $a_1, \dots, a_k$ are scalars. If the only possibility for (1.3.3) to hold is when $a_1 = 0, \dots, a_k = 0$ then the vectors $U_1, \dots, U_k$ are called *linearly independent*. If there exists at least one non-null vector $(a_1, \dots, a_k)$ such that (1.3.3) is satisfied then the system of vectors $U_1, \dots, U_k$ are *linearly dependent*.

If a non-null vector $(a_1, \dots, a_k)$ exists then at least one of the elements is nonzero. Let $a_1 \neq 0$. Then from (1.3.3)

$$U_1 = -\frac{a_2}{a_1} U_2 - \cdots - \frac{a_k}{a_1} U_k. \qquad (1.3.4)$$

That is, U_1 can be written as a linear function of $U_2, \dots, U_k$. Note that not all $a_2, \dots, a_k$ can be zeros. If they are all zeros then from (1.3.4) U_1 is a null vector. But a null vector is not included in our definition. Thus at least one of them can be written as a linear function of the others if $U_1, \dots, U_k$ are linearly dependent. If they are linearly independent then none can be written as a linear function of the others.

(i) A null vector is counted among dependent vectors. A set consisting of one non-null vector is counted as an independent system of vectors.

Example 1.3.1. Show that the basic unit vectors $e_1, \dots, e_n$ are linearly independent.

Solution 1.3.1. Consider the equation

$$a_1 e_1 + \cdots + a_n e_n = O \Rightarrow$$

$$a_1(1,0,\dots,0) + \dots + a_n(0,\dots,0,1) = (0,\dots,0) \implies$$
$$(a_1,\dots,a_n) = (0,\dots,0) \implies$$
$$a_1 = 0,\dots,a_n = 0$$

is the only solution, which means that $e_1,\dots,e_n$ are linearly independent.

Example 1.3.2. Show that a system of non-null mutually orthogonal vectors are linearly independent.

Solution 1.3.2. Let $V_1,\dots,V_k$ be a system of mutually orthogonal vectors. Consider the equation

$$a_1 V_1 + \dots + a_k V_k = O.$$

Take the dot product on both sides with respect to V_1. Then we have

$$a_1 V_1.V_1 + a_2 V_2.V_1 + \dots + a_k V_k.V_1 = O.V_1 = O.$$

But $V_j.V_1 = 0$ for $j \neq 1$ and $V_1.V_1 = \|V_1\|^2 \neq 0$. This means that $a_1 = 0$. Similarly $a_2 = 0,\dots,a_k = 0$ which means that $V_1,\dots,V_k$ are linearly independent. This is a very important result.

(ii) Every set of mutually orthogonal non-null vectors are linearly independent.
(iii) Any finite collection of vectors containing the null vector is counted as a linearly dependent system of vectors. If S_1 and S are two finite collections of vectors where S_1 is a subset of S, that is, $S_1 \subset S$, then the following hold: If S_1 is a linearly dependent system then S is also a linearly dependent system. If S is a linearly independent system then S_1 is also a linearly independent system.

Example 1.3.3. Check the linear dependence of the following sets of vectors:

(a) $U_1 = (1,2,1), \quad U_2 = (1,1,1)$;
(b) $U_1 = (1,-1,2), \quad U_2 = (1,1,0)$;
(c) $U_1 = (1,2,1), \quad U_2 = (1,-1,1), \quad U_3 = (3,3,3)$.

Solution 1.3.3. (a) For two vectors to be dependent one has to be a non-zero scalar multiple of the other. Hence U_1 and U_2 here are linearly independent.

(b) By inspection $U_1.U_2 = 0$ and hence they are orthogonal thereby linearly independent.

(c) U_1 and U_2 are evidently linearly independent, being not multiples of each other. By inspection $U_3 = 2U_1 + U_2$ and hence the set $\{U_1, U_2, U_3\}$ is a linearly dependent system.

(iv) Linear dependence or independence in a system of vectors is not altered by scalar multiplication of the vectors by non-zero scalars.

This result can be easily seen from the definition itself. Let the n-vectors $U_1, \ldots, U_k$ be linearly independent. Then

$$a_1 U_1 + \cdots + a_k U_k = O \Rightarrow a_1 = 0, \ldots, a_k = 0.$$

Let $c_1, \ldots, c_k$ be non-zero scalars. If $a_i = 0$ then $a_i c_i = 0$ and vice versa since $c_i \neq 0$, $i = 1, 2, \ldots, k$. Thus

$$a_1 (c_1 U_1) + \cdots + a_k (c_k U_k) = O \Rightarrow a_1 = 0, \ldots, a_k = 0.$$

On the other hand, if $U_1, \ldots, U_k$ are linearly dependent then at least one of them can be written as a linear function of the others. Let

$$U_1 = b_2 U_2 + \cdots + b_k U_k$$

where $b_2, \ldots, b_k$ are some constants, at least one of them nonzero. Then for $c_1 \neq 0, \ldots, c_k \neq 0$

$$c_1 U_1 = \frac{c_1 b_2}{c_2} (c_2 U_2) + \cdots + \frac{c_1 b_k}{c_k} (c_k U_k).$$

Thus $c_1 U_1, \ldots, c_k U_k$ are linearly dependent.

We have another important result on linear independence.

(v) Linear independence or dependence in a system of vectors is not altered by adding a scalar multiple of any vector in the system to any other vector in the system.

This result is easy to establish. Let the system $U_1, \ldots, U_k$ of n-vectors be linearly independent. Then

$$a_1 U_1 + \cdots + a_k U_k = O \Rightarrow a_1 = 0, \ldots, a_k = 0.$$

Now, consider a new system $U_1, c\, U_1 + U_2, \ldots, U_k$. [That is, U_2 is replaced by $c\, U_1 + U_2$, $c \neq 0$. In other words, $c\, U_1$ is added to U_2.] Consider the equation

$$a_1 U_1 + a_2 (c\, U_1 + U_2) + a_3 U_3 + \cdots + a_k U_k = O.$$

That is,

$$(a_1 + c a_2) U_1 + a_2 U_2 + \cdots + a_k U_k = O.$$

Then since $U_1, \ldots, U_k$ are linearly independent $a_1 + c a_2 = 0$, $a_2 = 0$, ..., $a_k = 0$ which means $a_1 = 0$ also which establishes that the system of vectors $U_1, c U_1 + U_2, U_3, \ldots, U_k$ is linearly independent. A similar procedure establishes that if the original system is linearly dependent then the new system is also linearly dependent.

By combining the results (iii) and (iv) above we can have the following result:

(vi) Consider a finite collection of n-vectors. If any number of vectors in this collection are multiplied by nonzero scalars or a linear function of any number of them is added to any member in the set, linear independence or dependence in the system is preserved. That is, if the original system is linearly independent then the new system is also linearly independent and if the original system is dependent then the new system is also linearly dependent.

Example 1.3.4. Check to see whether the following system of vectors is linearly independent or dependent:

$$U_1 = (1, 0, 2, -1, 5)$$
$$U_2 = (-1, 1, 1, -1, 2)$$
$$U_3 = (2, 1, 7, -4, 17)$$

Solution 1.3.4. Since nonzero scalar multiplication and addition do not alter independence or dependence let us create new systems of vectors. In what follows the following standard notations will be used:

A few standard notations

$$\text{“}\alpha(i) \Rightarrow \text{”} \text{ means the } i\text{-th vector multiplied by } \alpha \qquad (1.3.5)$$

In this operation the i-th vector in the set is replaced by α (Greek letter alpha) times the original i-th vector. For example "$-3(1) \Rightarrow$" means that "the first vector multiplied by -3 gives", that is, the new first vector is the original first vector multiplied by -3.

$$\text{“}\alpha(i) + (j) \Rightarrow \text{”} \text{ means } \alpha \text{ times the } i\text{-th vector added to the } j\text{-th vector} \qquad (1.3.6)$$

In this operation the original i-th vector remains the same whereas the new j-th vector is the original j-th vector plus α times the original i-th vector. Let us apply these types of operations on U_1, U_2, U_3, remembering that linear independence or dependence is preserved.

$$(1) + (2) \Rightarrow U_1 = (1, 0, 2, -1, 5)$$
$$V_2 = (0, 1, 3, -2, 7)$$
$$U_3 = (2, 1, 7, -4, 17)$$

In the above operation the second vector U_2 is replaced by $U_2 + U_1 = V_2$. Let us continue the operations.

$$-2(1) + (3) \Rightarrow U_1 = (1, 0, 2, -1, 5)$$

$$V_2 = (0, 1, 3, -2, 7)$$
$$V_3 = (0, 1, 3, -2, 7)$$

In the set U_1, V_2, V_3 we will do the next operation.

$$-(2) + (3) \implies U_1 = (1, 0, 2, -1, 5)$$
$$V_2 = (0, 1, 3, -2, 7)$$
$$W_3 = (0, 0, 0, 0, 0)$$

Here W_3 is obtained by adding (-1) times V_2 to V_3 or replacing V_3 by $V_3 - V_2 = W_3$. By the above sequence of operations W_3 has become a null vector which by definition is dependent. Hence the original system U_1, U_2, U_3 is a linearly dependent system.

Example 1.3.5. Check the linear dependence or independence of the following system of vectors:

$$U_1 = (2, -1, 1, 1, 3, 4)$$
$$U_2 = (5, 2, 1, -1, 2, 1)$$
$$U_3 = (1, -1, 1, 1, 1, 4)$$

Solution 1.3.5. Since linear dependence or independence is not altered by the order in which the vectors are selected we will write U_3 first and write only the elements in 3 rows and 6 columns as follows, rather than naming them as U_3, U_1, U_2:

$$
\begin{array}{rrrrrr}
1 & -1 & 1 & 1 & 1 & 4 \\
2 & -1 & 1 & 1 & 3 & 4 \\
5 & 2 & 1 & -1 & 2 & 1
\end{array}
$$

We have written them in the order U_3, U_1, U_2 to bring a convenient number, namely 1, at the first row first column position. This does not alter linear independence or dependence in the system. Now, we will carry out more than one operations at a time. [We add (-2) times the first row to the second row and (-5) times the first row to the third row. The first row remains the same. The result is the following:]

$$
-2(1) + (2); -5(1) + (3) \implies
\begin{array}{rrrrrr}
1 & -1 & 1 & 1 & 1 & 4 \\
0 & 1 & -1 & -1 & 1 & -4 \\
0 & 7 & -4 & -6 & -3 & -19
\end{array}
$$

[On the new configuration we add the second row to the first row and (-7) times the second row to the third row. The second row remains the same. The net result is the following:]

$$
(2) + (1); -7(2) + (3) \implies
\begin{array}{rrrrrr}
1 & 0 & 0 & 0 & 2 & 0 \\
0 & 1 & -1 & -1 & 1 & -4 \\
0 & 0 & 3 & 1 & -10 & 9
\end{array}
$$

[The third row is divided by 3. The third row changes.]

$$\frac{1}{3}(3) \Rightarrow \begin{array}{cccccc} 1 & 0 & 0 & 0 & 2 & 0 \\ 0 & 1 & -1 & -1 & 1 & -4 \\ 0 & 0 & 1 & \frac{1}{3} & -\frac{10}{3} & 3 \end{array}$$

[This operation is done to bring a convenient number at the third column position on the third row. Now we add the new third row to the second row. The new third row remains the same.]

$$(3)+(2) \Rightarrow \begin{array}{cccccc} 1 & 0 & 0 & 0 & 2 & 0 \\ 0 & 1 & 0 & -\frac{2}{3} & -\frac{7}{3} & -1 \\ 0 & 0 & 1 & \frac{1}{3} & -\frac{10}{3} & 3 \end{array}$$

The aim in the above sequences of operations is to bring a unity at all leading diagonal (the diagonal from the upper left-end corner down) positions, if possible. Interchanges of rows can be done if necessary to achieve the above aim, because interchanges do not alter the linear independence or dependence. During such a process if any row becomes null then automatically the original system, represented by the starting rows, is dependent. If no row becomes null during the process then at the end of the process look at the final first, second, etc columns. In our example above look at the first column. No non-zero linear combination of the second and third rows can create a 1 at the first position. Hence the first row cannot be written as a linear function of the second and third rows. Now look at the second column. By the same argument above the second row cannot be written as a linear function of the first and third rows. Now look at the third column. By the same argument the third row cannot be written as a linear combination of the first and second rows. Hence all the three rows are linearly independent or the original system $\{U_1, U_2, U_3\}$ is a linearly independent system.

The above procedure is called a *sweep-out procedure*. Then the principles to remember in a sweep-out procedure are the following: Assume that the system consists of m vectors, each is an n-vector.

Principles in a sweep-out procedure

(1) *Write the given vectors as rows, interchange if necessary to bring a convenient nonzero number, 1 if available, at the first row first column position. Do not interchange columns, the vectors will be altered.*
(2) *Add suitable multiples of the first row to the second, third, ..., m-th row to make the first column elements, except the first element, zeros.*
(3) *Start with the second row. Interchange 2nd, ..., m-th rows if necessary to bring a convenient nonzero number at the second position on the second row.*
(4) *Add suitable multiples of the second row to the first row, third row, ..., m-th row to make all elements in the second column, except the second element, zeros.*

(5) *Repeat the process with the third, fourth etc rows until all the leading diagonal elements are non-zeros, unities if possible.*

(6) *During the process if any row becomes null then shift it to the bottom position. If at any stage a vector has become null then the system is dependent. If all the leading diagonal elements are non-zeros when all other elements in the corresponding columns are wiped out (made zeros) by the above process then the system is linearly independent.*

(7) *If the first r, for some r, leading diagonal elements are non-zeros, none of the rows has become null so far and the (r + 1)th elements in all the remaining rows are zeros then continue the process with the (r + 2)th element on the (r + 1)th row and so on. If no row has become null by the end of the whole process then all the rows are linearly independent.*

(8) *Division of a row by a non-zero scalar usually brings in fractions. Hence multiply the rows with appropriate numbers to avoid fractions and to achieve the* sweep-out *process.*

The leading diagonal elements need not be brought to unities to check for linear dependence or independence. Only nonzero elements are to be brought to the diagonal positions, if possible. When doing the operations, try to bring the system to a triangular format by reducing all elements below the leading diagonal to zeros, if possible. When the system is in a triangular format all elements above nonzero diagonal elements can be simply put as zeros because this can always be achieved by operating with the last row first, wiping out all last column elements except the last column last row element, then last but one column elements and so on. Thus all elements above nonzero diagonal elements can be simply put as zeros once the matrix is in a triangular format.

(9) *If the vectors to be checked for linear independence or dependence are column vectors then write them as rows before executing a sweep-out process. This is done only for convenience because operations on rows are easier to visualize.*

(10) *When doing a sweep-out process always write first the row that you are operating with because this row is not changing and others can change as a result of the operations.*

Example 1.3.6. Check for linear independence or dependence in the following system of vectors:

$$U_1 = (2, 0, 1, 5)$$
$$U_2 = (1, -1, 1, 1)$$
$$U_3 = (4, 2, 2, 8)$$

Solution 1.3.6. For convenience write in the order U_2, U_1, U_3 and write only the elements and continue with the sweep-out process.

$$
\begin{array}{cccc}
1 & -1 & 1 & 1 \\
2 & 0 & 1 & 5 \\
4 & 2 & 2 & 8
\end{array}
\quad -2(1)+(2);\ -4(1)+(3) \Rightarrow
\begin{array}{cccc}
1 & -1 & 1 & 1 \\
0 & 2 & -1 & 3 \\
0 & 6 & -2 & 4
\end{array}
$$

$$
2(1) \Rightarrow
\begin{array}{cccc}
2 & -2 & 2 & 2 \\
0 & 2 & -1 & 3 \\
0 & 6 & -2 & 4
\end{array}
\quad (2)+(1);\ -3(2)+(3) \Rightarrow
\begin{array}{cccc}
2 & 0 & 1 & 5 \\
0 & 2 & -1 & 3 \\
0 & 0 & 1 & -5
\end{array}
$$

$$
(3)+(2);\ -(3)+(1) \Rightarrow
\begin{array}{cccc}
2 & 0 & 0 & 10 \\
0 & 2 & 0 & -2 \\
0 & 0 & 1 & -5
\end{array}
$$

The leading diagonal elements are $2, 2, 1$ which are non-zeros and hence the system is linearly independent. [During the process above the first row is multiplied by 2 in order to avoid fractions in the rest of the operations.]

Note that in the above operations the row that you are operating with remains the same and the other rows, to which constant multiples are added, change. In the last form above, are all the four columns linearly independent? Evidently not. The last column = 5 (column 1)−(column 2)−5(column 3).

If our aim is only to check for linear independence or dependence then we need to bring the original set to a triangular type format. In the second step above the operation 2(1) and in the third stage the operation (2) + (1) need not be done. That is,

$$
\begin{array}{cccc}
1 & -1 & 1 & 1 \\
0 & 2 & -1 & 3 \\
0 & 6 & -2 & 4
\end{array}
\quad -3(2)+(3) \Rightarrow
\begin{array}{cccc}
1 & -1 & 1 & 1 \\
0 & 2 & -1 & 3 \\
0 & 0 & 1 & -5
\end{array}
$$

Now we have the triangular type format with nonzero diagonal elements. Note that the first row cannot be written as a linear function of the second and third rows. Similarly no row can be written as a linear function of the other two. At this stage if we wish to create a diagonal format for the first three columns then by using the third row one can wipe out all other elements in the third column, then by using the second row we can wipe out all other elements in the second column. In other words, we can simply replace all those elements by zeros, then only the last column will change.

$$
\begin{array}{cccc}
1 & -1 & 1 & 1 \\
0 & 2 & -1 & 3 \\
0 & 0 & 1 & -5
\end{array}
\to
\begin{array}{cccc}
1 & -1 & 0 & 6 \\
0 & 2 & 0 & -2 \\
0 & 0 & 1 & -5
\end{array}
\ \text{(operating with the third row)}
$$

$$
\to
\begin{array}{cccc}
1 & 0 & 0 & 5 \\
0 & 2 & 0 & -2 \\
0 & 0 & 1 & -5
\end{array}
\ \text{(operating with the second row)} \to
\begin{array}{cccc}
1 & 0 & 0 & 5 \\
0 & 1 & 0 & -1 \\
0 & 0 & 1 & -5
\end{array}
$$

dividing the second row by 2. Thus, the first three columns are made basic unit vectors, the same procedure if we wish to create unit vectors in the first r columns and if there are r linearly independent rows.

At a certain stage, say the rth stage, suppose that all elements in the $(r+1)$th column below the rth row are zeros. Then start with a nonzero element in the remaining configuration of the columns in the remaining row and proceed to create a triangular format. For example, consider the following situation:

$$
\begin{array}{ccccccc}
1 & 1 & -1 & 1 & -1 & 1 & 1 \\
0 & 1 & 0 & 2 & 1 & -1 & 1 \\
0 & 0 & 0 & 0 & 2 & 0 & 1 \\
0 & 0 & 0 & 0 & 0 & 0 & 1 \\
0 & 0 & 0 & 1 & 0 & 0 & -1
\end{array}
\rightarrow
\begin{array}{ccccccc}
1 & 1 & -1 & 1 & -1 & 1 & 1 \\
0 & 1 & 0 & 2 & 1 & -1 & 1 \\
0 & 0 & 0 & 1 & 0 & 0 & -1 \\
0 & 0 & 0 & 0 & 2 & 0 & 1 \\
0 & 0 & 0 & 0 & 0 & 0 & 1
\end{array}
$$

The first two rows are evidently linearly independent. Our procedure of triangularization cannot proceed. Write the 5th row in the 3rd row position to get the matrix on the right above. Now we see that the new 3rd, 4th and 5th rows form a triangular type format. This shows that all the five rows are linearly independent. Note that by using the last row one can wipe out all other elements in the 7th column. Then by using the 4th row we can wipe out all other elements in the 5th column. Then by using the 3rd row we can wipe out all other elements in the 4th column. Then by using the second row we can wipe out all other elements in the second column. Now, one can see linear independence of all the five rows clearly. In the light of the above examples and discussions we can state the following result:

(vii) There cannot be more than n mutually orthogonal n-vectors and there cannot be more than n linearly independent n-vectors.

It is not difficult to establish this result. Consider the n-vectors $U_1, \ldots, U_n, U_{n+1}$, that is, $n+1$ vectors of n elements each. Write the $n+1$ vectors as $n+1$ rows and apply the above sweep-out process. If the first n vectors are linearly independent then all the n leading diagonal spots have nonzero entries with all elements in the corresponding columns zeros. Thus automatically the $(n+1)$th row becomes null. Hence the $(n+1)$th row depends on the other n rows or the maximum number of linearly independent n-vectors possible is n.

If possible, let $V_1, \ldots, V_{n+1}$ be mutually orthogonal n-vectors. From what we proved just above not all these $n+1$ vectors can be linearly independent. Then the $(n+1)$th can be written as a linear function of the other n vectors. Then there exists a non-null vector $b = (b_1, \ldots, b_n)$ such that

$$V_{n+1} = b_1 V_1 + \cdots + b_n V_n.$$

Take the dot product on both sides with respect to V_i. If all $V_1, \ldots, V_{n+1}$ are mutually orthogonal then we have

$$0 = 0 + b_i \|V_i\|^2 + 0 \implies b_i = 0, \quad i = 1, \ldots, n$$

since $\|V_i\| \neq 0$, $i = 1, \ldots, n$. This then contradicts the fact that b is a non-null vector. Thus they cannot be all mutually orthogonal. Since the orthogonal vectors are linearly independent, proved earlier, the maximum number of n-vectors which are mutually orthogonal is n.

1.3.1 A vector subspace

The vectors in our discussion so far are ordered n-tuples of real numbers. The notions of vector spaces, dimension etc will be introduced for such vectors. Then later we will generalize these ideas to cover some general objects called vectors satisfying some general postulates. Consider, for example, two given vectors

$$U_1 = (1, 0, -1) \quad \text{and} \quad U_2 = (2, 3, 1).$$

Evidently U_1 and U_2 are linearly independent. Two vectors being dependent means one is a multiple of the other. Consider a collection S_1 of vectors which are spanned by U_1 and U_2 by the following process. Every scalar multiple of U_1 as well as of U_2 is in S_1. For example

$$3U_1 = 3(1, 0, -1) = (3, 0, -3) \in S_1$$
$$-2U_2 = -2(2, 3, 1) = (-4, -6, -1) \in S_1$$
$$0U_1 = (0, 0, 0) \in S_1.$$

Every linear combination of U_1 and U_2 is also in S_1. For example,

$$2U_1 - 5U_2 = (2, 0, -2) + (-10, -15, -5) = (-8, -15, -7) \in S_1$$
$$U_1 + U_2 = (1, 0, -1) + (2, 3, 1) = (3, 3, 0) \in S_1$$
$$U_1 + 0U_2 = U_1 \in S_1.$$

Since a scalar multiplication and then addition will create a linear combination the basic operations are scalar multiplication and addition. Then every element in S_1, elements are vectors, can be written as a linear combination of U_1 and U_2. In this case we say that S_1 is *spanned* or *generated* or *created* by U_1 and U_2. Then we say that the collection $\{U_1, U_2\}$ is a *spanning set* of S_1.

Definition 1.3.2 (Vector subspace). Let S be a collection of vectors such that if $V_1 \in S$ then $cV_1 \in S$ where c is any scalar, including zero, and if $V_1 \in S$ and $V_2 \in S$ then $V_1 + V_2 \in S$. Then S is called a vector subspace.

Another way of defining S is that it is a collection which is closed under scalar multiplication and addition. When the elements of S are n-vectors (ordered set of n real numbers) then the operations "scalar multiplication" and "addition" are easily defined and many properties such as commutativity,

$$V_1 + V_2 = V_2 + V_1,$$

associativity

$$V_1 + (V_2 + V_3) = (V_1 + V_2) + V_3$$

and so on are easily established. But if the elements of S are some general objects then the operations "scalar multiplication" and "addition" are to be redefined and then all types of extra properties are to be double-checked before constructing such a collection which is closed under "scalar multiplication" and "addition". A more general definition of S will be introduced later. For the time being the elements in our S are all n-tuples of real numbers. The null vector is automatically an element of any such S. That is, $O \in S$. If $V \in S$ then $V + O = V$, $-V \in S$, $V - V = O$.

Definition 1.3.3 (A spanning set of a vector subspace). A collection of vectors which span the whole of a given vector subspace is called a spanning set of that vector subspace.

Note that there can be a number of spanning sets for a given subspace S. In our illustrative example $C_1 = \{U_1, U_2\}$, where $U_1 = (1, 0, -1)$, $U_2 = (2, 3, 1)$, spans the subspace S_1. The same subspace could be spanned by $C_2 = \{U_1, U_2, U_1 + U_2\}$ or $C_3 = \{U_1, U_2 + 3U_1\}$ or $C_4 = \{U_2, U_1 - U_2, 2U_1 + 5U_2, U_1\}$ and so on. Thus, for a given subspace there can be infinitely many spanning sets. In all the spanning sets, $C_1, \ldots, C_4$ above the smallest number of linearly independent vectors which can span S_1 or the maximum number of linearly independent vectors in all those spanning sets is 2.

Definition 1.3.4 (A basis for a vector subspace). A set of all linearly independent vectors in a spanning set of a vector subspace is called a basis for that vector subspace. That is, a basis is a spanning set consisting of only linearly independent vectors.

As there can be many spanning sets for a given vector subspace there can be infinitely many bases for a given vector subspace. In our illustrative example $B_1 = \{U_1, U_2\}$ is a basis, $B_2 = \{U_1, U_2 + 3U_1\}$ is another basis, $B_3 = \{U_2, U_1 - U_2\}$ is a third basis, but $B_4 = \{U_2, U_1 - U_2, 2U_1 + U_2\}$ is not a basis because one vector, namely

$$2U_1 + U_2 = 2(U_1 - U_2) + 3U_2,$$

is a linear function of the other two. B_4 is a spanning set but not a basis. We are imposing two conditions for a basis of a vector subspace. (i) A basis is a spanning set for that vector subspace; (ii) A basis consists of only linearly independent vectors.

Example 1.3.7. Construct 3 bases for the vector subspace spanned by the following set of vectors:

$$U_1 = (1, 1, 1), \quad U_2 = (1, -1, 2), \quad U_3 = (2, 0, 3).$$

Solution 1.3.7. Our first step is to determine the number of linearly independent vectors in the given set so that one set of the maximum number of linearly independent vectors can be collected. Let us apply the sweep-out process, writing the vectors as rows.

$$
\begin{array}{ccc}
1 & 1 & 1 \\
1 & -1 & 2 \\
2 & 0 & 3
\end{array}
\quad
-1(1)+(2);\quad -2(1)+(3) \Rightarrow
\begin{array}{ccc}
1 & 1 & 1 \\
0 & -2 & 1 \\
0 & -2 & 1
\end{array}
$$

$$
-1(2)+(3) \Rightarrow
\begin{array}{ccc}
1 & 1 & 1 \\
0 & -2 & 1 \\
0 & 0 & 0
\end{array}
$$

Thus the whole vector subspace S, which is spanned by $\{U_1, U_2, U_3\}$, can also be spanned by $\{V_1, V_2\}$ where

$$V_1 = (1,1,1), \quad \text{and} \quad V_2 = (0,-2,1).$$

Hence one basis for S is $B_1 = \{V_1, V_2\}$. Any set of 2 linearly independent vectors that can be constructed by using V_1 and V_2 is also a basis for S. For example,

$$B_2 = \{2V_1, 3V_2\}, \quad B_3 = \{V_1, V_2 + V_1\}$$

are two more bases for S. Infinitely many such bases can be constructed for the same vector subspace S. This means that if we start with V_1 only then we can span only a part of S or a subset of S, say S_1. This S_1 consists of all scalar multiples of V_1. Similarly if we start with only V_2 we can only span a part of S or a subset of S, say S_2. This S_2 consists of scalar multiples of V_2. Note that the union of S_1 and S_2, $S_1 \cup S_2$, is not S. All linear functions of V_1 and V_2 are also in S. Hence $S_1 \cup S_2$ is only a subset of S.

Definition 1.3.5 (Dimension of a vector subspace). The maximum number of linearly independent vectors in a spanning set of S or the smallest number of linearly independent vectors which can span the whole of S or the number of vectors in a basis of S is called the dimension of the subspace S.

In our illustrative Example 1.3.7 the dimension of S is 2. In general, observe that for a given subspace S there cannot be two different bases B_1 and B_2 where in B_1 the number of linearly independent vectors is m_1 whereas in B_2 that number is m_2 with $m_1 \neq m_2$. If possible let $m_1 < m_2$. Then every vector in S is a linear function of these m_1 vectors and hence by definition there cannot be a vector in S which is linearly independent of these m_1 vectors. That means m_1 must be equal to m_2.

One more point is worth observing. Since every 3-vector can be written as a linear function of the basic unit vectors, the vectors $U_1 = (1,0,-1)$ and $U_2 = (2,3,1)$ in our illustrative example can be written as linear functions of the basic unit vectors

$$e_1 = (1,0,0), \quad e_2 = (0,1,0), \quad e_3 = (0,0,1).$$

Note that

$$U_1 = e_1 - e_3 \quad \text{and} \quad U_2 = 2e_1 + 3e_2 + e_3.$$

In the set $B = \{e_1, e_2, e_3\}$ there are 3 linearly independent vectors. We have already seen that U_1 and U_2 can be written as linear functions of these unit vectors. Thus this set B could have spanned not only S of our Example 1.3.7, call it $\tilde{S}$, the vector subspace spanned by U_1 and U_2, but also a much larger space S where our $\tilde{S}$ is a subset or $\tilde{S}$ is contained in S or $\tilde{S} \subset S$ or $\tilde{S}$ is a subspace there. This is why we used the phrase "subspace" in our definitions. Incidently, since $S \subset S$ we can also call S itself a subspace.

Definition 1.3.6 (Orthogonal subspaces). Consider two subspaces, S and S^* of n-vectors such that for every vector $U \in S$ and every vector $V \in S^*$, $U.V = 0$. That is, vectors in S are orthogonal to the vectors in S^* and vice versa. Then S and S^* are called subspaces orthogonal to each other.

Obviously, since the same vector cannot be orthogonal to itself (except the null vector) the same non-null vector cannot be present in S as well as in S^*. For example, if $U_1 = (1,1,1)$ is in S then $V_1 = (1,-2,1)$ and $V_2 = (1,0,-1)$ are two possible vectors in S^* since $U_1.V_1 = 0$ and $U_1.V_2 = 0$. But V_1 or V_2 or both need not be present in S^*.

Example 1.3.8. If $U = (1,2,-1) \in S$ and if S is spanned by U itself then what is the maximum possible number of linearly independent vectors in a subspace S^* orthogonal to S? Construct a basis for such an S^*.

Solution 1.3.8. Let $X = (x_1, x_2, x_3)$ be in S^*. Then

$$U.X = 0 \implies x_1 + 2x_2 - x_3 = 0. \tag{1.3.7}$$

The maximum number of linearly independent 3-vectors possible is 3. Orthogonal vectors are linearly independent. Hence the maximum number of linearly independent X possible is $3 - 1 = 2$. In order to construct a basis we construct two linearly independent X from equation (1.3.7). For example, $X_1 = (-2,1,0)$ and $X_2 = (-1,1,1)$ are two linearly independent solutions of (1.3.7). Hence $\{X_1, X_2\}$ is a basis for the orthogonal space S^*. There can be many such bases for S^*, each basis will consist of two linearly independent solutions of (1.3.7). Note that the subspace spanned by $X_1 = (-2,1,0)$ alone will be orthogonal to S as well as the subspace spanned by $X_2 = (-1,1,1)$ alone will be orthogonal to S. But we were looking for that orthogonal subspace consisting of the maximum number of linearly independent solutions of (1.3.7).

Definition 1.3.7 (Orthogonal complement of a subspace). Let S be a vector subspace and S^* a subspace orthogonal to S. If all the maximum possible number of linearly independent vectors, orthogonal to S, are in S^* then S^* is the orthogonal complement of S and it is usually written as $S^* = S^{\perp}$.

(viii) If the dimension of a vector subspace S of n-vectors is $m < n$ and if S^* is the orthogonal complement of S then the dimension of S^* is $n - m$. If the dimension of S is n then the dimension of S^* is zero which means S^* contains only the null vector.

1.3.2 Gram–Schmidt orthogonalization process

From a given set $U_1, \dots, U_k$ of k linearly independent n-vectors can we create another set $V_1, \dots, V_k$ of vectors which form an orthonormal system and each V_j is a linear function of the U_j's? That is, $V_i.V_j = 0$, $i \neq j$ and $\|V_j\| = 1$, $j = 1, \dots, k$. The answer to this question is in the affirmative and the process by which we obtain the set $V_1, \dots, V_k$ from the set $U_1, \dots, U_k$ is known as the Gram–Schmidt orthogonalization process. This process can be described as follows: Take the normalized U_1 as V_1. Construct a V_2 where

$$V_2 = \frac{W_2}{\|W_2\|}, \quad W_2 = U_2 + aV_1$$

where a is a scalar quantity. Since we require V_1 to be orthogonal to W_2 we have $W_2.V_1 = 0$ or $U_2.V_1 + aV_1.V_1 = U_2.V_1 + a = 0$ since $V_1.V_1 = 1$. Then $a = -U_2.V_1$. That is, $W_2 = U_2 - (U_2.V_1)V_1$ where $U_2.V_1$ is the dot product of U_2 and V_1. Note that

$$W_2.V_1 = U_2.V_1 - (U_2.V_1)V_1.V_1 = U_2.V_1 - U_2.V_1 = 0$$

since $V_1.V_1 = \|V_1\|^2 = 1$. Thus V_1 and V_2 are orthogonal to each other and each one is a normalized vector. Now, consider the general formula

$$W_j = U_j - (U_j.V_1)V_1 - (U_j.V_2)V_2$$
$$- \cdots - (U_j.V_{j-1})V_{j-1} \quad \text{for } j = 2, \dots, k \text{ and} \tag{1.3.8}$$
$$V_j = \frac{W_j}{\|W_j\|}.$$

For example,

$$W_3 = U_3 - (U_3.V_1)V_1 - (U_3.V_2)V_2,$$
$$V_3 = \frac{W_3}{\|W_3\|}.$$

Let us see whether W_3 is orthogonal to both V_1 and V_2. Take the dot product

$$W_3.V_1 = U_3.V_1 - (U_3.V_1)V_1.V_1 - (U_3.V_2)V_2.V_1.$$

It is already shown that $V_2.V_1 = 0$ and $V_1.V_1 = \|V_1\|^2 = 1$. Hence

$$W_3.V_1 = U_3.V_1 - (U_3.V_1) = 0.$$

Now take,

$$W_3.V_2 = U_3.V_2 - (U_3.V_1)V_1.V_2 - (U_3.V_2)V_2.V_2$$
$$= U_3.V_2 - 0 - (U_3.V_2) = 0$$

since $V_2.V_2 = \|V_2\|^2 = 1$ and $V_2.V_1 = 0$.

The formula (1.3.8) is constructed by writing W_j as a linear function of $U_j, V_1, \ldots,$ V_{j-1} and then solving for the coefficients by using the conditions that the dot products of W_j with $V_1, \ldots, V_{j-1}$ are all zeros. One interesting observation can be made on (1.3.8). V_j is a linear function of $V_1, \ldots, V_{j-1}$ and U_j which implies that V_j is a linear function of $U_1, \ldots, U_j$ only. That is, V_1 is a function of U_1 only, V_2 is a function of U_1 and U_2 only and so on, a triangular format.

Example 1.3.9. Given the vectors

$$U_1 = (1,1,-1), \quad U_2 = (1,2,1), \quad U_3 = (2,3,4)$$

construct an orthonormal system by using U_1, U_2 and U_3, if possible.

Solution 1.3.9. Let

$$V_1 = \frac{U_1}{\|U_1\|}, \quad \|U_1\| = \sqrt{(1)^2 + (1)^2 + (-1)^2} = \sqrt{3} \implies$$
$$V_1 = \frac{1}{\sqrt{3}}(1,1,-1).$$

Let

$$W_2 = U_2 - (U_2.V_1)V_1$$

where

$$V_1.U_2 = \frac{1}{\sqrt{3}}(1,1,-1).(1,2,1)$$
$$= \frac{1}{\sqrt{3}}[(1)(1) + (1)(2) + (-1)(1)] = \frac{2}{\sqrt{3}}.$$
$$W_2 = U_2 - (U_2.V_1)V_1$$
$$= (1,2,1) - \frac{2}{\sqrt{3}}\frac{1}{\sqrt{3}}(1,1,-1) = \left(\frac{1}{3}, \frac{4}{3}, \frac{5}{3}\right) = \frac{1}{3}(1,4,5),$$
$$\|W_2\| = \frac{1}{3}\sqrt{(1)^2 + (4)^2 + (5)^2} = \frac{\sqrt{42}}{3} \implies$$
$$V_2 = \frac{W_2}{\|W_2\|} = \frac{1}{\sqrt{42}}(1,4,5).$$

Note that for any vector U and for any nonzero scalar a, $\|aU\| = |a|\,\|U\|$ and hence keep the constants outside when computing the lengths. Consider

$$W_3 = U_3 - (U_3.V_1)V_1 - (U_3.V_2)V_2,$$

where

$$V_1.U_3 = \frac{1}{\sqrt{3}}(1,1,-1).(2,3,4) = \frac{1}{\sqrt{3}},$$

$$(V_1.U_3)V_1 = \frac{1}{3}(1,1,-1),$$

$$V_2.U_3 = \frac{1}{\sqrt{42}}(1,4,5).(2,3,4) = \frac{34}{\sqrt{42}},$$

$$(V_2.U_3)V_2 = \frac{34}{42}(1,4,5).$$

Therefore

$$W_3 = (2,3,4) - \frac{1}{3}(1,1,-1) - \frac{34}{42}(1,4,5) = \frac{1}{7}(6,-4,2)$$

with

$$\|W_3\| = \frac{\sqrt{56}}{7} \ \Rightarrow \ V_3 = \frac{W_3}{\|W_3\|} = \frac{1}{\sqrt{56}}(6,-4,2).$$

Verification

$$V_1.V_2 = \left[\frac{1}{\sqrt{3}}(1,1,-1)\right].\left[\frac{1}{\sqrt{42}}(1,4,5)\right] = 0;$$

$$V_1.V_3 = \left[\frac{1}{\sqrt{3}}(1,1,-1)\right].\left[\frac{1}{\sqrt{56}}(6,-4,2)\right] = 0;$$

$$V_2.V_3 = \left[\frac{1}{\sqrt{42}}(1,4,5)\right].\left[\frac{1}{\sqrt{56}}(6,-4,2)\right] = 0.$$

Thus V_1, V_2, V_3 is the system of orthonormal vectors available from U_1, U_2, U_3.

Example 1.3.10. Given the vectors

$$U_1 = (1,1,-1), \quad U_2 = (1,2,1), \quad U_3 = (2,3,0)$$

construct an orthonormal system by using U_1, U_2, U_3, if possible.

Solution 1.3.10. Since U_1 and U_2 are the same as the ones in Example 1.3.9 we have

$$V_1 = \frac{1}{\sqrt{3}}(1,1,-1) \quad \text{and} \quad V_2 = \frac{1}{\sqrt{42}}(1,4,5).$$

Now, consider the equation

$$W_3 = U_3 - (U_3.V_1)V_1 - (U_3.V_2)V_2$$

where

$$V_1.U_3 = \left[\frac{1}{\sqrt{3}}(1,1,-1)\right].[2,3,0] = \frac{5}{\sqrt{3}},$$

$$(V_1.U_3)V_1 = \frac{5}{3}(1,1,-1)$$

$$V_2.U_3 = \left[\frac{1}{\sqrt{42}}(1,4,5)\right].[2,3,0] = \frac{14}{\sqrt{42}},$$

$$(V_2.U_3)V_2 = \frac{14}{42}(1,4,5).$$

Then

$$W_3 = (2,3,0) - \frac{5}{3}(1,1,-1) - \frac{14}{42}(1,4,5) = (0,0,0).$$

In this case the only orthogonal system possible is with a null vector and the non-null vectors V_1 and V_2. Here V_1 and V_2 are orthonormal but a null vector is orthogonal but not a normal vector. This situation arose because in the original set U_1, U_2, U_3, not all vectors are linearly independent. U_3 could have been written as a linear function of U_1 and U_2, in fact $U_3 = U_1 + U_2$, and that is why W_3 became null.

(ix) If there are m_1 dependent vectors and m_2 linearly independent vectors in a given system of $m_1 + m_2$ vectors of the same category then when the Gram–Schmidt orthogonalization process is applied on these $m_1 + m_2$ vectors we get only m_2 orthonormal vectors and the remaining m_1 will be null vectors.

When we start with a given set of vectors $U_1, \dots, U_k$ we do not know whether it is a linearly independent or dependent system. Hence, start with the orthogonalization process. If a W_j becomes null, ignore the corresponding U_j and proceed with the remaining to obtain a set of orthonormal vectors. This will be m_2 in number if in the original set $U_1, \dots, U_k$ only m_2 were linearly independent.

Note. For a more rigorous definition of a vector space we will wait until after the discussion of *matrices* so that these objects can also be included as elements in such a vector space.

Exercises 1.3

1.3.1. Check for linear dependence or independence in the following set of vectors:

(a) $\quad U_1 = \begin{bmatrix} 1 \\ -1 \\ 0 \\ 1 \\ 2 \end{bmatrix}, \quad U_2 = \begin{bmatrix} 2 \\ 0 \\ -1 \\ 1 \\ 5 \end{bmatrix}, \quad U_3 = \begin{bmatrix} 3 \\ 1 \\ 1 \\ -1 \\ 1 \end{bmatrix}, \quad U_4 = \begin{bmatrix} 1 \\ -1 \\ 0 \\ 0 \\ 1 \end{bmatrix};$

(b) $\quad U_1 = (2,0,1,-1), \quad U_2 = (3,0,-1,2), \quad U_3 = (5,0,0,1);$

(c) $\quad U_1 = (3,1,-1,1,2), \quad U_2 = (5,1,2,-1,0), \quad U_3 = (7,-1,1,-1,0).$

1.3.2. For each case in Exercise 1.3.1 find a basis for the vector subspace spanned by the vectors in the set.

1.3.3. For each of the subspaces spanned by the vectors in Exercise 1.3.1 construct a basis for the orthogonal complement and compute the dimensions of each of these orthogonal complements.

1.3.4. For each set of vectors in Exercise 1.3.1 construct a set of (i) mutually orthogonal vectors as linear functions of the given set of vectors, (ii) a set of orthonormal system of vectors as linear functions of the given set of vectors, if possible.

1.3.5. Let U_1 and U_2 be two linearly independent 2-vectors. Let V be an arbitrary 2-vector. Show that V can be written as a linear function of U_1 and U_2.

1.3.6. Illustrate the result in Exercise 1.3.5 geometrically.

1.3.7. Let U_1, U_2 and U_3 be three linearly independent 3-vectors and let V be an arbitrary 3-vector. Show that V can be written as a linear function of U_1, U_2 and U_3.

1.3.8. Treating vectors as arrowheads let $\vec{U}_1 = (1, 1, -1) = \vec{i} + \vec{j} - \vec{k}$ and $\vec{U}_2 = (2, 1, 0) = 2\vec{i} + \vec{j}$ give a geometric interpretation of a basis for the subspace orthogonal to the subspace spanned by $\vec{U}_1$ and $\vec{U}_2$.

1.3.9. In the language of analytical geometry two lines in a plane are perpendicular to each other if the product of their slopes is -1. Express this statement in terms of the dot product of two vectors being zero.

1.3.10. Find all vectors which are orthogonal to both $U_1 = (1, 1, 1, -1)$ and $U_2 = (2, 1, 3, 2)$.

1.3.11. If $U_1 = (1, 1, 1)$ and $U_2 = (1, 1, -1)$, are the following true? Prove your assertions by using the definition of linear independence. (i) U_1 and $2U_1 - U_2$, (ii) $U_1 + U_2$ and $U_1 - U_2$, (iii) $U_1 - U_2$ and $2U_1 + 2U_2$, (iv) $U_1 + U_2$ and $2U_1 - 2U_2$, are all linearly independent.

1.3.12. Consider a subspace spanned by the vectors U_1 and U_2 in Exercise 1.3.11. Is it true that the sets in (i) to (iv) there, are bases for that subspace. Justify your answer.

1.3.13. Let S be the vector subspace spanned by U_1 and U_2 of Exercise 1.3.11. Construct 2 bases for the orthogonal complement S^* of S. What are the dimensions of S and S^*?

1.3.14. Consider a 3-space and two planes passing through the origin. Consider the normals to these planes. Construct 3 bases for the subspace spanned by these normals if (1) the planes are parallel, (2) the planes are perpendicular to each other, (3) the planes are neither parallel nor perpendicular to each other.

1.3.15. In Exercise 1.3.14 construct the orthogonal complements of the subspaces spanned in the three cases and find 2 bases each for these orthogonal complements.

1.3.16. Let $V_j \in S$, $j = 1, 2, \ldots$ be n-vectors where S is a vector space of dimension n. Show that any set of n linearly independent V_j's is a basis of S.

1.3.17. Let $V_j \in S$, $j = 1, \ldots, r$, $1 \le r \le n - 1$ where the dimension of S is n and all V_j's are n-vectors. If $V_1, \ldots, V_r$ are linearly independent then show that there exist $n - r$ other elements $V_{r+1}, \ldots, V_n$ of S such that $V_1, \ldots, V_n$ is a basis of S.

1.3.18. Let S be the vector space of all 1×3 vectors. Let S_1 be spanned by $V_1 = (1, 1, 1)$, $V_2 = (1, 0, -1)$, $V_3 = (2, 1, 0)$ and S_2 be spanned by $U_1 = (2, 1, 1)$, $U_2 = (3, 1, -1)$. Show that (1) $S_1 \subset S$, $S_2 \subset S$, that is, S_1 and S_2 are subspaces in S. (2) $S_1 \cap S_2 \ne O$, that is, the intersection is not empty. (3) Determine the dimensions of S_1 and S_2. (4) Construct the subspace S_3 such that if $W \in S_3$ then $W = V + U$ where $V \in S_1$ and $U \in S_2$. [This $S_3 \subset S$ is called a simple sum of S_1 and S_2 and it is usually written as $S_3 = S_1 + S_2$.]

1.3.19. Consider the same S as in Exercise 1.3.18. Let

$$e_1 = (1, 0, 0), \quad e_2 = (0, 1, 0), \quad e_3 = (0, 0, 1).$$

Let S_1 be spanned by e_1 and e_2 and S_2 be spanned by e_3. Show that (1) $S_1 \subset S$ and $S_2 \subset S$. (2) $S_1 \cap S_2 = O$. (3) Construct S_3 as in Exercise 1.3.18.

1.3.20. *Direct sum of subspaces.* Let S be a finite dimensional linear space (vector space) and let S_1 and S_2 be subspaces of S. Then the *simple sum* of S_1 and S_2, denoted by $S_1 + S_2$, is the set of all sums of the type $U + V$ where $U \in S_1$ and $V \in S_2$. Note that $S_1 + S_2$ is also a subspace of S. In addition, if $S_1 \cap S_2 = O$, that is, the intersection of S_1 and S_2 is null or empty then the simple sum is called a *direct sum*, and it will be denoted by $S_1 \hat{+} S_2$. Show that for the simple sums,

$$\dim(S_1 + S_2) + \dim(S_1 \cap S_2) = \dim(S_1) + \dim(S_2)$$

where $\dim(\cdot)$ denotes the dimension of $(\cdot)$ and $+$ the simple sum.

1.3.21. Let S_j, $j = 1, \ldots, k$ be subspaces of a finite dimensional space S. Show that, for the simple sums,

$$\dim(S_1 + \cdots + S_k) \le \sum_{i=1}^{k} \dim(S_i).$$

1.3.22. Let S_1 and S_2 be as in Exercise 1.3.20. Then show that every element $W \in (S_1 + S_2)$ can be written as $W = U + V$, $U \in S_1$, $V \in S_2$ and that this decomposition $W = U + V$ is unique if and only if $S_1 \cap S_2 = O$ where O means a null set.

1.3.23. Let $S_0, S_1, \ldots, S_k$ be subspaces of a finite dimensional linear space S. Show that the subspace S_0 can be written as a direct sum of the subspaces $S_1, \ldots, S_k$ if and only if the union of the bases for $S_1, \ldots, S_k$ forms a basis for S_0.

1.3.24. Let $S_j \in S$, $j = 0, 1, \ldots, k$ where S is a finite dimensional linear space. Show that

$$S_0 = S_1 \hat{+} \cdots \hat{+} S_k$$

if and only if

$$\dim(S_0) = \sum_{j=1}^{k} \dim(S_j).$$

1.3.25. Let S_j, $j = 0, 1, \ldots, k$ be as in Exercise 1.3.24. Show that

$$S_0 = S_1 \dotplus \cdots \dotplus S_k$$

if and only if

$$S_i \cap (S_1 + \cdots + S_{i-1}) = O, \quad i = 1, \ldots, k$$

where O is a null set.

1.3.26. By using vector methods prove that the segment joining the midpoints of two sides of any triangle is parallel to the third side and half as long.

1.3.27. By using vector methods prove that the medians of a triangle (the line segments joining the vertices to the midpoints of opposite sides) intersect in a point of trisection of each.

1.3.28. By using vector methods prove that the midpoints of the sides of any plane convex quadrilateral are the vertices of a parallelogram.

1.3.29. By using vector methods prove that the lines from any vertex of a parallelogram to the midpoints of the opposite sides trisect the diagonal they intersect.

1.3.30. If $U_1, \ldots, U_k$ is a finite collection of vectors and if $\|U_j\|$ denotes the length of U_j then show that

$$\|U_1 + \cdots + U_k\| \le \|U_1\| + \|U_2\| + \cdots + \|U_k\|.$$

1.4 Some applications

We will explore a few applications of vector methods in multivariable calculus, statistical problems, model building and other related areas. The students who are not familiar with multivariable calculus may skip this section.

1.4.1 Partial differential operators

Consider a scalar function (as opposed to a vector function) of many real scalar (as opposed to vector) variables, $f(x_1, \ldots, x_n)$, where $x_1, \ldots, x_n$ are functionally independent (no variable can be written as a function of the other variables), or distinct, real variables. For example,

(i) $f = 2x_1^3 + x_2^2 - 3x_1x_2 + x_2 - 5x_1 + 8$
(ii) $f = 3x_1^2 + 2x_2^2 - x_1x_2 - x_2 - 2x_1 + 10$

are two such functions of two real scalar variables x_1 and x_2. Consider the vector of partial differential operators. Let us use the following notations:

$$X = \begin{bmatrix} x_1 \\ x_2 \\ \vdots \\ x_n \end{bmatrix}, \quad \frac{\partial}{\partial X} = \begin{bmatrix} \frac{\partial}{\partial x_1} \\ \frac{\partial}{\partial x_2} \\ \vdots \\ \frac{\partial}{\partial x_n} \end{bmatrix}, \quad \frac{\partial}{\partial X}f = \frac{\partial f}{\partial X} = \begin{bmatrix} \frac{\partial f}{\partial x_1} \\ \frac{\partial f}{\partial x_2} \\ \vdots \\ \frac{\partial f}{\partial x_n} \end{bmatrix},$$

$$\frac{\partial}{\partial X'} = \left(\frac{\partial}{\partial x_1}, \ldots, \frac{\partial}{\partial x_n} \right),$$

$$\frac{\partial}{\partial X'}f = \frac{\partial f}{\partial X'} = \left(\frac{\partial f}{\partial x_1}, \ldots, \frac{\partial f}{\partial x_n} \right). \tag{1.4.1}$$

For example, $\frac{\partial f}{\partial x_1}$ means to differentiate f with respect to x_1 partially which means assuming all other variables $x_2, \ldots, x_n$ to be constants. In (ii) above $\frac{\partial}{\partial x_1}$ operating on f gives

$$\frac{\partial f}{\partial x_1} = \frac{\partial}{\partial x_1}(3x_1^2 + 2x_2^2 - x_1x_2 - x_2 - 2x_1 + 10)$$

$$= \frac{\partial}{\partial x_1}(3x_1^2) + \frac{\partial}{\partial x_1}(2x_2^2) + \frac{\partial}{\partial x_1}(-x_1x_2)$$

$$+ \frac{\partial}{\partial x_1}(-x_2) + \frac{\partial}{\partial x_1}(-2x_1) + \frac{\partial}{\partial x_1}(10)$$

$$= 6x_1 + 0 - x_2 + 0 - 2 + 0$$

$$= 6x_1 - x_2 - 2.$$

Similarly $\frac{\partial}{\partial x_2}$ operating on this f gives

$$\frac{\partial f}{\partial x_2} = \frac{\partial}{\partial x_2}(3x_1^2 + 2x_2^2 - x_1x_2 - x_2 - 2x_1 + 10)$$

$$= 0 + 4x_2 - x_1 - 1 - 0 + 0$$

$$= 4x_2 - x_1 - 1.$$

Then $\frac{\partial}{\partial X}$ operating on f is a column vector, namely,

$$\frac{\partial}{\partial X}f = \begin{pmatrix} \frac{\partial f}{\partial x_1} \\ \frac{\partial f}{\partial x_2} \end{pmatrix} = \begin{pmatrix} 6x_1 - x_2 - 2 \\ 4x_2 - x_1 - 1 \end{pmatrix}.$$

The transpose of this vector is denoted by $\frac{\partial f}{\partial X'}$ ($\frac{\partial}{\partial X'}$ operating on f). That is,

$$\frac{\partial f}{\partial X'} = (6x_1 - x_2 - 2, \; 4x_2 - x_1 - 1).$$

1.4.2 Maxima/minima of a scalar function of many real scalar variables

When looking for points where the function may have local maximum or local minimum we differentiate the function partially with respect to each variable, equate to zero and solve the system of equations to determine the critical points or turning points or points where the function may have local maximum or local minimum or saddle points. These steps, in vector notation, are equivalent to solving the equation

$$\frac{\partial f}{\partial X} = O \tag{1.4.2}$$

where O denotes the null vector. In our illustrative example

$$\frac{\partial f}{\partial X} = O \Rightarrow \begin{pmatrix} 6x_1 - x_2 - 2 \\ 4x_2 - x_1 - 1 \end{pmatrix} = \begin{pmatrix} 0 \\ 0 \end{pmatrix}.$$

That is,
(a) $6x_1 - x_2 - 2 = 0$,
(b) $4x_2 - x_1 - 1 = 0$.

When solving $\frac{\partial f}{\partial X} = O$ we need not write down the individual equations as in (a) and (b) above. One can use matrix methods, which will be discussed in the next chapter, and solve (1.4.2) directly. Solving (a) and (b) we have

$$\begin{pmatrix} x_1 \\ x_2 \end{pmatrix} = \begin{pmatrix} 9/23 \\ 8/23 \end{pmatrix}.$$

In our illustrative example there is only one critical point

$$(x_1, x_2) = \left(\frac{9}{23}, \frac{8}{23} \right).$$

This critical point may correspond to a maximum or a minimum or it may be a saddle point. In order to check for maxima/minima we look for the whole configuration of the matrix of second order partial derivatives and look for definiteness of matrices. This aspect will be considered after introducing matrices in the next chapter.

1.4.3 Derivatives of linear and quadratic forms

Some obvious results when we use the operator $\frac{\partial}{\partial X}$ on linear and quadratic forms will be examined here. A linear form is available by taking a dot product of X with a constant vector. For example if

$$X = \begin{pmatrix} x_1 \\ \vdots \\ x_n \end{pmatrix}, \quad a = \begin{pmatrix} a_1 \\ \vdots \\ a_n \end{pmatrix}$$

then

$$X.a = a.X = a_1x_1 + \cdots + a_nx_n \tag{1.4.3}$$

is a linear form. For example,

$$y_1 = 2x_1 - x_2 + 3x_3 + x_4$$
$$y_2 = x_1 + x_1 + x_3 + x_4 - 2x_5 + 7x_6$$

are two linear forms. In a linear form each term is of degree one and all terms are of degree one each or a linear form is homogeneous of degree 1 in the variables. For example, the degree of a term is determined as follows: $3x^5$ (degree $0 + 5 = 5$), $x_1^5 + 3x_2^5$ (each term is of degree 5), $2x_1^4x_2$ (degree $0 + 4 + 1 = 5$), $6x_1$ (degree $0 + 1 = 1$, linear), 5 (degree 0, constant).

What will be the result if a linear form is operated with the operator $\frac{\partial}{\partial X}$? Let $y = X.a$ then

$$\frac{\partial}{\partial X}y = \frac{\partial y}{\partial X} = \begin{bmatrix} \frac{\partial y}{\partial x_1} \\ \vdots \\ \frac{\partial y}{\partial x_n} \end{bmatrix} = \begin{bmatrix} a_1 \\ \vdots \\ a_n \end{bmatrix} = a.$$

Hence we have the following important result:

(i) Consider the operator $\frac{\partial}{\partial X}$ and the linear form $X.a$ where a is a constant vector. Then

$$y = X.a = a.X \implies \frac{\partial}{\partial X}y = \frac{\partial y}{\partial X} = a$$

where a is the column vector of the coefficients in $X.a$.

Example 1.4.1. Evaluate $\frac{\partial y}{\partial X}$ if

$$y = x_1 - 5x_2 + x_3 - 2x_4.$$

Solution 1.4.1.

$$\frac{\partial y}{\partial x_1} = 1, \quad \frac{\partial y}{\partial x_2} = -5, \quad \frac{\partial y}{\partial x_3} = 1, \quad \frac{\partial y}{\partial x_4} = -2$$

and hence

$$\frac{\partial}{\partial X}y = \frac{\partial y}{\partial X} = \begin{bmatrix} 1 \\ -5 \\ 1 \\ -2 \end{bmatrix}.$$

Now, let us examine a simple quadratic form. Consider the sum of squares of a number of variables. Let

$$X = \begin{bmatrix} x_1 \\ \vdots \\ x_n \end{bmatrix} \quad \text{then} \quad X.X = x_1^2 + x_2^2 + \cdots + x_n^2.$$

This is a special case of a quadratic form. In a quadratic form, every term is of degree 2 each or it is a homogeneous function of many variables of degree 2. For the time being, we consider the above simple quadratic form. More general quadratic forms will be considered after introducing matrices in the next chapter. What will happen if a sum of squares is operated with the operator $\frac{\partial}{\partial X}$? Proceeding as in the linear case the result is the following:

(ii) Let $y = X.X = x_1^2 + \cdots + x_n^2$ then

$$y = X.X \implies \frac{\partial y}{\partial X} = \begin{pmatrix} 2x_1 \\ \vdots \\ 2x_n \end{pmatrix} = 2X.$$

1.4.4 Model building

Suppose that a gardener suspects that the growth of a particular species of plant (growth measured in terms of the height of the plant) is linearly related to the amount of a certain fertilizer used. Let the amount of the fertilizer used be denoted by x and the corresponding growth (height) be y. Then the gardener's suspicion is that

$$y = a + bx$$

where a and b are some constants, that is, y and x are linearly related. What exactly is this linear relationship? The gardener conducts an experiment to estimate the values of a and b. Suppose that the gardener applies the amounts $x_1, \ldots, x_n$ of the fertilizer x on different plants of the same species, in a carefully planned experiment, and take the corresponding measurements $y_1, \ldots, y_n$ on y. Thus the gardener has the following pairs of values (x_i, y_i), $i = 1, \ldots, n$. For example, when one spoon of fertilizer (measured in spoon units) is applied the growth (measured after a fixed time) noted is 3 inches (growth measured in inches) then the corresponding pair is $(x_1, y_1) = (1, 3)$. If $y = a + bx$ is a mathematical relationship then every pair (x, y) should satisfy the equation $y = a + bx$. Then we need only two pairs of values on (x, y) to exactly evaluate a and b and then every other value on (x, y) must satisfy the relationship. But this is not the situation here. The gardener is thinking that there may be a relationship between x and y, that relationship may be a linear relationship and that she will be able to estimate y at a preassigned value of x. Then the error in estimating y by using such a relationship at a given value of x is $y - (a + bx)$. Denoting the error in the i-th pair by ϵ_i we have

$$\epsilon_i = y_i - a - bx_i.$$

One way of estimating the unknown parameters a and b is to minimize the sum of squares of the errors (error = observed value minus the modeled value, whatever be the model, linear or not). Such a method of estimating the parameters in a model by minimizing the error sum of squares is known as the *method of least squares*. The error vector and the error sum of squares in our linear model are given by

$$\epsilon = \begin{pmatrix} \epsilon_1 \\ \vdots \\ \epsilon_n \end{pmatrix},$$

$$\epsilon.\epsilon = \epsilon_1^2 + \cdots + \epsilon_n^2 = \sum_{i=1}^{n}(y_i - a - bx_i)^2. \tag{1.4.4}$$

Equation (1.4.4) can be written in a more elegant way as a quadratic form after discussing matrices. Let the vector of unknowns be denoted by $\alpha = \binom{a}{b}$. Then the method of least squares implies that $\epsilon.\epsilon$ is minimized with respect to α. It is obvious that the maximum of $\epsilon.\epsilon$, being a non-negative arbitrary quantity, is at $+\infty$. Then the minimizing equations, often known as the *normal equations* in least square analysis, are the following:

$$\frac{\partial}{\partial \alpha}(\epsilon.\epsilon) = O \implies \begin{pmatrix} \frac{\partial}{\partial a} \\ \frac{\partial}{\partial b} \end{pmatrix}(\epsilon.\epsilon) = O \implies$$

$$\begin{pmatrix} -2\sum_{i=1}^{n}(y_i - a - bx_i) \\ -2\sum_{i=1}^{n}x_i(y_i - a - bx_i) \end{pmatrix} = \begin{pmatrix} 0 \\ 0 \end{pmatrix} \implies$$

$$\sum_{i=1}^{n}(y_i - a - bx_i) = 0 \tag{a}$$

and

$$\sum_{i=1}^{n}x_i(y_i - a - bx_i) = 0 \tag{b}$$

since $-2 \neq 0$. Opening up the sum we have, from (a) and (b),

$$\left(\sum_{i=1}^{n}y_i\right) - na - b\left(\sum_{i=1}^{n}x_i\right) = 0 \tag{c}$$

and

$$\left(\sum_{i=1}^{n}x_iy_i\right) - a\left(\sum_{i=1}^{n}x_i\right) - b\left(\sum_{i=1}^{n}x_i^2\right) = 0. \tag{d}$$

Denoting

$$\bar{y} = \sum_{i=1}^{n}\frac{y_i}{n} \quad \text{and} \quad \bar{x} = \sum_{i=1}^{n}\frac{x_i}{n}$$

and solving (c) and (d) we get the values of a and b. Let us denote these estimates by $\hat{a}$ and $\hat{b}$ respectively. Then we have

$$\hat{b} = \frac{\sum_{i=1}^{n}(x_i - \bar{x})(y_i - \bar{y})}{\sum_{i=1}^{n}(x_i - \bar{x})^2} = \frac{(\sum_{i=1}^{n} x_i y_i) - n(\bar{x}\bar{y})}{(\sum_{i=1}^{n} x_i^2) - n(\bar{x})^2} \tag{1.4.5}$$

and

$$\hat{a} = \bar{y} - \hat{b}\bar{x}. \tag{1.4.6}$$

From (1.4.5) and (1.4.6) we have the estimates for a and b, and the estimated linear model by using the method of least squares is then

$$y = \hat{a} + \hat{b}x. \tag{1.4.7}$$

Example 1.4.2. In a feeding experiment with beef cattle the farmer suspects that the increase in weight is linearly related to the quantity of a particular combination of feed. The farmer has obtained the following data. Construct the estimating function by the method of least squares and then estimate the weight if the quantity of feed is 2.2 kg.

Data:
y = (gain in weight in kg)	0.5	0.8	1.5	2.0
x = (quantity of feed in kg)	1.2	1.5	2.0	2.5

Solution 1.4.2.

$$\bar{x} = \frac{1.2 + 1.5 + 2.0 + 2.5}{4} = 1.8, \quad \bar{y} = \frac{0.5 + 0.8 + 1.5 + 2.0}{4} = 1.2.$$

For convenience of computations let us form the following table: [Use a calculator or computer to compute $\hat{a}$ and $\hat{b}$ directly.]

x	y	$x - \bar{x}$	$y - \bar{y}$	$(x - \bar{x})^2$	$(x - \bar{x})(y - \bar{y})$
1.2	0.5	−0.6	−.7	0.36	0.42
1.5	0.8	−0.3	−0.4	0.09	0.12
2.0	1.5	0.2	0.3	0.04	0.06
2.5	2.0	0.7	0.8	0.49	0.56
				−−−−	−−−−
				0.98	1.16

$$\hat{b} = \frac{1.16}{0.98} \approx 1.1837, \quad \hat{a} = 1.2 - \frac{(1.16)}{(0.98)}(1.8) \approx -0.9306.$$

The estimated model is

$$y = -0.9306 + 1.1837x.$$

Then the predicted value of y at $x = 2.2$ is

$$\hat{y} = -0.9306 + 1.1837(2.2) \approx 1.6735 \text{ kg}.$$

Exercises 1.4

1.4.1. Find the critical points for the following functions and then check to see whether these correspond to maxima or minima or something else:
(a) $f = 2x_1^2 + x_2^2 - 3x_2 + 5x_1x_2 - x_1 + 5$.
(b) $f = x_1^2 + x_2^2 - 2x_1x_2 - 5x_1 - 2x_2 + 8$.

1.4.2. Evaluate $\frac{\partial f}{\partial X}$ and write the results in vector notations:
(a) $f = 3x_1 - x_2 + 5x_3 - x_4 + 10$.
(b) $f = x_1^2 + 2x_1x_2 + x_1x_3 - x_2^2 + 3x_3^2$.
(c) $f = 2x_1^2 + x_2^2 + x_3^2 - 5x_1x_2 + x_2x_3$.

1.4.3. Write the operator $\frac{\partial}{\partial X'}$. Then on each element of this vector apply the operator $\frac{\partial}{\partial X}$. Explain what you have in this configurations of n rows and n columns.

1.4.4. Apply the operator $\frac{\partial}{\partial X} \frac{\partial}{\partial X'}$ on f in each of (a), (b), (c) in Exercise 1.4.2.

1.4.5. Fit linear models of the type $y = a + bx$ for the following data:
(a) $(x,y) = \{(0,2),(1,5),(2,6),(3,9)\}$.
(b) $(x,y) = \{(-1,1),(-2,-2),(0,3),(1,6)\}$.

1.4.6. Fit a model of the type $y = a + bx + cx^2$ to the following data:

$$(x,y) = \{(-1,2),(0,1),(1,5),(2,7),(3,21)\}.$$

1.4.7. In statistical distribution theory the moment generating function of a real vector $X' = (x_1,\ldots,x_k)$ random variable is denoted by $M(T)$, $T' = (t_1,\ldots,t_k)$ where T is a vector of parameters. When $M(T)$ is evaluated for the real multivariate Gaussian distribution we obtain

$$M(T) = e^{\phi(T)}$$

where

$$\phi(T) = t_1\mu_1 + \cdots + t_k\mu_k + \frac{1}{2}\left[\sum_{i,j=1}^{k} \sigma_{ij}t_it_j\right]$$

where $\mu_1,\ldots,\mu_k$ as well as σ_{ij}, $i = 1,\ldots,k$, $j = 1,\ldots,k$ are constants, free of T. When $M(T)$ is available and differentiable, then the expected value of X or the first moment of X, denoted by $\mu = E(X)$, is obtained as $\frac{\partial}{\partial T}M(T)|_{T=O}$, that is the first derivative evaluated at $T = O$, and the variance–covariance matrix is $\frac{\partial}{\partial T}\frac{\partial}{\partial T'}M(T)|_{T=O} - \mu\mu'$. Evaluate $E(X)$ and the variance–covariance matrix for the multivariate Gaussian distribution.

1.4.8. The exponential series is

$$e^y = y^0 + \frac{y}{1!} + \frac{y^2}{2!} + \cdots, \quad y^0 = 1.$$

Consider the operator $D = \frac{d}{dx}$. Then

$$e^{xD} = (xD)^0 + \frac{xD}{1!} + \frac{x^2 D^2}{2!} + \cdots.$$

where, for example, $D^r = DD\ldots D$ stands for D operating repeatedly r times. Let $e^{xD} f_0$ denote e^{xD} operating on f and then $D^r f$ is evaluated at $x = 0$, $r = 0, 1, \ldots$. Then

$$e^{xD} f_0 = f(0) + \frac{x}{1!}\left(\frac{d}{dx}f\right)_{x=0} + \frac{x^2}{2!}\left(\frac{d^2}{dx^2}f\right)_{x=0} + \cdots.$$

This is Taylor series in one variable. Now consider a two variable case. Let

$$\nabla = \begin{pmatrix} D_1 \\ D_2 \end{pmatrix}, \quad D_i = \frac{\partial}{\partial x_i}, \quad i = 1, 2$$

and the increment vector at the point (a_1, a_2) is $\Delta' = (x_1 - a_1, x_2 - a_2)$. Then the dot product is given by

$$\nabla.\Delta = (x_1 - a_1)D_1 + (x_2 - a_2)D_2.$$

As before, let $e^{\nabla.\Delta} f_0$ denote $e^{\nabla.\Delta}$ operating on f where the various derivatives are evaluated at the point (a_1, a_2). Write down the Taylor series expansion for two variables (x_1, x_2) at the point (a_1, a_2) explicitly up to the terms involving all the second order derivatives.

1.4.9. By using the operator ∇ in Exercise 1.4.8 expand the following functions by using Taylor expansion, at the specified points:
(a) $x_1^2 + 2x_1 x_2^2 + x_2^3 + 5x_1 - x_2 + 7$ at $(1, -1)$.
(b) $2x_1^2 + x_2^2 - 3x_1 x_2 + 8$ at $(-2, -3)$.
(c) $x_1^4 + x_1^3 x_2 + 3x_2^4 - x_1 x_2 + 4$ at $(2, 0)$.

1.4.10. Extend the ideas in Exercise 1.4.8 to a scalar function $f(x_1, x_2, x_3)$ of 3 real variables x_1, x_2, x_3, at the point (a_1, a_2, a_3). In this case $D_i = \frac{\partial}{\partial x_i}$, $i = 1, 2, 3$. Evaluate the first few terms of the series explicitly up to the terms involving $(\nabla.\Delta)^3$.

1.4.11. Apply the result in Exercise 1.4.10 to expand the following function up to terms involving $(\nabla.\Delta)^3$, and at the point $(1, 0, -1)$:

$$x_1^2 e^{-x_1 - x_2 - x_3} + 5x_1^3 x_2^2 x_3 - e^{-2x_1 + 3x_2}.$$

1.4.12. For Exercise 1.4.5 (a) estimate y at (i) $x = 2.7$, (ii) $x = 3.1$. Is it reasonable to use the model to estimate y at $x = 10$?

1.4.13. For Exercise 1.4.6 estimate y at (i) $x = 0.8$, (ii) $x = 3.1$. Is it reasonable to predict y at (iii) $x = -4$, (iv) $x = 8$ by using the same model?

1.4.14. Use the method of least squares to fit the model

$$y = a_0 + a_1 x_1^2 + a_2 x_1 x_2 + a_3 x_2^2$$

to the following data:

$$(x_1, x_2, y) = (0,0,1), (0,1,0), (0,2,-2), (1,-1,-1), (2,1,8), (1,2,3).$$

1.4.15. Can the method of least squares, as minimizing the error sum of squares with error defined as "observed minus the modeled value", be used to fit the model $y = ab^x$ to the data

$$(x,y) = (x_1, y_1), (x_2, y_2), \dots, (x_n, y_n)$$

and if not what are the difficulties encountered?

2 Matrices

2.0 Introduction

One of the most elegant tools in simplifying matters and dealing with systems of objects is the entity called a matrix. Plural of the word matrix is matrices. Suppose we have a set of mn objects, such as mn real numbers, and if these objects are arranged in m rows and n column we get the configuration called a matrix. For example if 6 numbers are arranged in 3 rows and 2 columns we get a matrix, if the same numbers are arranged in 2 rows and 3 columns we get another matrix, one row and 6 columns we get a third matrix and so on:

$$A_1 = \begin{bmatrix} 5 & -1 & 2 \\ 1 & 4 & 0 \end{bmatrix} = \text{a matrix,}$$

$$A_2 = \begin{bmatrix} 5 & 1 \\ -1 & 4 \\ 2 & 0 \end{bmatrix} = \text{another matrix,}$$

$$A_3 = \begin{bmatrix} 5 & 1 & -1 & 4 & 2 & 0 \end{bmatrix} = \text{another matrix,}$$

$$A_4 = \begin{bmatrix} 5 & -1 & 2 & 1 \\ 4 & 0 & & \end{bmatrix} = \text{not a matrix.}$$

In the last representation two positions are empty and hence it is not a matrix. Here A_1 has 2 rows and 3 columns whereas A_3 has one row and 6 columns.

Notation 2.0.1. If a matrix has m rows and n columns it is called an $m \times n$ (m by n) matrix.

Here m represents the number of rows and n represents the number of columns. In our illustrative examples A_1 is 2×3 (not 6) matrix, A_2 is a 3×2 matrix and A_3 is a 1×6 matrix. The symbol $\times$ (cross) simply separates the numbers m and n and it is not used as a multiplication symbol in this notation. $m.n$ or $m * n$ are not appropriate notations in this respect. Obviously, a $1 \times n$ matrix is a row vector of n elements and an $n \times 1$ matrix is a column vector of n elements. Thus all items in Chapter 1 become special cases of the various properties of matrices.

Example 2.0.1 (Grades of students). Let the following tables give the grades, in percentages, obtained by 3 students in four class tests in two courses:

Course 1

	test 1	test 2	test 3	test 4
Student 1	80	85	90	82
Student 2	65	60	70	72
Student 3	75	72	74	78

Course 2

	test 1	test 2	test 3	test 4
Student 1	99	92	90	95
Student 2	60	62	65	63
Student 3	80	72	77	81

The number at each position signifies something. The first row second column entry for Course 1, namely 85, is the grade of student 1 in test 2 in course 1. Thus the entries cannot be arbitrarily interchanged. The interchanged arrangement will signify something different from the original arrangement.

A convenient notation that can be used to denote a matrix is by denoting the element at the i-th row, j-th column position in a matrix A by a_{ij}. In this case we write a general matrix in the form

$$A = (a_{ij}) = \begin{bmatrix} a_{11} & a_{12} & \cdots & a_{1n} \\ a_{21} & a_{22} & \cdots & a_{2n} \\ \vdots & \vdots & \cdots & \vdots \\ a_{m1} & a_{m2} & \cdots & a_{mn} \end{bmatrix}. \tag{2.0.1}$$

The elements are enclosed by square brackets [·] or by ordinary brackets (·). The notation $A = (a_{ij})$ means the matrix where the i-th row, j-th column element or the (i,j)-th element is a_{ij} for all i and j. In Example 2.0.1 for course 1, for example,

$$a_{11} = 80, \quad a_{13} = 90, \quad a_{22} = 60, \quad a_{24} = 72, \quad a_{31} = 75, \quad a_{34} = 78.$$

The elements are usually separated by spaces. If there is possibility of confusion then the elements in the configuration are separated by commas. When some of the elements are numbers, some are long expressions involving some variables etc there is possibility of confusion. In this case we will separate the elements by commas.

If the matrix of grades in course 2 is denoted by $B = (b_{ij})$ then

$$A = \begin{bmatrix} 80 & 85 & 90 & 82 \\ 65 & 60 & 70 & 72 \\ 75 & 72 & 74 & 78 \end{bmatrix}, \quad B = \begin{bmatrix} 99 & 92 & 90 & 95 \\ 60 & 62 & 65 & 63 \\ 80 & 72 & 77 & 81 \end{bmatrix}.$$

For example, $b_{13} = 90$, $b_{24} = 63$, $b_{32} = 72$. Here $a_{11} = 80 \neq b_{11} = 99$ whereas $a_{32} = 72 = b_{32}$. If the student had exactly the same profiles of grades in the two courses then the corresponding entries would have been all equal, that is $a_{ij} = b_{ij}$ for all i and j.

2.1 Various definitions

A lot of technical terms and definitions will be introduced, all at once, since they can be recognized easily and the properties can be memorized without difficulty.

Definition 2.1.1 (Equality of matrices). Two matrices A and B are said to be equal if (i) they are of the same category, that is, if A is $m \times n$ then B has to be $m \times n$, (ii) element-wise they must be equal, that is, $a_{ij} = b_{ij}$ for all i and j.

Example 2.1.1. Let

$$C = \begin{bmatrix} 1 & -1 & 2 \\ 0 & 3 & x \end{bmatrix}, \quad D = \begin{bmatrix} a & b & 2 \\ 0 & 3 & 5 \end{bmatrix}.$$

Here both C and D are 2×3 matrices. $C = D$ if and only if (iff) $1 = a$, $-1 = b$, $x = 5$. If all elements in C are equal to the corresponding elements in D, except for one pair, still $C \neq D$.

In Example 2.0.1 if the grades are to be represented out of 20 points each, rather than as percentages, then each grade is to be divided by 5. Then the whole configuration of grades is to be divided by 5, or each element there, is to be divided by 5. In this case we say that the whole matrix is divided by 5. For example, for A, the configuration of grades out of 20 points will then be

$$\begin{bmatrix} \frac{80}{5} & \frac{85}{5} & \frac{90}{5} & \frac{82}{5} \\ \frac{65}{5} & \frac{60}{5} & \frac{70}{5} & \frac{72}{5} \\ \frac{75}{5} & \frac{72}{5} & \frac{74}{5} & \frac{78}{5} \end{bmatrix} = \frac{1}{5} \begin{bmatrix} 80 & 85 & 90 & 82 \\ 65 & 60 & 70 & 72 \\ 75 & 72 & 74 & 78 \end{bmatrix} = \frac{1}{5} A.$$

Definition 2.1.2 (Scalar multiplication of matrices).

$$cA = (ca_{ij})$$

where c is a scalar quantity (1×1 matrix).

That is, if every element in A is multiplied by c then we say that the matrix A is multiplied by c. As a convention the scalar quantity c is written on the left of A as cA, and not on the right as Ac, when writing a scalar multiple of A.

Example 2.1.2. Let

$$A = \begin{bmatrix} 1 & -1 \\ 2 & 5 \end{bmatrix}.$$

Then

$$2A = \begin{bmatrix} 2 & -2 \\ 4 & 10 \end{bmatrix}, \quad -A = \begin{bmatrix} -1 & 1 \\ -2 & -5 \end{bmatrix}, \quad 0A = \begin{bmatrix} 0 & 0 \\ 0 & 0 \end{bmatrix}.$$

Definition 2.1.3 (A null matrix). If all the elements in a matrix are zeros it is called a null matrix and it is denoted by a big O.

Definition 2.1.4 (Square and rectangular matrices). In an $m \times n$ matrix if $m = n$, that is, the number of rows is equal to the number of columns, then it is called a square matrix. Non-square matrices are called rectangular matrices:

$$\begin{bmatrix} 1 & -1 \\ 1 & 0 \end{bmatrix} = \text{a } 2 \times 2 \text{ square matrix,}$$

$$\begin{bmatrix} 2 & 1 & 5 \\ 1 & 1 & -1 \end{bmatrix} = \text{a } 2 \times 3 \text{ rectangular matrix,}$$

$$[1 \quad -1 \quad 2 \quad 3] = \text{a } 1 \times 4 \text{ rectangular matrix or a row vector,}$$

$$\begin{bmatrix} 1 \\ -7 \end{bmatrix} = \text{a } 2 \times 1 \text{ rectangular matrix or a column vector.}$$

Note that vectors of Chapter 1 are all either $1 \times n$ (row vector) or $n \times 1$ (column vector) rectangular matrices.

Definition 2.1.5 (A diagonal matrix). In an $n \times n$ square matrix $A = (a_{ij})$ if $a_{ij} = 0$ for all i and j, $i \neq j$ and at least one $a_{ii} \neq 0$, $i = 1, \dots, n$ then A is called a diagonal matrix.

That is, the matrix has to be square, and all elements other than the ones on the leading diagonal (the diagonal starting from the top left corner and going down; in a square matrix this diagonal ends at the bottom right corner) are zeros and there is at least one nonzero element on the diagonal. If all the elements on the diagonal are also zeros then obviously it is a null matrix. A null matrix is not counted among diagonal matrices even if the null matrix is a square matrix. For example,

$$\begin{bmatrix} 0 & 0 & 0 \\ 0 & 2 & 0 \\ 0 & 0 & 1 \end{bmatrix}, \quad \begin{bmatrix} 1 & 0 \\ 0 & 1 \end{bmatrix}, \quad \begin{bmatrix} 0 & 0 \\ 0 & -5 \end{bmatrix}$$

are all diagonal matrices. A convenient notation for a diagonal matrix is the following:

$$D = \text{diag}(d_1, \dots, d_n) \tag{2.1.1}$$

which means D is a diagonal matrix with the diagonal elements $d_1, \dots, d_n$ respectively or to be more specific, to indicate rows and columns, we may write

$$D = \text{diag}(d_{11}, \dots, d_{nn}).$$

Definition 2.1.6 (A triangular matrix). A square matrix with all elements above the leading diagonal zeros (there may be some zeros on the diagonal and below the diagonal also) then it is called a *lower triangular matrix* and if all elements below the leading diagonal are zeros then it is called an *upper triangular matrix*. (A null matrix is not counted as a triangular matrix or a diagonal matrix.)

For example,

$$
\begin{bmatrix} 1 & 0 & 0 \\ 3 & 0 & 0 \\ -1 & 1 & 5 \end{bmatrix} \text{ is lower triangular,} \quad
\begin{bmatrix} 7 & 1 & 0 \\ 0 & -1 & 1 \\ 0 & 0 & 2 \end{bmatrix} \text{ is upper triangular.}
$$

Definition 2.1.7 (Identity and scalar matrices). A diagonal matrix with all diagonal elements equal to d, $d \neq 1$, $d \neq 0$, is called a scalar matrix and if $d = 1$, that is, all diagonal elements are equal to 1 then it is called an identity matrix and an identity matrix is denoted by I, or I_n if the order is to be indicated that it is an $n \times n$ matrix.

For example,

$$
I_2 = \begin{bmatrix} 1 & 0 \\ 0 & 1 \end{bmatrix}, \quad
I_3 = \begin{bmatrix} 1 & 0 & 0 \\ 0 & 1 & 0 \\ 0 & 0 & 1 \end{bmatrix}, \quad
I_4 = \begin{bmatrix} 1 & 0 & 0 & 0 \\ 0 & 1 & 0 & 0 \\ 0 & 0 & 1 & 0 \\ 0 & 0 & 0 & 1 \end{bmatrix}.
$$

Definition 2.1.8 (The transpose of a matrix). If the i-th row of an $m \times n$ matrix A is written as the i-th column for all $i = 1, \ldots, m$ then the new matrix thus obtained is called the transpose of A and it is usually denoted as A' (A prime) or A^T (A transpose), transpose of the matrix A.

Thus when A is $m \times n$ then A' is $n \times m$. For example,

$$
A = \begin{bmatrix} 1 & 1 & 1 \\ 1 & -1 & 2 \end{bmatrix} \Rightarrow A' = \begin{bmatrix} 1 & 1 \\ 1 & -1 \\ 1 & 2 \end{bmatrix};
$$

$$
B = (0, 1, 5) \Rightarrow B' = \begin{pmatrix} 0 \\ 1 \\ 5 \end{pmatrix};
$$

$$
C = \begin{pmatrix} 1 \\ -1 \end{pmatrix} \Rightarrow C' = (1 \quad -1).
$$

(i) The transpose of a 1×1 matrix (scalar quantity) is itself.

Definition 2.1.9 (A symmetric matrix). If a square matrix $A = (a_{ij})$ is such that $a_{ij} = a_{ji}$, that is, the element in the (i, j)-th position is equal to the element in the (j, i)-th position for all i and j then A is called a symmetric matrix. That is, if A is symmetric then $A = A'$.

For example,

$$A = \begin{bmatrix} 1 & 7 \\ 7 & -3 \end{bmatrix} = A', \quad B = \begin{bmatrix} 1 & 2 & 3 \\ 2 & 0 & -4 \\ 3 & -4 & 7 \end{bmatrix} = B', \quad C = \begin{bmatrix} 2 & 0 \\ 0 & -5 \end{bmatrix} = C'.$$

The following properties are immediate:

(ii) $I = I'$; $D = D'$ (D a diagonal matrix); (lower triangular)$'$ = upper triangular and vice versa; $O' = O$ (when O is a square null matrix).

Definition 2.1.10 (A skew symmetric matrix). If a square matrix $A = (a_{ij})$ is such that $a_{ij} = -a_{ji}$ for all i and j then A is called a skew symmetric matrix. That is, $A' = -A$.

For example,

$$A = \begin{bmatrix} 0 & -4 \\ 4 & 0 \end{bmatrix} \Rightarrow A' = -A;$$

$$B = \begin{bmatrix} 0 & 1 & 3 \\ -1 & 0 & -2 \\ -3 & 2 & 0 \end{bmatrix} \Rightarrow B' = -B.$$

Note that when $A = (a_{ij})$ is skew symmetric then $a_{ii} = -a_{ii}$ which means $a_{ii} = 0$. That is, all the diagonal elements are zeros.

(iii) All the leading diagonal elements of a skew symmetric matrix are zeros.

Example 2.1.3 (Consumption profiles). Suppose that the following tables give the quantities (all in kg (kilograms)) of food items consumed by 3 families over two weeks.

week 1

	beef	pork	chicken	beans
family 1	10	5	10	10
family 2	8	7	8	10
family 3	10	15	10	12

week 2

	beef	pork	chicken	beans
family 1	10	15	15	5
family 2	8	10	10	12
family 3	12	15	12	10

The two matrices of weekly consumption, for all the 3 families, are therefore

$$A = \begin{bmatrix} 10 & 5 & 10 & 10 \\ 8 & 7 & 8 & 10 \\ 10 & 15 & 10 & 12 \end{bmatrix} \quad \text{and} \quad B = \begin{bmatrix} 10 & 15 & 15 & 5 \\ 8 & 10 & 10 & 12 \\ 12 & 15 & 12 & 10 \end{bmatrix}.$$

If we want to find the profile of total consumption in the two weeks together then we add the corresponding elements. That is,

$$A + B = \begin{bmatrix} 10+10 & 5+15 & 10+15 & 10+5 \\ 8+8 & 7+10 & 8+10 & 10+12 \\ 10+12 & 15+15 & 10+12 & 12+10 \end{bmatrix}$$

$$= \begin{bmatrix} 20 & 20 & 25 & 15 \\ 16 & 17 & 18 & 22 \\ 22 & 30 & 22 & 22 \end{bmatrix}.$$

Definition 2.1.11 (Sum of two matrices). It is defined only for matrices of the same category, both must be $m \times n$ (same m, same n for both). Let $A = (a_{ij})$ and $B = (b_{ij})$. Then the sum is defined as

$$A + B = (a_{ij} + b_{ij})$$

or the matrix obtained by adding the corresponding elements as in the illustrative example.

For example,

$$(1,1,-1) + (2,0,1) = (3,1,0);$$

$$(1,1,-1) + \begin{pmatrix} 2 \\ 0 \\ 1 \end{pmatrix} = \text{not defined};$$

$$\begin{bmatrix} 1 & 1 & 0 \\ 2 & -1 & 0 \end{bmatrix} + \begin{bmatrix} 0 & 0 & 0 \\ 0 & 0 & 0 \end{bmatrix} = \begin{bmatrix} 1 & 1 & 0 \\ 2 & -1 & 0 \end{bmatrix};$$

$$\begin{pmatrix} 0 \\ 1 \\ -1 \end{pmatrix} + \begin{pmatrix} 8 \\ 1 \\ 2 \end{pmatrix} = \begin{pmatrix} 8 \\ 2 \\ 1 \end{pmatrix};$$

$$\begin{bmatrix} 5 & 2 & -1 \\ 1 & 0 & -1 \\ 2 & 2 & 4 \end{bmatrix} + \begin{bmatrix} 0 & 1 & 0 \\ 0 & 2 & 0 \\ 0 & 0 & 1 \end{bmatrix} = \begin{bmatrix} 5 & 3 & -1 \\ 1 & 2 & -1 \\ 2 & 2 & 5 \end{bmatrix}$$

$$= \begin{bmatrix} 0 & 1 & 0 \\ 0 & 2 & 0 \\ 0 & 0 & 1 \end{bmatrix} + \begin{bmatrix} 5 & 2 & -1 \\ 1 & 0 & -1 \\ 2 & 2 & 4 \end{bmatrix}.$$

We can extend this definition to any number of matrices of the same category. Combining with the definition of scalar multiplication we can define a linear function of two matrices (or of several matrices) of the same category. That is,

$$\alpha A + \beta B = (\alpha a_{ij} + \beta b_{ij}) \tag{2.1.2}$$

where α, β are scalars and $A = (a_{ij})$ and $B = (b_{ij})$. That is, the corresponding linear functions of the elements are taken. Now, we can establish the following properties easily. For matrix addition we use the symbol +.

(vi)
$$-A = (-1)A; \quad A - A = O; \quad A + O = A;$$
$$A + B = B + A; \quad A + (B + C) = (A + B) + C;$$
$$\alpha(A + B) = \alpha A + \alpha B \tag{2.1.3}$$

where α is a scalar.

Example 2.1.4. In Example 2.1.3 if the price per kg for beef, pork, chicken and beans for week 1 are respectively $(2, 1, 0.5, 3)$ dollars and those for week 2 are respectively $(2.1, 1.2, 0.8, 3.2)$ dollars then construct the expense profiles for week 1 and week 2 for the 3 families.

Solution 2.1.4. If the price vectors are

$$U = \begin{pmatrix} 2 \\ 1 \\ 0.5 \\ 3 \end{pmatrix} \quad \text{and} \quad V = \begin{pmatrix} 2.1 \\ 1.2 \\ 0.8 \\ 3.2 \end{pmatrix}$$

respectively then the money value, in dollars, of the expense profiles for the two weeks are the following: For the first week, writing it as AU,

$$AU = \begin{bmatrix} 10 & 5 & 10 & 10 \\ 8 & 7 & 8 & 10 \\ 10 & 15 & 10 & 12 \end{bmatrix} \begin{bmatrix} 2 \\ 1 \\ 0.5 \\ 3 \end{bmatrix}$$

$$= \begin{bmatrix} (10)(2) + (5)(1) + (10)(0.5) + (10)(3) \\ (8)(2) + (7)(1) + (8)(0.5) + (10)(3) \\ (10)(2) + (15)(1) + (10)(0.5) + (12)(3) \end{bmatrix}$$

$$= \begin{bmatrix} 60 \\ 57 \\ 76 \end{bmatrix}$$

and that for the second week, writing it as BV, we have

$$BV = \begin{bmatrix} 10 & 15 & 15 & 5 \\ 8 & 10 & 10 & 12 \\ 12 & 15 & 12 & 10 \end{bmatrix} \begin{bmatrix} 2.1 \\ 1.2 \\ 0.8 \\ 3.2 \end{bmatrix}$$

$$= \begin{bmatrix} (10)(2.1) + (15)(1.2) + (15)(0.8) + (5)(3.2) \\ (8)(2.1) + (10)(1.2) + (10)(0.8) + (12)(3.2) \\ (12)(2.1) + (15)(1.2) + (12)(0.8) + (10)(3.2) \end{bmatrix}$$

$$= \begin{bmatrix} 67 \\ 75.2 \\ 84.8 \end{bmatrix}.$$

The total expenses for the two weeks combined is then

$$\begin{bmatrix} 60 \\ 57 \\ 76 \end{bmatrix} + \begin{bmatrix} 67 \\ 75.2 \\ 84.8 \end{bmatrix} = \begin{bmatrix} 127 \\ 132.2 \\ 160.8 \end{bmatrix}.$$

Some sort of multiplication and addition of matrices is involved in calculating the combined expense profile of the three families for two weeks. We will define matrix multiplication in a formal way.

The matrix $A = (a_{ij})$ postmultiplied by $B = (b_{ij})$, denoted as AB (or B premultiplied by A) is defined only when A and B are of the following types: A is $m \times n$ and B is $n \times r$, that is, the number of columns of A is equal to the number of rows of B. For example,

if A is 2×5, B is 5×4 then AB is defined but BA is not defined;

if A is 3×3, B is 3×3 then AB is defined and BA is also defined;

if A is $1 \times n$ and B is $n \times 1$ then AB and BA are defined;

if A is 3×4 and B is 2×3 then AB is not defined but BA is defined.

Definition 2.1.12 (Multiplication of matrices). Let $A = (a_{ij})$ be $m \times n$ and $B = (b_{ij})$ be $n \times r$ then AB is an $m \times r$ matrix where the (i,j)-th element in AB is the dot product of the i-th row vector of A with the j-th column vector of B.

The i-th row vector of A, denoted by α_i, is the following:

$$\alpha_i = (a_{i1}, a_{i2}, \dots, a_{in}).$$

The j-th column vector of B, denoted by β_j, is

$$\beta_j = \begin{pmatrix} b_{1j} \\ b_{2j} \\ \vdots \\ b_{nj} \end{pmatrix}.$$

Then writing element-wise multiplication and addition as

$$\alpha_i \beta_j = (a_{i1}, \ldots, a_{in}) \begin{pmatrix} b_{1j} \\ \vdots \\ b_{nj} \end{pmatrix}$$

$$= a_{i1} b_{1j} + a_{i2} b_{2j} + \cdots + a_{in} b_{nj}$$

$$= \sum_{k=1}^{n} a_{ik} b_{kj}.$$

If AB is denoted by $AB = C = (c_{ij})$ then

$$c_{ij} = \alpha_i \beta_j = \sum_{k=1}^{n} a_{ik} b_{kj}. \qquad (2.1.4)$$

Symbolically the multiplication can be expressed as follows:

The first row of A dot product with the various columns of B gives the first row of $C = AB$, the second row of A dot product with the various columns of B gives the second row of C, and so on.

Example 2.1.5. Evaluate the product of the matrices A and B to obtain AB, wherever AB is defined:

(a) $\quad A = (1 \quad -1 \quad 1 \quad 2), \quad B = \begin{pmatrix} 0 \\ 1 \\ 2 \\ -1 \end{pmatrix}$;

(b) $\quad A = \begin{bmatrix} 0 & 1 & -1 & 2 \\ -1 & 1 & 5 & 4 \end{bmatrix}, \quad B = \begin{bmatrix} 1 \\ 2 \\ -1 \\ 0 \end{bmatrix}$;

(c) $\quad A = \begin{bmatrix} 2 & 1 & -1 \\ 3 & 1 & 1 \\ 1 & 1 & 1 \end{bmatrix}, \quad B = \begin{bmatrix} -1 & 1 \\ 1 & -1 \\ 0 & 0 \end{bmatrix}$;

(d) $\quad A = \begin{bmatrix} 2 & 1 & -1 \\ 3 & 1 & 1 \\ 1 & 1 & 1 \end{bmatrix}, \quad B = \begin{bmatrix} 2 & 1 & 1 & 2 \\ 1 & 1 & 0 & 0 \end{bmatrix}$;

(e) $A = I_3, \quad B = (b_{ij})$;

(f) $A = \begin{bmatrix} 0 & 0 & 0 \\ 0 & 0 & 0 \end{bmatrix}, \quad B = (b_{ij})$ is 3×3.

Solutions 2.1.5.

(a) $AB = (1 \quad -1 \quad 1 \quad 2) \begin{pmatrix} 0 \\ 1 \\ 2 \\ -1 \end{pmatrix}$

$= (1)(0) + (-1)(1) + (1)(2) + (2)(-1) = -1.$

(b) $AB = \begin{bmatrix} 0 & 1 & -1 & 2 \\ -1 & 1 & 5 & 4 \end{bmatrix} \begin{bmatrix} 1 \\ 2 \\ -1 \\ 0 \end{bmatrix}$

where A is 2×4 and B is 4×1. Thus AB is 2×1 which can be remembered symbolically as $2 \times 1 = 2 \times (4 : 4) \times 1$. The first row element in AB is

$$(0 \quad 1 \quad -1 \quad 2) \begin{pmatrix} 1 \\ 2 \\ -1 \\ 0 \end{pmatrix} = (0)(1) + (1)(2) + (-1)(-1) + (2)(0) = 3$$

and the second row element in AB is

$$(-1 \quad 1 \quad 5 \quad 4) \begin{pmatrix} 1 \\ 2 \\ -1 \\ 0 \end{pmatrix} = (-1)(1) + (1)(2) + (5)(-1) + (4)(0) = -4.$$

Hence

$$AB = \begin{bmatrix} 3 \\ -4 \end{bmatrix}.$$

(c) $AB = \begin{bmatrix} 2 & 1 & -1 \\ 3 & 1 & 1 \\ 1 & 1 & 1 \end{bmatrix} \begin{bmatrix} -1 & 1 \\ 1 & -1 \\ 0 & 0 \end{bmatrix}$

$= \begin{bmatrix} (2)(-1) + (1)(1) + (-1)(0), (2)(1) + (1)(-1) + (-1)(0) \\ (3)(-1) + (1)(1) + (1)(0), (3)(1) + (1)(-1) + (1)(0) \\ (1)(-1) + (1)(1) + (1)(0), (1)(1) + (1)(-1) + (1)(0) \end{bmatrix}$

$= \begin{bmatrix} -1 & 1 \\ -2 & 2 \\ 0 & 0 \end{bmatrix}.$

(d) Since A is 3×3 and B is 2×4, AB is not defined.

(e) Here A is an identity matrix and B is a general matrix.

$$AB = \begin{bmatrix} 1 & 0 & 0 \\ 0 & 1 & 0 \\ 0 & 0 & 1 \end{bmatrix} \begin{bmatrix} b_{11} & b_{12} & b_{13} \\ b_{21} & b_{22} & b_{23} \\ b_{31} & b_{32} & b_{33} \end{bmatrix}$$

$$= \begin{bmatrix} b_{11} & b_{12} & b_{13} \\ b_{21} & b_{22} & b_{23} \\ b_{31} & b_{32} & b_{33} \end{bmatrix} = B.$$

(f) $\quad AB = \begin{bmatrix} 0 & 0 & 0 \\ 0 & 0 & 0 \end{bmatrix} \begin{bmatrix} b_{11} & b_{12} & b_{13} \\ b_{21} & b_{22} & b_{23} \\ b_{31} & b_{32} & b_{33} \end{bmatrix} = \begin{bmatrix} 0 & 0 & 0 \\ 0 & 0 & 0 \end{bmatrix} = O.$

It is interesting to observe the following general properties. As long as the products are defined

(v) $\qquad (A')' = A; \ IA = A; \ BI = B; \ OC = O; \ DO = O; \ AB \neq BA$

$\qquad ABC \neq ACB$ $\hfill (2.1.5)$

where A, B, C, D are arbitrary matrices and I and O denote the identity matrix and the null matrix respectively, and when the products are defined.

Also note that if X is an $n \times 1$ column vector then

$$X = \begin{bmatrix} x_1 \\ \vdots \\ x_n \end{bmatrix} \Rightarrow X'X = (x_1, \ldots, x_n) \begin{pmatrix} x_1 \\ \vdots \\ x_n \end{pmatrix}$$

$$= x_1^2 + x_2^2 + \ldots + x_n^2 \hfill (2.1.6)$$

whereas

$$XX' = \begin{pmatrix} x_1 \\ \vdots \\ x_n \end{pmatrix} (x_1, \ldots, x_n)$$

$$= \begin{bmatrix} x_1^2 & x_1 x_2 & \cdots & x_1 x_n \\ x_2 x_1 & x_2^2 & \cdots & x_2 x_n \\ \vdots & \vdots & \cdots & \vdots \\ x_n x_1 & x_n x_2 & \cdots & x_n^2 \end{bmatrix}. \hfill (2.1.7)$$

That is, $X'X$ is a scalar $(1 \times 1$ matrix) whereas XX' is an $n \times n$ matrix.

Example 2.1.6. Write the following systems of linear equations in matrix notation:

$$
\begin{aligned}
\text{(a)} \quad 2x_1 - x_2 + x_3 &= 4 \\
x_1 + x_2 - x_3 &= 2;
\end{aligned}
$$

$$
\begin{aligned}
\text{(b)} \quad 3x_1 + x_2 + x_3 &= 1 \\
x_1 - 2x_2 + x_3 &= 3 \\
2x_1 + x_2 - x_3 &= 2;
\end{aligned}
$$

$$
\begin{aligned}
\text{(c)} \quad 5x_1 - x_2 + x_3 - x_4 &= 0 \\
2x_1 + x_2 - 3x_3 + x_4 &= 0.
\end{aligned}
$$

Solution 2.1.6. (a) The first equation in the first set can be written as

$$
(2 \quad -1 \quad 1) \begin{pmatrix} x_1 \\ x_2 \\ x_3 \end{pmatrix} = 4
$$

and the second as

$$
(1, 1, -1) \begin{pmatrix} x_1 \\ x_2 \\ x_3 \end{pmatrix} = 2
$$

and combining the two we have

$$
\begin{bmatrix} 2 & -1 & 1 \\ 1 & -1 & 1 \end{bmatrix} \begin{bmatrix} x_1 \\ x_2 \\ x_3 \end{bmatrix} = \begin{bmatrix} 4 \\ 2 \end{bmatrix} \quad \text{or} \quad AX = b
$$

where the coefficient matrix is A,

$$
A = \begin{bmatrix} 2 & -1 & 1 \\ 1 & -1 & 1 \end{bmatrix}, \quad X = \begin{bmatrix} x_1 \\ x_2 \\ x_3 \end{bmatrix} \quad \text{and} \quad b = \begin{bmatrix} 4 \\ 2 \end{bmatrix}.
$$

(b) The coefficient matrix in the three equations is

$$
A = \begin{bmatrix} 3 & 1 & 1 \\ 1 & -2 & 1 \\ 2 & 1 & -1 \end{bmatrix}
$$

and hence the three equations together is $AX = b$ where A is given above

$$
X = \begin{pmatrix} x_1 \\ x_2 \\ x_3 \end{pmatrix} \quad \text{and} \quad b = \begin{pmatrix} 1 \\ 3 \\ 2 \end{pmatrix}.
$$

(c) Writing the two equations together as $AX = b$ we have

$$A = \begin{bmatrix} 5 & -1 & 1 & -1 \\ 2 & 1 & -3 & 1 \end{bmatrix}, \quad X = \begin{bmatrix} x_1 \\ x_2 \\ x_3 \\ x_4 \end{bmatrix},$$

$$b = \begin{pmatrix} 0 \\ 0 \end{pmatrix} \quad \text{or} \quad AX = O$$

where O is a null vector.

Example 2.1.7 (Linear forms). Write the following linear forms (all terms are homogeneous of degree 1 each) in matrix notation:

(a) $y = x_1 - 3x_2 + x_3$

(b) $y = x_1 - x_2 - x_3 + 2x_4$

(c) $y_1 = 2x_1 - x_2 + x_3$

  $y_2 = x_1 + x_2$

(d) $y_1 = a_{11}x_1 + a_{12}x_2 + \ldots + a_{1n}x_n$

  $\vdots$

  $y_m = a_{m1}x_1 + a_{m2}x_2 + \ldots + a_{mn}x_n$

where a_{ij}'s are constants.

Solution 2.1.7.

(a) $y = (1, -3, 1) \begin{pmatrix} x_1 \\ x_2 \\ x_3 \end{pmatrix} = (x_1, x_2, x_3) \begin{pmatrix} 1 \\ -3 \\ 1 \end{pmatrix} = a'X = X'a$

where

$$a = \begin{pmatrix} 1 \\ -3 \\ 1 \end{pmatrix} \quad \text{and} \quad X = \begin{pmatrix} x_1 \\ x_2 \\ x_3 \end{pmatrix}.$$

(b) Writing it as $y = a'X = X'a$ we have

$$a = \begin{bmatrix} 1 \\ -1 \\ -1 \\ 2 \end{bmatrix} \quad \text{and} \quad X = \begin{bmatrix} x_1 \\ x_2 \\ x_3 \\ x_4 \end{bmatrix}.$$

(c) Writing the two linear forms together as one equation $Y = AX$ we have

$$Y = \begin{bmatrix} y_1 \\ y_2 \end{bmatrix}, \quad A = \begin{bmatrix} 2 & -1 & 1 \\ 1 & 1 & 0 \end{bmatrix}, \quad X = \begin{bmatrix} x_1 \\ x_2 \\ x_3 \end{bmatrix}.$$

(d) Writing the m linear forms together as one equation $Y = AX$ we have

$$Y = \begin{bmatrix} y_1 \\ \vdots \\ y_m \end{bmatrix}, \quad A = \begin{bmatrix} a_{11} & a_{12} & \cdots & a_{1n} \\ \vdots & \vdots & \cdots \vdots & \vdots \\ a_{m1} & a_{m2} & \cdots & a_{mn} \end{bmatrix}, \quad X = \begin{bmatrix} x_1 \\ \vdots \\ x_n \end{bmatrix}.$$

The representation in (d) is a general linear form or it can also be considered as a linear transformation of the vector X into the vector Y. It is linear in the sense that each y_j is a linear function (of the first degree in every term) of $x_1, \ldots, x_n$.

Example 2.1.8 (Quadratic forms). Write the following quadratic forms (all terms are homogeneous of degree 2 each) in matrix notations:

(a) $y = x_1^2 + x_2^2 + x_3^2$

(b) $y = x_1^2 - x_2^2 + x_3^2$

(c) $y = 2x_1^2 + x_2^2 - x_3^2 + 5x_1x_2 - 2x_1x_3 + x_2x_3$

(d) $y = x_2^2 + 4x_1x_2 - 2x_2x_3$

(e) $y = \sum_{i=1}^{n} a_{ii}x_i^2 + 2 \sum_{i<j=1}^{n} a_{ij}x_ix_j = \sum_{ij} a_{ij}x_ix_j.$

Solutions 2.1.8. This is a simple sum of squares.

$$\text{(a)} \quad y = X'X, \quad X = \begin{bmatrix} x_1 \\ x_2 \\ x_3 \end{bmatrix}.$$

(b) Here the coefficients of x_1^2, x_2^2 and x_3^2 are different. This format can be created by a diagonal matrix. That is,

$$y = X'AX, \quad X = \begin{bmatrix} x_1 \\ x_2 \\ x_3 \end{bmatrix}, \quad A = \begin{bmatrix} 1 & 0 & 0 \\ 0 & -1 & 0 \\ 0 & 0 & 1 \end{bmatrix}.$$

(c) Here the product terms are also present. Hence A has nonzero off-diagonal elements.

$$y = X'AX, \quad X = \begin{bmatrix} x_1 \\ x_2 \\ x_3 \end{bmatrix}, \quad A = \begin{bmatrix} 2 & 5 & -2 \\ 0 & 1 & 1 \\ 0 & 0 & -1 \end{bmatrix}.$$

Note that the coefficient of $x_1 x_2$ is $a_{12} = 5$. But $x_1 x_2 = x_2 x_1$. Hence we could have written 5 as a_{21} instead as a_{12} or we could have distributed 5 equally to a_{12} and a_{21}, that is, $a_{12} = a_{21} = \frac{5}{2}$ which is a symmetric format. Writing a symmetric format for A we have the same quadratic form written as

$$y = X'AX, \quad A = \begin{bmatrix} 2 & \frac{5}{2} & -1 \\ \frac{5}{2} & 1 & \frac{1}{2} \\ -1 & \frac{1}{2} & -1 \end{bmatrix}.$$

In this format, the diagonal elements of A are the coefficients of the square terms, that is a_{ii} is the coefficient of x_i^2, and the non-diagonal elements are the coefficients of the corresponding product terms where the coefficients are distributed equally as $a_{ij} = a_{ji}$, that is the coefficients of $x_i x_j$ as well as that of $x_j x_i$ is $(a_{ij} + a_{ji})/2$ in order to write A as a symmetric matrix for elegance.

$$(d) \quad y = X'AX, \quad X = \begin{bmatrix} x_1 \\ x_2 \\ x_3 \end{bmatrix}, \quad A = \begin{bmatrix} 0 & 2 & 0 \\ 2 & 1 & -1 \\ 0 & -1 & 0 \end{bmatrix}.$$

Here A is written as a symmetric matrix. We could have written the quadratic form in many different ways if we did not want A to be symmetric. For example, the same quadratic form

$$y = X'BX = X'CX$$

where

$$B = \begin{bmatrix} 0 & 4 & 0 \\ 0 & 1 & -2 \\ 0 & 0 & 0 \end{bmatrix} \quad \text{and} \quad C = \begin{bmatrix} 0 & 1 & 0 \\ 3 & 1 & 0 \\ 0 & -2 & 0 \end{bmatrix}.$$

The following is a general quadratic form:

$$(e) \quad y = X'AX, \quad X = \begin{bmatrix} x_1 \\ \vdots \\ x_n \end{bmatrix}, \quad A = (a_{ij})$$

and without loss of generality we can take A to be symmetric, that is, $A = A'$.

(vi) The matrix A of the quadratic form $X'AX$ can be taken as symmetric, that is, $A = A'$, without loss of generality.

If A is not symmetric then the quadratic form can be rewritten equivalently as

$$X'AX = X'BX, \quad \text{where } B = \frac{1}{2}(A + A') = B'.$$

Thus B is symmetric.

The following general properties can be observed for transposes and products. The student may verify them by taking arbitrary 3×3 matrices:

$$(A')' = A; \quad (A + B)' = A' + B'; \quad (AB)' = B'A'; \quad (AA')' = AA';$$
$$I' = I; \quad O' = O; \quad (A_1 A_2 \ldots A_k)' = A_k' \ldots A_2' A_1'. \tag{2.1.8}$$

Whenever the sums and products are defined,
  the transpose of a lower triangular matrix is upper triangular;
  the transpose of an upper triangular matrix is lower triangular;
  the transpose of a diagonal matrix is diagonal.

Whenever the product is defined,
  the product of two identity matrices is an identity matrix;
  the product of two diagonal matrices is a diagonal matrix;
  the product of a null matrix with any other matrix is null;
  the product of two lower triangular matrices is lower triangular;
  the product of two upper triangular matrices is upper triangular.

Whenever the sum is defined,
  the sum of two lower triangular matrices is lower triangular;
  the sum of two upper triangular matrices is upper triangular;
  the sum of two diagonal matrices is diagonal;
  the sum of two identity matrices is a scalar matrix;
  the sum of two symmetric matrices is symmetric;
  the sum of two skew symmetric matrices is skew symmetric.

2.1.1 Some more practical situations

Many situations from various disciplines can be listed where systems of entities can be written in nice elegant simplified forms with the help of matrices. A few more situations will be listed here where only the sums and products of matrices are involved.

Example 2.1.9 (The Jacobian matrix). Consider the following system of linear equations:

$$y_1 = 2x_1 + x_2 - x_3 + x_4$$
$$y_2 = x_1 + 3x_2 + x_3 + 2x_4$$
$$y_3 = -x_1 + x_2 + x_3 - x_4$$
$$y_4 = x_1 + x_2 + x_3 + x_4$$

which can be written in matrix notation as,

$$Y = AX, \quad Y = \begin{bmatrix} y_1 \\ y_2 \\ y_3 \\ y_4 \end{bmatrix}, \quad A = \begin{bmatrix} 2 & 1 & -1 & 1 \\ 1 & 3 & 1 & 2 \\ -1 & 1 & 1 & -1 \\ 1 & 1 & 1 & 1 \end{bmatrix}, \quad X = \begin{bmatrix} x_1 \\ x_2 \\ x_3 \\ x_4 \end{bmatrix}.$$

Consider the partial derivative of y_i with respect to x_j for all i and j and let the matrix of these partial derivatives be denoted by

$$\frac{\partial Y}{\partial X} = \left(\frac{\partial y_i}{\partial x_j} \right) = \begin{bmatrix} \frac{\partial y_1}{\partial x_1} & \cdots & \frac{\partial y_1}{\partial x_4} \\ \vdots & \cdots & \vdots \\ \frac{\partial y_4}{\partial x_1} & \cdots & \frac{\partial y_4}{\partial x_4} \end{bmatrix}.$$

Then the matrix $\frac{\partial Y}{\partial X}$ is called the *Jacobian matrix* of this transformation X to Y. Evaluate the Jacobian matrix in the above transformation.

Solution 2.1.9.

$$\frac{\partial y_1}{\partial x_1} = 2, \quad \frac{\partial y_1}{\partial x_2} = 1, \quad \frac{\partial y_1}{\partial x_3} = -1, \quad \frac{\partial y_1}{\partial x_4} = 1, \quad \text{etc.}$$

Then

$$\frac{\partial Y}{\partial X} = \left(\frac{\partial y_i}{\partial x_j} \right) = \begin{bmatrix} 2 & 1 & -1 & 1 \\ 1 & 3 & 1 & 2 \\ -1 & 1 & 1 & -1 \\ 1 & 1 & 1 & 1 \end{bmatrix} = A =$$

the coefficient matrix in $Y = AX$.

Instead of linear functions if

$$y_i = f_i(x_1, \ldots, x_k), \quad i = 1, \ldots, k$$

then the Jacobian matrix is still

$$\frac{\partial Y}{\partial X} = \left(\frac{\partial y_i}{\partial x_j} \right)$$

where $\frac{\partial y_i}{\partial x_j}$ is the partial derivative of y_i with respect to x_j. The Jacobian matrices are relevant only when the number of $x_1, \ldots, x_k$, that is k, is the same as the number of $y_1, \ldots, y_k$, that is k. These numbers are equal, and further, we should be able to write $x_1, \ldots, x_k$ uniquely in terms of $y_1, \ldots, y_k$ and vice versa. In this case we say that there is a one-to-one transformation.

Example 2.1.10 (Derivative of a quadratic form). Let

$$u = 3x_1^2 + x_2^2 - 2x_3^2 - 2x_1x_2 + x_1x_3$$

and consider the differential operator discussed in Chapter 1, namely,

$$\frac{\partial}{\partial X} = \begin{pmatrix} \frac{\partial}{\partial x_1} \\ \vdots \\ \frac{\partial}{\partial x_n} \end{pmatrix}. \tag{2.1.9}$$

Here $n = 3$. Evaluate $\frac{\partial u}{\partial X}$.

Solution 2.1.10. Writing A as a symmetric matrix

$$u = X'AX \implies X' = (x_1, x_2, x_3), \quad A = \begin{bmatrix} 3 & -1 & \frac{1}{2} \\ -1 & 1 & 0 \\ \frac{1}{2} & 0 & -2 \end{bmatrix}$$

$$\frac{\partial u}{\partial X} = \begin{bmatrix} \frac{\partial u}{\partial x_1} \\ \frac{\partial u}{\partial x_2} \\ \frac{\partial u}{\partial x_3} \end{bmatrix} = \begin{bmatrix} 6x_1 - 2x_2 + x_3 \\ -2x_1 + 2x_2 \\ x_1 - 4x_3 \end{bmatrix}$$

$$= \begin{bmatrix} 6 & -2 & 1 \\ -2 & 2 & 0 \\ 1 & 0 & -4 \end{bmatrix} \begin{bmatrix} x_1 \\ x_2 \\ x_3 \end{bmatrix}$$

$$= 2 \begin{bmatrix} 3 & -1 & \frac{1}{2} \\ -1 & 1 & 0 \\ \frac{1}{2} & 0 & -2 \end{bmatrix} \begin{bmatrix} x_1 \\ x_2 \\ x_3 \end{bmatrix} = 2AX.$$

This, in fact, is a general result.

(vii) Let X be a $k \times 1$ vector of real variables, $A = A'$ a symmetric $k \times k$ matrix of constants, $\frac{\partial}{\partial X}$ the $k \times 1$ vector of partial derivative operator then

$$u = X'AX, \quad A = A', \implies \frac{\partial u}{\partial X} = 2AX, \tag{2.1.10}$$

and

$$\frac{\partial}{\partial X} \frac{\partial u}{\partial X'} = 2A. \tag{2.1.11}$$

If $u = X'AX$ where A is not assumed to be symmetric then

$$u = X'AX \implies \frac{\partial u}{\partial X} = (A + A')X \tag{2.1.12}$$

and

$$\frac{\partial}{\partial X}\frac{\partial u}{\partial X'} = (A + A').$$ (2.1.13)

Before concluding this section we will define powers of square matrices.

Definition 2.1.13 (Powers of a square matrix). Let A be an $n \times n$ matrix. Then the r-th power of A for $r = 0, 1, \ldots$ (non-negative integers) is defined as

$$A^r = AA \cdots A$$

(product taken r times), with $A^0 = I$.

As examples,

$$A = \begin{bmatrix} 1 & 0 \\ 0 & 1 \end{bmatrix} \Rightarrow A^2 = \begin{bmatrix} 1 & 0 \\ 0 & 1 \end{bmatrix};$$

$$B = \begin{bmatrix} 0 & 1 \\ 1 & 0 \end{bmatrix} \Rightarrow B^2 = \begin{bmatrix} 0 & 1 \\ 1 & 0 \end{bmatrix}\begin{bmatrix} 0 & 1 \\ 1 & 0 \end{bmatrix} = \begin{bmatrix} 1 & 0 \\ 0 & 1 \end{bmatrix};$$

$$C = \begin{bmatrix} 0 & -1 \\ -1 & 0 \end{bmatrix} \Rightarrow C^2 = \begin{bmatrix} 1 & 0 \\ 0 & 1 \end{bmatrix},$$
$$C^3 = C, \quad C^4 = I, \quad C^{2m} = I, \quad C^{2m+1} = C;$$

$$A_1 = \begin{bmatrix} 0 & 1 \\ 0 & 0 \end{bmatrix} \Rightarrow A_1^2 = \begin{bmatrix} 0 & 0 \\ 0 & 0 \end{bmatrix} = O;$$

$$A_2 = \begin{bmatrix} d_1 & 0 \\ 0 & d_2 \end{bmatrix} \Rightarrow A_2^m = \begin{bmatrix} d_1^m & 0 \\ 0 & d_2^m \end{bmatrix}.$$

Definition 2.1.14 (Idempotent matrices). If $A = A^2$ then A is called an idempotent matrix when A is non-null.

As examples we have,

(i) $I = I^2 \Rightarrow I$ is an idempotent matrix.

Consider the $n \times 1$ vector of unities, denoted by J. Then

$$J = \begin{pmatrix} 1 \\ \vdots \\ 1 \end{pmatrix} \Rightarrow J'J = n, \quad JJ' = \begin{bmatrix} 1 & 1 & \cdots & 1 \\ \vdots & \vdots & \cdots & \vdots \\ 1 & 1 & \cdots & 1 \end{bmatrix}.$$

Let $A = \frac{1}{n}JJ'$ then

(ii) $\quad A^2 = \left(\frac{1}{n}\right)^2 JJ'JJ' = \left(\frac{1}{n}\right)^2 J(J'J)J = \frac{1}{n}JJ' = A.$

(iii) $\quad A = \begin{bmatrix} 0 & 0 \\ 1 & 1 \end{bmatrix}, \quad A^2 = \begin{bmatrix} 0 & 0 \\ 1 & 1 \end{bmatrix}\begin{bmatrix} 0 & 0 \\ 1 & 1 \end{bmatrix} = \begin{bmatrix} 0 & 0 \\ 1 & 1 \end{bmatrix} = A.$

Thus A is idempotent. We can construct many such examples of idempotent matrices.

Definition 2.1.15 (Nilpotent matrix of order r). If a matrix $B \neq O$ is such that $B^r = O$, $B^{r-1} \neq O$, for some fixed r, $r = 2, 3, \ldots$ then B is called a nilpotent matrix of order r, where r is the smallest integer where B^r becomes null.

We can construct many examples of nilpotent matrices of various orders. For example,

(i) $\quad B = \begin{bmatrix} 0 & 1 \\ 0 & 0 \end{bmatrix} \Rightarrow B^2 = \begin{bmatrix} 0 & 1 \\ 0 & 0 \end{bmatrix}\begin{bmatrix} 0 & 1 \\ 0 & 0 \end{bmatrix} = \begin{bmatrix} 0 & 0 \\ 0 & 0 \end{bmatrix} = O,$

nilpotent of order 2;

(ii) $\quad C = \begin{bmatrix} 0 & 0 \\ 1 & 0 \end{bmatrix} \Rightarrow C^2 = O, \quad$ nilpotent of order 2.

Exercises 2.1

2.1.1. Compute $2A - B + \frac{1}{2}C$ for the following matrices, wherever it is defined:

(a) $\quad A = (1, -1, 2, 3), \quad B = (2, 5, 0), \quad C = (-1, 1);$

(b) $\quad A = (1, -1, 2, 3), \quad B = \begin{bmatrix} 2 \\ 0 \\ 1 \\ -1 \end{bmatrix}, \quad C = \begin{bmatrix} 0 \\ 0 \\ 0 \\ 0 \end{bmatrix};$

(c) $\quad A = \begin{bmatrix} 1 & 1 & -1 \\ 0 & 1 & 0 \end{bmatrix}, \quad B = \begin{bmatrix} 0 & 1 & 2 \\ -2 & 1 & 5 \end{bmatrix}, \quad C = \begin{bmatrix} 0 & 4 & 6 \\ 1 & -1 & 0 \end{bmatrix};$

(d) $\quad A = \begin{bmatrix} 1 & 0 \\ 1 & 1 \end{bmatrix}, \quad B = \begin{bmatrix} 0 & 1 \\ 1 & 0 \end{bmatrix}, \quad C = \begin{bmatrix} 1 & 0 \\ 0 & 1 \end{bmatrix}.$

2.1.2. Compute AB and BA, wherever defined, for the matrices A and B in Exercise 2.1.1. Are $AB = BA$ in general?

2.1.3. By taking an arbitrary $m \times n$ matrix $A = (a_{ij})$ show that $OA = O$, $AO = O$ whenever the products are defined, where O indicates a null matrix.

2.1.4. By taking an arbitrary $m \times n$ matrix $A = (a_{ij})$ show that $I_m A = A$ and $AI_n = A$ where I_m and I_n are $m \times m$ and $n \times n$ identity matrices.

2.1.5. Construct a 2×2 matrix A such that (a) $A^3 = O$ but $A \neq O$, (b) $A^2 = A$, $A \neq O$, $A \neq I$.

2.1.6. Consider a general 2×2 matrix A and consider

$$P_1 = \begin{pmatrix} 0 & 1 \\ 1 & 0 \end{pmatrix}, \quad P_2 = \begin{pmatrix} 0 & -1 \\ 1 & 0 \end{pmatrix},$$

$$P_3 = \begin{pmatrix} 0 & 1 \\ -1 & 0 \end{pmatrix}, \quad P_4 = \begin{pmatrix} 0 & -1 \\ -1 & 0 \end{pmatrix}.$$

(a) Premultiply A with each of $P_1, \dots, P_4$ and explain in each case what happens to the matrix A.
(b) Postmultiply A with each of $P_1, \dots, P_4$ and explain in each case what happens to A.

2.1.7. Consider the 2×2 matrix A in Exercise 2.1.6 as two ordered points in a plane. Explain geometrically what is seen in (a) and (b) of Exercise 2.1.6.

2.1.8. Let

$$A = \begin{bmatrix} \cos\theta & -\sin\theta \\ \sin\theta & \cos\theta \end{bmatrix}.$$

Compute AA' and $A'A$.

2.1.9. Construct two examples each of two matrices A and B where (i) $AB \neq BA$, (ii) $AB = BA$. Exclude trivial cases involving identity, null and diagonal matrices.

2.1.10. Let

$$A = \begin{bmatrix} 1 & 1 \\ 1 & -1 \end{bmatrix}.$$

Construct a matrix B such that $AB = I, BA = I$ where I is a 2×2 identity matrix.

2.1.11. Write the following systems of linear equations in matrix notation:

$$\text{(a)} \quad \begin{aligned} x_1 - x_2 + x_3 &= 2 \\ 2x_1 + x_2 - 5x_3 &= 4; \end{aligned}$$

$$\text{(b)} \quad \begin{aligned} 2x_1 + 3x_2 + x_3 - x_4 &= 1 \\ x_1 + x_2 + x_3 + x_4 &= 7 \\ 3x_1 + 2x_2 - x_3 - x_4 &= 5 \\ x_1 - x_2 - x_3 + x_4 &= 4. \end{aligned}$$

2.1.12. If the right sides in (a) and (b) in Exercise 2.1.11 are replaced by variables $y_1, y_2, \ldots$ then write the transformation in the form $Y = AX$ and identity Y, X, A in each case.

2.1.13. Evaluate the Jacobian matrix in Exercise 2.1.12 (b).

2.1.14. Write the following quadratic forms in the form $u = X'AX$ where (i) $A = A'$, (ii) $A \neq A'$:
(a) $u = 2x_1^2 + x_2^2 - x_3^2 + x_4^2 - 2x_1x_2 - x_2x_4 + 3x_3x_4$;
(b) $u = -x_1^2 + x_4^2 + 2x_1x_3 - 2x_1x_4 + x_1x_2$.

2.1.15. If $\frac{\partial}{\partial X}$ denotes the column vector of partial differential operators evaluate $\frac{\partial u}{\partial X}$ and $\frac{\partial}{\partial X} \frac{\partial u}{\partial X'}$ for (a) and (b) in Exercise 2.1.14. Write the final forms in matrix notations (one case each where the matrix of the quadratic form is (i) symmetric, (ii) not symmetric).

2.1.16. Let

$$A = \begin{bmatrix} 1 & 1 & 1 \\ 1 & 0 & -1 \end{bmatrix}.$$

(a) Construct a 3×1 non-null vector B such that $AB = O$.
(b) Construct a 3×2 non-null matrix B such that $AB = O$, if possible.

2.1.17. Let A and B be two matrices where AB is defined. Suppose that the second row of A is 5 times the first row. Then show that, whatever be B, the second row of AB is 5 times the first row of AB.

2.2 More properties of matrices

A scalar function of the elements of a square matrix called the *trace* of the matrix is very useful in some practical applications.

Definition 2.2.1 (The trace of a matrix). It is defined only for square matrices. Let $A = (a_{ij})$ be $n \times n$. Then the trace of A, denoted by $\text{tr}(A)$, is the sum of the leading diagonal elements of A. That is,

$$\text{tr}(A) = a_{11} + a_{22} + \cdots + a_{nn}.$$

For example,

$$A = \begin{bmatrix} 1 & -1 & 1 \\ 2 & 3 & 0 \\ 1 & 4 & -5 \end{bmatrix} \Rightarrow \text{tr}(A) = 1 + 3 + (-5) = -1;$$

$$A = \begin{bmatrix} 2 & 0 \\ 0 & 3 \end{bmatrix} \Rightarrow \operatorname{tr}(A) = 2 + 3 = 5;$$

$$A = \begin{bmatrix} 0 & 0 \\ 0 & 0 \end{bmatrix} \Rightarrow \operatorname{tr}(A) = 0.$$

The following properties follow from the definition itself:

$$\operatorname{tr}(A') = \operatorname{tr}(A) \tag{2.2.1}$$

$$\operatorname{tr}(AB) = \operatorname{tr}(BA) \tag{2.2.2}$$

whenever AB and BA are defined. Note that, in general, AB need not be equal to BA but their traces are equal. Extending (2.2.2) we have

$$\operatorname{tr}(ABC) = \operatorname{tr}(CAB) = \operatorname{tr}(BCA) \tag{2.2.3}$$

even though $ABC \neq CAB \neq BCA$.

Let A be a square matrix. For some matrices A we can construct another matrix B such that

$$AB = I, \quad BA = I$$

where I is the identity matrix.

Definition 2.2.2 (Regular inverse). If there exists a matrix, denoted by A^{-1}, such that

$$AA^{-1} = I, \quad A^{-1}A = I$$

then A^{-1} is called the regular inverse, or simply the inverse, of A.

For example,

$$A = I_2 = \begin{bmatrix} 1 & 0 \\ 0 & 1 \end{bmatrix} \Rightarrow A^{-1} = I_2, \quad AA^{-1} = I_2, \quad A^{-1}A = I_2;$$

$$A = \begin{bmatrix} 2 & 0 \\ 0 & -3 \end{bmatrix} \Rightarrow A^{-1} = \begin{bmatrix} \frac{1}{2} & 0 \\ 0 & -\frac{1}{3} \end{bmatrix}, \quad AA^{-1} = I, \quad A^{-1}A = I;$$

$$A = \begin{bmatrix} 1 & 1 \\ 1 & 2 \end{bmatrix} \Rightarrow A^{-1} = \begin{bmatrix} 2 & -1 \\ -1 & 1 \end{bmatrix}, \quad AA^{-1} = I, \quad A^{-1}A = I;$$

$$A = \begin{bmatrix} 1 & 0 \\ -1 & 1 \end{bmatrix} \Rightarrow A^{-1} = \begin{bmatrix} 1 & 0 \\ 1 & 1 \end{bmatrix}, \quad AA^{-1} = I, \quad A^{-1}A = I.$$

Later we will discuss a systematic way of evaluating the inverse of a given matrix, whenever the inverse exists. The following properties are evident from the definition itself.

(i) If A is an $n \times n$ diagonal matrix with the diagonal elements $d_1 \neq 0, d_2 \neq 0, \dots, d_n \neq 0$ then A^{-1} is an $n \times n$ diagonal matrix with the diagonal elements $\frac{1}{d_1}, \dots, \frac{1}{d_n}$.

(ii) A diagonal matrix A with at least one of the diagonal elements zero has no inverse A^{-1}.

(iii) A triangular matrix A (lower or upper) with at least one of the diagonal elements zero has no inverse A^{-1}.

(iv) For a given matrix A if A^{-1} exists then it is unique. That is, if $AB = I, BA = I$ as well as $AC = I, CA = I$ then $B = C = A^{-1}$.

(v) A null matrix has no inverse.

(vi) If A and B are square matrices with A^{-1} and B^{-1} existing then

$$(AB)^{-1} = B^{-1}A^{-1}.$$

This result is easily proved by evaluating

$$(AB)(B^{-1}A^{-1}) \quad \text{as well as} \quad (B^{-1}A^{-1})(AB).$$

Note that

$$(AB)(B^{-1}A^{-1}) = A(BB^{-1})A^{-1} = AIA^{-1} = AA^{-1} = I$$

and similarly

$$(B^{-1}A^{-1})(AB) = B^{-1}(A^{-1}A)B = B^{-1}IB = B^{-1}B = I.$$

This result can be extended to any number of $n \times n$ matrices having inverses. That is,

$$(A_1 A_2 \cdots A_k)^{-1} = A_k^{-1} \cdots A_2^{-1} A_1^{-1}. \tag{2.2.4}$$

We have already seen a similar result on transposes. That is,

$$(A_1 A_2 \cdots A_k)' = A_k' \cdots A_2' A_1'. \tag{2.2.5}$$

An application of the regular inverse in solving systems of linear equations can be stated as follows: Consider a system of n linear equations in n unknowns, written in matrix notation as, $AX = b$. If the coefficient matrix A has a regular inverse then premultiplying both sides by A^{-1} we get the unique solution of the system.

$$AX = b \quad \text{with } A^{-1} \text{ existing means} \quad X = A^{-1}b. \tag{2.2.6}$$

Example 2.2.1. By using the illustrative example to Definition 2.2.2 solve the following systems of linear equations by inspection:

(a) $\begin{aligned} x_1 + x_2 &= 2 \\ x_1 + 2x_2 &= 0. \end{aligned}$
(b) $\begin{aligned} 2x_1 - x_2 &= 1 \\ -x_1 + x_2 &= 3. \end{aligned}$

Solution 2.2.1. (a) Writing the system as $AX = b$ we have

$$A = \begin{bmatrix} 1 & 1 \\ 1 & 2 \end{bmatrix}, \quad X = \begin{bmatrix} x_1 \\ x_2 \end{bmatrix}, \quad b = \begin{bmatrix} 2 \\ 0 \end{bmatrix}.$$

But from the illustrative example

$$A^{-1} = \begin{bmatrix} 2 & -1 \\ -1 & 1 \end{bmatrix}.$$

Hence

$$A^{-1}b = \begin{bmatrix} 2 & -1 \\ -1 & 1 \end{bmatrix} \begin{bmatrix} 2 \\ 0 \end{bmatrix} = \begin{bmatrix} 4 \\ -2 \end{bmatrix} \Rightarrow x_1 = 4 \quad \text{and} \quad x_2 = -2.$$

(b) Writing the system as $AX = b$ we have

$$A = \begin{bmatrix} 2 & -1 \\ -1 & 1 \end{bmatrix}, \quad X = \begin{pmatrix} x_1 \\ x_2 \end{pmatrix}, \quad b = \begin{pmatrix} 1 \\ 3 \end{pmatrix}.$$

From the above (a) itself

$$A^{-1} = \begin{bmatrix} 1 & 1 \\ 1 & 2 \end{bmatrix} \Rightarrow$$

$$A^{-1}b = \begin{bmatrix} 1 & 1 \\ 1 & 2 \end{bmatrix} \begin{bmatrix} 1 \\ 3 \end{bmatrix} = \begin{bmatrix} 4 \\ 7 \end{bmatrix} \Rightarrow x_1 = 4 \quad \text{and} \quad x_2 = 7.$$

Computing A^{-1} first and then solving the system of linear equations by using the formula $X = A^{-1}b$ is not the easiest way of solving the system even when A^{-1} exists. We can see from the above examples that if A is an $n \times n$ matrix with $n \geq 3$ then by inspection we may not be able to come up with A^{-1} even when A^{-1} exists. Another simpler way of solving systems of linear equations by using a procedure called *elementary transformations* will be considered later on.

A result on trace which will be useful in many applied problems is on the trace of a product of the type AA' where A need not be a square matrix. If A is $m \times n$ then A' is $n \times m$ and AA' is $m \times m$. The trace of an $m \times m$ matrix is defined. Similarly A' is $n \times m$ which makes $A'A$ an $n \times n$ matrix. Trace is again defined. By straight multiplication and then summing up the leading diagonal elements we can establish the following result:

(vii) Let A be any $m \times n$ matrix. Then

$$\text{tr}(AA') = \text{tr}(A'A) = \sum_i \sum_j a_{ij}^2 =$$

sum of squares of all elements in A.

The following results can be established easily.

(viii)
$$\mathrm{tr}(AA^{-1}) = \mathrm{tr}(I) = n$$

when A is an $n \times n$ matrix with A^{-1} existing.

(ix) If A is idempotent and if $A \neq I$ then A^{-1} does not exist.

(x) If A is a nilpotent matrix of order r, for some r, then A^{-1} does not exist. (A null matrix is not counted among nilpotent matrices.)

Example 2.2.2. Let the elements of an $m \times n$ matrix X be all functionally independent (distinct) real variables x_{ij}. Let $\int_X$ denote the multiple integral over all the variables x_{ij}'s and dX the wedge product of all differentials in X. Then evaluate the integral

$$y = \int_X e^{-\mathrm{tr}(XX')} dX.$$

Solution 2.2.2. Since

$$\mathrm{tr}(XX') = \sum_{i=1}^{m} \sum_{j=1}^{n} x_{ij}^2$$

(sum of squares of all elements in X), we have

$$y = \int \cdots \int e^{-\sum_{i=1}^{m} \sum_{j=1}^{n} x_{ij}^2} dx_{11} \wedge \cdots \wedge dx_{mn}.$$

Since all the integrals over the individual variables are identical we need to evaluate only one integral. Let

$$\delta = \int_{-\infty}^{\infty} e^{-x^2} dx = 2 \int_{0}^{\infty} e^{-x^2} dx$$

(since e^{-x^2} is an even function and since the integral exists)

$$\delta = \int_{0}^{\infty} y^{\frac{1}{2}-1} e^{-y} dy \quad \left(\text{put } y = x^2 \Rightarrow x = y^{\frac{1}{2}} \Rightarrow dx = \frac{1}{2} y^{\frac{1}{2}-1} dy \right)$$
$$= \Gamma\left(\frac{1}{2}\right) = \sqrt{\pi}.$$

Therefore

$$y = (\sqrt{\pi})^{mn} = \pi^{mn/2}.$$

Note. The integral representation of a gamma function is the following: [A gamma function is defined in different ways.]

$$\Gamma(\alpha) = \int_{0}^{\infty} x^{\alpha-1} e^{-x} dx \quad \text{for } \Re(\alpha) > 0 \tag{2.2.7}$$

where $\mathbb{R}(\cdot)$ denotes the real part of $(\cdot)$. The notation $\Gamma(\alpha)$ (gamma alpha) is a standard notation. It is a function notation, function of α, gamma of alpha. The integral in (2.2.7) exists even for complex parameter values of α provided the real part of this complex parameter α is positive. Then, of course, it exists for real α such that $\alpha > 0$. For defining a gamma function this condition is not necessary. Only for representing a gamma function as an integral the condition $\mathbb{R}(\alpha) > 0$ is needed, otherwise $\Gamma(\alpha)$ exists for all values of $\alpha \neq 0, -1, -2, \ldots$. Two immediate properties of a gamma function are the following:

$$\Gamma(\alpha) = (\alpha - 1)\Gamma(\alpha - 1) \quad \text{for } \mathbb{R}(\alpha - 1) > 0 \tag{2.2.8}$$

and

$$\Gamma\left(\frac{1}{2}\right) = \sqrt{\pi}. \tag{2.2.9}$$

Observe that (2.2.8) can be recursively applied to evaluate $\Gamma(\alpha)$ when α is a positive integer. That is,

$$\Gamma(n) = (n - 1)! \quad \text{for } n = 1, 2, \ldots . \tag{2.2.10}$$

If the inverses of A and B exist and if A and B are $n \times n$ matrices then the following properties hold:

(xi) $(A^m)^{-1} = (A^{-1})^m = A^{-1}A^{-1} \cdots A^{-1}, \quad m = 0, 1, 2, \ldots .$

(xii) $(A^m B^r)^{-1} = (B^r)^{-1}(A^m)^{-1} = (B^{-1})^r (A^{-1})^m, \quad m, r = 0, 1, 2, \ldots .$

We have seen that the positive integer powers of a square matrix are defined. It is natural to ask the question: is the square root of a square matrix A defined? Can we find a matrix B such that $BB = B^2 = A$? Then, naturally, we can say that B is a square root of A. Can we find such a square root for a given matrix A and when we can find one such B, is that B going to be unique? Let us examine this a little bit further. Consider the following matrices:

$$A_1 = \begin{bmatrix} 1 & 0 \\ 0 & 1 \end{bmatrix}, \quad A_2 = \begin{bmatrix} 0 & 1 \\ 1 & 0 \end{bmatrix},$$

$$A_3 = \begin{bmatrix} 0 & -1 \\ -1 & 0 \end{bmatrix}, \quad A_4 = \begin{bmatrix} -1 & 0 \\ 0 & -1 \end{bmatrix}.$$

Note that

$$A_1^2 = \begin{bmatrix} 1 & 0 \\ 0 & 1 \end{bmatrix}\begin{bmatrix} 1 & 0 \\ 0 & 1 \end{bmatrix} = \begin{bmatrix} 1 & 0 \\ 0 & 1 \end{bmatrix};$$

$$A_2^2 = \begin{bmatrix} 1 & 0 \\ 0 & 1 \end{bmatrix}, \quad A_3^2 = \begin{bmatrix} 1 & 0 \\ 0 & 1 \end{bmatrix}, \quad A_4^2 = \begin{bmatrix} 1 & 0 \\ 0 & 1 \end{bmatrix}.$$

The squares of A_1, A_2, A_3, A_4 all are equal to the 2×2 identity matrix I_2. Hence $A_1, \dots, A_4$ all qualify to be a square root of I_2. One of the simplest matrices that we can consider is an identity matrix. We see that when we can find a square root the square root is not unique. In general, there need not exist a matrix B such that $B^2 = A$ for a given matrix A, and when such a B exists it need not be unique. Hence we will not deal with fractional powers of matrices in the following sections. The square root will be explored further after discussing something called the *eigenvalues* of a matrix later on.

2.2.1 Some more practical situations

A number of practical situations, where matrices come in naturally, are already discussed. Some more will be listed here which involve only sums and products of matrices. The student is urged to take note of all the practical situations listed so far, and also the ones to be listed later, because we are going to enlarge on each of them later on.

Example 2.2.3 (Center of gravity and moment of inertia). Some concepts connected with physics and statistics are the mean values and variances. Some of these will be examined here. Consider a set of numbers $x_1, \dots, x_n$ (such as the heights of students in a class, incomes of wage-earners in a city and so on). The average

$$\bar{x} = \frac{x_1 + \cdots + x_n}{n} = \frac{1}{n}J'X = \frac{1}{n}X'J \tag{2.2.11}$$

where $J' = (1, 1, \dots, 1)$ and $X' = (x_1, \dots, x_n)$. Then

$$(\bar{x})^2 = \bar{x}\bar{x} = \bar{x}(\bar{x})'$$

(note that $\bar{x}' = \bar{x}$ since $\bar{x}$ is 1×1)

$$(\bar{x})^2 = \left(\frac{1}{n}\right)^2 X'J(J'X) = \frac{1}{n^2}X'JJ'X = \frac{1}{n}X'BX \tag{2.2.12}$$

(a quadratic form in X) where

$$B = \frac{1}{n}\begin{bmatrix} 1 & 1 & \cdots & 1 \\ 1 & 1 & \cdots & 1 \\ \vdots & \vdots & \cdots & \vdots \\ 1 & 1 & \cdots & 1 \end{bmatrix}.$$

One interesting property of B is that

$$B = B^2$$

which means that B is idempotent.

If $x_1, \ldots, x_n$ are the values taken by a discrete real random variable x with the corresponding probabilities $p_1, \ldots, p_n$, $p_i > 0$, $i = 1, \ldots, n$, $p_1 + \cdots + p_n = 1$ then the mean value of the random variable x, denoted as $\mu = E(x)$, is given by

$$\mu = \sum_{j=1}^{n} x_j p_j = X'P = P'X$$

$$= \frac{\sum_{i=1}^{n} x_i p_i}{\sum_{i=1}^{n} p_i} \quad \left(\text{since } \sum_{i=1}^{n} p_i = 1 \right) \tag{2.2.13}$$

where $X' = (x_1, \ldots, x_n)$ and $P' = (p_1, \ldots, p_n)$. The expression in (2.2.13) is the mean value of the random variable x in statistical literature and it is the center of gravity of the system X when P is the vector of weights or forces and so on, in physics. The variance of the real discrete random variable x, denoted by σ^2, is defined as

$$\sigma^2 = \sum_{i=1}^{n} (x_i - \mu)^2 p_i$$

which is also the moment of inertia of the system X and P in physics. When $p_1 = \cdots = p_n = \frac{1}{n}$ we have

$$\sigma^2 = \frac{1}{n} \sum_{i=1}^{n} (x_i - \mu)^2 = \frac{1}{n}(X - \tilde{\mu})'(X - \tilde{\mu})$$

$$= \frac{1}{n} \sum_{i=1}^{n} x_i^2 - (\bar{x})^2$$

where $X' = (x_1, \ldots, x_n)$, $\tilde{\mu}' = (\mu, \ldots, \mu)$,

$$\mu = \sum_{i=1}^{n} \frac{x_i}{n} = \bar{x} = \frac{1}{n} X' J.$$

But note that $\frac{1}{n} \sum_{i=1}^{n} x_i^2 = \frac{1}{n} X'X$ and $\bar{x}^2 = \frac{1}{n} X'BX$ where B is defined in (2.2.12). Therefore

$$\sigma^2 = \frac{1}{n} X'X - \frac{1}{n} X'BX = \frac{1}{n} X'[I - B]X \tag{2.2.14}$$

where

$$I - B = \begin{bmatrix} 1 - \frac{1}{n} & -\frac{1}{n} & \cdots & -\frac{1}{n} \\ -\frac{1}{n} & 1 - \frac{1}{n} & \cdots & -\frac{1}{n} \\ \vdots & \vdots & \cdots & \vdots \\ -\frac{1}{n} & -\frac{1}{n} & \cdots & 1 - \frac{1}{n} \end{bmatrix}. \tag{2.2.15}$$

Observe that

$$(I - B)^2 = (I - B)(I - B) = I - B - B + B^2 = I - 2B + B = I - B$$

since $B = B^2$ is idempotent. (The student may verify by directly computing $(I - B)^2$ also.) This means that $I - B$ is also idempotent. Further,

$$(I - B)B = B - B^2 = B - B = O$$

since B is idempotent. The properties that B and $I - B$ are idempotent and that $(I - B)B = O$ are the fundamental results in the analysis of variance principle, regression analysis, design of experiments, independence of quadratic forms and in many other similar topics in statistics, econometrics and related areas.

Definition 2.2.3 (Orthogonal matrices). If two non-null matrices A and B are such that $AB = O$ then A and B are said to be orthogonal to each other.

This is a generalization of the concept of orthogonal vectors in Chapter 1. For example,

$$\text{(a)} \quad A = \begin{bmatrix} 1 & 1 & 1 \\ 1 & -2 & 1 \end{bmatrix}, \quad B = \begin{bmatrix} 1 \\ 0 \\ -1 \end{bmatrix} \Rightarrow AB = O;$$

$$\text{(b)} \quad A = \begin{bmatrix} 1 & 1 & 1 \\ 1 & -2 & 1 \end{bmatrix}, \quad B = \begin{bmatrix} 1 & 3 \\ 0 & 0 \\ -1 & -3 \end{bmatrix} \Rightarrow AB = O.$$

Note that when $AB = O$ every column vector in B is orthogonal to every row vector in A or the angle between these vectors is $\pi/2$. As an immediate consequence of this definition we can observe the following result:

(xiii) In a system of linear equations $AX = O$ every solution vector X is orthogonal to the rows of A, where A is $m \times n$ and X is $n \times 1$.

Definition 2.2.4 (Orthonormal matrix). If a square matrix A is such that $AA' = I$, $A'A = I$ then A is called an *orthonormal matrix* or an element of the orthogonal group.

For example,

$$A = \begin{bmatrix} \cos\theta & -\sin\theta \\ \sin\theta & \cos\theta \end{bmatrix},$$

$$AA' = \begin{bmatrix} \cos\theta & -\sin\theta \\ \sin\theta & \cos\theta \end{bmatrix} \begin{bmatrix} \cos\theta & \sin\theta \\ -\sin\theta & \cos\theta \end{bmatrix} = \begin{bmatrix} 1 & 0 \\ 0 & 1 \end{bmatrix} = I$$

since $\cos^2\theta + \sin^2\theta = 1$. Hence A is an orthonormal matrix. Consider

$$B = \begin{bmatrix} \frac{1}{\sqrt{2}} & \frac{1}{\sqrt{2}} \\ \frac{1}{\sqrt{2}} & -\frac{1}{\sqrt{2}} \end{bmatrix} \Rightarrow BB' = I, \quad B'B = I$$

which means B is orthonormal. Now, take

$$C = \begin{bmatrix} \frac{1}{\sqrt{3}} & \frac{1}{\sqrt{3}} & \frac{1}{\sqrt{3}} \\ \frac{1}{\sqrt{6}} & -\frac{2}{\sqrt{6}} & \frac{1}{\sqrt{6}} \\ \frac{1}{\sqrt{2}} & 0 & -\frac{1}{\sqrt{2}} \end{bmatrix} \Rightarrow CC' = I, \quad C'C = I$$

which means C is orthonormal.

The following properties are immediate:

(xiv) If A is an $n \times n$ orthonormal matrix then (a) the length of each row vector is unity and the row vectors are orthogonal to each other, (b) the length of each column vector is unity and the column vectors are orthogonal to each other.

(xv) If A is $n \times n$ and if $AA' = I$ then $A^{-1} = A'$ and $A'A = I$.

This is a very important result which has consequences in linear transformations, orthogonal transformations, reductions of quadratic forms and so on.

(xvi) An $n \times n$ orthonormal matrix A represents a rotation of the coordinate axes.

This can be easily noticed geometrically as well as algebraically by looking at the transformation

$$Y = AX,$$

where

$$A = \begin{bmatrix} \cos\theta & -\sin\theta \\ \sin\theta & \cos\theta \end{bmatrix},$$

$$X = \begin{pmatrix} x_1 \\ x_2 \end{pmatrix}, \quad Y = \begin{pmatrix} y_1 \\ y_2 \end{pmatrix}, \quad AA' = I, \quad A'A = I.$$

If the coordinate axes are rotated through the angle θ then the point $\begin{pmatrix} x_1 \\ x_2 \end{pmatrix}$ in the original axes of coordinates becomes $\begin{pmatrix} y_1 \\ y_2 \end{pmatrix}$ in the new axes of coordinates.

Definition 2.2.5 (A semi-orthonormal matrix). If an $m \times n, m \leq n$ rectangular matrix A is such that $AA' = I_m$ then A is called a semi-orthonormal matrix or an element of the Stiefel manifold, and if $m = n$ then it is a full orthonormal matrix.

For example,

$$A = (\cos\theta, -\sin\theta) \Rightarrow AA' = I_1 = 1 \Rightarrow A \text{ is semi-orthonormal};$$

$$B = \begin{bmatrix} \frac{1}{\sqrt{3}} & \frac{1}{\sqrt{3}} & \frac{1}{\sqrt{3}} \\ \frac{1}{\sqrt{2}} & 0 & -\frac{1}{\sqrt{2}} \end{bmatrix} \Rightarrow BB' = I_2 \Rightarrow B \text{ is semi-orthonormal}.$$

Observe that when A is $m \times n$ such that $AA' = I_m$ then each row has length 1 and the rows are orthogonal whereas the columns do not have these properties when $m < n$ which means that when $m < n$, $AA' = I_m$ does not imply that $A'A = I_n$ which can be verified from the above illustrative examples.

Example 2.2.4 (The covariance matrix). The *covariance matrix* or the *variance–co-variance matrix* in statistical theory, denoted by V, is defined as follows:

$$V = E[(X - \mu)(X - \mu)'] = E(XX') - \mu\mu'$$

where E denotes the expected value, $\mu = E(X)$ and X is a $p \times 1$ vector of real random variables. This matrix represents the configuration of variances and covariances in the vector random variable X. For example, let the joint density of x_1 and x_2, $X' = (x_1, x_2)$, be

$$f(X') = f(x_1, x_2) = x_1 + x_2, \quad 0 \le x_1 \le 1, \, 0 \le x_2 \le 1$$

and $f(X') = 0$ elsewhere. Then

$$X = \begin{pmatrix} x_1 \\ x_2 \end{pmatrix}, \quad E(X) = \begin{pmatrix} E(x_1) \\ E(x_2) \end{pmatrix} = \mu = \begin{pmatrix} \mu_1 \\ \mu_2 \end{pmatrix};$$

$$E(XX') = E\begin{bmatrix} x_1^2 & x_1 x_2 \\ x_2 x_1 & x_2^2 \end{bmatrix} = \begin{bmatrix} E(x_1^2) & E(x_1 x_2) \\ E(x_2 x_1) & E(x_2^2) \end{bmatrix};$$

$$E(x_1) = \int_0^1 \int_0^1 x_1(x_1 + x_2) dx_1 \wedge dx_2$$

$$= \int_0^1 x_1\left(x_1 + \frac{1}{2}\right) dx_1 = \frac{7}{12} = E(x_2)$$

due to symmetry, and

$$E(x_1 x_2) = \int_0^1 \int_0^1 x_1 x_2(x_1 + x_2) dx_1 \wedge dx_2$$

$$= \int_0^1 x_1\left\{\int_0^1 [x_1 x_2 + x_2^2] dx_2\right\} dx_1 = \frac{1}{3};$$

$$E(x_1^2) = \int_0^1 \int_0^1 x_1^2(x_1 + x_2) dx_1 \wedge dx_2 = \frac{5}{12} = E(x_2^2)$$

due to symmetry. Then the covariance matrix

$$V = \begin{bmatrix} E(x_1^2) & E(x_1 x_2) \\ E(x_2 x_1) & E(x_2^2) \end{bmatrix} - \begin{bmatrix} \mu_1^2 & \mu_1 \mu_2 \\ \mu_2 \mu_1 & \mu_2^2 \end{bmatrix}$$

$$= \begin{bmatrix} \frac{5}{12} & \frac{1}{3} \\ \frac{1}{3} & \frac{5}{12} \end{bmatrix} - \begin{bmatrix} (\frac{7}{12})^2 & (\frac{7}{12})(\frac{7}{12}) \\ (\frac{7}{12})(\frac{7}{12}) & (\frac{7}{12})^2 \end{bmatrix}$$

$$= \begin{bmatrix} \frac{11}{144} & -\frac{1}{144} \\ -\frac{1}{144} & \frac{11}{144} \end{bmatrix}.$$

Example 2.2.5 (Maxima/minima). When a scalar function

$$f = f(x_1, \ldots, x_n)$$

of many real variables $x_1, \ldots, x_n$ is considered the critical points are available by solving $\frac{\partial f}{\partial X} = O$ where $\frac{\partial}{\partial X} = \begin{pmatrix} \frac{\partial}{\partial x_1} \\ \vdots \\ \frac{\partial}{\partial x_n} \end{pmatrix}$, as seen in Chapter 1. One can check for maxima/minima at these critical points by evaluating the matrix $\frac{\partial}{\partial X}\frac{\partial f}{\partial X'}$ at these critical points. For example, let

$$f = x_1^2 + 2x_2^2 + 2x_1x_2 - x_1 - 2x_2 + 8.$$

Then

$$\frac{\partial}{\partial X} = \begin{pmatrix} \frac{\partial}{\partial x_1} \\ \frac{\partial}{\partial x_2} \end{pmatrix} \Rightarrow \frac{\partial f}{\partial X} = \begin{pmatrix} \frac{\partial f}{\partial x_1} \\ \frac{\partial f}{\partial x_2} \end{pmatrix} = \begin{bmatrix} 2x_1 + 2x_2 - 1 \\ 2x_1 + 4x_2 - 2 \end{bmatrix}.$$

Hence

$$\frac{\partial f}{\partial X} = O \Rightarrow \begin{bmatrix} 2x_1 + 2x_2 - 1 \\ 2x_1 + 4x_2 - 2 \end{bmatrix} = \begin{bmatrix} 0 \\ 0 \end{bmatrix} \Rightarrow$$

$$\begin{pmatrix} x_1 \\ x_2 \end{pmatrix} = \begin{pmatrix} 0 \\ \frac{1}{2} \end{pmatrix}.$$

There is only one critical point $\begin{pmatrix} x_1 \\ x_2 \end{pmatrix} = \begin{pmatrix} 0 \\ \frac{1}{2} \end{pmatrix}$. Now, consider the matrix operator

$$\frac{\partial}{\partial X}\frac{\partial f}{\partial X'} = \begin{bmatrix} \frac{\partial^2}{\partial x_1^2} & \cdots & \frac{\partial^2}{\partial x_1 \partial x_n} \\ \vdots & \cdots & \vdots \\ \frac{\partial^2}{\partial x_n \partial x_1} & \cdots & \frac{\partial^2}{\partial x_n^2} \end{bmatrix},$$

which in our example is, with $n = 2$, operating on f.

$$\frac{\partial}{\partial X}\frac{\partial f}{\partial X'} = \frac{\partial}{\partial X}(2x_1 + 2x_2 - 1, 2x_1 + 4x_2 - 2)$$

$$= \begin{bmatrix} 2 & 2 \\ 2 & 4 \end{bmatrix}.$$

Since this matrix here is free of x_1 and x_2 this matrix evaluated at the critical point is itself. This matrix is positive definite and hence the critical point corresponds to a minimum. [Definiteness of matrices will be considered later on.]

Example 2.2.6 (Transition probability matrix). A worker finds that his boss has three stages of her mood, "pleasant", "tolerable", "intolerable". If the boss is in a given mood in the morning it can change to one of the three stages by the evening. That is, for example, "pleasant" can change into "pleasant" or "tolerable" or "intolerable". Let p_{ij} be the chance (probability) that the i-th stage of the mood in the morning changes to the j-th stage of the mood by the evening. Then the 3×3 matrix (p_{ij}) is a transition probability matrix. For example, suppose

$$P = (p_{ij}) = \begin{bmatrix} 0.5 & 0.4 & 0.1 \\ 0.3 & 0.6 & 0.1 \\ 0.1 & 0.2 & 0.7 \end{bmatrix}$$

which in terms of the various stages of the mood is the following:

		evening		
		pleasant	tolerable	intolerable
	pleasant	0.5	0.4	0.1
morning	tolerable	0.3	0.6	0.1
	intolerable	0.1	0.2	0.7

For example the chance of going from "tolerable" to "pleasant" is 0.3 or 30%, whereas going from "tolerable" to "intolerable" is 0.1 or 10%. The chance of going from "pleasant" to "pleasant" is 0.5 and that from "intolerable" to "intolerable" is 0.7. In general, if there are k stages and if p_{ij} is the probability of going from stage i to stage j then the transition probability matrix is

$$P = (p_{ij}) = \begin{bmatrix} p_{11} & p_{12} & \cdots & p_{1k} \\ p_{21} & p_{22} & \cdots & p_{2k} \\ \vdots & \vdots & \cdots & \vdots \\ p_{k1} & p_{k2} & \cdots & p_{kk} \end{bmatrix}$$

where the sum of each row is unity, $\sum_{j=1}^{k} p_{ij} = 1$ for each $i = 1, 2, \ldots, k$ and each entry $p_{ij} \geq 0$. If the probability of transition from the j-th stage to the i-th stage is denoted by p_{ij} then we have the transpose of the P above. In this case the sum of the elements in each column will be 1. Such a matrix P where, either the elements in each row sum to 1 or the elements in each column sum to 1, but not both, is also called a *singly stochastic matrix*. If both, the elements in each row and each column sum to 1, that is, $\sum_{j=1}^{k} p_{ij} = 1$ for each i as well as $\sum_{i=1}^{k} p_{ij} = 1$ for each j then such a matrix is called a *doubly stochastic matrix*.

For our example of the boss let us examine one interesting aspect. Suppose that there is no mood change in the night. She will be in the same mood in the morning as the one of the previous evening. If she is in a "pleasant" mood in the first day morning what is the chance that she will be in a "pleasant" mood in the second day evening? Let the three stages "pleasant", "tolerable", and "intolerable" be denoted by 1, 2, 3 respectively. If she is in stage 1 in the first day morning it can go from 1 to 1 or 1 to 2 or 1 to 3 by the evening. Then on the second day it can go from 1 to 1, given that she was in stage 1 by the previous evening, or 2 to 1, given that she was in stage 2 by the previous evening, or 3 to 1, given that she was in stage 3 by the previous evening. Thus the chance of finding her in stage 1 on the second evening is given by

$$(0.5)(0.5) + (0.4)(0.3) + (0.1)(0.1) = 0.38.$$

From stage 1 in the first morning to stage 2 in the second evening has the probability

$$(0.5)(0.4) + (0.4)(0.6) + (0.1)(0.2) = 0.46,$$

and so on. [Probability of the intersection of two events = conditional probability multiplied by the marginal probability.] The transition probability matrix for the second evening starting with P in the first day morning is then

$$PP = P^2 = \begin{bmatrix} 0.38 & 0.46 & 0.16 \\ 0.34 & 0.50 & 0.16 \\ 0.18 & 0.30 & 0.52 \end{bmatrix}.$$

The transition probability matrix from the first day morning to the k-th day evening is then P^k. There are several interesting aspects that can be studied by using transition probability matrices. Some of these will be considered later on in the coming chapters.

Definition 2.2.6. If A and B are two non-null matrices such that $AB = BA$ then A and B are said to commute or are said to be commutative.

Note that in general, whenever AB and BA are defined $AB \neq BA$. But in some cases $AB = BA$. For example,

$$IA = AI \implies$$

that any $n \times n$ matrix A and the $n \times n$ identity matrix are commutative. Let D_1 and D_2 be two diagonal matrices of the same order (which includes identity and scalar matrices) then

$$D_1 D_2 = D_2 D_1.$$

(xvii) Diagonal matrices of the same order are commutative.

If A is an arbitrary matrix and D a diagonal matrix (not equal to a scalar matrix) then $AD \neq DA$. Let

$$A = \begin{bmatrix} 1 & 1 & \cdots & 1 \\ 1 & 1 & \cdots & 1 \\ \vdots & \vdots & \cdots & \vdots \\ 1 & 1 & \cdots & 1 \end{bmatrix}, \quad B = (b_{ij}) = B'$$

then $AB = BA$ (or A and B commute).

2.2.2 Pre and post multiplications by diagonal matrices

Let $A = (a_{ij})$ be an $n \times n$ matrix and $D = \text{diag}(d_1, \ldots, d_n)$ be a diagonal matrix. The effects of pre and post multiplications of the matrix A by the diagonal matrix D are something very interesting. The student must memorize these results because in many structural decompositions of matrices these results will come in handy:

$$DA = \begin{bmatrix} d_1 & 0 & \cdots & 0 \\ 0 & d_2 & \cdots & 0 \\ \vdots & \vdots & \cdots & \vdots \\ 0 & 0 & \cdots & d_n \end{bmatrix} \begin{bmatrix} a_{11} & a_{12} & \cdots & a_{1n} \\ a_{21} & a_{22} & \cdots & a_{2n} \\ \vdots & \vdots & \cdots & \vdots \\ a_{n1} & a_{n2} & \cdots & a_{nn} \end{bmatrix}$$

$$= \begin{bmatrix} d_1 a_{11} & d_1 a_{12} & \cdots & d_1 a_{1n} \\ d_2 a_{21} & d_2 a_{22} & \cdots & d_2 a_{2n} \\ \vdots & \vdots & \cdots & \vdots \\ d_n a_{n1} & d_n a_{n2} & \cdots & d_n a_{nn} \end{bmatrix}.$$

(xviii) Premultiplication of a matrix A by a diagonal matrix D is equivalent to multiplying each row of A by the corresponding diagonal elements in D.

That is, the first row of A is multiplied by the first diagonal element in D, the second row of A is multiplied by the second diagonal element in D and so on. Now, let us see what happens if we postmultiply A with D:

$$AD = \begin{bmatrix} a_{11} & a_{12} & \cdots & a_{1n} \\ a_{21} & a_{22} & \cdots & a_{2n} \\ \vdots & \vdots & \cdots & \vdots \\ a_{n1} & a_{n2} & \cdots & a_{nn} \end{bmatrix} \begin{bmatrix} d_1 & 0 & \cdots & 0 \\ 0 & d_2 & \cdots & 0 \\ \vdots & \vdots & \cdots & \vdots \\ 0 & 0 & \cdots & d_n \end{bmatrix}$$

$$= \begin{bmatrix} d_1 a_{11} & d_2 a_{12} & \cdots & d_n a_{1n} \\ d_1 a_{21} & d_2 a_{22} & \cdots & d_n a_{2n} \\ \vdots & \vdots & \cdots & \vdots \\ d_1 a_{n1} & d_2 a_{n2} & \cdots & d_n a_{nn} \end{bmatrix}.$$

(xix) Postmultiplication of A by the diagonal matrix D is equivalent to multiplying each column of A by the corresponding diagonal elements in D.

That is, the first column of A is multiplied by the first diagonal element in D, the second column of A is multiplied by the second diagonal element in D and so on. For example,

$$\begin{bmatrix} 2 & 0 \\ 0 & 3 \end{bmatrix} \begin{bmatrix} 1 & 1 \\ 2 & 4 \end{bmatrix} = \begin{bmatrix} 2 & 2 \\ 6 & 12 \end{bmatrix};$$

$$\begin{bmatrix} 1 & 1 \\ 2 & 4 \end{bmatrix} \begin{bmatrix} 2 & 0 \\ 0 & 3 \end{bmatrix} = \begin{bmatrix} 2 & 3 \\ 4 & 12 \end{bmatrix}.$$

It is worth observing that the same properties hold if an $m \times n$ matrix A is premultiplied by an $m \times m$ diagonal matrix and postmultiplied by an $n \times n$ diagonal matrix.

Example 2.2.7 (Covariance and correlation matrices). In statistical analysis the covariance between two real scalar random variables x_i and x_j, denoted by σ_{ij}, has the following representation:

$$\sigma_{ij} = \rho_{ij}\sigma_i\sigma_j, \quad \sigma_i \neq 0, \quad \sigma_j \neq 0$$

where ρ_{ij} (Greek letter rho) is the *correlation* between x_i and x_j, σ_i and σ_j are the *standard deviations* (measures of scatter) in x_i and x_j respectively (σ_i^2 is the variance of x_i). Then the covariance matrix or the variance–covariance matrix can be written in the following structural form:

$$V = (\sigma_{ij}) = \begin{bmatrix} \rho_{11}\sigma_1^2 & \rho_{12}\sigma_1\sigma_2 & \cdots & \rho_{1n}\sigma_1\sigma_n \\ \vdots & \vdots & \cdots & \vdots \\ \rho_{n1}\sigma_n\sigma_1 & \rho_{n2}\sigma_n\sigma_2 & \cdots & \rho_{nn}\sigma_n^2 \end{bmatrix} = DPD$$

where

$$D = \begin{bmatrix} \sigma_1 & 0 & \cdots & 0 \\ 0 & \sigma_2 & \cdots & 0 \\ \vdots & \vdots & \cdots & \vdots \\ 0 & 0 & \cdots & \sigma_n \end{bmatrix} \quad \text{and} \quad P = \begin{bmatrix} \rho_{11} & \rho_{12} & \cdots & \rho_{1n} \\ \rho_{21} & \rho_{22} & \cdots & \rho_{2n} \\ \vdots & \vdots & \cdots & \vdots \\ \rho_{n1} & \rho_{n2} & \cdots & \rho_{nn} \end{bmatrix}$$

where V is the covariance matrix, D is a diagonal matrix of standard deviations and P is the matrix of correlations. (Incidently $\rho_{ii} = 1$, $i = 1, \ldots, n$ which follows from the definition itself.)

Exercises 2.2

2.2.1. Compute the traces of the following matrices:

(a) $\quad A = \begin{bmatrix} 0 & -1 & 2 \\ 2 & 3 & 1 \\ 0 & 0 & 7 \end{bmatrix}, \quad B = \begin{bmatrix} -1 & 0 & 1 \\ 0 & 2 & 1 \\ 0 & 0 & -2 \end{bmatrix}, \quad C = \begin{bmatrix} 2 & 0 & 0 \\ 1 & 1 & 2 \\ 0 & 1 & -1 \end{bmatrix};$

(b) AB, BA; (c) ABC, CAB, BCA; (d) $2A - 5B$, $2\,\mathrm{tr}(A) - 5\,\mathrm{tr}(B)$.

2.2.2. If $A = (a_{ij})$ is a 3×3 matrix obtain (a) $\mathrm{tr}(A^2)$, (b) $\mathrm{tr}(AA')$ and compare the results in (a) and (b).

2.2.3. If X is a $p \times p$ matrix which can be written as $X = TT'$ where T is a lower triangular matrix, (a) compute the trace of X in terms of the elements of T; (b) can you represent every element in T as a function of the elements in X, if so, is the representation unique? (c) What are the conditions on the elements of T so that the transformation $X = TT'$ (it is a nonlinear transformation) is one-to-one, that is, every element in T can be uniquely written as a function of the elements in X and vice versa?

2.2.4. By inspection, write down the regular inverses, if they exist, for the following matrices:

$$A = \begin{bmatrix} 2 & 1 \\ 3 & 5 \end{bmatrix}, \quad B = \begin{bmatrix} 2 & 0 & 0 & 0 \\ 1 & 0 & 0 & 0 \\ -1 & 2 & 1 & 0 \\ 5 & -7 & 2 & 8 \end{bmatrix},$$

$$C = \begin{bmatrix} 5 & 0 & 0 & 0 \\ 0 & -2 & 0 & 0 \\ 0 & 0 & 0 & 0 \\ 0 & 0 & 0 & 9 \end{bmatrix}, \quad D = \begin{bmatrix} -2 & 0 & 0 \\ 0 & 1 & 0 \\ 0 & 0 & 4 \end{bmatrix}.$$

2.2.5. Show that if A is lower (upper) triangular then its regular inverse, when it exists, is also lower (upper) triangular. Verify the result for a general 3×3 lower (upper) triangular matrix.

2.2.6. Taking the matrices in Exercise 2.2.1 (a) verify the following results:

(a) $\quad (AB)' = B'A'$, $\qquad$ (b) $\quad (ABC)' = C'B'A'$.

2.2.7. Let

$$A = \frac{1}{n}\begin{bmatrix} 1 & \cdots & 1 \\ \vdots & \cdots & \vdots \\ 1 & \cdots & 1 \end{bmatrix}, \quad B = \begin{bmatrix} 1 & 0 \\ 1 & -1 \end{bmatrix}, \quad C = \frac{1}{2}\begin{bmatrix} 1 & -1 \\ 1 & -1 \end{bmatrix},$$

where A is $n \times n$. Compute $A^2, A^3, A^{100}, A^{111}, B^2, B^3, B^{100}, B^{121}, C^2, C^3, C^{30}, C^{43}$. What are A^k, B^k, C^k for a general k?

2.2.8. $(AB)^2 \neq A^2B^2$ in general. But for some special situations $(AB)^2 = A^2B^2$. Construct non-null, non-diagonal 3×3 matrices A and B such that (a) $(AB)^2 \neq A^2B^2$, (b) $(AB)^2 = A^2B^2$.

2.2.9. Is $(A + B)^2 = A^2 + 2AB + B^2$ for general $n \times n$ matrices, $n > 1$? If not, what is the correct expansion formula? Derive the expansion formula for $(A + B)^3$.

2.2.10. The product $AB = C = (c_{ij})$ is defined in such a way that the (i,j)-th element in C, namely c_{ij}, is the dot product of the i-th row of A with the j-th column of B. Now, let $\alpha_1, \ldots, \alpha_n$ be the columns of A and $\beta_1, \ldots, \beta_n$ be the rows of B. Then obviously $\alpha_i\beta_j$ is an $n \times n$ matrix. Show that the product AB can also be written as a sum of such matrices in the following form:

$$AB = \alpha_1\beta_1 + \alpha_2\beta_2 + \cdots + \alpha_n\beta_n.$$

2.2.11. Construct different 3×3 matrices A, B, C, other than the ones in the text, such that (a) $AA' = I$, (b) B is semiorthonormal, (c) $BC = O$.

2.2.12. Let the illustrative orthonormal matrix in the text involving θ be denoted by

$$P(\theta) = \begin{bmatrix} \cos\theta & -\sin\theta \\ \sin\theta & \cos\theta \end{bmatrix}.$$

Let $P(\theta_i)$ be the same $P(\theta)$ with θ replaced by θ_i. Then $P(\theta_1)$ and $P(\theta_2)$ represent rotations of the coordinate axes through angles θ_1 and θ_2 respectively and $P(\theta_1)P(\theta_2)$ represents the situation of first rotating through an angle θ_2 and then rotating through an angle θ_1. Show that
(a) $P(\theta_1)P(\theta_2) = P(\theta_1 + \theta_2)$.
(b) What is the geometrical interpretation of $P(-\theta)$?

2.2.13. Construct different 2×2 non-null matrices A, B, C with real elements such that

$$\text{(a)} \quad A^2 = -I, \qquad \text{(b)} \quad BC = -CB.$$

2.2.14. Let A be a given $n \times n$ matrix and X an arbitrary $n \times n$ matrix such that $AX = XA$ for all X. Then show that A is a scalar multiple of an identity matrix (scalar matrix).

2.2.15. Show that in Example 2.2.7 the inverse of the covariance matrix, V^{-1}, whenever it exists, can be computed by the formula

$$V^{-1} = \begin{bmatrix} \frac{1}{\sigma_1} & 0 & \cdots & 0 \\ \vdots & \vdots & \cdots & \vdots \\ 0 & 0 & \cdots & \frac{1}{\sigma_n} \end{bmatrix} P^{-1} \begin{bmatrix} \frac{1}{\sigma_1} & 0 & \cdots & 0 \\ \vdots & \vdots & \cdots & \vdots \\ 0 & 0 & \cdots & \frac{1}{\sigma_n} \end{bmatrix}.$$

2.2.16. (a) If A is an $m \times n$ matrix and if $A = -A$ then show that A is a null matrix.

(b) If c is a scalar (1×1 matrix) then the scalar multiplication cA is defined. Show that the matrix multiplication cA is defined only if $m = 1$.

2.2.17. If A is an $n \times n$ matrix and if A commutes with every $n \times n$ matrix then show that A is an identity matrix.

2.2.18. Prove that A is symmetric iff A' is symmetric and vice versa.

2.2.19. Give two examples of different symmetric matrices A and B such that (1) AB is symmetric; (2) AB is not symmetric.

2.2.20. If A is symmetric, is (1) A^k symmetric, where k is a positive integer? (2) Is A^{-1} symmetric if A is nonsingular also? (3) For any matrix A if A^2 is symmetric is A symmetric? Justify your answer by examples (2 each) and counter examples (2 each).

2.2.21. If A is a skew symmetric matrix then show that $B = (I + A)(I - A)^{-1}$ is orthonormal, that is, $BB' = I, B'B = I$, and that $B^{-1} = B'$.

2.2.22. For any matrix A show that $\mathrm{tr}(B^{-1}AB) = \mathrm{tr}(A)$ where B^{-1} exists and the product is defined.

2.2.23. If $A = (a_{ij}(t))$ where each element of A is a function of the real variable t, then show that

$$\frac{d}{dt}(A^{-1}) = -A^{-1}\left(\frac{dA}{dt}\right)A^{-1}.$$

2.2.24. Right and left inverses. Let A be any $m \times n$ matrix. Any matrix B for which $BA = I$ is called a left-inverse of A and any matrix C for which $AC = I$ is called a right-inverse of A. Any matrix X such that $AX = I, XA = I$ is called the regular inverse of A. Compute a left inverse B, and all possible right inverses C for the matrix

$$A = \begin{bmatrix} 1 & 1 & -1 \\ 1 & 1 & 1 \end{bmatrix}.$$

2.2.25. Show that, in general, if L is a left inverse of any matrix A then L' is a right inverse of A' and that if R is a right inverse of A then R' is a left inverse of A'.

2.2.26. Let A be $m \times n$ and B be $n \times r$ with $\alpha_1, \dots, \alpha_m$ being the rows of A and $\beta_1, \dots, \beta_r$ being the columns of B. Then show that

$$AB = \begin{pmatrix} \alpha_1 B \\ \vdots \\ \alpha_m B \end{pmatrix} = (A\beta_1, \dots, A\beta_r).$$

2.2.27. For any square matrix A show that $B = A + A'$ is symmetric, $C = A - A'$ is skew symmetric, $A = \frac{1}{2}B + \frac{1}{2}C = $ sum of a symmetric and a skew symmetric matrices and that this representation is unique.

2.3 Elementary matrices and elementary operations

Elementary matrices and elementary operations (multiplications by elementary matrices) have wide spread applications in solving systems of linear equations, in determining linear independence and dependence of vectors, in determining the rank of a matrix, in obtaining a basis for a vector space, in evaluating the inverse of a matrix, and so on. Here we will define the basic elementary matrices and then will look into various types of operations with elementary matrices.

Definition 2.3.1 (The basic elementary matrices). The two basic elementary matrices are the following: Consider an $n \times n$ identity matrix I_n. If any row (column) of I_n is multiplied by a nonzero scalar then the resulting matrix is called an elementary matrix. This is one basic type of an elementary matrix. We will call this an E type elementary matrix. If any row (column) of I_n is added to any other row (column) of I_n then the resulting matrix is the second basic type of an elementary matrix. We will call this an F type elementary matrix.

The E and F types of elementary matrices are the basic types of elementary matrices. For example, consider a 3×3 identity matrix $I = I_3$:

$$E_1 = \begin{bmatrix} 5 & 0 & 0 \\ 0 & 1 & 0 \\ 0 & 0 & 1 \end{bmatrix}$$

is an elementary matrix (the first row of I_3 is multiplied by 5);

$$E_2 = \begin{bmatrix} 1 & 0 & 0 \\ 0 & -2 & 0 \\ 0 & 0 & 1 \end{bmatrix}$$

is an elementary matrix (the second row of I_3 is multiplied by -2);

$$E_3 = \begin{bmatrix} 1 & 0 & 0 \\ 0 & 1 & 0 \\ 0 & 0 & x \end{bmatrix}$$

is an elementary matrix (the third row of I_3 is multiplied by x for $x \neq 0$);

$$F_1 = \begin{bmatrix} 1 & 0 & 0 \\ 1 & 1 & 0 \\ 0 & 0 & 1 \end{bmatrix}$$

is an elementary matrix (the first row of I_3 is added to the second row);

$$F_2 = \begin{bmatrix} 1 & 0 & 0 \\ 0 & 1 & 0 \\ 1 & 0 & 1 \end{bmatrix}$$

is an elementary matrix (the first row of I_3 is added to the third row);

$$F_3 = \begin{bmatrix} 1 & 0 & 1 \\ 0 & 1 & 0 \\ 0 & 0 & 1 \end{bmatrix}$$

is an elementary matrix (the third row of I_3 is added to the first row);

$$F_4 = \begin{bmatrix} 1 & 0 & 0 \\ 0 & 1 & 0 \\ 0 & 1 & 1 \end{bmatrix}$$

is an elementary matrix (the second row of I_3 is added to the third row).

The row which is added remains the same. The net effect is on the row to which another row is added. Before we start operating with these basic elementary matrices let us look at the inverses of these. What is that matrix which nullifies the effect on I_n or I_n is regained by premultiplication of a given elementary matrix with the new matrix? For example, consider E_1. What is E_1^{-1} such that $E_1^{-1}E_1 = I_3$? Multiplication of a certain row of an identity matrix by a nonzero scalar can be nullified by dividing the same row by that nonzero scalar. Therefore

$$E_1^{-1} = \begin{bmatrix} \frac{1}{5} & 0 & 0 \\ 0 & 1 & 0 \\ 0 & 0 & 1 \end{bmatrix} \quad \text{so that} \quad E_1^{-1}E_1 = I_3 = E_1 E_1^{-1};$$

$$E_2^{-1} = \begin{bmatrix} 1 & 0 & 0 \\ 0 & -\frac{1}{2} & 0 \\ 0 & 0 & 1 \end{bmatrix} \Rightarrow E_2^{-1}E_2 = I_3 = E_2 E_2^{-1};$$

$$E_3^{-1} = \begin{bmatrix} 1 & 0 & 0 \\ 0 & 1 & 0 \\ 0 & 0 & \frac{1}{x} \end{bmatrix} \Rightarrow E_3^{-1}E_3 = I_3 = E_3 E_3^{-1}, \quad x \neq 0.$$

Thus the inverses for the E series of elementary matrices are easily obtained. Similar is the situation whatever be the order n in I_n. Now, look at the F series of elementary matrices. How can the effect in F_1 be nullified so that F_1 multiplied by a matrix, denoted by F_1^{-1}, gives back I_3? F_1 is obtained by adding the first row to the second row. Naturally the effect can be nullified by adding (-1) times the first row to the second row. That is,

$$F_1^{-1} = \begin{bmatrix} 1 & 0 & 0 \\ -1 & 1 & 0 \\ 0 & 0 & 1 \end{bmatrix} \quad \text{so that} \quad F_1^{-1}F_1 = I_3 = F_1 F_1^{-1};$$

$$F_2^{-1} = \begin{bmatrix} 1 & 0 & 0 \\ 0 & 1 & 0 \\ -1 & 0 & 0 \end{bmatrix} \Rightarrow F_2^{-1}F_2 = F_2 F_2^{-1} = I_3;$$

$$F_3^{-1} = \begin{bmatrix} 1 & 0 & -1 \\ 0 & 1 & 0 \\ 0 & 0 & 1 \end{bmatrix} \Rightarrow F_3^{-1}F_3 = I_3 = F_3 F_3^{-1};$$

$$F_4^{-1} = \begin{bmatrix} 1 & 0 & 0 \\ 0 & 1 & 0 \\ 0 & -1 & 1 \end{bmatrix} \Rightarrow F_4^{-1}F_4 = I_3 = F_4 F_4^{-1}.$$

From the way we defined the basic elementary matrices it is evident that the regular inverses exist for all elementary matrices.

(i) Regular inverses exist for all basic elementary matrices or elementary matrices, thereby the products of elementary matrices, are nonsingular.

Definition 2.3.2. For a given square matrix A if there exists a matrix B such that $AB = I, BA = I$ then the regular inverse exists and it is denoted by $B = A^{-1}$ and in this case A is called a *nonsingular* matrix and square matrices for which regular inverses do not exist are called *singular* matrices.

(ii) A square null matrix is a singular matrix.

(iii) A diagonal matrix with all nonzero diagonal elements is nonsingular and if there is at least one zero diagonal element then the diagonal matrix is singular.

(iv) A triangular matrix (lower or upper) with all nonzero diagonal elements is nonsingular and it is singular if there is at least one zero diagonal element.

2.3.1 Premultiplication of a matrix by elementary matrices

Consider an arbitrary 3×3 matrix $A = (a_{ij})$ and consider the E_1 of the previous section. Then

$$E_1 A = \begin{bmatrix} 5 & 0 & 0 \\ 0 & 1 & 0 \\ 0 & 0 & 1 \end{bmatrix} \begin{bmatrix} a_{11} & a_{12} & a_{13} \\ a_{21} & a_{22} & a_{23} \\ a_{31} & a_{32} & a_{33} \end{bmatrix}$$

$$= \begin{bmatrix} 5a_{11} & 5a_{12} & 5a_{13} \\ a_{21} & a_{22} & a_{23} \\ a_{31} & a_{32} & a_{33} \end{bmatrix}.$$

Note that E_1 is created by multiplying the first row of an identity matrix by 5. When we premultiply any 3×3 matrix A by E_1 the effect is exactly the same, that is, the first row of A is multiplied by 5.

(v) If E is an elementary matrix created from I_n by multiplying the i-th row by a nonzero scalar c and if any $n \times n$ matrix $A = (a_{ij})$ is premultiplied by E, that is EA, the net effect is that the i-th row of A is multiplied by c.

Note that E^{-1} in this case is a diagonal matrix with the i-th diagonal element $\frac{1}{c}$ and all other elements unities. If we again premultiply EA with this matrix E^{-1} then we get back A. That is,

$$E^{-1}(EA) = A.$$

Now, consider premultiplication of a 3×3 matrix $A = (a_{ij})$ by the F_1 of the previous section. That is,

$$F_1 A = \begin{bmatrix} 1 & 0 & 0 \\ 1 & 1 & 0 \\ 0 & 0 & 1 \end{bmatrix} \begin{bmatrix} a_{11} & a_{12} & a_{13} \\ a_{21} & a_{22} & a_{23} \\ a_{31} & a_{32} & a_{33} \end{bmatrix}$$

$$= \begin{bmatrix} a_{11} & a_{12} & a_{13} \\ a_{21} + a_{11} & a_{22} + a_{12} & a_{23} + a_{13} \\ a_{31} & a_{32} & a_{33} \end{bmatrix}.$$

Observe that F_1 is created by adding the first row to the second row of an identity matrix. (First row remains the same, the second row becomes the original second row plus the original first row.) The net effect of premultiplication of A by F_1 is exactly the same, the first row of A is added to the second row of A. In general, we have the following result:

(vi) If F is an elementary matrix created by adding the i-th row to the j-th row in I_n and if an arbitrary $n \times n$ matrix A is premultiplied by F, that is FA, the net effect is the same, that is, the i-th row of A is added to the j-th row of A.

Example 2.3.1. What are the net effects of the following operations? (a) $E_2 F_1 E_1 A$, (b) $E_4 F_2 E_3 E_2 F_1 E_1 A$ where

$$E_1 = \begin{bmatrix} -2 & 0 & 0 \\ 0 & 1 & 0 \\ 0 & 0 & 1 \end{bmatrix}, \quad F_1 = \begin{bmatrix} 1 & 0 & 0 \\ 1 & 1 & 0 \\ 0 & 0 & 1 \end{bmatrix}, \quad E_2 = E_1^{-1},$$

$$E_3 = \begin{bmatrix} -3 & 0 & 0 \\ 0 & 1 & 0 \\ 0 & 0 & 1 \end{bmatrix}, \quad F_2 = \begin{bmatrix} 1 & 0 & 0 \\ 0 & 1 & 0 \\ 1 & 0 & 1 \end{bmatrix}, \quad E_4 = E_3^{-1},$$

$$A = \begin{bmatrix} 1 & 1 & -1 \\ 2 & 0 & -1 \\ 3 & 1 & 4 \end{bmatrix}.$$

Solution 2.3.1.

$$E_1 A = \begin{bmatrix} -2 & 0 & 0 \\ 0 & 1 & 0 \\ 0 & 0 & 1 \end{bmatrix} \begin{bmatrix} 1 & 1 & -1 \\ 2 & 0 & -1 \\ 3 & 1 & 4 \end{bmatrix} = \begin{bmatrix} -2 & -2 & 2 \\ 2 & 0 & -1 \\ 3 & 1 & 4 \end{bmatrix},$$

$$F_1 E_1 A = F_1 (E_1 A)$$

$$= \begin{bmatrix} 1 & 0 & 0 \\ 1 & 1 & 0 \\ 0 & 0 & 1 \end{bmatrix} \begin{bmatrix} -2 & -2 & 2 \\ 2 & 0 & -1 \\ 3 & 1 & 4 \end{bmatrix} = \begin{bmatrix} -2 & -2 & 2 \\ 0 & -2 & 1 \\ 3 & 1 & 4 \end{bmatrix},$$

$$E_2 F_1 E_1 A = E_2 (F_1 E_1 A)$$

$$= \begin{bmatrix} -\frac{1}{2} & 0 & 0 \\ 0 & 1 & 0 \\ 0 & 0 & 1 \end{bmatrix} \begin{bmatrix} -2 & -2 & 2 \\ 0 & -2 & 1 \\ 3 & 1 & 4 \end{bmatrix} = \begin{bmatrix} 1 & 1 & -1 \\ 0 & -2 & 1 \\ 3 & 1 & 4 \end{bmatrix}.$$

The net effect of the operations so far is that the $(2, 1)$-th element in A is made zero.

$$E_3 E_2 F_1 E_1 A = E_3 (E_2 F_1 E_1 A)$$

$$= \begin{bmatrix} -3 & 0 & 0 \\ 0 & 1 & 0 \\ 0 & 0 & 1 \end{bmatrix} \begin{bmatrix} 1 & 1 & -1 \\ 0 & -2 & 1 \\ 3 & 1 & 4 \end{bmatrix} = \begin{bmatrix} -3 & -3 & 3 \\ 0 & -2 & 1 \\ 3 & 1 & 4 \end{bmatrix},$$

$$F_2 (E_3 E_2 F_1 E_1 A) = \begin{bmatrix} 1 & 0 & 0 \\ 0 & 1 & 0 \\ 1 & 0 & 1 \end{bmatrix} \begin{bmatrix} -3 & -3 & 3 \\ 0 & -2 & 1 \\ 3 & 1 & 4 \end{bmatrix} = \begin{bmatrix} -3 & -3 & 3 \\ 0 & -2 & 1 \\ 0 & -2 & 7 \end{bmatrix}.$$

The net effect of the operations so far is that the first column elements, except the first one, are reduced to zeros. Then operation on the left with $E_4 = E_3^{-1}$ gives

$$E_4 (F_2 E_3 E_2 F_1 E_1 A) = \begin{bmatrix} 1 & 1 & -1 \\ 0 & -2 & 1 \\ 0 & -2 & 7 \end{bmatrix}.$$

Example 2.3.2. Reduce the matrix A in Example 2.3.1 to a triangular form by premultiplication with elementary matrices (by elementary operations).

Solution 2.3.2. Part of the work is already done in Example 2.3.1. Now we continue. Consider the elementary matrices

$$E_5 = \begin{bmatrix} 1 & 0 & 0 \\ 0 & -1 & 0 \\ 0 & 0 & 1 \end{bmatrix}, \quad F_3 = \begin{bmatrix} 1 & 0 & 0 \\ 0 & 1 & 0 \\ 0 & 1 & 1 \end{bmatrix}.$$

Then E_5 operating on the reduced form from Example 2.3.1 gives

$$E_5 \begin{bmatrix} 1 & 1 & -1 \\ 0 & -2 & 1 \\ 0 & -2 & 7 \end{bmatrix} = \begin{bmatrix} 1 & 0 & 0 \\ 0 & -1 & 0 \\ 0 & 0 & 1 \end{bmatrix} \begin{bmatrix} 1 & 1 & -1 \\ 0 & -2 & 1 \\ 0 & -2 & 7 \end{bmatrix}$$

$$= \begin{bmatrix} 1 & 1 & -1 \\ 0 & 2 & -1 \\ 0 & -2 & 7 \end{bmatrix}.$$

Now, F_3 operating on the above form gives

$$F_3 \begin{bmatrix} 1 & 1 & -1 \\ 0 & 2 & -1 \\ 0 & -2 & 7 \end{bmatrix} = \begin{bmatrix} 1 & 0 & 0 \\ 0 & 1 & 0 \\ 0 & 1 & 1 \end{bmatrix} \begin{bmatrix} 1 & 1 & -1 \\ 0 & 2 & -1 \\ 0 & -2 & 7 \end{bmatrix}$$

$$= \begin{bmatrix} 1 & 1 & -1 \\ 0 & 2 & -1 \\ 0 & 0 & 6 \end{bmatrix}.$$

This is an upper triangular matrix. Hence the solution is complete.

Note that if A is to be recovered from the final form then write the final equation as,

$$F_3 E_5 E_4 F_2 E_3 E_2 F_1 E_1 A = \begin{bmatrix} 1 & 1 & -1 \\ 0 & 2 & -1 \\ 0 & 0 & 6 \end{bmatrix}.$$

Premultiply both sides by the inverses F_3^{-1}, E_5^{-1}, and so on, in that order. Then

$$A = E_1^{-1} F_1^{-1} E_2^{-1} E_3^{-1} F_2^{-1} E_4^{-1} E_5^{-1} F_3^{-1} \begin{bmatrix} 1 & 1 & -1 \\ 0 & 2 & -1 \\ 0 & 0 & 6 \end{bmatrix}.$$

But

$$F_3^{-1} = \begin{bmatrix} 1 & 0 & 0 \\ 0 & 1 & 0 \\ 0 & -1 & 1 \end{bmatrix}, \quad E_5^{-1} = \begin{bmatrix} 1 & 0 & 0 \\ 0 & -1 & 0 \\ 0 & 0 & 1 \end{bmatrix},$$

$$E_4^{-1} = \begin{bmatrix} -3 & 0 & 0 \\ 0 & 1 & 0 \\ 0 & 0 & 1 \end{bmatrix}, \quad F_2^{-1} = \begin{bmatrix} 1 & 0 & 0 \\ 0 & 1 & 0 \\ -1 & 0 & 1 \end{bmatrix},$$

$$E_3^{-1} = \begin{bmatrix} -\frac{1}{3} & 0 & 0 \\ 0 & 1 & 0 \\ 0 & 0 & 1 \end{bmatrix}, \quad E_2^{-1} = \begin{bmatrix} -2 & 0 & 0 \\ 0 & 1 & 0 \\ 0 & 0 & 1 \end{bmatrix},$$

$$F_1^{-1} = \begin{bmatrix} 1 & 0 & 0 \\ -1 & 1 & 0 \\ 0 & 0 & 1 \end{bmatrix}, \quad E_1^{-1} = \begin{bmatrix} -\frac{1}{2} & 0 & 0 \\ 0 & 1 & 0 \\ 0 & 0 & 1 \end{bmatrix}.$$

Multiplications can be carried out by inspection, remembering that if any matrix is premultiplied by an elementary matrix the same effect is there on that matrix. For example, $E_5^{-1}F_3^{-1}$ will have the effect that the second row of F_3^{-1} is multiplied by (-1). That is,

$$E_5^{-1}F_3^{-1} = \begin{bmatrix} 1 & 0 & 0 \\ 0 & -1 & 0 \\ 0 & -1 & 1 \end{bmatrix}.$$

Premultiplying this by E_4^{-1} will have the effect that the first row is multiplied by (-3). That is,

$$E_4^{-1}E_5^{-1}F_3^{-1} = \begin{bmatrix} -3 & 0 & 0 \\ 0 & -1 & 0 \\ 0 & -1 & 1 \end{bmatrix}.$$

Premultiplying this by F_2^{-1} has the effect that (-1) times the first row is added to the third row, and so on:

$$F_2^{-1}E_4^{-1}E_5^{-1}F_3^{-1} = \begin{bmatrix} -3 & 0 & 0 \\ 0 & -1 & 0 \\ 3 & -1 & 1 \end{bmatrix};$$

$$E_3^{-1}(F_2^{-1}E_4^{-1}E_5^{-1}F_3^{-1}) = \begin{bmatrix} 1 & 0 & 0 \\ 0 & -1 & 0 \\ 3 & -1 & 1 \end{bmatrix};$$

$$E_2^{-1}(E_3^{-1}F_2^{-1}E_4^{-1}E_5^{-1}F_3^{-1}) = \begin{bmatrix} -2 & 0 & 0 \\ 0 & -1 & 0 \\ 3 & -1 & 1 \end{bmatrix}.$$

Premultiplying this with F_1^{-1} has the effect that (-1) times the first row is added to the second row. That is,

$$F_1^{-1}(E_2^{-1}E_3^{-1}F_2^{-1}E_4^{-1}E_5^{-1}F_3^{-1}) = \begin{bmatrix} -2 & 0 & 0 \\ 2 & -1 & 0 \\ 3 & -1 & 1 \end{bmatrix}.$$

Then

$$E_1^{-1}(F_1^{-1}E_2^{-1}E_3^{-1}F_2^{-1}E_4^{-1}E_5^{-1}F_3^{-1}) = \begin{bmatrix} 1 & 0 & 0 \\ 2 & -1 & 0 \\ 3 & -1 & 1 \end{bmatrix}.$$

Thus

$$A = \begin{bmatrix} 1 & 0 & 0 \\ 2 & -1 & 0 \\ 3 & -1 & 1 \end{bmatrix} \begin{bmatrix} 1 & 1 & -1 \\ 0 & 2 & -1 \\ 0 & 0 & 6 \end{bmatrix}.$$

It can be verified by straight multiplication of the two matrices on the right and then comparing it with the given matrix in the example. The whole process is done in detail in Examples 2.3.1 and 2.3.2 so that the student can clearly understand the effects of premultiplications by elementary matrices, how to write down the inverses of elementary matrices or products of elementary matrices without doing any computation etc so that at later stages many such operations can be carried out simultaneously instead of doing it one at a time. The final solution is that A is written as a product of a lower triangular matrix, say L, and an upper triangular matrix, say U. That is,

$$A = LU. \tag{2.3.1}$$

Is the reduction of any given $n \times n$ matrix A to a product of $n \times n$ lower and upper triangular matrices possible? We can answer this question by making a series of observations.

(vii) Elementary matrices of the E category are always diagonal.

(viii) Elementary matrices of the F category are always lower triangular if the elementary matrices are created by adding the i-th row to the j-th row of an identity matrix, with $i < j$. [If $i > j$ the elementary matrix is no longer lower triangular.]

(ix) Product of a lower triangular matrix with a lower triangular matrix or with a diagonal matrix remains lower triangular.

(x) Regular inverse of a lower triangular matrix is lower triangular (write it as a product of elementary matrices and prove the result) and that of a diagonal matrix is diagonal.

Therefore in attempting to reduce a given matrix A to the form LU if only the steps in (vii) to (x) are involved we have a possibility of obtaining the form LU. If during the process, at any stage, the lower triangular nature of an elementary matrix of the F category is violated then we cannot expect the form LU unless the effect of that elementary operation is nullified by another elementary matrix during the process. Still we may not be able to get the form LU. Note from the examples that we could reduce the elements below the leading diagonal to zeros because we had a nonzero diagonal element sitting there at that stage of the operations. If the first row first column element in A was zero then by adding suitable multiples of the first row to the other rows we could not reduce nonzero elements in the same column to zeros. We can bring in a nonzero element to the $(1,1)$-th position by using an elementary matrix of the F category provided there is at least one nonzero element in the first column. But this F will not be lower triangular since an i-th row will be added to the j-th row with $i > j$. After reducing all elements in the first column to zeros except the first element, we can use the second diagonal element, $(2,2)$-th element, in the resulting matrix. If this resulting $(2,2)$-th element is zero the same situations as described above will arise. That is, at every stage, the leading diagonal elements must be nonzero in the sense that when we come to reducing the elements below the diagonal on the j-th column to zeros the (j,j)-th element must be nonzero for $j = 1, \ldots, n-1$. Then we end up with the form $A = LU$ where L is lower triangular as well as nonsingular, being product of elementary matrices, and U may or may not be nonsingular. If A itself is nonsingular then both L and U will be nonsingular and in this case,

$$A = LU \implies A^{-1} = U^{-1}L^{-1}$$

where U^{-1} and L^{-1} are easier to evaluate, being triangular, compared to the evaluation of the inverse of a general $n \times n$ matrix.

Example 2.3.3. Reduce the following matrix A to a triangular form by elementary operations on the left of A, where

$$A = \begin{bmatrix} 0 & 1 & -1 \\ 2 & -1 & 2 \\ 1 & 3 & -2 \end{bmatrix}.$$

Solution 2.3.3. Since the $(1,1)$-th position has a zero element we add the third row to the first row to bring in a nonzero entry at the $(1,1)$-th position. This can be achieved by premultiplying with

$$F_1 = \begin{bmatrix} 1 & 0 & 1 \\ 0 & 1 & 0 \\ 0 & 0 & 1 \end{bmatrix}.$$

That is,

$$F_1 A = \begin{bmatrix} 1 & 0 & 1 \\ 0 & 1 & 0 \\ 0 & 0 & 1 \end{bmatrix} \begin{bmatrix} 0 & 1 & -1 \\ 2 & -1 & 2 \\ 1 & 3 & -2 \end{bmatrix} = \begin{bmatrix} 1 & 4 & -3 \\ 2 & -1 & 2 \\ 1 & 3 & -2 \end{bmatrix}.$$

Note that the net effect is on the first row and not on the third row. Instead of the third row we could have added the second row to the first row which would have brought in a 2 at the $(1,1)$-th position. But 1 is easier to handle than 2. Instead of operating with elementary matrices of the E and F categories we may do two or more such operations together and write the resulting products of elementary matrices as G category matrices. Let

$$G_1 = \begin{bmatrix} 1 & 0 & 0 \\ -2 & 1 & 0 \\ 0 & 0 & 1 \end{bmatrix}$$

which is evidently a product of the basic elementary matrices. The net effect of premultiplying with G_1 is that (-2) times the first row is added to the second row. Later on, we will make a statement of such a premultiplication as follows:

$$\alpha(i) + (j) \implies \tag{2.3.2}$$

which means that α times the i-th row is added to the j-th row, where the i-th row remains the same whereas the j-th row changes to the original j-th row plus α times the original i-th row. In terms of the notation in (2.3.2) we can write G_1 as equivalent to the operation

$$-2(1) + (2) \;\Rightarrow\; G_1 F_1 A = \begin{bmatrix} 1 & 4 & -3 \\ 0 & -9 & 8 \\ 1 & 3 & -2 \end{bmatrix}.$$

Let

$$G_2 = \begin{bmatrix} 1 & 0 & 0 \\ 0 & 1 & 0 \\ -1 & 0 & 1 \end{bmatrix} \quad \text{or} \quad -1(1) + (3) \Rightarrow .$$

Then

$$G_2 G_1 F_1 A = \begin{bmatrix} 1 & 4 & -3 \\ 0 & -9 & 8 \\ 0 & -1 & 1 \end{bmatrix}.$$

Since F_1 is no longer a lower triangular form we do not expect A to be written as LU, product of lower and upper triangular matrices. Now to make $(3, 2)$-th element zero we have two choices, either divide the second row by -9, which produces a fraction at the $(2, 3)$-th position, then operate with the resulting second row, or interchange the third and second rows, which is a product of elementary operations, and then operate with the new second row. This last procedure avoids fractions. Hence consider

$$G_3 = \begin{bmatrix} 1 & 0 & 0 \\ 0 & 0 & 1 \\ 0 & 1 & 0 \end{bmatrix}$$

and then

$$G_3(G_2 G_1 F_1 A) = \begin{bmatrix} 1 & 0 & 0 \\ 0 & 0 & 1 \\ 0 & 1 & 0 \end{bmatrix} \begin{bmatrix} 1 & 4 & -3 \\ 0 & -9 & 8 \\ 0 & -1 & 1 \end{bmatrix}$$

$$= \begin{bmatrix} 1 & 4 & -3 \\ 0 & -1 & 1 \\ 0 & -9 & 8 \end{bmatrix}.$$

Let

$$G_4 = \begin{bmatrix} 1 & 0 & 0 \\ 0 & 1 & 0 \\ 0 & -9 & 1 \end{bmatrix} \quad \text{or} \quad -9(2) + (3) \Rightarrow$$

(−9 times the second row is added to the third row). Then

$$G_4 G_3 G_2 G_1 F_1 A = \begin{bmatrix} 1 & 4 & -3 \\ 0 & -1 & 1 \\ 0 & 0 & -1 \end{bmatrix}.$$

This is the desired triangular form. Since G_1, G_2, G_3, G_4 are products of the basic elementary matrices they are nonsingular and their regular inverses exist. Again, these inverses can be written down by inspection. In fact,

$$F_1^{-1} = \begin{bmatrix} 1 & 0 & -1 \\ 0 & 1 & 0 \\ 0 & 0 & 1 \end{bmatrix}, \quad G_1^{-1} = \begin{bmatrix} 1 & 0 & 0 \\ 2 & 1 & 0 \\ 0 & 0 & 1 \end{bmatrix},$$

$$G_2^{-1} = \begin{bmatrix} 1 & 0 & 0 \\ 0 & 1 & 0 \\ 1 & 0 & 1 \end{bmatrix}, \quad G_3^{-1} = \begin{bmatrix} 1 & 0 & 0 \\ 0 & 0 & 1 \\ 0 & 1 & 0 \end{bmatrix},$$

$$G_4^{-1} = \begin{bmatrix} 1 & 0 & 0 \\ 0 & 1 & 0 \\ 0 & 9 & 1 \end{bmatrix}.$$

From the above representation

$$A = F_1^{-1} G_1^{-1} G_2^{-1} G_3^{-1} G_4^{-1} \begin{bmatrix} 1 & 4 & -3 \\ 0 & -1 & 1 \\ 0 & 0 & -1 \end{bmatrix}.$$

But G_3^{-1} operating on G_4^{-1} will make the second and third rows interchanged in the original G_4^{-1}. That is,

$$G_3^{-1} G_4^{-1} = \begin{bmatrix} 1 & 0 & 0 \\ 0 & 9 & 1 \\ 0 & 1 & 0 \end{bmatrix};$$

$$G_2^{-1} G_3^{-1} G_4^{-1} = \begin{bmatrix} 1 & 0 & 0 \\ 0 & 1 & 0 \\ 1 & 0 & 1 \end{bmatrix} \begin{bmatrix} 1 & 0 & 0 \\ 0 & 9 & 1 \\ 0 & 1 & 0 \end{bmatrix} = \begin{bmatrix} 1 & 0 & 0 \\ 0 & 9 & 0 \\ 1 & 1 & 0 \end{bmatrix};$$

$$G_1^{-1} G_2^{-1} G_3^{-1} G_4^{-1} = \begin{bmatrix} 1 & 0 & 0 \\ 2 & 9 & 1 \\ 1 & 1 & 0 \end{bmatrix};$$

$$F_1^{-1} G_1^{-1} G_2^{-1} G_3^{-1} G_4^{-1} = \begin{bmatrix} 0 & -1 & 0 \\ 2 & 9 & 1 \\ 1 & 1 & 0 \end{bmatrix}.$$

Thus A has the decomposition

$$A = \begin{bmatrix} 0 & -1 & 0 \\ 2 & 9 & 1 \\ 1 & 1 & 0 \end{bmatrix} \begin{bmatrix} 1 & 4 & -3 \\ 0 & -1 & 1 \\ 0 & 0 & -1 \end{bmatrix}$$

where the first matrix on the right is not lower triangular but nonsingular (being product of elementary matrices) but the second matrix on the right is upper triangular. Thus, in general, we can have a decomposition of an $n \times n$ matrix A to the form

$$A = BU \qquad\qquad (2.3.3)$$

where B is nonsingular and U is upper triangular. This U will be nonsingular if A is nonsingular. In our example, U is nonsingular since none of the diagonal elements in U is zero.

(xi) Interchange of two rows (columns) is a product of elementary operations.

2.3.2 Reduction of a square matrix into a diagonal form

Here we consider the reduction of a square matrix into a diagonal form by premultiplication with elementary matrices, that is, by premultiplication alone. Later we will consider the reduction to a diagonal form by postmultiplication alone, and then reduction to a diagonal form by pre and post multiplications.

Example 2.3.4. Reduce the following matrix A to a diagonal form by premultiplication alone, where

$$A = \begin{bmatrix} 0 & 1 & -2 \\ 1 & 2 & 5 \\ 3 & -1 & 0 \end{bmatrix}.$$

Solution 2.3.4. Let

$$F_1 = \begin{bmatrix} 1 & 1 & 0 \\ 0 & 1 & 0 \\ 0 & 0 & 1 \end{bmatrix},$$

that is, add the second row to the first row $((2) + (1) \Rightarrow)$.

$$F_1 A = \begin{bmatrix} 1 & 1 & 0 \\ 0 & 1 & 0 \\ 0 & 0 & 1 \end{bmatrix} \begin{bmatrix} 0 & 1 & -2 \\ 1 & 2 & 5 \\ 3 & -1 & 0 \end{bmatrix} = \begin{bmatrix} 1 & 3 & 3 \\ 1 & 2 & 5 \\ 3 & -1 & 0 \end{bmatrix}.$$

Let

$$G_1 = \begin{bmatrix} 1 & 0 & 0 \\ -1 & 1 & 0 \\ 0 & 0 & 1 \end{bmatrix}, \quad G_2 = \begin{bmatrix} 1 & 0 & 0 \\ 0 & 1 & 0 \\ -3 & 0 & 1 \end{bmatrix} \quad \text{or}$$

$$G_1 : \ -1(1) + (2) \Rightarrow; \quad G_2 : \ -3(1) + (3) \Rightarrow.$$

$$G_2 G_1 F_1 A = \begin{bmatrix} 1 & 3 & 3 \\ 0 & -1 & 2 \\ 0 & -10 & -9 \end{bmatrix}.$$

Let

$$G_3 = \begin{bmatrix} 1 & 3 & 0 \\ 0 & 1 & 0 \\ 0 & 0 & 1 \end{bmatrix} \quad \text{or} \quad 3(2) + (1) \Rightarrow \quad \text{and}$$

$$G_4 = \begin{bmatrix} 1 & 0 & 0 \\ 0 & 1 & 0 \\ 0 & -10 & 1 \end{bmatrix} \quad \text{or} \quad -10(2) + (3) \Rightarrow.$$

Then

$$G_4 G_3 G_2 G_1 F_1 A = \begin{bmatrix} 1 & 0 & 9 \\ 0 & -1 & 2 \\ 0 & 0 & -29 \end{bmatrix}.$$

Let

$$E_1 = \begin{bmatrix} 1 & 0 & 0 \\ 0 & 1 & 0 \\ 0 & 0 & -\frac{1}{29} \end{bmatrix} \quad \text{or} \quad -\frac{1}{29}(3) \Rightarrow$$

or the third row is divided by (-29). Then

$$E_1 G_4 G_3 G_2 G_1 F_1 A = \begin{bmatrix} 1 & 0 & 9 \\ 0 & -1 & 2 \\ 0 & 0 & 1 \end{bmatrix}.$$

Let

$$G_5 = \begin{bmatrix} 1 & 0 & 0 \\ 0 & 1 & -2 \\ 0 & 0 & 1 \end{bmatrix} \quad \text{or} \quad -2(3) + (2) \Rightarrow;$$

$$G_6 = \begin{bmatrix} 1 & 0 & -9 \\ 0 & 1 & 0 \\ 0 & 0 & 1 \end{bmatrix} \quad \text{or} \quad -9(3) + (1) \Rightarrow.$$

Then

$$G_6 G_5 E_1 G_4 G_3 G_2 G_1 F_1 A = \begin{bmatrix} 1 & 0 & 0 \\ 0 & -1 & 0 \\ 0 & 0 & 1 \end{bmatrix}.$$

That is, BA is a diagonal matrix where

$$B = G_6 G_5 E_1 G_4 G_3 G_2 G_1 F_1$$

is a nonsingular matrix, being products of elementary matrices. Therefore, in this representation

$$A = B^{-1} \text{ multiplied by a diagonal matrix,}$$
$$B^{-1} = F_1^{-1} G_1^{-1} G_2^{-1} G_3^{-1} G_4^{-1} E_1^{-1} G_5^{-1} G_6^{-1}.$$

Note that if A is nonsingular we can expect the diagonal matrix to have all nonzero diagonal elements. If A is singular then at least one diagonal element in the diagonal matrix will be zero. Thus for any $n \times n$ matrix A we have the representation

$$CA = D \quad \text{or} \quad A = C^{-1}D \tag{2.3.4}$$

where C is nonsingular and D is diagonal. If A is singular then D is singular and if A is nonsingular then D is nonsingular. We can also represent A in the forms

$$A = DB, \quad A = C_1 D_1 B_1$$

where C, C_1 and B, B_1 are nonsingular matrices and D, D_1 are diagonal matrices. These will be considered later after looking into the solution of a system of linear equations.

2.3.3 Solving a system of linear equations

As we have already seen that a system of m linear equations in n real scalar variables can be written in the form

$$AX = b$$

where A is the $m \times n$ known coefficient matrix, X is the $n \times 1$ vector of variables or unknowns and b is an $m \times 1$ known vector. For example,

$$\begin{aligned} x_1 - x_2 + x_3 - 2x_4 &= 5 \\ 2x_1 + x_2 - x_3 + x_4 &= 2 \implies AX = b \\ x_1 + x_2 + x_3 - 3x_4 &= 4 \end{aligned}$$

where

$$A = \begin{bmatrix} 1 & -1 & 1 & -2 \\ 2 & 1 & -1 & 1 \\ 1 & 1 & 1 & -3 \end{bmatrix}, \quad X = \begin{bmatrix} x_1 \\ \vdots \\ x_4 \end{bmatrix}, \quad b = \begin{bmatrix} 5 \\ 2 \\ 4 \end{bmatrix}.$$

Here $m = 3$, $n = 4$. One method of solving this system is by successive elimination, which the student may be familiar with. This will be quite tedious when the number of variables and the number of equations are large. We will solve this system by elementary operations. If there is a solution for $AX = b$, that is, if there exists a vector X such that $AX = b$ is satisfied then

$$BAX = Bb$$

has the same solution as the original equation $AX = b$ as long as B is a nonsingular matrix. Note that, by premultiplying both sides by B^{-1}, when B is nonsingular B^{-1} exists, we get back the original equation

$$B^{-1}(BAX) = B^{-1}(Bb) \implies AX = b.$$

Since elementary matrices or products of them are nonsingular we may operate both sides of $AX = b$ by elementary matrices. Premultiplication by elementary matrices will be stated by using our notation in (2.3.2). When premultiplying both sides of $AX = b$ by elementary matrices the effects will be on A and b. Hence we need to look only at the effects rather than writing the whole system of equations each time. A convenient way of writing A and b is to write A and then b separated by a vertical line or two vertical lines to indicate that A and b are on separate sides of the equation $AX = b$. For the illustrative example, this representation is then,

$$(a) \quad \begin{array}{cccc|c} 1 & -1 & 1 & -2 & 5 \\ 2 & 1 & -1 & 1 & 2 \\ 1 & 1 & 1 & -3 & 4 \end{array}$$

Now the idea is to get rid off the elements in the first column, except the $(1,1)$-th element, the elements in the second column, except the element at the $(2,2)$-th position, and so on or to reduce the elements below the leading diagonal of A to zeros or to reduce the elements below as well as above the leading diagonal to zeros. Instead of doing one operation at a time we can do several operations simultaneously:

$$-2(1) + (2); \quad -1(1) + (3) \implies$$

This means that (-2) times the first row (on both sides of the vertical line) is added to the second row and then (-1) times the first row is added to the third row. The first row, the one we are operating with, remains the same and the other rows change. The net result of these operations will give the following configurations: [Always write the row that you are operating with first, since it is not going to change, and then write the result of the operations on the other rows, on both sides.]

$$
(b) \quad
\begin{array}{cccc|c}
1 & -1 & 1 & -2 & 5 \\
0 & 3 & -3 & 5 & -8 \\
0 & 2 & 0 & -1 & -1
\end{array}
$$

For a triangular type reduction we will try to get rid off the $(3,2)$-th element by using the $(2,2)$-th element. This can be done by first dividing the second row by 3 and then adding (-2) times the second row to the third row:

$$
(c) \quad \frac{1}{3}(2); \quad -2(2)+(3) \Rightarrow
\begin{array}{cccc|c}
1 & -1 & 1 & -2 & 5 \\
0 & 1 & -1 & \frac{5}{3} & -\frac{8}{3} \\
0 & 0 & 2 & -\frac{13}{3} & \frac{13}{3}
\end{array}
$$

In this triangular type reduction we cannot go further. One way of solving the system is to write the whole system at this stage and then solve starting from the last equation. Translating (c) in terms of the original variables we have the following system of equations:

$$
\begin{aligned}
x_1 - x_2 + x_3 - 2x_4 &= 5 \\
x_2 - x_3 + \frac{5}{3}x_4 &= -\frac{8}{3} \\
2x_3 - \frac{13}{3}x_4 &= \frac{13}{3}.
\end{aligned}
$$

There are infinitely many solutions in this case because any one variable, for example, x_4, can be free. We can assign any value to x_4 and can solve for the remaining variables. For example let $x_4 = 0$. Then we have $2x_3 = \frac{13}{3}$ or $x_3 = \frac{13}{6}$. Then from the next line above

$$
x_2 = x_3 - \frac{5}{3}x_4 - \frac{8}{3} = \frac{13}{6} - 0 - \frac{8}{3} = -\frac{1}{2}
$$

and finally

$$
x_1 = x_2 - x_3 + 2x_4 + 5 = -\frac{1}{2} - \frac{13}{6} + 0 + 5 = \frac{7}{3}.
$$

Therefore one solution is

$$
(x_1, x_2, x_3, x_4) = \left(\frac{7}{3}, -\frac{1}{2}, \frac{13}{6}, 0\right).
$$

By taking other values for x_4 we get other solutions. Note that when solving a system such as the one above it is wiser to verify the final answer by substituting in the original system because it is likely that we may have made some computational errors during the process.

If we wish to reduce our system to a diagonal type format then at the stage (c) above, add the second row to the first row to obtain

$$
(d) \qquad \begin{array}{cccc|c}
1 & 0 & 0 & -1/3 & 7/3 \\
0 & 1 & -1 & 5/3 & -8/3 \\
0 & 0 & 2 & -13/3 & 13/3
\end{array}
$$

Now divide the third row by 2 and then add the third row to the second row:

$$
(e) \qquad \frac{1}{2}(3); \quad (3)+(2) \Rightarrow
\begin{array}{cccc|c}
1 & 0 & 0 & -1/3 & 7/3 \\
0 & 1 & 0 & -1/2 & -1/2 \\
0 & 0 & 1 & -13/6 & 13/6
\end{array}
$$

Now with $x_4 = 0$ we can read off values of the other variables from the right side itself. That is, $x_3 = \frac{13}{6}$, $x_2 = -\frac{1}{2}$, $x_1 = \frac{7}{3}$. Thus one solution is

$$
(x_1, x_2, x_3, x_4) = \left(\frac{7}{3}, -\frac{1}{2}, \frac{13}{6}, 0 \right).
$$

When doing elementary operations to solve a system of equations the following points are worth observing:

(xii) Interchange the rows (changing the order of the equations does not affect the solutions), if necessary, to bring a nonzero number at the $(1,1)$-th position to start with. Repeat the same technique when dealing with the (i,i)-th position on the way, $i = 1, 2, \dots$.

(xiii) If a division creates fractions at any stage then multiply the equations with appropriate numbers to avoid fractions when adding a constant multiple of a row to another row.

(xiv) At any stage of the operations if any equation results in an impossible statement, such as on the one side of the vertical line there is a zero only whereas on the other side there is a nonzero number, then stop the process. There is no solution for the system.

Example 2.3.5. Solve the following system of linear equations if there exists a solution:

$$
\begin{aligned}
x_1 - x_2 + 2x_3 &= 2 \\
2x_1 + 2x_2 - x_3 &= 3 \\
3x_1 + x_2 + x_3 &= 5.
\end{aligned}
$$

Solution 2.3.5. To start with we do not know whether the system has a solution or not. Hence start the process and continue. Writing as before

$$\begin{array}{ccc|c} 1 & -1 & 2 & 2 \\ 2 & 2 & -1 & 3 \\ 3 & 1 & 1 & 5 \end{array}$$

$$-2(1) + (2); \quad -3(1) + (3) \Rightarrow \begin{array}{ccc|c} 1 & -1 & 2 & 2 \\ 0 & 4 & -5 & -1 \\ 0 & 4 & -5 & -1 \end{array}$$

$$-1(2) + (3) \Rightarrow \begin{array}{ccc|c} 1 & -1 & 2 & 2 \\ 0 & 4 & -5 & -1 \\ 0 & 0 & 0 & 0 \end{array}$$

The last equation resulted in a statement $0 = 0$ which is a valid statement. Thus the last row disappears. We can start solving from this triangular type format or try to get rid off the element at the $(1, 2)$-th position. This can be done by two steps, at the same time avoiding fractions also. Multiply the first row by 4 and then add the second row to the first row. That is,

$$4(1); \quad (2) + (1) \Rightarrow \begin{array}{ccc|c} 4 & 0 & 3 & 7 \\ 0 & 4 & -5 & -1 \end{array}$$

Writing the resulting equations we have

$$4x_1 + 3x_3 = 7$$
$$4x_2 - 5x_3 = -1.$$

There are several solutions. We can assign an arbitrary value to x_3. For example let $x_3 = 0$ then one solution is

$$(x_1, x_2, x_3) = \left(\frac{7}{4}, -\frac{1}{4}, 0\right).$$

Example 2.3.6. Solve the following system of linear equations if there exists a solution:

$$2x_2 - x_3 = 1$$
$$x_1 - x_2 + 3x_3 = 2$$
$$x_1 + x_2 + 2x_3 = 5.$$

Solution 2.3.6. We will interchange the first and second equations to bring a nonzero number at the $(1, 1)$-th position. The resulting configuration is the following:

$$\begin{array}{ccc|c} 1 & -1 & 3 & 2 \\ 0 & 2 & -1 & 1 \\ 1 & 1 & 2 & 5 \end{array}$$

Now we start the elementary operations:

$$-1(1) + (3) \Rightarrow \begin{array}{ccc|c} 1 & -1 & 3 & 2 \\ 0 & 2 & -1 & 1; \\ 0 & 2 & -1 & 3 \end{array} \qquad -1(2) + (3) \Rightarrow \begin{array}{ccc|c} 1 & -1 & 3 & 2 \\ 0 & 2 & -1 & 1. \\ 0 & 0 & 0 & 2 \end{array}$$

The last equation has resulted in an inconsistent statement that $2 = 0$ and hence the system has no solution, the system is inconsistent.

Definition 2.3.3. A system of linear equations $AX = b$ is said to be *consistent* if there exists at least one vector X (at least one solution) such that the equation $AX = b$ is satisfied. If there is no such X the system is said to be *inconsistent*.

In Example 2.3.6 the system is inconsistent whereas the system in Example 2.3.5 is consistent. When the system is consistent we may have just one solution (unique solution) or many solutions.

(xv) If A in $AX = b$ is a square and nonsingular matrix then there is a unique solution and the solution is $X = A^{-1}b$.

(xvi) If $AX = b$, with A a square matrix, and if the system is consistent with A singular then there are many solutions.

(xvii) If A is $m \times n$, $m < n$ the system $AX = b$ may not have a solution. Consistency of the system does not go with $m < n$ or $m = n$.

Exercises 2.3

2.3.1. Write the following matrices as products of the basic elementary matrices of the E and F types, if possible (see Definition 2.3.1):

$$A_1 = \begin{bmatrix} 1 & 0 & 0 & 0 \\ -1 & 1 & 0 & 0 \\ -2 & 0 & 1 & 0 \\ 0 & 0 & 0 & 1 \end{bmatrix}, \quad A_2 = \begin{bmatrix} 0 & 0 & 1 & 0 \\ 0 & 1 & 0 & 0 \\ 1 & 0 & 0 & 0 \\ 0 & 0 & 0 & 0 \end{bmatrix}, \quad A_3 = \begin{bmatrix} 1 & -3 & 0 & 0 \\ 0 & 1 & 0 & 0 \\ 0 & 2 & 1 & 0 \\ 0 & 0 & 0 & 1 \end{bmatrix}.$$

2.3.2. Prove that the interchange of the i-th and the j-th rows of I_n is a product of elementary matrices of the E and F types.

2.3.3. Evaluate the regular inverses of the matrices in Exercise 2.3.1, if they exist, by first writing them as product of elementary matrices and then inverting them.

2.3.4. Let A be the matrix obtained by interchanging the i-th and j-th rows of I_n. Evaluate the regular inverse of A.

2.3.5. Reduce the following matrices into the form LU, wherever possible, where L is lower triangular and U is upper triangular:

$$A = \begin{bmatrix} 1 & -1 & 2 \\ -2 & 0 & 1 \\ 3 & 1 & -1 \end{bmatrix}, \quad B = \begin{bmatrix} 0 & 1 & 1 \\ -1 & 1 & 5 \\ 2 & 0 & 4 \end{bmatrix}, \quad C = \begin{bmatrix} 1 & 0 & 1 \\ 2 & 3 & 1 \\ 4 & 3 & 3 \end{bmatrix}.$$

2.3.6. Reduce the matrices in Exercise 2.3.5 to the form QU where Q is nonsingular and U is upper triangular.

2.3.7. Reduce the matrices in Exercise 2.3.5 to the form SD where D is diagonal and S is nonsingular.

2.3.8. Under what conditions the following matrices A and B nonsingular?

$$A = \begin{bmatrix} 2 & 0 & 0 \\ 0 & c & 0 \\ 3 & 2 & 1 \end{bmatrix} \begin{bmatrix} d_1 & 0 & 0 \\ 0 & d_2 & 0 \\ 0 & 0 & d_3 \end{bmatrix}, \quad B = A \begin{bmatrix} 1 & 2 & 1 \\ 0 & 2 & 3 \\ 0 & 0 & x \end{bmatrix}.$$

2.3.9. Solve the systems of equations by reducing the coefficient matrices to triangular type forms:

$$\begin{aligned}
\text{(a)} \quad & x_2 - 3x_3 + x_4 = 1 \\
& 2x_1 - x_2 + x_3 + x_4 = 2 \\
& 3x_1 - 2x_2 + x_3 - x_4 = 1; \\
\text{(b)} \quad & x_2 - 2x_3 + 2x_4 = 1 \\
& x_1 - 2x_2 + x_3 + x_4 = 2 \\
& x_1 - x_2 - x_3 + 3x_4 = 4; \\
\text{(c)} \quad & x_1 + x_2 - x_3 + x_4 = 1 \\
& 2x_1 - x_2 + x_3 + 2x_4 = 2 \\
& x_1 + x_2 + x_3 + x_4 = 2 \\
& 3x_1 - 2x_2 + x_3 - 2x_4 = 4.
\end{aligned}$$

2.3.10. Solve the same systems of equations in Exercise 2.3.9 by reducing the coefficient matrices to the diagonal type forms.

2.3.11. Solve the system of equations

$$\begin{aligned}
x_1 + x_2 + x_3 + x_4 &= 0 \\
x_1 - x_2 + x_3 - x_4 &= 0 \\
-x_1 - x_2 + x_3 + x_4 &= 0 \\
x_1 - x_2 - x_3 + x_4 &= 0.
\end{aligned}$$

2.3.12. Solve the system of equations

$$x_1 + x_2 + x_3 + x_4 = 1$$
$$x_1 - x_2 + x_3 - x_4 = 2$$
$$-x_1 - x_2 + x_3 + x_4 = 1$$
$$2x_1 + 2x_3 = 3.$$

2.3.13. Solve the system of equations

$$x_1 + x_2 + x_3 + x_4 = 2$$
$$x_1 - x_2 + x_3 - x_4 = 1$$
$$x_1 + 2x_2 + x_3 + x_4 = 3$$
$$2x_1 + x_2 + 2x_3 = 2.$$

2.3.14. Writing the equations in Exercises 2.3.11, 2.3.12 and 2.3.13 as $AX = b$ and then reducing A to triangular type forms determine whether or not (a) A is nonsingular in each case, (b) A can be represented as LU in each case where L is lower triangular and U is upper triangular, (c) A can be written as BD in each case where B is nonsingular and D is diagonal.

2.3.15. Writing the equations in Exercises 2.3.11, 2.3.12 and 2.3.13 in the form $AX = b$ write $A = BDC$ where D is diagonal, B and C nonsingular, $B \neq I_4$, $C \neq I_4$.

2.3.16. Solve the system of equations

$$x_1 + 2x_2 + x_3 - x_4 = 2$$
$$x_2 + 5x_3 + x_4 = 4$$
$$x_1 - x_2 + x_3 = 2$$
$$x_1 + 2x_2 + x_3 - x_4 = 5.$$

2.3.17. If there is a solution for the system in Exercise 2.3.16 what is the geometric interpretation of a solution? If there is no solution in Exercise 2.3.16 explain the geometry.

2.3.18. If there is a system of n linear equations in n unknowns $x_1, \ldots, x_n$, that is, $AX = b$ where A is $n \times n$ and $X' = (x_1, \ldots, x_n)$, and if A is orthonormal, is the system consistent? If so how many solutions are there? Obtain a solution without using elementary operations.

2.3.19. Suppose A, B and $A + B$ are all nonsingular $n \times n$ matrices. Show that $A^{-1} + B^{-1}$ is nonsingular and that

$$(A^{-1} + B^{-1})^{-1} = A(A + B)^{-1}B = B(A + B)^{-1}A.$$

2.3.20. Verify the results in Exercise 2.3.19 for

$$A = \begin{bmatrix} 1 & 0 & 0 \\ 1 & 2 & 0 \\ 2 & 1 & 3 \end{bmatrix}, \quad B = \begin{bmatrix} 1 & 2 & 1 \\ 0 & 1 & 1 \\ 0 & 0 & 4 \end{bmatrix}.$$

2.4 Inverse, linear independence and ranks

In this section we shall consider a method of evaluating the regular inverse of a given nonsingular matrix by elementary operations, checking for linear dependence of a set of vectors by elementary operations, the concepts of row and column ranks of a matrix and the rank of a matrix.

First we deal with a technique of evaluating the regular inverse of a given matrix whenever the inverse exists. There are several ways of doing this. One method based on elementary operations will be discussed here.

2.4.1 Inverse of a matrix by elementary operations

Let $A = (a_{ij})$ be a given $n \times n$ matrix. If the regular inverse of A exists let us denote it by A^{-1}. Then from the definition itself

$$AA^{-1} = I \tag{2.4.1}$$

where I is the identity matrix. Equation (2.4.1) is the same equation if both sides are multiplied by the same nonsingular matrix, say B, in the sense

$$AA^{-1} = I \implies BAA^{-1} = B \implies AA^{-1} = I.$$

Also note that $B(AA^{-1}) = (BA)A^{-1}$ or we can premultiply A with a nonsingular matrix B and it will be equivalent to premultiplying (AA^{-1}) with B. Since elementary matrices are nonsingular we will premultiply on both sides by elementary matrices and try to reduce A to an identity matrix. If this is possible then A^{-1} is the product of the elementary matrices on the right.

Example 2.4.1. Evaluate the regular inverse of the following matrix A if it exists.

$$A = \begin{bmatrix} 1 & 0 & -1 \\ 2 & 1 & 3 \\ 3 & 1 & 4 \end{bmatrix}.$$

Solution 2.4.1. If A was a rectangular matrix we would not have attempted to evaluate the regular inverse since regular inverses do not exist for rectangular matrices. Our matrix A is a square matrix. It may or may not have the regular inverse A^{-1}. We start with the assumption that A^{-1} exists and proceed with elementary operations on the left on both sides. If A^{-1} does not exist then some inconsistency will arise during the process. Then we stop. If no inconsistency arises during the process then A^{-1} will be available on the right side when A reduces to an identity matrix. Hence consider the equation

$$\begin{bmatrix} 1 & 0 & -1 \\ 2 & 1 & 3 \\ 3 & 1 & 4 \end{bmatrix} A^{-1} = \begin{bmatrix} 1 & 0 & 0 \\ 0 & 1 & 0 \\ 0 & 0 & 1 \end{bmatrix}.$$

We premultiply both sides by elementary matrices. In the first stage our aim is to reduce the first column elements of A to zeros except the first element. This can be achieved by a sequence of operations with elementary matrices which in our notation can be stated as follows:

$$-2(1) + (2); \quad -3(1) + (3) \Rightarrow$$

$[(-2)$ times the first row added to the second row and (-3) times the first row added to the third row give]

$$\begin{bmatrix} 1 & 0 & -1 \\ 0 & 1 & 5 \\ 0 & 1 & 7 \end{bmatrix} A^{-1} = \begin{bmatrix} 1 & 0 & 0 \\ -2 & 1 & 0 \\ -3 & 0 & 1 \end{bmatrix}.$$

Note that we do the same operations on both sides (same premultiplications by elementary matrices on both sides). The net effect of these multiplications on the left is on the matrix A itself. The final result of this stage of operations is given above. Our next aim is to get rid off the elements in the second column of the resulting A by operating with the second row (or with the help of the element at the $(2,2)$-th position). We have the $(1,2)$-th element already zero and hence we need to get rid off only the $(3,2)$-th element.

$$-1(2) + (3) \Rightarrow \begin{bmatrix} 1 & 0 & -1 \\ 0 & 1 & 5 \\ 0 & 0 & 2 \end{bmatrix} A^{-1} = \begin{bmatrix} 1 & 0 & 0 \\ -2 & 1 & 0 \\ -1 & -1 & 1 \end{bmatrix}.$$

The next stage is to get rid off the elements in the third column and at the same time make the elements at $(3,3)$-th position 1. This can be done first by dividing the third row by 2 and with the help of this new third row get rid off the elements at the $(2,3)$-th and $(1,3)$-th positions. The operations are the following:

$$\frac{1}{2}(3); \quad -5(3) + (2); \quad (3) + (1) \Rightarrow$$

$$\begin{bmatrix} 1 & 0 & 0 \\ 0 & 1 & 0 \\ 0 & 0 & 1 \end{bmatrix} A^{-1} = \begin{bmatrix} \frac{1}{2} & -\frac{1}{2} & \frac{1}{2} \\ \frac{1}{2} & \frac{7}{2} & -\frac{5}{2} \\ -\frac{1}{2} & -\frac{1}{2} & \frac{1}{2} \end{bmatrix}.$$

Hence, writing with $\frac{1}{2}$ outside,

$$A^{-1} = \frac{1}{2} \begin{bmatrix} 1 & -1 & 1 \\ 1 & 7 & -5 \\ -1 & -1 & 1 \end{bmatrix}.$$

The student may verify by multiplying this with A to see whether the product is an identity matrix, to make sure that no computational error is made during the process.

Instead of writing the reduced matrix, A^{-1} and the reduced right side each time, we may simply write the configurations of the elements in A first then put a vertical line and write an identity matrix. Continue with the operations with the aim of reducing A to an identity matrix, each time doing the same operations on both sides of the vertical line, [always premultiplications by elementary matrices on both sides of the vertical line]. If A reduces to an identity matrix then what is obtained on the right of the vertical line at this stage is A^{-1}.

Example 2.4.2. Evaluate the regular inverse of A if it exists, where

$$A = \begin{bmatrix} 1 & 0 & -1 \\ -3 & 2 & 1 \\ -2 & 2 & 0 \end{bmatrix}.$$

Solution 2.4.2. We start with the equation $AA^{-1} = I$ assuming that A^{-1} exists. Let us write the configuration in A, a vertical line, the configuration in an identity matrix, in that order. That is,

$$\begin{array}{ccc|ccc} 1 & 0 & -1 & 1 & 0 & 0 \\ -3 & 2 & 1 & 0 & 1 & 0 \\ -2 & 2 & 0 & 0 & 0 & 1 \end{array}$$

Now we start with the operations on both sides, using the same notations as before:

$$3(1) + (2); \quad 2(1) + (3) \Rightarrow \begin{array}{ccc|ccc} 1 & 0 & -1 & 1 & 0 & 0 \\ 0 & 2 & -2 & 3 & 1 & 0 \\ 0 & 2 & -2 & 2 & 0 & 1 \end{array}$$

$$-1(2) + (3) \Rightarrow \begin{array}{ccc|ccc} 1 & 0 & -1 & 1 & 0 & 0 \\ 0 & 2 & -2 & 3 & 1 & 0 \\ 0 & 0 & 0 & -1 & -1 & 1 \end{array}$$

The last row on the left is null. Whatever linear operations we do on the left this form cannot be reduced to an identity matrix. Hence A^{-1} does not exist in this case.

(i) While doing elementary operations on the left if any row of the reduced matrix on the left becomes null at any stage then there is no regular inverse for the given matrix. Hence stop the process then and there when a null vector is obtained.

(ii) If the given matrix is rectangular then there is no regular inverse and hence do not start elementary operations if the aim is to find the regular inverse.

(iii) While doing elementary operations on the left multiply the rows by appropriate numbers (do the same multiplications on both sides of the vertical line) in order to avoid fractions. Then at the very end of the operations divide the rows by appropriate numbers to create an identity matrix on the left of the vertical line. This will make the computations much easier.

2.4.2 Checking linear independence through elementary operations

Consider ordered sets of real numbers defined as vectors. Consider m such n-vectors. These m vectors can also be looked upon as the m rows of an $m \times n$ matrix. In such a case we will be checking the linear dependence of the rows of a matrix also. Write the m vectors as a matrix and apply elementary operations on the left, that is premultiply by elementary matrices. As seen from Chapter 1, linear independence or dependence in a set of vectors is not altered by nonzero scalar multiplications or additions, the two basic elementary operations corresponding to the two types of basic elementary matrices. Also interchanges of rows will not alter the linear independence or dependence in the set. An interchange of rows can be looked upon as a product of elementary operations. Let $m \leq n$ for convenience. By elementary operations on the left, with interchanges if necessary, bring $A = (a_{ij})$ the matrix representing the m vectors to the following form:

$$A \rightarrow \begin{bmatrix} I_r & C \\ O & O \end{bmatrix}$$

where I_r is an identity matrix, $r \leq m$, O indicates a null matrix and C is a matrix which may or may not be null. Such a reduction is always possible provided there are no null column vectors in the first r columns. If $r = m$ then the null matrices will not be present. Since I_r is an $r \times r$ identity matrix, $r \leq m \leq n$ the first r rows of the reduced matrix are linearly independent. Thus the maximum number of linearly independent rows is r or the maximum number of linearly independent vectors in the given set of m vectors is r. This process will be clear from the following example.

Example 2.4.3. Determine the maximum number of linearly independent row vectors and the maximum number of linearly independent column vectors in the following matrix A where,

$$A = \begin{bmatrix} 2 & -3 & 1 & 5 \\ 1 & 2 & -1 & 1 \\ 3 & -1 & 0 & 6 \end{bmatrix}.$$

Solution 2.4.3. Since our aim is to check for linear independence or dependence we can interchange the rows, if necessary. Let us interchange rows 1 and 2 to bring a 1 at the $(1,1)$-th position. That is,

$$\begin{matrix} 1 & 2 & -1 & 1 \\ 2 & -3 & 1 & 5 \\ 3 & -1 & 0 & 6 \end{matrix}$$

Now we do elementary operations, using our notations introduced earlier:

$$-2(1)+(2); \quad -3(1)+(3) \Rightarrow \begin{matrix} 1 & 2 & -1 & 1 \\ 0 & -7 & 3 & 3 \\ 0 & -7 & 3 & 3 \end{matrix}$$

$$-1(2)+(3) \Rightarrow \begin{matrix} 1 & 2 & -1 & 1 \\ 0 & -7 & 3 & 3 \\ 0 & 0 & 0 & 0 \end{matrix}$$

Linear independence can be determined at this stage itself without bringing the matrix A to the form $\begin{bmatrix} I_r & C \\ O & O \end{bmatrix}$. The first two rows are linearly independent. Hence the maximum number of linearly independent rows is 2. If we wish to bring the matrix to the above form then do the following operations. Divide the second row by (-7) and then add (-2) times the second row to the first row:

$$-\frac{1}{7}(2); \quad -2(2)+(1) \Rightarrow \begin{bmatrix} 1 & 0 & -\frac{1}{7} & \frac{13}{7} \\ 0 & 1 & -\frac{3}{7} & -\frac{3}{7} \\ 0 & 0 & 0 & 0 \end{bmatrix} = \begin{bmatrix} I_r & C \\ O & O \end{bmatrix}$$

$$C = \begin{bmatrix} -\frac{1}{7} & \frac{13}{7} \\ -\frac{3}{7} & -\frac{3}{7} \end{bmatrix}.$$

Let

$$e_1 = \begin{pmatrix} 1 \\ 0 \end{pmatrix}, \quad e_2 = \begin{pmatrix} 0 \\ 1 \end{pmatrix}.$$

Then

$$\begin{pmatrix} -1/7 \\ -3/7 \end{pmatrix} = -\frac{1}{7}e_1 - \frac{3}{7}e_2$$

and

$$\begin{pmatrix} 13/7 \\ -3/7 \end{pmatrix} = \frac{13}{7}e_1 - \frac{3}{7}e_2.$$

Hence the last two columns are linearly dependent on the first two columns. Thus the maximum number of linearly independent columns is also 2. This in fact is a general result. Before stating the general results we may observe one aspect. In the above example we had interchanged two rows to start with. Thus the columns are disturbed. Is linear independence of columns affected by such an interchange? Note that at the last stage we may interchange the rows back to the original order. Still we have the two unit vectors e_2 and e_1 and still the last columns or the columns in C are linear functions of e_1 and e_2. Hence linear dependence in the columns is not affected by such an interchange of rows.

Definition 2.4.1. The maximum number of linearly independent row vectors in a matrix A is called the *row rank* of A. The maximum number of linearly independent column vectors in a matrix A is called the *column rank* of A. It can be proved that the row rank equals the column rank. Then this common rank is called the *rank of the matrix*.

(iv) In any matrix the row rank is equal to the column rank.

This result can be proved without much difficulty. Premultiplication of a matrix by elementary matrices does not alter the linear independence of the system of row vectors or column vectors. By elementary operations, and row interchanges if necessary, bring the given $m \times n$ matrix, $m \leq n$, to the form

$$\text{(a)} \qquad A \rightarrow \begin{bmatrix} I_r & C \\ O & O \end{bmatrix}$$

where O indicates a null matrix and C may or may not be null. Let $e_1, \dots, e_r$ be the r basic unit column vectors. Since C is an $r \times (n - r)$ matrix every column there is an r-vector and hence can be written as a linear combination of the basic unit vectors $e_1, \dots, e_r$. Since every column vector in C is dependent on $e_1, \dots, e_r$ the column rank is r. Since I_r is present as the first block the first r row vectors are linearly independent, the remaining are null vectors. Hence the row rank is also r. A similar argument holds for the case $m \geq n$.

Definition 2.4.2. If the rank of an $m \times n$ or $n \times m$ matrix with $m \leq n$ is m then the matrix is said to be *a full rank matrix*.

Definition 2.4.3. An $n \times n$ matrix A with rank n is said to be a *nonsingular matrix* with a regular inverse A^{-1} such that $AA^{-1} = I_n, A^{-1}A = I_n$ and if the rank is $r < n$ then it is called a *singular matrix* with no regular inverse A^{-1}.

In Example 2.4.3 the rank of the matrix is 2. It is rectangular. It is not a full rank matrix. If the rank was 3 then we would have called it a full rank matrix. Some of the immediate consequences of the concept of rank are the following:

(v) The rank of an $m \times n$ matrix cannot exceed m or n. The maximum value possible is the smaller of m and n. If the maximum is attained then it is a full rank matrix. A nonsingular matrix is also a full rank matrix.

(vi) An $n \times n$ matrix is nonsingular iff its rank is n.

(vii) The rank of a null matrix is zero.

(viii) The rank of an $n \times n$ orthonormal matrix is n.

(ix) The rank of an $n \times n$ idempotent matrix, not equal to I_n, is less than n.

(x) The rank of an $n \times n$ nilpotent matrix is less than n.

(xi) The rank of any matrix A and the rank of cA are the same where c is a nonzero scalar.

Exercises 2.4

2.4.1. Evaluate the ranks of the following matrices:

$$A = \begin{bmatrix} 2 & 1 & -1 & 2 \\ 1 & 1 & -1 & 3 \\ 1 & 0 & 0 & 1 \\ 4 & 2 & -2 & 6 \end{bmatrix}, \quad B = \begin{bmatrix} 2 & 1 & 0 & 0 \\ 2 & -1 & 1 & 2 \\ 3 & 1 & 1 & -1 \\ 4 & 0 & 1 & -1 \end{bmatrix}, \quad C = \begin{bmatrix} 0 & 1 & -1 & 1 \\ 2 & 1 & 0 & -1 \\ 3 & -1 & 0 & 1 \end{bmatrix}.$$

2.4.2. Evaluate the ranks of AB and CB where A, B, C are given in Exercise 2.4.1. What can you say about the rank of a product of two matrices in terms of the ranks of the individual matrices?

2.4.3. Show that the rank of AB, where A and B are general matrices with AB defined, cannot exceed the rank of A or the rank of B.

2.4.4. Compute the ranks of (a) $A + B$, (b) $2A + 3B$ for the A and B given in Exercise 2.4.1.

2.4.5. Show that for two arbitrary matrices A and B such that $\alpha A + \beta B$ is defined, where α and β are scalars, the rank of $\alpha A + \beta B$ cannot exceed the rank of A plus the rank of B.

2.4.6. Show that the only idempotent matrix with full rank is the identity matrix.

2.4.7. If A is a square matrix what can you say about the rank of A^2 in terms of the rank of A. Verify your result for the A of Exercise 2.4.1.

2.4.8. Let A be $n \times n$, X an $n \times 1$ vector then show that for the system of linear equations $AX = O$ to have a non-null solution (at least one X satisfying $AX = O$ is such that $X \neq O$) the rank of A must be less than n. If the rank is n then the only solution is $X = O$.

2.4.9. Let

$$A = \begin{bmatrix} 1 & 0 & -1 \\ 1 & 1 & 2 \\ 0 & 1 & -1 \end{bmatrix}, \quad X = \begin{bmatrix} x_1 \\ x_2 \\ x_3 \end{bmatrix}$$

and consider the equation $AX = \lambda X$. Find the values of λ so that the equation $AX = \lambda X$ has a non-null solution $(X \neq O)$.

2.4.10. Evaluate the ranks of the following $n \times n$ matrices:

$$A = \begin{bmatrix} a & b & \ldots & b \\ b & a & \ldots & b \\ \vdots & \vdots & \ldots & \vdots \\ b & b & \ldots & a \end{bmatrix}, \quad a \neq 0, b \neq 0, a \neq b; \quad B = \begin{bmatrix} 1 & 1 & \ldots & 1 \\ 1 & 1 & \ldots & 1 \\ \vdots & \vdots & \ldots & \vdots \\ 1 & 1 & \ldots & 1 \end{bmatrix};$$

$$C = \begin{bmatrix} 1 - \frac{1}{n} & -\frac{1}{n} \ldots & -\frac{1}{n} \\ \vdots & \ldots & \vdots \\ -\frac{1}{n} & -\frac{1}{n} & \ldots & 1 - \frac{1}{n} \end{bmatrix}; \quad G = \begin{bmatrix} 1 & a_1 & a_1^2 & \ldots & a_1^{n-1} \\ 1 & a_2 & a_2^2 & \ldots & a_2^{n-1} \\ \vdots & \vdots & \vdots & \ldots & \vdots \\ a & a_n & a_n^2 & \ldots & a_n^{n-1} \end{bmatrix},$$

a_j's are distinct and nonzero.

2.4.11. What can you say about the rank of a singly stochastic matrix (sum of the elements in each row or each column is 1). Verify your result for a 2×2 and 3×3 matrices.

2.4.12. Let A and B be nonsingular $n \times n$ matrices. Show that

$$AB, \quad AB^{-1}, \quad A^{-1}B, \quad A^{-1}B^{-1}$$

are nonsingular matrices.

2.4.13. Let A and B be $n \times n$ nonsingular matrices. Show that $A + B$ and $A - B$ need not be nonsingular. Give two such examples of A and B.

2.4.14. If A is $m \times n$, $m < n$ and with rank m then show that AA' is nonsingular.

2.4.15. Show that the rank of AB is zero iff $AB = O$.

2.4.16. If A is $m \times n$, $m < n$ and with rank m and if B is $m \times m$ nonsingular and C is $n \times n$ nonsingular is BA of rank m? Is AC of rank m or n or something else?

2.5 Row and column subspaces and null spaces

Here we examine the subspaces generated by the rows of a given matrix, the columns of a given matrix and subspaces which are orthogonal to these. Then we examine the bases and dimensions of these subspaces.

2.5.1 The row and column subspaces

Consider the subspaces generated by the row vectors of a given matrix A by the following basic operations of scalar multiplication and addition. Let this subspace be denoted by S_1. Then S_1 is called the *row subspace of the matrix A*. Thus every linear combination of the row vectors of A is in S_1. That is, all the rows are in S_1, every scalar multiple of every row vector is in S_1, every sum of such scalar multiples is also in S_1. Now, consider the subspace generated by the column of A. Then this subspace, denoted by S_2, is called the *column subspace of A*. Thus every linear combination of the columns of A is in S_2.

If A is an $m \times n$ matrix then every vector in S_1 is an n-vector whereas every vector in S_2 is an m-vector. Since the maximum number of linearly independent row vectors is r, the rank of A, which is also equal to the maximum number of linearly independent column vectors of A, the dimension of S_1 is r which is also equal to the dimension of S_2. Consider the equation

$$AX = O \tag{2.5.1}$$

where A is the given $m \times n$ matrix and X is an $n \times 1$ vector of unknowns (variables or parameters) and O is the null vector. [$AX = O$ is also called the homogeneous system of linear equations.] In (2.5.1) each row vector is orthogonal to each solution vector X. Consider the set of all such solutions, $\{X\}$, that is the set of all possible X satisfying (2.5.1).

Definition 2.5.1 (The null space). The set of all possible solutions, $\{X\}$, of the equations in (2.5.1) is called the *null space* or the *right null space* of the matrix A.

Definition 2.5.2 (The left null space). The solution space $\{Y\}$ of the equation $A'Y = O$, where A' is the transpose of A, is called the *left null space* of A.

Let us denote the null space or the right null space by S_3 and the left null space by S_4. Then each vector in S_3 is an n-vector and it is orthogonal to each vector in the row subspace S_1. Similarly, each vector in S_4 is an m-vector and it is orthogonal to each vector in the column subspace S_2. The subspaces S_1 and S_3 are orthogonal to each other. Also the subspaces S_2 and S_4 are orthogonal to each other. Not only that, S_3 is the orthogonal complement of S_1 and S_4 is the orthogonal complement of S_2. (See Chapter 1 for the definitions.)

Definition 2.5.3 (Orthogonal complements). If two subspaces of n-vectors S and S^* are such that the dimension of S is r, the dimension of S^* is $n - r$ and further, S and S^* are orthogonal to each other then S^* is called the orthogonal complement of S, and vice versa.

We will use the notation $\perp$ to denote orthogonality. Then we have

$$S_1 \perp S_3 \quad \text{or} \quad S_3 \perp S_1 \quad \text{(orthogonal to each other).}$$
$$S_2 \perp S_4 \quad \text{or} \quad S_4 \perp S_2 \quad \text{(orthogonal to each other).}$$

The maximum number of linearly independent n-vectors is n. Also from Chapter 1 we know that orthogonal vectors are linearly independent. The dimension of S_1 is r, the row rank of A. Since S_3 is the orthogonal complement of S_1 the dimension of S_3 is $n - r$. Similarly the dimension of S_4 is $m - r$, the number of rows minus the column rank of A. Thus we have the following results:

(i) The row subspace of an $m \times n$ matrix A and the null space (the right null space) S_3 are of dimensions r and $n - r$ respectively, where r is the rank of A, and further, S_1 and S_3 are orthogonal complements of each other. Similarly the column subspace S_2 and the left null space S_4 are of dimensions r and $m - r$ respectively and further, S_2 and S_4 are orthogonal complements of each other.

(ii)
$$\text{dimension of } S_1 + \text{dimension of } S_3 = n$$
$$\text{dimension of } S_2 + \text{dimension of } S_4 = m.$$

Example 2.5.1. Obtain 2 bases each for the row subspace S_1, the column subspace S_2 and the null space of A, where

$$A = \begin{bmatrix} 1 & 1 & 0 & 1 \\ 2 & -1 & 1 & 2 \\ 4 & 1 & 1 & 4 \end{bmatrix}.$$

Solutions 2.5.1. In order to determine the rank or establish a basis for the row or column subspace we proceed as before, namely do elementary operations since the linear independence or dependence in the set of row or column vectors is unaltered by these operations. Also we use the same notations as before:

$$-2(1) + (2); \quad -4(1) + (3) \Rightarrow \begin{matrix} 1 & 1 & 0 & 1 \\ 0 & -3 & 1 & 0 \\ 0 & -3 & 1 & 0 \end{matrix}$$

$$-1(2) + (3) \Rightarrow \begin{matrix} 1 & 1 & 0 & 1 \\ 0 & -3 & 1 & 0 \\ 0 & 0 & 0 & 0 \end{matrix} \qquad (a)$$

At this stage we know that the row rank $= r = 2 =$ the column rank. Hence the dimension of the row subspace S_1 is 2, that of the null space S_3 is $4 - 2 = 2$, that of the column subspace S_2 is 2 and that of the left null space S_4 is $3 - 2 = 1$. Any two linearly independent row vectors is a basis for the row subspace S_1. Two such bases are then two such collections of two linearly independent row vectors of A. Many such sets can be constructed. One such set is already in (α). Two such bases are the following:

$$(1) \quad \left\{ \begin{matrix} (1,1,0,1) \\ (2,-1,1,2) \end{matrix} \right\}, \qquad (2) \quad \left\{ \begin{matrix} (1,1,0,1) \\ (4,1,1,4) \end{matrix} \right\}.$$

Any two linearly independent column vectors of A is a basis for the column subspace S_2. Two such bases are the following:

$$(1) \quad \left\{ \begin{pmatrix} 1 \\ 2 \\ 4 \end{pmatrix}, \begin{pmatrix} 1 \\ -1 \\ 1 \end{pmatrix} \right\}, \qquad (2) \quad \left\{ \begin{pmatrix} 1 \\ 2 \\ 4 \end{pmatrix}, \begin{pmatrix} 0 \\ 1 \\ 1 \end{pmatrix} \right\}.$$

Every vector in the null space S_3 is orthogonal to every vector in S_1. There are sets of two such linearly independent vectors. A general method of constructing a basis for S_3 would be to take a basis in S_1 and construct two orthogonal vectors. In our example here a simpler way of doing it is to reduce the reduced form of A, given in (α) above, further. Divide the second row by (-3) and add (-1) times the second row to the first row to obtain

$$\begin{matrix} 1 & 0 & \frac{1}{3} & 1 \\ 0 & 1 & -\frac{1}{3} & 0 \\ 0 & 0 & 0 & 0 \end{matrix} \qquad (\beta)$$

Then a vector in the null space S_3 must be orthogonal to the vectors $(1,0,\frac{1}{3},1)$ as well as to $(0,1,-\frac{1}{3},0)$. Let V, $V' = (a,b,c,d)$ be a vector in S_3. Then

$$\left(1,0,\frac{1}{3},1\right)V = 0 \Rightarrow a + \frac{1}{3}c + d = 0$$

and

$$\left(0,1,-\frac{1}{3},0\right)V = 0 \Rightarrow b - \frac{1}{3}c = 0.$$

Two solutions are

$$(a,b,c,d) = (-2,1,3,1), \ (-3,1,3,2).$$

Two other solutions are

$$(a,b,c,d) = (-4,1,3,3), \ (-5,1,3,4).$$

Hence two bases for the null space S_3 are the following:

$$(1) \quad \left\{ \begin{bmatrix} -2 \\ 1 \\ 3 \\ 1 \end{bmatrix}, \begin{bmatrix} -3 \\ 1 \\ 3 \\ 2 \end{bmatrix} \right\}, \qquad (2) \quad \left\{ \begin{bmatrix} -4 \\ 1 \\ 3 \\ 3 \end{bmatrix}, \begin{bmatrix} -5 \\ 1 \\ 3 \\ 4 \end{bmatrix} \right\}.$$

If two bases for the left null space S_4 are needed then we can start with A. Since we have done row operations to reduce A to (β) hence (β) is not a suitable starting point here. Take any arbitrary column vector U, $U' = (a, b, c)$. Then U must be orthogonal to the columns of A. Take any two linearly independent columns, say the second and the third columns in A. Then take the dot products with U. That is,

$$a - b + c = 0, \quad b + c = 0.$$

Two solutions are the following:

$$\begin{pmatrix} a \\ b \\ c \end{pmatrix} = \begin{pmatrix} 2 \\ 1 \\ -1 \end{pmatrix}, \begin{pmatrix} -2 \\ -1 \\ 1 \end{pmatrix}.$$

Note that the dimension of S_4 is 1 and hence all vectors in S_4 will be scalar multiples of U. Then, for example, two bases for S_4 are the two vectors given above. [A basis consists of only one vector in this case.]

Example 2.5.2. Show that every row vector in A of Example 2.5.1 can be written as a linear function of the vectors in each basis for the row subspace S_1 constructed in Example 2.5.1 and that each column vector in A can be written as a linear function of the vectors in each basis of S_2 there.

Solutions 2.5.2. Let us start with the basis (1) of the row subspace S_1. Then the vector

$$(4, 1, 1, 4) = 2(1, 1, 0, 1) + (2, -1, 1, 2).$$

Hence all the row vectors are expressible as linear functions of the vectors in each basis of S_1. Now let us consider the basis (1) of the column subspace S_2.

$$a \begin{pmatrix} 1 \\ 2 \\ 4 \end{pmatrix} + b \begin{pmatrix} 1 \\ -1 \\ 1 \end{pmatrix} = \begin{pmatrix} 0 \\ 1 \\ 1 \end{pmatrix} \Rightarrow a = \frac{1}{3}, \quad b = -\frac{1}{3}, \quad \text{or}$$

$$\frac{1}{3} \begin{pmatrix} 1 \\ 2 \\ 4 \end{pmatrix} - \frac{1}{3} \begin{pmatrix} 1 \\ -1 \\ 1 \end{pmatrix} = \begin{pmatrix} 0 \\ 1 \\ 1 \end{pmatrix}.$$

$$c \begin{pmatrix} 1 \\ -1 \\ 1 \end{pmatrix} + d \begin{pmatrix} 0 \\ 1 \\ 1 \end{pmatrix} = \begin{pmatrix} 1 \\ 2 \\ 4 \end{pmatrix} \Rightarrow c = 1, \quad d = 3, \quad \text{or}$$

$$\begin{pmatrix} 1 \\ -1 \\ 1 \end{pmatrix} + 3 \begin{pmatrix} 0 \\ 1 \\ 1 \end{pmatrix} = \begin{pmatrix} 1 \\ 2 \\ 4 \end{pmatrix}.$$

2.5.2 Consistency of a system of linear equations

A system of linear equations can be written as $AX = b$, that is,

$$\begin{bmatrix} a_{11} & \cdots & a_{1n} \\ \vdots & \cdots & \vdots \\ a_{m1} & \cdots & a_{mn} \end{bmatrix} \begin{bmatrix} x_1 \\ \vdots \\ x_n \end{bmatrix} = \begin{bmatrix} b_1 \\ \vdots \\ b_m \end{bmatrix} \tag{2.5.2}$$

where A and b are known and X is the unknown part. Writing the equations in (2.5.2) in a slightly different equivalent form we have,

$$x_1 \begin{bmatrix} a_{11} \\ \vdots \\ a_{m1} \end{bmatrix} + x_2 \begin{bmatrix} a_{12} \\ \vdots \\ a_{m2} \end{bmatrix} + \ldots + x_n \begin{bmatrix} a_{1n} \\ \vdots \\ a_{mn} \end{bmatrix} = \begin{bmatrix} b_1 \\ \vdots \\ b_m \end{bmatrix}. \tag{2.5.3}$$

Note that (2.5.2) and (2.5.3) are equivalent representations of $AX = b$. If there is a solution for $AX = b$ then we have a set of numbers for

$$X' = (x_1, \ldots, x_n) = (c_1, \ldots, c_n)$$

and for this set (2.5.3) becomes

$$c_1 \begin{bmatrix} a_{11} \\ \vdots \\ a_{m1} \end{bmatrix} + c_2 \begin{bmatrix} a_{12} \\ \vdots \\ a_{m2} \end{bmatrix} + \ldots + c_n \begin{bmatrix} a_{1n} \\ \vdots \\ a_{mn} \end{bmatrix} = \begin{bmatrix} b_1 \\ \vdots \\ b_m \end{bmatrix}.$$

In other words, the vector b is a linear combination of the column vectors of the matrix A or b is an element of the column subspace S_2 of A. Therefore for the system $AX = b$ to have a solution, b must be an element of S_2.

(iii) The system of linear equations $AX = b$ is consistent (have at least one solution) if and only if b is an element of the column subspace S_2 of A.

Now, let us examine the general solution for the system $AX = b$. Consider the homogeneous system

$$AX = O \tag{2.5.4}$$

and consider all solutions of this homogeneous system. There are $n - r$ linearly independent solutions for (2.5.4) when A is $m \times n$ and when r is the rank of A. Let one set of such linearly independent solutions be denoted by $X_{(1)}, \ldots, X_{(n-r)}$. That is,

$$AX_{(i)} = O, \quad \text{for } i = 1, 2, \ldots, n - r.$$

Take a general linear combination of $X_{(i)}$, $i = 1, \ldots, n - r$, say

$$Y = d_1 X_{(1)} + \cdots + d_{n-r} X_{(n-r)} \tag{2.5.5}$$

where $d_1, \dots, d_{n-r}$ are arbitrary constants. Then

$$AY = O. \tag{2.5.6}$$

Let X_0 be a particular solution of $AX = b$. That is,

$$AX_0 = b. \tag{2.5.7}$$

Combining (2.5.6) and (2.5.7) we have

$$A(Y + X_0) = O + b = b \tag{2.5.8}$$

or $Y + X_0$ is a solution of the starting system $AX = b$. Thus a general solution of $AX = b$ is available as follows:

(iv) The general solution of the linear system $AX = b$ is $Y + X_0$ where X_0 is a particular solution of $AX = b$ and Y is the general solution of $AX = O$.

Example 2.5.3. Consider the matrix A in Example 2.5.1 and consider the system of linear equations

$$\begin{bmatrix} 1 & 1 & 0 & 1 \\ 2 & -1 & 1 & 2 \\ 4 & 1 & 1 & 4 \end{bmatrix} \begin{bmatrix} x_1 \\ x_2 \\ x_3 \\ x_4 \end{bmatrix} = \begin{bmatrix} b_1 \\ b_2 \\ b_3 \end{bmatrix} \quad \text{or} \quad AX = b.$$

(i) Find one vector b_0 so that the system is consistent.
(ii) Find the general solution of the system $AX = b_0$.

Solution 2.5.3. (i) For the system to be consistent b must be an element of the column subspace S_2 of A. That is, b must be a linear function of the vectors in a basis of S_2. Since the rank of A is 2 any two linearly independent columns of A is a basis of S_2. For example take

$$\begin{pmatrix} 1 \\ 2 \\ 4 \end{pmatrix}, \quad \begin{pmatrix} 1 \\ -1 \\ 1 \end{pmatrix}$$

and b as the sum,

$$b = \begin{pmatrix} 1 \\ 2 \\ 4 \end{pmatrix} + \begin{pmatrix} 1 \\ -1 \\ 1 \end{pmatrix} = \begin{pmatrix} 2 \\ 1 \\ 5 \end{pmatrix}.$$

Many such b's can be constructed. (ii) A particular solution of this system

$$\begin{bmatrix} 1 & 1 & 0 & 1 \\ 2 & -1 & 1 & 2 \\ 4 & 1 & 1 & 4 \end{bmatrix} \begin{bmatrix} x_1 \\ x_2 \\ x_3 \\ x_4 \end{bmatrix} = \begin{bmatrix} 2 \\ 1 \\ 5 \end{bmatrix} \tag{2.5.9}$$

can be taken either from the representation (α) or from (β) of Example 2.5.1. Taking from (β) a particular solution $X_0' = (0, 2, 3, 0)$. The general solution of the system $AX = O$ is a general linear combination of the vectors in a basis for the null space S_3. Taking the basis (1) of S_3 in Example 2.5.1 a general linear combination is

$$c \begin{pmatrix} -2 \\ 1 \\ 3 \\ 1 \end{pmatrix} + d \begin{pmatrix} -3 \\ 1 \\ 3 \\ 2 \end{pmatrix}$$

where c and d are arbitrary constants. Therefore the general solution of the system (2.5.9), denoted by Z, is

$$Z = \begin{bmatrix} 0 \\ 2 \\ 3 \\ 0 \end{bmatrix} + c \begin{bmatrix} -2 \\ 1 \\ 3 \\ 1 \end{bmatrix} + d \begin{bmatrix} -3 \\ 1 \\ 3 \\ 2 \end{bmatrix} = \begin{bmatrix} -2c - 3d \\ c + d + 2 \\ 3c + 3d + 3 \\ c + 2d \end{bmatrix}$$

where c and d are arbitrary constants. [The student may substitute back in (2.5.9) and verify the result.]

Definition 2.5.4. A linear system of equations $AX = b$, where A is $n \times n$, is said to be *a singular system* if A is singular, and *a nonsingular system* if A is nonsingular.

Exercises 2.5

2.5.1. Show that interchange of two rows in an $m \times n$ matrix can be effected by pre-multiplying it by a product of elementary matrices.

2.5.2. Compute the ranks of the following matrices:

$$A = \begin{bmatrix} 3 & 1 & 0 & -1 & 2 \\ 4 & 0 & 1 & -1 & 1 \\ 1 & 1 & 2 & -2 & 1 \\ 2 & 4 & 1 & 0 & 1 \\ 1 & 0 & 1 & 1 & -1 \end{bmatrix}, \quad B = \begin{bmatrix} 1 & -1 & 1 & 2 \\ 2 & 1 & -1 & 1 \\ 1 & 2 & -1 & 1 \\ 4 & 2 & -1 & 4 \end{bmatrix},$$

$$C = \begin{bmatrix} 2 & 0 & 1 & -1 \\ 1 & 2 & 1 & 1 \\ 3 & 2 & 2 & 0 \\ 5 & 2 & 3 & -1 \end{bmatrix}.$$

2.5.3. Construct 3 bases (not scalar multiples of each other) for (1) the row subspace S_1 and (2) the column subspace S_2 for each of the matrices in Exercise 2.5.2.

2.5.4. Construct 3 bases for the null space S_3 and the left null space S_4 for each of the matrices in Exercise 2.5.2.

2.5.5. Check for linear independence and determine the maximum number of linearly independent vectors in each of the following sets:

(a) $U_1 = (1, 1, -1, 2)$, $U_2 = (1, 1, 0, 1)$,
      $U_3 = (2, -1, 1, 4)$, $U_4 = (1, 2, 1, 2)$,
      $U_5 = (0, 1, -1, 1)$.

(b) $V_1 = \begin{bmatrix} 1 \\ -1 \\ 2 \\ 1 \\ 1 \end{bmatrix}$, $V_2 = \begin{bmatrix} 2 \\ -1 \\ 1 \\ 2 \\ 1 \end{bmatrix}$, $V_3 = \begin{bmatrix} 3 \\ 1 \\ -1 \\ 1 \\ 2 \end{bmatrix}$, $V_4 = \begin{bmatrix} 0 \\ 1 \\ 1 \\ 0 \\ 1 \end{bmatrix}$.

2.5.6. For each of the matrices in Exercise 2.5.2 write the systems of linear equations as $AX = b_1$, $BY = b_2$, $CZ = b_3$. (a) Construct two different vectors for each of b_1, b_2, b_3 so that the systems are consistent. (b) Construct two vectors for each of b_1, b_2, b_3 so that the systems are not consistent. (c) For your answers in (a) evaluate the most general solutions for each of the systems.

2.5.7. Determine the product of the basic elementary matrices of the E and F types so that the matrix

$$\begin{bmatrix} 0 & 1 & 2 & 3 \\ 1 & -1 & 2 & 4 \\ 2 & 1 & -1 & 1 \\ 3 & 2 & 1 & 4 \end{bmatrix} \rightarrow \begin{bmatrix} 0 & 1 & 2 & 3 \\ 2 & 1 & -1 & 1 \\ 3 & 2 & 1 & 4 \\ 1 & -1 & 2 & 4 \end{bmatrix}.$$

2.5.8. Let U and V be $n \times 1$ vectors and let $A = VV'$, $B = UV'$. Show that the $n \times n$ matrices A and B have rank 1.

2.5.9. Show that the Vandermonde's matrix

$$V = \begin{bmatrix} 1 & a_1 & a_1^2 & \cdots & a_1^{n-1} \\ 1 & a_2 & a_2^2 & \cdots & a_2^{n-1} \\ \vdots & \vdots & \vdots & \cdots & \vdots \\ 1 & a_n & a_n^2 & \cdots & a_n^{n-1} \end{bmatrix}$$

is nonsingular, where $a_i \neq 0$, $i = 1, \dots, n$ and a_j's are distinct.

2.5.10. For a 3×3 non-null matrix A, each row dot product with the vector $(1, -1, 1)$ is zero. What can you say about the rank of A? What can you say about the dimension of the null space?

2.5.11. Determine the rank and construct a basis each for (1) row subspace, (2) column subspace, (3) right null space, (4) left null space of the matrix

$$A = \begin{bmatrix} a & b & \cdots & b \\ b & a & \cdots & b \\ \vdots & \vdots & \ddots & \vdots \\ b & b & \cdots & a \end{bmatrix}, \quad a \ne b,\ a \ne 0,\ b \ne 0.$$

2.5.12. In a non-null 3×3 matrix A each row is orthogonal to the vectors

$$V_1 = \begin{pmatrix} 1 \\ 0 \\ 1 \end{pmatrix}, \quad V_2 = \begin{pmatrix} 1 \\ 1 \\ 1 \end{pmatrix}.$$

Determine the rank of A and construct a basis for the row subspace of A.

2.5.13. Consider the linear system of equations $AX = b$ where

$$A = \begin{bmatrix} 1 & 1 & 1 \\ 2 & 1 & 2 \\ 1 & 0 & 1 \end{bmatrix}.$$

(a) Construct two examples of b where the system $AX = b$ is not consistent; (b) Construct two examples of b where the system is consistent; (c) In (b) construct one basis each for the right null space of A; (d) In (b) construct the general solution in each case.

2.5.14. Let J be the $n \times 1$ column vector of unities, that is, $J' = (1, 1, \ldots, 1)$. Let

$$A = \frac{1}{n} JJ' \quad \text{and} \quad B = I_n - \frac{1}{n} JJ'.$$

Evaluate the ranks of A and B.

2.5.15. Construct one basis each for the right null space of the matrices A and B in Exercise 2.5.14.

2.5.16. Show that any matrix A of rank r can be written as

$$A = R \begin{bmatrix} I_r & O \\ O & O \end{bmatrix} S$$

where R and S are nonsingular matrices.

2.5.17. If A and B are rectangular matrices of the same rank then show that there exist two nonsingular matrices R and S such that

$$B = RAS.$$

2.5.18. If A and B are nonsingular matrices and C is any matrix then show that the matrices C, AC, CB, ACB have the same rank as long as the multiplications are defined.

2.6 Permutations and elementary operations on the right

So far we were dealing exclusively with operating on the left with elementary matrices, that is, premultiplications of a matrix with elementary matrices. By this time the student may be very clear about the effects of such premultiplications on a given matrix. Now we consider postmultiplications by elementary matrices.

2.6.1 Permutations

It is already seen that if we wish to permute or interchange rows we can effect that with a product of the basic elementary matrices. Consider the matrices

$$G = \begin{bmatrix} 1 & 0 & 0 \\ 0 & 0 & 1 \\ 0 & 1 & 0 \end{bmatrix}, \quad F_1 = \begin{bmatrix} 1 & 0 & 0 \\ 0 & 1 & 0 \\ 0 & 1 & 1 \end{bmatrix}, \quad E_1 = \begin{bmatrix} 1 & 0 & 0 \\ 0 & 1 & 0 \\ 0 & 0 & -1 \end{bmatrix},$$

$$F_2 = \begin{bmatrix} 1 & 0 & 0 \\ 0 & 1 & 1 \\ 0 & 0 & 1 \end{bmatrix}, \quad E_2 = \begin{bmatrix} 1 & 0 & 0 \\ 0 & -1 & 0 \\ 0 & 0 & 1 \end{bmatrix},$$

$$F_3 = \begin{bmatrix} 1 & 0 & 0 \\ 0 & 1 & 0 \\ 0 & 1 & 1 \end{bmatrix}, \quad E_{3=} \begin{bmatrix} 1 & 0 & 0 \\ 0 & 1 & 0 \\ 0 & 0 & -1 \end{bmatrix}.$$

Then it is easily seen that

$$G = E_3 F_3 E_2 F_2 E_1 F_1,$$

a product of the basic elementary matrices. If we premultiply an arbitrary matrix with G then we have

$$\begin{bmatrix} 1 & 0 & 0 \\ 0 & 0 & 1 \\ 0 & 1 & 0 \end{bmatrix} \begin{bmatrix} a_{11} & a_{12} & a_{13} \\ a_{21} & a_{22} & a_{23} \\ a_{31} & a_{32} & a_{33} \end{bmatrix} = \begin{bmatrix} a_{11} & a_{12} & a_{13} \\ a_{31} & a_{32} & a_{33} \\ a_{21} & a_{32} & a_{33} \end{bmatrix}.$$

That is, the second and the third rows are interchanged as it is the case in G or G can be looked upon as an identity matrix with the second and the third rows interchanged. Thus, in general, permutations of the rows can be achieved by premultiplication with a product of elementary matrices.

2.6.2 Postmultiplications by elementary matrices

The technique of postmultiplications is postponed this far mainly to give the student time to have the premultiplication ideas to sink in clearly. Otherwise there is great chance of getting confused about the effects of pre and post multiplications. Let us start with the basic elementary matrices of the E and F types. Let

$$E_1 = \begin{bmatrix} 1 & 0 & 0 \\ 0 & 4 & 0 \\ 0 & 0 & 1 \end{bmatrix}$$

a matrix obtained by multiplying the second row of an identity matrix with 4. E_1 operating on the left has the effect that the second row is multiplied by 4. Observe that E_1 can also be looked upon as created by multiplying the second column of an identity matrix by 4. Let us see what happens when E_1 operates on the right:

$$AE_1 = \begin{bmatrix} a_{11} & a_{12} & a_{13} \\ a_{21} & a_{22} & a_{23} \\ a_{31} & a_{32} & a_{33} \end{bmatrix} \begin{bmatrix} 1 & 0 & 0 \\ 0 & 4 & 0 \\ 0 & 0 & 1 \end{bmatrix} = \begin{bmatrix} a_{11} & 4a_{12} & a_{13} \\ a_{21} & 4a_{22} & a_{23} \\ a_{31} & 4a_{32} & a_{33} \end{bmatrix}.$$

That is, the second column is multiplied by 4. Thus we have the following result:

(i) If E is a basic elementary matrix created by multiplying the i-th column of an identity matrix by the nonzero scalar c then any matrix A postmultiplied by E will have the effect that the i-th column of A is multiplied by c.

Now, let us consider an elementary matrix of the F type. Let

$$F_1 = \begin{bmatrix} 1 & 0 & 0 \\ 1 & 1 & 0 \\ 0 & 0 & 1 \end{bmatrix},$$

obtained by adding the first row of I_3 to the second row, which can also be considered as obtained by adding the second column of I_3 to the first column. Let us see what happens if A is postmultiplied by F_1:

$$AF_1 = \begin{bmatrix} a_{11} & a_{12} & a_{13} \\ a_{21} & a_{22} & a_{23} \\ a_{31} & a_{32} & a_{33} \end{bmatrix} \begin{bmatrix} 1 & 0 & 0 \\ 1 & 1 & 0 \\ 0 & 0 & 1 \end{bmatrix} = \begin{bmatrix} a_{11} + a_{12} & a_{12} & a_{13} \\ a_{21} + a_{22} & a_{22} & a_{23} \\ a_{31} + a_{32} & a_{32} & a_{33} \end{bmatrix}.$$

That is, the second column is added to the first column, exactly as F_1 is obtained by adding the second column of I_3 to the first column. Now see the net effect of operating on the right with F_1' the transpose of F_1. This transpose could also be looked upon as obtained by adding the first column of I_3 to the second column:

$$AF_1' = \begin{bmatrix} a_{11} & a_{12} & a_{13} \\ a_{21} & a_{22} & a_{23} \\ a_{31} & a_{32} & a_{33} \end{bmatrix} \begin{bmatrix} 1 & 1 & 0 \\ 0 & 1 & 0 \\ 0 & 0 & 1 \end{bmatrix} = \begin{bmatrix} a_{11} & a_{12} + a_{11} & a_{13} \\ a_{21} & a_{22} + a_{21} & a_{13} \\ a_{31} & a_{32} + a_{31} & a_{33} \end{bmatrix}.$$

Now the first column is added to the second column. Thus F_1' is the matrix to operate on the right if the same type of effects on the columns is needed as the effects on the rows when F_1 operates on the left.

(ii) Let F be an elementary matrix obtained by adding the i-th row to the j-th row in I_n. Let A be an $n \times n$ arbitrary matrix. Then FA has the effect that the i-th row of A is added to the j-th row of A. AF has the effect that the j-th column is added to the i-th column. AF' has the effect that the i-th column is added to the j-th column. [F' is also the elementary matrix obtained by adding the i-th column to the j-th column of I_n.]

(iii) If F_2 is an elementary matrix obtained by adding the i-th column of I_n to the j-th column of I_n then any arbitrary $n \times n$ matrix A postmultiplied by F_2 has the effect that the i-th column of A is added to the j-th column of A.

When premultiplying a matrix A by an elementary matrix then create the elementary matrix by operating on the rows of the identity matrix I. The effect will be exactly the same on the rows of A. When postmultiplying A with an elementary matrix then create the elementary matrix by operating on the columns of I. The effect will be exactly the same on the columns of A. The student is urged to memorize these properties. These will be helpful when trying to reduce a matrix to a triangular or diagonal form.

Example 2.6.1. Write the following symmetric matrix A in the form $A = QDQ'$ where D is a diagonal matrix, Q is a nonsingular matrix and Q' the transpose of Q, where

$$A = \begin{bmatrix} 1 & 0 & -1 \\ 0 & 2 & 4 \\ -1 & 4 & 4 \end{bmatrix}.$$

Solution 2.6.1. Let

$$F_1 = \begin{bmatrix} 1 & 0 & 0 \\ 0 & 1 & 0 \\ 1 & 0 & 1 \end{bmatrix} \quad \text{or} \quad (1) + (3) \Rightarrow$$

$$F_1 A = \begin{bmatrix} 1 & 0 & -1 \\ 0 & 2 & 4 \\ 0 & 4 & 3 \end{bmatrix}, \quad F_1 A F_1' = \begin{bmatrix} 1 & 0 & 0 \\ 0 & 2 & 4 \\ 0 & 4 & 3 \end{bmatrix}.$$

Observe that by operating on the left by F_1 and on the right by its transpose F_1' the symmetric nature of the resulting matrix is maintained. If we had operated on the right first and then on the left we would have got the same result. [The student may verify this aspect.] Now our aim is to get rid off the elements at the $(3, 2)$-th and $(2, 3)$-th positions and at the same time maintaining symmetry. This can be achieved by doing the following operations. Let

$$G_1 = \begin{bmatrix} 1 & 0 & 0 \\ 0 & 1 & 0 \\ 0 & -2 & 1 \end{bmatrix} \quad \text{or} \quad -2(2) + (3) \Rightarrow$$

$$G_1(F_1AF_1') = \begin{bmatrix} 1 & 0 & 0 \\ 0 & 2 & 4 \\ 0 & 0 & -5 \end{bmatrix},$$

$$G_1(F_1AF_1')G_1' = \begin{bmatrix} 1 & 0 & 0 \\ 0 & 2 & 0 \\ 0 & 0 & -5 \end{bmatrix}.$$

That is,

$$A = F_1^{-1}G_1^{-1} \begin{bmatrix} 1 & 0 & 0 \\ 0 & 2 & 0 \\ 0 & 0 & -5 \end{bmatrix} (F_1^{-1}G_1^{-1})'.$$

But, by inspection

$$F_1^{-1} = \begin{bmatrix} 1 & 0 & 0 \\ 0 & 1 & 0 \\ -1 & 0 & 1 \end{bmatrix}, \quad G_1^{-1} = \begin{bmatrix} 1 & 0 & 0 \\ 0 & 1 & 0 \\ 0 & 2 & 1 \end{bmatrix} \Rightarrow$$

$$F_1^{-1}G_1^{-1} = \begin{bmatrix} 1 & 0 & 0 \\ 0 & 1 & 0 \\ -1 & 0 & 1 \end{bmatrix} \begin{bmatrix} 1 & 0 & 0 \\ 0 & 1 & 0 \\ 0 & 2 & 1 \end{bmatrix} = \begin{bmatrix} 1 & 0 & 0 \\ 0 & 1 & 0 \\ -1 & 2 & 1 \end{bmatrix} = Q.$$

Therefore

$$A = QDQ', \quad Q = \begin{bmatrix} 1 & 0 & 0 \\ 0 & 1 & 0 \\ -1 & 2 & 1 \end{bmatrix}, \quad D = \begin{bmatrix} 1 & 0 & 0 \\ 0 & 2 & 0 \\ 0 & 0 & -5 \end{bmatrix}.$$

[The student may verify the result by straight multiplication of Q, D and Q'.]

Note that when the matrix A is symmetric we need to consider only a triangulariza-tion of A by premultiplication with elementary matrices. The transpose of the product of the elementary matrices on the left is going to be the matrix on the right. Hence there is no need to evaluate the inverses of the transposes of the elementary matrices also. In the above example we obtained $Q = F_1^{-1}G_1^{-1}$ and Q' is the matrix on the right. Also, Q being a product of elementary matrices will be nonsingular whereas the diagonal matrix D need not be nonsingular. If the matrix A is singular then there will be at least one zero diagonal element in D and if A is nonsingular then all diagonal elements in D will be nonzeros.

Example 2.6.2. Reduce the following matrix A to the form $A = PDQ$ where P and Q are nonsingular matrices and D is a diagonal matrix, where

$$A = \begin{bmatrix} 1 & 0 & -1 \\ 0 & 2 & 4 \\ -1 & 2 & 4 \end{bmatrix}.$$

Solution 2.6.2. Note that A is not symmetric here. Let

$$F_1 = \begin{bmatrix} 1 & 0 & 0 \\ 0 & 1 & 0 \\ 1 & 0 & 1 \end{bmatrix}.$$

Then

$$F_1 A F_1' = \begin{bmatrix} 1 & 0 & 0 \\ 0 & 1 & 0 \\ 1 & 0 & 1 \end{bmatrix} \begin{bmatrix} 1 & 0 & -1 \\ 0 & 2 & 4 \\ -1 & 2 & 4 \end{bmatrix} \begin{bmatrix} 1 & 0 & 1 \\ 0 & 1 & 0 \\ 0 & 0 & 1 \end{bmatrix} = \begin{bmatrix} 1 & 0 & 0 \\ 0 & 2 & 4 \\ 0 & 2 & 3 \end{bmatrix}.$$

Let

$$G_2 = \begin{bmatrix} 1 & 0 & 0 \\ 0 & 1 & 0 \\ 0 & -1 & 1 \end{bmatrix}, \quad G_3 = \begin{bmatrix} 1 & 0 & 0 \\ 0 & 1 & -2 \\ 0 & 0 & 1 \end{bmatrix}.$$

Then

$$G_2(F_1 A F_1') = \begin{bmatrix} 1 & 0 & 0 \\ 0 & 1 & 0 \\ 0 & -1 & 1 \end{bmatrix} \begin{bmatrix} 1 & 0 & 0 \\ 0 & 2 & 4 \\ 0 & 2 & 3 \end{bmatrix} = \begin{bmatrix} 1 & 0 & 0 \\ 0 & 2 & 4 \\ 0 & 0 & -1 \end{bmatrix},$$

$$G_2(F_1 A F_1') G_3 = \begin{bmatrix} 1 & 0 & 0 \\ 0 & 2 & 4 \\ 0 & 0 & -1 \end{bmatrix} \begin{bmatrix} 1 & 0 & 0 \\ 0 & 1 & -2 \\ 0 & 0 & 1 \end{bmatrix} = \begin{bmatrix} 1 & 0 & 0 \\ 0 & 2 & 0 \\ 0 & 0 & -1 \end{bmatrix}.$$

Hence

$$A = F_1^{-1} G_2^{-1} D G_3^{-1} (F_1')^{-1}.$$

The inverses are available by inspection. That is,

$$F_1^{-1} = \begin{bmatrix} 1 & 0 & 0 \\ 0 & 1 & 0 \\ -1 & 0 & 1 \end{bmatrix}, \quad G_2^{-1} = \begin{bmatrix} 1 & 0 & 0 \\ 0 & 1 & 0 \\ 0 & 1 & 1 \end{bmatrix},$$

$$G_3^{-1} = \begin{bmatrix} 1 & 0 & 0 \\ 0 & 1 & 2 \\ 0 & 0 & 1 \end{bmatrix}, \quad D = \begin{bmatrix} 1 & 0 & 0 \\ 0 & 2 & 0 \\ 0 & 0 & -1 \end{bmatrix}.$$

Then

$$P = F_1^{-1} G_2^{-1} = \begin{bmatrix} 1 & 0 & 0 \\ 0 & 1 & 0 \\ -1 & 1 & 1 \end{bmatrix},$$

$$Q = G_3^{-1}(F_1^{-1})'$$

$$= \begin{bmatrix} 1 & 0 & 0 \\ 0 & 1 & 2 \\ 0 & 0 & 1 \end{bmatrix} \begin{bmatrix} 1 & 0 & -1 \\ 0 & 1 & 0 \\ 0 & 0 & 1 \end{bmatrix} = \begin{bmatrix} 1 & 0 & -1 \\ 0 & 1 & 2 \\ 0 & 0 & 1 \end{bmatrix},$$

$$A = PDQ.$$

[In order to check for possible computational errors verify the result by straight multiplication.]

(iv) If there is a zero at the $(1,1)$-th position then add the i-th row to the first row as well as the i-th column to the first column if the element at the $(i,1)$-th position is nonzero. This will bring a nonzero element to the $(1,1)$-th position as well as keep the symmetry. If symmetry is not to be maintained then do only the first operation above. Repeat a similar process at every stage when dealing with a zero diagonal element.

Example 2.6.3. Reduce the same A in Example 2.6.2 to the form $A = DQ$ where D is diagonal and Q is nonsingular.

Solution 2.6.3. Since we want a form DQ we operate only on the right of A with elementary matrices. Let

$$F_1 = \begin{bmatrix} 1 & 0 & 1 \\ 0 & 1 & 0 \\ 0 & 0 & 1 \end{bmatrix}, \quad E_1 = \begin{bmatrix} 1 & 0 & 0 \\ 0 & \frac{1}{2} & 0 \\ 0 & 0 & 1 \end{bmatrix}, \quad F_2 = \begin{bmatrix} 1 & 0 & 0 \\ 1 & 1 & 0 \\ 0 & 0 & 1 \end{bmatrix},$$

$$G_3 = \begin{bmatrix} 1 & 0 & 0 \\ 0 & 1 & -4 \\ 0 & 0 & 1 \end{bmatrix}, \quad F_4 = \begin{bmatrix} 1 & 0 & 0 \\ 0 & 1 & 0 \\ 0 & 1 & 1 \end{bmatrix}, \quad G_5 = \begin{bmatrix} 1 & 0 & 0 \\ -1 & 1 & 0 \\ 0 & 0 & 1 \end{bmatrix}.$$

These are created from an identity matrix as follows:
  F_1: first column is added to the third column;
  E_1: second column is divided by 2;
  F_2: second column is added to the first column;
  G_3: (-4) times the second column is added to the third column;
  F_4: the third column is added to the second column;
  G_5: (-1) times the second column is added to the first column.

When postmultiplying A with these elementary matrices the effects will be exactly the same on the columns of A:

$$AF_1 = \begin{bmatrix} 1 & 0 & -1 \\ 0 & 2 & 4 \\ -1 & 2 & 4 \end{bmatrix} \begin{bmatrix} 1 & 0 & 1 \\ 0 & 1 & 0 \\ 0 & 0 & 1 \end{bmatrix} = \begin{bmatrix} 1 & 0 & 0 \\ 0 & 2 & 4 \\ -1 & 2 & 3 \end{bmatrix};$$

$$AF_1E_1 = \begin{bmatrix} 1 & 0 & 0 \\ 0 & 1 & 4 \\ -1 & 1 & 3 \end{bmatrix}; \quad AF_1E_1F_2 = \begin{bmatrix} 1 & 0 & 0 \\ 1 & 1 & 4 \\ 0 & 1 & 3 \end{bmatrix};$$

$$AF_1E_1F_2G_3 = \begin{bmatrix} 1 & 0 & 0 \\ 1 & 1 & 0 \\ 0 & 1 & -1 \end{bmatrix};$$

$$AF_1E_1F_2G_3F_4 = \begin{bmatrix} 1 & 0 & 0 \\ 1 & 1 & 0 \\ 0 & 0 & -1 \end{bmatrix}.$$

Hence

$$AF_1E_1F_2G_3F_4G_5 = \begin{bmatrix} 1 & 0 & 0 \\ 0 & 1 & 0 \\ 0 & 0 & -1 \end{bmatrix}.$$

Therefore,

$$A = DG_5^{-1}F_4^{-1}G_3^{-1}F_2^{-1}E_1^{-1}F_1^{-1}$$

where

$$D = \begin{bmatrix} 1 & 0 & 0 \\ 0 & 1 & 0 \\ 0 & 0 & -1 \end{bmatrix}, \quad F_1^{-1} = \begin{bmatrix} 1 & 0 & -1 \\ 0 & 1 & 0 \\ 0 & 0 & 1 \end{bmatrix}, \quad E_1^{-1} = \begin{bmatrix} 1 & 0 & 0 \\ 0 & 2 & 0 \\ 0 & 0 & 1 \end{bmatrix},$$

$$F_2^{-1} = \begin{bmatrix} 1 & 0 & 0 \\ -1 & 1 & 0 \\ 0 & 0 & 1 \end{bmatrix}, \quad G_3^{-1} = \begin{bmatrix} 1 & 0 & 0 \\ 0 & 1 & 4 \\ 0 & 0 & 1 \end{bmatrix}, \quad F_4^{-1} = \begin{bmatrix} 1 & 0 & 0 \\ 0 & 1 & 0 \\ 0 & -1 & 1 \end{bmatrix},$$

$$G_5^{-1} = \begin{bmatrix} 1 & 0 & 0 \\ 1 & 1 & 0 \\ 0 & 0 & 1 \end{bmatrix}, \quad E_1^{-1}F_1^{-1} = \begin{bmatrix} 1 & 0 & -1 \\ 0 & 2 & 0 \\ 0 & 0 & 1 \end{bmatrix}.$$

$$F_2^{-1}E_1^{-1}F_1^{-1} = F_2^{-1}\begin{bmatrix} 1 & 0 & -1 \\ 0 & 2 & 0 \\ 0 & 0 & 1 \end{bmatrix} = \begin{bmatrix} 1 & 0 & -1 \\ -1 & 2 & 1 \\ 0 & 0 & 1 \end{bmatrix}.$$

(Remember that this is a premultiplication; the effect is on the rows.)

$$G_3^{-1}F_2^{-1}E_1^{-1}F_1^{-1} = G_3^{-1}\begin{bmatrix} 1 & 0 & -1 \\ -1 & 2 & 1 \\ 0 & 0 & 1 \end{bmatrix} = \begin{bmatrix} 1 & 0 & -1 \\ -1 & 2 & 5 \\ 0 & 0 & 1 \end{bmatrix},$$

$$F_4^{-1}G_3^{-1}F_2^{-1}E_1^{-1}F_1^{-1} = F_4^{-1}\begin{bmatrix} 1 & 0 & -1 \\ -1 & 2 & 5 \\ 0 & 0 & 1 \end{bmatrix} = \begin{bmatrix} 1 & 0 & -1 \\ -1 & 2 & 5 \\ 1 & -2 & -4 \end{bmatrix},$$

$$G_5^{-1}F_4^{-1}G_3^{-1}F_2^{-1}E_1^{-1}F_1^{-1} = G_5^{-1}\begin{bmatrix} 1 & 0 & -1 \\ -1 & 2 & 5 \\ 1 & -2 & -4 \end{bmatrix} = \begin{bmatrix} 1 & 0 & -1 \\ 0 & 2 & 4 \\ 1 & -2 & -4 \end{bmatrix}$$

$$= Q.$$

Thus we have written

$$A = DQ$$

where

$$D = \begin{bmatrix} 1 & 0 & 0 \\ 0 & 1 & 0 \\ 0 & 0 & -1 \end{bmatrix} \quad \text{and} \quad Q = \begin{bmatrix} 1 & 0 & -1 \\ 0 & 2 & 4 \\ 1 & -2 & -4 \end{bmatrix}.$$

By operating on the right as well as on the left of a square matrix A by elementary matrices, as well as operating on the left alone (considered in the previous sections) and operating on the right alone we can reduce a given matrix A to the following forms:

$A = PD$, P nonsingular, D diagonal;   (2.6.1)

$A = DQ$, D diagonal, Q nonsingular;   (2.6.2)

$A = RDS$, D diagonal, R, S nonsingular;   (2.6.3)

$A = ZDZ'$, Z nonsingular, D diagonal, when $A = A'$ (2.6.4)

$A = LU$, L lower and U upper triangular,   (2.6.5)

$A = L_1DU_1$, L_1 lower and U_1 upper triangular, D diagonal,   (2.6.6)

where the representations in (2.6.5) and (2.6.6) are not always possible. They depend on the nature of A. As an application of (2.6.4) we can consider reduction of quadratic forms to their canonical forms.

2.6.3 Reduction of quadratic forms to their canonical forms

Let

$$u = X'AX, \quad A = A'$$

be a general quadratic form, where X is an $n \times 1$ vector, A is an $n \times n$ matrix of known elements and, as we have seen before, A can be taken as a symmetric matrix without any loss of generality. By using (2.6.4) write A as

$$A = PDP'$$

where P is a nonsingular matrix, D is a diagonal matrix and P' is the transpose of P. Then the quadratic form

$$u = X'AX = X'PDP'X = Y'DY$$

where

$$Y = P'X = \begin{pmatrix} y_1 \\ \vdots \\ y_n \end{pmatrix}.$$

Let the diagonal elements in D be $d_1, \ldots, d_n$ (some of these may be zeros depending upon the singularity of A). Then

$$u = (y_1, \ldots, y_n) \begin{bmatrix} d_1 & 0 & \cdots & 0 \\ \vdots & \vdots & \cdots & \vdots \\ 0 & 0 & \cdots & d_n \end{bmatrix} \begin{bmatrix} y_1 \\ \vdots \\ y_n \end{bmatrix}$$

$$= d_1 y_1^2 + \cdots + d_n y_n^2 \tag{2.6.7}$$

a linear combination of the squares of y_j's. This form in (2.6.7) is known as the canonical form of the quadratic form. This reduction has many applications in different fields. Many such applications are given in the book *Quadratic Forms in Random Variables: Theory and Applications* [7]. We may also observe one interesting aspect in (2.6.7). All the y_j, $j = 1, \ldots, n$ are linear functions of the original x_j's (the elements in X) since P' is a matrix of constants.

Example 2.6.4. Reduce the following quadratic form to its canonical form:

$$u = x_1^2 - 2x_1x_3 + 2x_2^2 + 8x_2x_3 + 4x_3^2.$$

Solution 2.6.4. Writing a symmetric matrix A the quadratic form can be written as

$$u = (x_1, x_2, x_3) \begin{bmatrix} 1 & 0 & -1 \\ 0 & 2 & 4 \\ -1 & 4 & 4 \end{bmatrix} \begin{bmatrix} x_1 \\ x_2 \\ x_3 \end{bmatrix},$$

$$A = \begin{bmatrix} 1 & 0 & -1 \\ 0 & 2 & 4 \\ -1 & 4 & 4 \end{bmatrix}.$$

This matrix is already reduced to the form $A = PDP'$ in Example 2.6.1, where

$$D = \begin{bmatrix} 1 & 0 & 0 \\ 0 & 2 & 0 \\ 0 & 0 & -5 \end{bmatrix}, \quad P = \begin{bmatrix} 1 & 0 & 0 \\ 0 & 1 & 0 \\ -1 & 2 & 1 \end{bmatrix} \Rightarrow$$

$$P' = \begin{bmatrix} 1 & 0 & -1 \\ 0 & 1 & 2 \\ 0 & 0 & 1 \end{bmatrix}.$$

and

$$P'X = \begin{bmatrix} 1 & 0 & -1 \\ 0 & 1 & 2 \\ 0 & 0 & 1 \end{bmatrix} \begin{bmatrix} x_1 \\ x_2 \\ x_3 \end{bmatrix} = \begin{pmatrix} x_1 - x_3 \\ x_2 + 2x_3 \\ x_3 \end{pmatrix}.$$

Writing

$$u = X'AX = X'PDP'X = Y'DY$$
$$= y_1^2 + 2y_2^2 - 5y_3^2$$

we have

$$Y' = (x_1 - x_3, x_2 + 2x_3, x_3) \Rightarrow$$
$$y_1 = x_1 - x_3, \quad y_2 = x_2 + 2x_3, \quad y_3 = x_3.$$

Observe that D in the canonical reduction above is not unique. By further elementary operations we could have taken out various factors from the diagonal elements. Thus D in the representation PDP' is not unique.

2.6.4 Rotations

Here we look at stretching, rotations and projections. The basic ideas will be illustrated in a 2-space. Permutation matrices are already considered in the beginning of Section 2.6. Consider a 2×2 scalar matrix

$$A = cI_2 = \begin{pmatrix} c & 0 \\ 0 & c \end{pmatrix}.$$

If we premultiply a 2×2 matrix with A then every row vector there is multiplied by c, or we say, stretched by c. Then the above A is a *stretching operator*. Let

$$B = \begin{pmatrix} 0 & -1 \\ 1 & 0 \end{pmatrix}, \quad X = \begin{pmatrix} x_1 \\ x_2 \end{pmatrix}.$$

Then B operating on the left of X gives

$$BX = \begin{bmatrix} 0 & -1 \\ 1 & 0 \end{bmatrix} \begin{bmatrix} x_1 \\ x_2 \end{bmatrix} = \begin{bmatrix} -x_2 \\ x_1 \end{bmatrix}.$$

For example a point $(2,1)$ or the vector $\vec{a} = 2\vec{i} + \vec{j}$, $\vec{i} = (1,0)$, $\vec{j} = (0,1)$ goes to $\vec{b} = -\vec{i} + 2\vec{j}$. The dot product is

$$(x_1, x_2) \begin{pmatrix} -x_2 \\ x_1 \end{pmatrix} = -x_1 x_2 + x_2 x_1 = 0.$$

This means our matrix B rotates the vector $\vec{x} = x_1\vec{i} + x_2\vec{j}$ through a 90° angle. Then B above is a *rotation operator*. Let

$$C = \begin{pmatrix} 0 & 1 \\ 1 & 0 \end{pmatrix}, \quad X = \begin{pmatrix} x_1 \\ x_2 \end{pmatrix}$$

then

$$CX = \begin{pmatrix} 0 & 1 \\ 1 & 0 \end{pmatrix} \begin{pmatrix} x_1 \\ x_2 \end{pmatrix} = \begin{pmatrix} x_2 \\ x_1 \end{pmatrix}.$$

The two points $\begin{pmatrix} x_1 \\ x_2 \end{pmatrix}$ and $\begin{pmatrix} x_2 \\ x_1 \end{pmatrix}$ are the mirror images on both sides of the line $x_1 = x_2$. Here we say that C is a *reflection operator*. Let

$$D = \begin{bmatrix} 1 & 0 \\ 0 & 0 \end{bmatrix} \Rightarrow DX = \begin{bmatrix} 1 & 0 \\ 0 & 0 \end{bmatrix} \begin{bmatrix} x_1 \\ x_2 \end{bmatrix} = \begin{bmatrix} x_1 \\ 0 \end{bmatrix}.$$

This gives the projection of the vector $\vec{x} = x_1\vec{i} + x_2\vec{j}$ onto the x-axis, namely $x_1\vec{i}$. Then D above is called a *projection operator*, see Figure 2.6.1. These ideas can be generalized to the n-space, $n = 3, 4, \dots$.

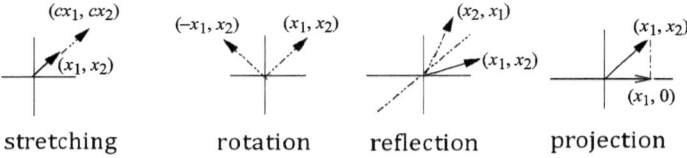

| stretching | rotation | reflection | projection |

Figure 2.6.1: Stretching, rotation, reflection and projection.

2.6.5 Linear transformations

Consider a vector X in n-space, an n-vector, or a point in the Euclidean n-space. $X' = (x_1, \dots, x_n)$. Let A be an $m \times n$ matrix. Then we have the general properties
(a) $AO = O$, where O is a null vector;
(b) $A(cX) = cAX$, c is a scalar;
(c) $A(X + Y) = AX + AY$ where Y is another n-vector.

Property (c) says that the operation is linear. $U = AX$ in general represents a transformation of X going to U where every element of U is a linear function of X. That is, if $U' = (u_1, \dots, u_m)$ and $A = (a_{ij})$ an $m \times n$ matrix of constants then

$$u_i = a_{i1}x_1 + \cdots + a_{in}x_n, \quad i = 1, \dots, m.$$

It is a linear transformation. Later we will give a more general definition of a linear transformation after introducing a more general definition for a vector. In that case X will be an object called *vector* and "A" will stand for an operator and AX will be A operating on X. In that case "scalar multiplication" and "addition" will also be re-defined. In terms of general objects called *vectors* and an operator denoted by A the transformation satisfying (a), (b), (c) above will be called a linear transformation. For the time being we will confine our discussion to ordered n-tuples of real numbers as vectors and A representing an $m \times n$ matrix of constants, c a scalar and addition and scalar multiplication as defined before.

Definition 2.6.1 (Linear and orthogonal transformations). $Y = AX$ where A is an $m \times n$ matrix of constants and X is an $n \times 1$ vector of real variables, will be called a linear transformation. When A is $n \times n$ and orthonormal, $AA' = I, A'A = I$, then the transformation is called an *orthogonal transformation*. When A is $m \times n$, $m < n$ and $AA' = I_m$ it is called a *semiorthonormal transformation*.

(v) Geometrically, an orthogonal transformation represents a rotation of the axes of coordinates.

Example 2.6.5. Show that the following transformations are orthogonal transformations: $Y = AX$ where

$$(a) \qquad A = \begin{bmatrix} \cos\theta & -\sin\theta \\ \sin\theta & \cos\theta \end{bmatrix}.$$

(This rotates the axes through an angle θ.)

$$(b) \qquad A = \begin{bmatrix} \frac{1}{\sqrt{3}} & \frac{1}{\sqrt{3}} & \frac{1}{\sqrt{3}} \\ \frac{1}{\sqrt{6}} & -\frac{2}{\sqrt{6}} & \frac{1}{\sqrt{6}} \\ \frac{1}{\sqrt{2}} & 0 & -\frac{1}{\sqrt{2}} \end{bmatrix}.$$

Solution 2.6.5. (a) This transformation is

$$\begin{pmatrix} y_1 \\ y_2 \end{pmatrix} = \begin{pmatrix} \cos\theta & -\sin\theta \\ \sin\theta & \cos\theta \end{pmatrix} \begin{pmatrix} x_1 \\ x_2 \end{pmatrix} \Rightarrow$$

$$y_1 = x_1 \cos\theta - x_2 \sin\theta, \quad y_2 = x_1 \sin\theta + x_2 \cos\theta;$$

$$AA' = \begin{bmatrix} \cos\theta & -\sin\theta \\ \sin\theta & \cos\theta \end{bmatrix} \begin{bmatrix} \cos\theta & \sin\theta \\ -\sin\theta & \cos\theta \end{bmatrix} = \begin{bmatrix} 1 & 0 \\ 0 & 1 \end{bmatrix} = A'A.$$

Hence this linear transformation is an orthogonal transformation.

$$(b) \qquad \begin{bmatrix} y_1 \\ y_2 \\ y_3 \end{bmatrix} = \begin{bmatrix} \frac{1}{\sqrt{3}} & \frac{1}{\sqrt{3}} & \frac{1}{\sqrt{3}} \\ \frac{1}{\sqrt{6}} & -\frac{2}{\sqrt{6}} & \frac{1}{\sqrt{6}} \\ \frac{1}{\sqrt{2}} & 0 & -\frac{1}{\sqrt{2}} \end{bmatrix} \begin{bmatrix} x_1 \\ x_2 \\ x_3 \end{bmatrix} \Rightarrow$$

$$y_1 = \frac{1}{\sqrt{3}}(x_1 + x_2 + x_3), \quad y_2 = \frac{1}{\sqrt{6}}(x_1 - 2x_2 + x_3),$$

$$y_3 = \frac{1}{\sqrt{2}}(x_1 - x_3); \quad AA' = I, A'A = I$$

and hence this linear transformation is an orthogonal transformation. [If, for example, the last row in (b) is deleted we have a semiorthogonal transformation.]

Suppose we transform an n-vector to an m-vector by the linear transformation $Y = AX$ where A is $m \times n$. Suppose then we transform this m-vector Y to a p-vector Z by the linear transformation $Z = BY$ where B is $p \times m$. What is the net result of transforming X to Z?

$$Z = BY = BAX.$$

Here BA is still a matrix of constants, BA is $p \times n$. Hence $X \to Z$ is also a linear transformation.

(vi) Product of two linear transformations, in the above sense, is again a linear transformation.

Let us see what happens to the shape, angles etc under a linear transformation. In order to illustrate the changes we will examine a simple linear transformation. Let $0 \leq x_1 \leq 2$, $0 \leq x_2 \leq 1$. Consider the linear transformation

$$y_1 = x_1 + x_2, \quad y_2 = x_1 \quad \text{or} \quad \begin{bmatrix} y_1 \\ y_2 \end{bmatrix} = A \begin{bmatrix} x_1 \\ x_2 \end{bmatrix}, \quad A = \begin{bmatrix} 1 & 1 \\ 1 & 0 \end{bmatrix} \Rightarrow$$

$$x_1 = y_2, \quad x_2 = y_1 - y_2.$$

Under this transformation the rectangle $OACB$, with angles $\frac{\pi}{2}$ each or the angle between OA and OB is $\frac{\pi}{2}$ in Figure 2.6.2 and the lengths $OA = 2$ and $OB = 1$, is transformed into a parallelogram with angle between OA and OB changed to $\frac{\pi}{4}$ and lengths of OA changed to 1 and OB changed to $2\sqrt{2}$, see illustrations in Figure 2.6.3. Thus, in this case the shape is not preserved, the lengths are not preserved and the angles are not preserved.

Now, consider the transformation $y_1 = \frac{1}{\sqrt{2}}x_1 + \frac{1}{\sqrt{2}}x_2$ and $y_2 = \frac{1}{\sqrt{2}}x_1 - \frac{1}{\sqrt{2}}x_2$ or

$$\begin{bmatrix} y_1 \\ y_2 \end{bmatrix} = \begin{bmatrix} \frac{1}{\sqrt{2}} & \frac{1}{\sqrt{2}} \\ \frac{1}{\sqrt{2}} & -\frac{1}{\sqrt{2}} \end{bmatrix} \begin{bmatrix} x_1 \\ x_2 \end{bmatrix},$$

$$A = \begin{bmatrix} \frac{1}{\sqrt{2}} & \frac{1}{\sqrt{2}} \\ \frac{1}{\sqrt{2}} & -\frac{1}{\sqrt{2}} \end{bmatrix}, \quad AA' = I, \quad A'A = I.$$

Here A is an orthonormal matrix. Let us see what happens to the rectangle $OACB$ in the (x_1, x_2)-plane. Under this transformation

$$x_1 = \frac{1}{\sqrt{2}}y_1 + \frac{1}{\sqrt{2}}y_2, \quad x_2 = \frac{1}{\sqrt{2}}y_1 - \frac{1}{\sqrt{2}}y_2$$

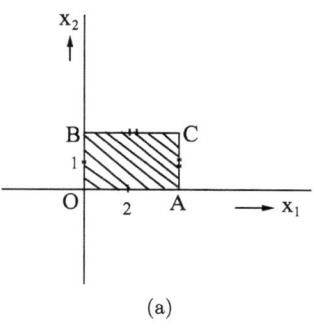 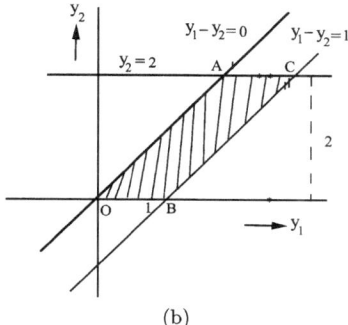

Figure 2.6.2: Linear transformations.

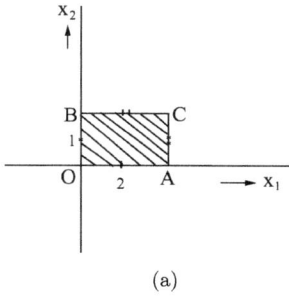 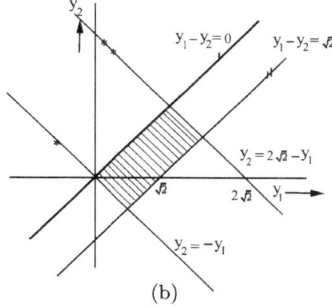

Figure 2.6.3: Additional linear transformations.

and

$$x_1 = 0 \implies y_2 = -y_1, \quad x_1 = 2 \implies y_1 + y_2 = 2\sqrt{2},$$
$$x_2 = 0 \implies y_1 = y_2, \quad x_2 = 1 \implies y_1 - y_2 = \sqrt{2}.$$

Here the angle between OA and OB is preserved as $\frac{\pi}{2}$. The lengths of the sides are preserved. The shape is also preserved. The net effect is the rotation of the axes of coordinates through an angle $\theta = \frac{\pi}{4}$ here. The general orthogonal transformation is of the form

$$\begin{bmatrix} y_1 \\ y_2 \end{bmatrix} = \begin{bmatrix} \cos\theta & \sin\theta \\ \sin\theta & -\cos\theta \end{bmatrix} \begin{bmatrix} x_1 \\ x_2 \end{bmatrix} \implies A = \begin{bmatrix} \cos\theta & \sin\theta \\ \sin\theta & -\cos\theta \end{bmatrix}, \quad AA' = I, \quad A'A = I.$$

Orthogonal transformations are simply rotations of the axes of coordinates through an angle θ where the angles, lengths and shapes in the original region in the (x_1, x_2)-plane are preserved in the (y_1, y_2)-plane.

2.6.6 Orthogonal bases for a vector subspace

Suppose we have located a basis for a given vector subspace. How can we convert this basis to an orthogonal system of vectors or to come up with an orthogonal basis for a given vector subspace? Recall the Gram–Schmidt orthogonalization process from Chapter 1. This is one method of selecting a linear function of the given vectors so that the new set will be mutually orthonormal. Then, after locating a basis transform them to an orthonormal system by Gram–Schmidt orthogonalization process to obtain an orthogonal basis.

Example 2.6.6. Construct an orthonormal basis for the row subspace of the matrix

$$A = \begin{bmatrix} 1 & 1 & 1 & 1 \\ 1 & -1 & 1 & 1 \\ 1 & -1 & -1 & 1 \\ 3 & -1 & 1 & 3 \end{bmatrix}.$$

Solution 2.6.6. Through elementary operations on the left try to determine the rank and a basis. Writing the operations by using our standard notations we have the following:

$$-1(1) + (2); \quad -1(1) + (3); \quad -3(1) + (4) \Rightarrow$$

$$A \rightarrow \begin{bmatrix} 1 & 1 & 1 & 1 \\ 0 & -2 & 0 & 0 \\ 0 & -2 & -2 & 0 \\ 0 & -4 & -2 & 0 \end{bmatrix} = B;$$

$$-1(2) + (4); \quad -1(3) + (4) \Rightarrow$$

$$B \rightarrow \begin{bmatrix} 1 & 1 & 1 & 1 \\ 0 & -2 & 0 & 0 \\ 0 & -2 & -2 & 0 \\ 0 & 0 & 0 & 0 \end{bmatrix} = C;$$

$$-1(2) + (3) \Rightarrow$$

$$C \rightarrow \begin{bmatrix} 1 & 1 & 1 & 1 \\ 0 & -2 & 0 & 0 \\ 0 & 0 & -2 & 0 \\ 0 & 0 & 0 & 0 \end{bmatrix} = C_1.$$

Hence the rank is 3 and a basis, from C_1 above, is

$$U_1 = (0, 0, -1, 0), \quad U_2 = (0, -1, 0, 0), \quad U_3 = (1, 1, 1, 1).$$

Now apply Gram–Schmidt process. Let

$$V_1 = \frac{U_1}{\|U_1\|} = (0,0,-1,0)$$

is the normalized U_1. Consider

$$W_2 = U_2 - (U_2 V_1')V_1$$

$$= (0,-1,0,0) - (0,-1,0,0)\begin{bmatrix} 0 \\ 0 \\ -1 \\ 0 \end{bmatrix} V_1 = (0,-1,0,0);$$

$$V_2 = \frac{W_2}{\|W_2\|} = (0,-1,0,0);$$

$$W_3 = U_3 - (U_3 V_1')V_1 - (U_3 V_2')V_2$$

$$= (1,1,1,1) - (1,1,1,1)\begin{bmatrix} 0 \\ 0 \\ -1 \\ 0 \end{bmatrix}(0,0,-1,0)$$

$$- (1,1,1,1)\begin{bmatrix} 0 \\ -1 \\ 0 \\ 0 \end{bmatrix}(0,-1,0,0)$$

$$= (1,1,1,1) + (0,0,-1,0) + (0,-1,0,0) = (1,0,0,1);$$

$$V_3 = \frac{W_3}{\|W_3\|} = \frac{1}{\sqrt{2}}(1,0,0,1).$$

Evidently

$$\|V_i\| = 1, \quad i = 1,2,3 \quad \text{and} \quad V_1 V_2' = 0, \quad V_1 V_3' = 0, \quad V_2 V_3' = 0.$$

Verification. Can we write all the row vectors in A as linear functions of V_1, V_2 and V_3? Note that

$$(1,1,1,1) = -V_1 - V_2 + \sqrt{2}V_3;$$
$$(1,-1,1,1) = -V_1 + V_2 + \sqrt{2}V_3;$$
$$(1,-1,-1,1) = V_1 + V_2 + \sqrt{2}V_3$$

and the fourth row is already a linear function of the first three rows. Hence $\{V_1, V_2, V_3\}$ is an orthonormal basis for the row subspace of A.

2.6.7 A vector subspace, a more general definition

Now, we are in a better position to give a more general definition to a vector subspace. The elements of the subspace are some general objects satisfying some conditions. In the definition that we are going to give here the operations "scalar multiplication" and "addition" are as defined for vectors (as n-tuples) and matrices before. A still more abstract definition can be given by defining "scalar multiplication" and "addition" as well. Let S be a set of some objects on which one can define scalar multiplication and addition. Suppose S and its elements satisfy the following conditions:

(a) If $V \in S$ then $cV \in S$ where c is any scalar, including zero.

(b) If $U \in S$, $V \in S$ then $U + V \in S$.

That is, S is closed under scalar multiplication and addition. Then S will be called a vector subspace.

Note that the same definitions introduced for linear dependence, independence, rank or dimension etc go through for this general definition also.

Example 2.6.7. Check whether the following sets satisfy the conditions for a vector subspace and if so construct a basis for the subspace:

(a) The set consisting only of the null vector.

(b) The set of all polynomials in t of degree ≤ 5.

(c) The set of all 2×2 matrices with real numbers as elements.

(d) The set of all 2×1, 2×2 and 2×3 matrices.

Solution 2.6.7. (a) It is a trivial case of a vector subspace. Since a null vector is not counted in the set of linearly independent vectors we take the dimension of this subspace as zero.

(b) Let this set be denoted by S. Then $2t + t^2$ and t^5 are two such polynomials in S. Thus, for example, $5(2t + t^2) = 10t + 5t^2$ is in S, $12t^5 \in S$. If

$$a_0 + a_1 t + \cdots + a_5 t^5 \quad \text{and} \quad b_0 + b_1 t + \cdots + b_5 t^5$$

are two polynomials in S then their sum and scalar multiples are also in S. These two operations cannot create a polynomial not in S. We cannot create a 6-th degree or higher degree polynomial by addition and scalar multiplication. Hence S is a vector subspace. Note that any general polynomial of degree up to 5 can be generated by the following quantities

$$1, t, t^2, t^3, t^4, t^5.$$

Obviously these are in S and linearly independent. None can be written as a linear combination of the others. Hence a basis is $\{1, t, t^2, t^3, t^4, t^5\}$ and the dimension of this vector subspace is 6. [Note that, for example, $t^2 = t \times t$ but this is not a scalar multiple or a constant multiple of t. Hence t^2 cannot be written in terms of other elements through the two operations of scalar multiplication and addition. Thus, these elements $1, t, ..., t^5$ are linearly independent.]

(c) Let this set be denoted by S. All elements are 2×2 matrices. Three typical elements are

$$V = \begin{bmatrix} a & b \\ c & d \end{bmatrix} \in S, \quad V_1 = \begin{bmatrix} a_1 & b_1 \\ c_1 & d_1 \end{bmatrix} \in S \quad \text{and} \quad V_2 = \begin{bmatrix} a_2 & b_2 \\ c_2 & d_2 \end{bmatrix} \in S.$$

Then obviously $V_1 + V_2$ is a 2×2 matrix and hence in S. Also $cV_1 \in S$ for any scalar c. For $c = 0$ it is a null matrix and hence the null matrix is also in S.

$$O = \begin{bmatrix} 0 & 0 \\ 0 & 0 \end{bmatrix} \in S, \quad -V = \begin{bmatrix} -a & -b \\ -c & -d \end{bmatrix} \in S,$$

$$V - V = O \in S, \quad I_2 V = V = V I_2 \in S, \quad I_2 \in S,$$

$$V + O = V, \quad c_1 c_2 V = c_1 (c_2 V) = c_2 (c_1 V)$$

where c_1 and c_2 are scalars. Obviously the following general conditions are satisfied by the elements of S in this case:

(1) $V_1 + V_2 = V_2 + V_1, \quad V_1 \in S, \quad V_2 \in S$

(2) $V_1 + (V_2 + V_3) = (V_1 + V_2) + V_3, \quad V_3 \in S$

(3) $V + O = V \quad \text{for all} \quad V, \quad V \in S, \quad O \in S$

(4) $-V \in S, \quad V - V = V + (-1)V = O$

(5) $IV = V, \quad I \in S$

(6) $(c_1 c_2)V = c_1 (c_2 V), \quad c_1, c_2 \text{ scalars}$

(7) $c(V_1 + V_2) = cV_1 + cV_2, \quad c \text{ a scalar}$

(8) $(c_1 + c_2)V = c_1 V + c_2 V.$

These eight properties are in fact the conditions that we will impose when we have a more abstract definition of a vector subspace where we will also define what is meant by "+" and "cV". But we will not make the definition more abstract in this book.

What is a basis for our vector subspace in this case? Note that a general matrix of the form $V = \begin{bmatrix} a & b \\ c & d \end{bmatrix}$ can be generated as a linear function of the following four matrices:

$$U_1 = \begin{bmatrix} 1 & 0 \\ 0 & 0 \end{bmatrix}, \quad U_2 = \begin{bmatrix} 0 & 1 \\ 0 & 0 \end{bmatrix}, \quad U_3 = \begin{bmatrix} 0 & 0 \\ 1 & 0 \end{bmatrix}, \quad U_4 = \begin{bmatrix} 0 & 0 \\ 0 & 1 \end{bmatrix}.$$

Hence a basis is $\{U_1, U_2, U_3, U_4\}$ and the dimension is 4 since these four are linearly independent.

(d) Let the set be denoted by S. Let

$$U = \begin{pmatrix} a \\ b \end{pmatrix}, \quad V = \begin{pmatrix} c & d \\ d & f \end{pmatrix}.$$

Then $U \in S$, $V \in S$ whereas $U + V$ is not defined and hence S is not a vector subspace in this case.

2.6.8 A linear transformation, a more general definition

Taking the general definition of a vector subspace in Section 2.6.7 we can define a linear transformation. Let the elements of a vector subspace S of Section 2.6.7 be some general objects where S is closed under addition and scalar multiplication. Let A represent some operator operating on the elements of S such that the following conditions (a), (b), (c) are satisfied:

(a) $AO = O$, where O is a null vector.
(b) $A(cX) = cAX$, c is a scalar and $X \in S$.
(c) $A(X + Y) = AX + AY$ where $X \in S$, $Y \in S$.

Then $Y = AX$, $X \in S$ is called a linear transformation. Of course this general definition also covers the case when X is an ordered n-tuple and A an $m \times n$ matrix. Let us see what are the general operators that we can include under this general definition of a linear transformation.

Example 2.6.8. Consider a vector subspace S of all real polynomials in the real scalar variable θ. Then a typical vector in this subspace S, for example a polynomial of degree 3, will be of the form

$$X = a_0 + a_1\theta + a_2\theta^2 + a_3\theta^3.$$

Consider the operator $A = \frac{d}{d\theta}$. Then A operating on X, namely AX, will be to differentiate X with respect to θ. Show that A is a linear operator.

Solution 2.6.8.

$$AX = \frac{d}{d\theta}X = a_1 + 2a_2\theta + 3a_3\theta^2 \in S;$$
$$A(cX) = c\left(\frac{d}{d\theta}X\right) = c(a_1 + 2a_2\theta + 3a_3\theta^2) \in S,$$

where c is a constant, free of θ. If

$$U = b_0 + b_1\theta + b_2\theta^2 + b_3\theta^3 \in S$$

then obviously

$$A(U + X) = AU + AX = (a_1 + b_1) + 2(a_2 + b_2)\theta + 3(a_3 + b_3)\theta^2 \in S.$$

Hence $A = \frac{d}{d\theta}$ is a linear operator. What is the null space here? This means the set of all X satisfying the condition

$$AX = O \implies \frac{d}{d\theta}X = O \implies X = \text{constant}.$$

The null space or the right null space is the set of all constants here.

Example 2.6.9. Consider the same vector subspace S of Example 2.6.8. Let A be the integration operator, $\int_\theta$. Show that A is a linear operator.

Solution 2.6.9. Let X and U be as defined in Example 2.6.8. Then

$$AX = \int [a_0 + a_1\theta + a_2\theta^2 + a_3\theta^3]d\theta$$

$$= a_0\theta + a_1\frac{\theta^2}{2} + a_2\frac{\theta^3}{3} + a_3\frac{\theta^4}{4} + c_1$$

where c_1 is a constant. Note that $A(cX) = cAX$ here as well as

$$A(X + U) = AX + AU.$$

Hence $A = \int_\theta$ is a linear operator. $AX = O \implies \int Xd\theta = O$. This has no solution here and hence the null space for this vector subspace with respect to the operator $A = \int_\theta$ is empty.

Example 2.6.10. Let S be the vector subspace of all real polynomials in the real scalar variable t. Let A be the operator $AX = X^2$. (A operating on an element gives the square of that element.) Is A linear?

Solution 2.6.10. Two typical vectors in this S, for example of degree 1 each, are of the form $U = a_0 + a_1t$ and $V = b_0 + b_1t$. Then

$$AU = (a_0 + a_1t)^2 = a_0^2 + 2a_0a_1t + a_1^2t^2$$
$$AV = (b_0 + b_1t)^2 = b_0^2 + 2b_0b_1t + b_1^2t^2$$
$$A(U + V) = A[(a_0 + b_0) + (a_1 + b_1)t] = [(a_0 + b_0) + (a_1 + b_1)t]^2$$
$$\neq (a_0 + a_1t)^2 + (b_0 + b_1t)^2 = AU + AV.$$

Hence this operator A is not linear.

Exercises 2.6

2.6.1. Construct the matrix G, product of 4×4 elementary matrices, so that GA effects the permutation of the rows to the order $2, 3, 4, 1$ where A is a general 4×4 matrix.

2.6.2. Write the following matrices in the form QDQ' where Q is a nonsingular matrix and D is a diagonal matrix:

$$A = \begin{bmatrix} 1 & 0 & 2 \\ 0 & 1 & 1 \\ 2 & 1 & 4 \end{bmatrix}, \quad B = \begin{bmatrix} 1 & 1 & 1 & 1 \\ 1 & 1 & 1 & 1 \\ 1 & 1 & 1 & 1 \\ 1 & 1 & 1 & 1 \end{bmatrix}, \quad C = \begin{bmatrix} 2 & 1 & -1 \\ 1 & 3 & 2 \\ -1 & 2 & -1 \end{bmatrix}.$$

2.6.3. Write the following matrices in the form PDQ where P and Q are nonsingular matrices and D is a diagonal matrix:

$$A = \begin{bmatrix} 2 & 1 & -1 \\ 1 & 1 & 2 \\ -1 & 0 & 4 \end{bmatrix}, \quad B = \begin{bmatrix} 3 & 1 & -1 \\ 2 & 0 & 2 \\ 2 & 1 & -1 \end{bmatrix}, \quad C = \begin{bmatrix} -2 & 2 & -1 \\ 0 & -1 & 0 \\ 2 & 1 & 5 \end{bmatrix}.$$

2.6.4. Write the matrices in Exercise 2.6.2 in the form (a) DQ, (b) PD where P and Q are nonsingular matrices and D is a diagonal matrix.

2.6.5. Reduce the following quadratic forms to their canonical forms:
(a) $u_1 = 2x_1^2 - x_2^2 + 3x_1x_2 + x_3^2 - 4x_2x_3$
(b) $u_2 = 3x_1^2 + x_2^2 - 2x_1x_2 + 2x_1x_3 - 2x_2x_3 + 2x_3^2$.

2.6.6. Construct two orthonormal bases each for the row subspaces of the matrices in Exercise 2.6.3.

2.6.7. Using the general definition in Section 2.6.7 are the following sets vector subspaces?
(a) The set of all $n \times n$ lower triangular matrices for a fixed n. If so, what is its dimension?
(b) The set of all $n \times n$ diagonal matrices for a fixed n?
(c) The set of all integers, including zero.
(d) The set of all possible scalar functions $f(x)$ of the real scalar variable x defined on the interval $[3,5]$.
(e) All couplets of real numbers (a,b), $a \geq 0$, $b \geq 0$.

2.6.8. Taking the general definition of linear transformation in Section 2.6.8 are the following linear transformations?
(a) Let S be a vector subspace of polynomials in t of degree ≤ 3. For $V \in S$ let $AV = (1+t)V$.
(b) Let S be the same subspace in (a) above. Let A be such that $AV = V + 5$.
(c) Let S be the subspace of 3×3 matrices. For $V \in S$ let $AV = V + I_3$ where I_3 is the identity matrix.

2.6.9. If $A + B = I_n$, $A = A'$, $B = B'$, $AB = O$, where A and B are $n \times n$ matrices, then show that A and B are idempotent with ranks r and $n - r$ respectively.

2.6.10. If $A + B = I_n$, $A = A'$, $B = B'$, $AB = O$ then show that rank of A plus rank of B is n, where A and B are $n \times n$ matrices.

2.6.11. If $A + B = I_n$, $A = A'$, $B = B'$, where A and B are $n \times n$ matrices, then show that both A and B can be reduced to diagonal forms by the same orthonormal matrix, say P.

2.6.12. Show that every nonsingular matrix can be written as a product of the basic elementary matrices.

2.6.13. Write the matrices in Exercise 2.6.3 into the forms (a) DQ, (b) PD where D is a diagonal matrix and P and Q are nonsingular matrices.

2.6.14. Construct one orthonormal basis each for the right null spaces of the matrices in Exercise 2.6.3, if possible.

2.6.15. Construct two examples each for the following: (a) sum of two nonsingular matrices is singular; (b) sum of two nonsingular matrices is nonsingular; (c) sum of two singular matrices is singular; (d) sum of two singular matrices is nonsingular.

2.6.16. (a) Can the product of two nonsingular matrices be singular? (b) Can the product of two singular matrices be nonsingular? Prove your assertions.

2.6.17. Let A and B be rectangular matrices where AB is defined. (a) If A and B are of full ranks can AB be of less than full rank? (b) If A and B are of less than full ranks each can AB be of full rank? (c) If A or B is of full rank and the other is of less than full rank can AB be of full rank? Prove your assertions.

2.6.18. Generalized inverse of a matrix. Consider an $m \times n$ matrix A of any rank. A generalized inverse (there can be many such inverses for a given matrix) or *g-inverse* of A is an $n \times m$ matrix, denoted by A^-, such that $X = A^- b$ is a solution of the consistent system of linear equations $AX = b$ or equivalently A^- is a *g*-inverse iff $AA^-A = A$. Evaluate a *g*-inverse of A where

$$A = \begin{bmatrix} 1 & 1 & -1 \\ 1 & 0 & 1 \end{bmatrix}.$$

2.6.19. Let A^- be a *g*-inverse of A and let $H = A^-A$. Show that (a) H is idempotent, (b) $AH = A$, (c) rank of A = rank of H = trace of H.

2.6.20. Let $AX = b$ be a consistent system of linear equations. Show that a general solution of this system can be written as

$$A^- b + (H - I)Z$$

where A^- is a *g*-inverse of A, $H = A^-A$ and Z is arbitrary.

2.6.21. Consider the space S of all $n \times n$ matrices. Construct three subspaces S_1, S_2, S_3 of this space S.

2.6.22. Consider the vector space S of 3×3 matrices. Let

$$A_1 = \begin{bmatrix} 1 & 1 & 0 \\ 0 & 1 & 0 \\ 0 & 0 & 0 \end{bmatrix}, \quad A_2 = \begin{bmatrix} 0 & 0 & 0 \\ 0 & 0 & 1 \\ 0 & 0 & 1 \end{bmatrix}, \quad A_3 = \begin{bmatrix} 0 & 0 & 0 \\ 0 & 0 & 0 \\ 0 & 1 & 0 \end{bmatrix}.$$

Are A_1, A_2, A_3, as elements in the vector space S, linearly independent? Prove your assertion.

2.6.23. Consider the vector space S of all polynomials of degree less than or equal to 2 in the real variable t. Let

$$p_1(t) = 1 + 2t, \quad p_2(t) = t^2, \quad p_3(t) = 3 + 5t + 2t^2.$$

Are these linearly independent? Prove your assertion.

2.6.24. Show that any matrix of rank r can be written as a sum of r matrices of rank one each.

2.6.25. Let S_1 and S_2 be nontrivial subspaces of a vector space S. Show that S_1 and S_2 are orthogonal to each other if and only if each basis element of S_1 is orthogonal to all basis elements of S_2 or vice versa.

2.6.26. Let S be the space of all polynomials in x of degree not exceeding n, for a fixed n, with the inner product

$$\int_{-1}^{1} p(x)q(x)\mathrm{d}x, \quad p(x) \in S, \ q(x) \in S$$

defined. Show that (Cauchy–Schwartz inequality)

$$\left| \int_{-1}^{1} p(x)q(x)\mathrm{d}x \right| \leq \left[\int_{-1}^{1} p^2(x)\mathrm{d}x \right]^{\frac{1}{2}} \left[\int_{-1}^{1} q^2(x)\mathrm{d}x \right]^{\frac{1}{2}}.$$

2.6.27. Let S be the same vector space of polynomials in Exercise 2.6.26. Observe that $1, x, x^2, \ldots, x^n$ is a basis of S. Obtain an orthonormal basis by applying the Gram–Schmidt orthogonalization process.

2.6.28. Show that the orthonormal basis obtained in Exercise 2.6.27 can be written in terms of the Legendre polynomials

$$L_k(x) = \frac{1}{2^k k!} \frac{\mathrm{d}^k}{\mathrm{d}x^k}(x^2 - 1)^k.$$

2.7 Partitioning of matrices

This is a very convenient way of doing matrix multiplications and computations of inverses when we have large matrices [number of rows or columns or both large]. The ideas will be introduced by looking at some special cases.

2.7.1 Partitioning and products

Consider the multiplication involving two vectors

$$(1, -1, 2) \begin{bmatrix} 3 \\ 4 \\ -5 \end{bmatrix} = [(1)(3) + (-1)(4)] + [(2)(-5)]$$

$$= [-1] + [-10] = -11.$$

Let

$$A_1 = (1, -1), \quad A_2 = (2), \quad B_1 = \begin{bmatrix} 3 \\ 4 \end{bmatrix}, \quad B_2 = (-5).$$

Then the above multiplication can be written as

$$(A_1, A_2) \begin{bmatrix} B_1 \\ B_2 \end{bmatrix} = A_1 B_1 + A_2 B_2,$$

$$A_1 B_1 = (1)(3) + (-1)(4) = -1,$$

$$A_2 B_2 = (2)(-5) = -10.$$

Here the multiplication is carried out as if A_1, A_2, B_1, B_2 are scalars, but keeping the order in which the submatrices occur. This means that instead of taking in the order $A_1 B_1$ we are not allowed to take in the order $B_1 A_1$ when writing down the products. In general, let a and b be $n \times 1$ vectors. Let them be partitioned as follows:

$$a = \begin{bmatrix} A_1 \\ A_2 \end{bmatrix}, \quad b = \begin{bmatrix} B_1 \\ B_2 \end{bmatrix} \Rightarrow a' = (A_1', A_2'),$$

$$a = \begin{bmatrix} a_1 \\ \vdots \\ a_n \end{bmatrix}, \quad b = \begin{bmatrix} b_1 \\ \vdots \\ b_n \end{bmatrix}, \quad A_1 = \begin{bmatrix} a_1 \\ \vdots \\ a_r \end{bmatrix}, \quad B_1 = \begin{bmatrix} b_1 \\ \vdots \\ b_r \end{bmatrix}.$$

Then

$$a'b = (a_1 b_1 + \cdots + a_r b_r) + (a_{r+1} b_{r+1} + \cdots + a_n b_n)$$

$$= A_1' B_1 + A_2' B_2.$$

This is possible as long as all the products are defined. If A_1 is $r \times 1$ and B_1 is $s \times 1$, $r \neq s$ then such a partitioning will not produce a simplified form, $A_1 B_1$ is not defined here. Consider the product

$$(1, -1, | \, 2) \begin{bmatrix} 3 & 1 & 1 \\ 4 & 0 & 1 \\ --- & --- & --- \\ -5 & -2 & 1 \end{bmatrix} = (A_1, A_2) \begin{bmatrix} B_1 \\ B_2 \end{bmatrix},$$

$$A_1 = (1, -1), \quad A_2 = (2), \quad B_1 = \begin{bmatrix} 3 & 1 & 1 \\ 4 & 0 & 1 \end{bmatrix}, \quad B_2 = (-5, -2, 1).$$

Let us multiply A_i's and B_j's as if they are scalars but keeping the order. Then

$$(A_1, A_2)\begin{pmatrix} B_1 \\ B_2 \end{pmatrix} = A_1 B_1 + A_2 B_2$$

$$= (1, -1)\begin{bmatrix} 3 & 1 & 1 \\ 4 & 0 & 1 \end{bmatrix} + (2)(-5, -2, 1)$$

$$= (-1, 1, 0) + (-10, -4, 2) = (-11, -3, 2).$$

We get the same final answer if they are multiplied element-wise or directly. Consider the partition

$$\begin{bmatrix} 1 & -1 & | & 2 \\ 0 & 1 & | & 1 \end{bmatrix}\left[\begin{array}{ccc} 3 & 1 & 1 \\ 4 & 0 & 1 \\ --- & --- & --- \\ -5 & -2 & 1 \end{array}\right] = (A_1, A_2)\begin{pmatrix} B_1 \\ B_2 \end{pmatrix}$$

$$= A_1 B_1 + A_2 B_2$$

where

$$A_1 = \begin{bmatrix} 1 & -1 \\ 0 & 1 \end{bmatrix}, \quad A_2 = \begin{bmatrix} 2 \\ 1 \end{bmatrix}, \quad B_1 = \begin{bmatrix} 3 & 1 & 1 \\ 4 & 0 & 1 \end{bmatrix},$$

$$B_2 = (-5, -2, 1).$$

Then

$$A_1 B_1 = \begin{bmatrix} 1 & -1 \\ 0 & 1 \end{bmatrix}\begin{bmatrix} 3 & 1 & 1 \\ 4 & 0 & 1 \end{bmatrix} = \begin{bmatrix} -1 & 1 & 0 \\ 4 & 0 & 1 \end{bmatrix},$$

$$A_2 B_2 = \begin{bmatrix} 2 \\ 1 \end{bmatrix}[-5, -2, 1] = \begin{bmatrix} -10 & -4 & 2 \\ -5 & -2 & 1 \end{bmatrix}$$

and

$$A_1 B_1 + A_2 B_2 = \begin{bmatrix} -1 & 1 & 0 \\ 4 & 0 & 1 \end{bmatrix} + \begin{bmatrix} -10 & -4 & 2 \\ -5 & -2 & 1 \end{bmatrix}$$

$$= \begin{bmatrix} -11 & -3 & 2 \\ -1 & -2 & 2 \end{bmatrix}.$$

If we multiply element-wise we get the same answer.

Definition 2.7.1 (Conformal partitioning of matrices). Two matrices A and B are said to be partitioned conformally for the product AB, when A and B are partitioned into submatrices and if the multiplication AB is carried out treating the submatrices as if they are scalars, but keeping the order, and when all products and sums of submatrices involved are defined.

Example 2.7.1. Check whether the following partitioning is conformal for the product AB:

(a)
$$A = \begin{bmatrix} 1 & 0 & | & 2 \\ -1 & 1 & | & 1 \\ 1 & 1 & | & 1 \end{bmatrix} = (A_1, A_2),$$

$$B = \begin{bmatrix} 1 & 1 & -1 & 2 \\ 0 & 0 & 1 & 1 \\ -- & -- & -- & -- \\ 1 & 1 & 0 & 1 \end{bmatrix} = \begin{bmatrix} B_1 \\ B_2 \end{bmatrix};$$

(b)
$$A = \begin{bmatrix} 1 & | & 0 & 2 \\ -1 & | & 1 & 1 \\ r1 & | & 1 & 1 \end{bmatrix} = (A_1, A_2),$$

$$B = \begin{bmatrix} 1 & 1 & -1 & 2 \\ 0 & 0 & 1 & 1 \\ -- & -- & -- & -- \\ 1 & 1 & 0 & 1 \end{bmatrix} = \begin{bmatrix} B_1 \\ B_2 \end{bmatrix}.$$

Solution 2.7.1. (a) If the submatrices are treated as scalars and if the multiplication is carried out we get

$$AB = A_1 B_1 + A_2 B_2$$

where

$$A_1 = \begin{bmatrix} 1 & 0 \\ -1 & 1 \\ 1 & 1 \end{bmatrix}, \quad B_1 = \begin{bmatrix} 1 & 1 & -1 & 2 \\ 0 & 0 & 1 & 1 \end{bmatrix},$$

$$A_2 = \begin{bmatrix} 2 \\ 1 \\ 1 \end{bmatrix}, \quad B_2 = (1, 1, 0, 1).$$

$A_1 B_1$, $A_2 B_2$ and $A_1 B_1 + A_2 B_2$ are all defined and hence the partitioning is conformal for the product AB.

(b) Here

$$A_1 = \begin{bmatrix} 1 \\ -1 \\ 1 \end{bmatrix}, \quad B_1 = \begin{bmatrix} 1 & 1 & -1 & 2 \\ 0 & 0 & 1 & 1 \end{bmatrix}$$

but $A_1 B_1$ is not defined and hence the partition is not conformal. If the partition was after the first row that is, if $B_1 = (1, 1, -1, 2)$ then the partitioning would be conformal for the product AB.

2.7.2 Partitioning of quadratic forms

In statistical theory, regression analysis, econometrics, model building and in many other areas one requires to study a part of a quadratic form or to study different parts separately. This leads to partitioning of a quadratic form. Let X be an $n \times 1$ vector of real variables and let A be an $n \times n$ symmetric matrix of constants. We have already seen that a quadratic form can be written in the form

$$u = X'AX, \quad A = A'. \tag{2.7.1}$$

Consider the partitions

$$X = \begin{pmatrix} X_1 \\ X_2 \end{pmatrix}, \quad X_1 = \begin{bmatrix} x_1 \\ \vdots \\ x_r \end{bmatrix}, \quad X_2 = \begin{bmatrix} x_{r+1} \\ \vdots \\ x_n \end{bmatrix},$$

$$A = \begin{bmatrix} A_{11} & A_{12} \\ A_{21} & A_{22} \end{bmatrix},$$

where A_{11} is $r \times r$, A_{22} is $(n-r) \times (n-r)$, A_{12} is $r \times (n-r)$ and A_{21} is $(n-r) \times r$. When $A = A'$ we have $A'_{21} = A_{12}$. Then in the partitioned form the quadratic form is the following:

$$u = (X'_1, X'_2) \begin{pmatrix} A_{11} & A_{12} \\ A_{21} & A_{22} \end{pmatrix} \begin{pmatrix} X_1 \\ X_2 \end{pmatrix}.$$

It is easy to note that the partitioning is conformal to carry out all the multiplications. Treating the submatrices as if they are scalars and completing the multiplications we have

$$u = (X'_1 A_{11} + X'_2 A_{21}, X'_1 A_{12} + X'_2 A_{22}) \begin{pmatrix} X_1 \\ X_2 \end{pmatrix}$$

$$= X'_1 A_{11} X_1 + X'_2 A_{21} X_1 + X'_1 A_{12} X_2 + X'_2 A_{22} X_2. \tag{2.7.2}$$

We obtain two quadratic forms $X'_1 A_{11} X_1$ and $X'_2 A_{22} X_2$ and two bilinear forms $X'_2 A_{21} X_1$ and $X'_1 A_{12} X_2$. When $A'_{21} = A_{12}$ we have an interesting property.

$$(X'_2 A_{21} X_1)' = X'_1 A'_{21} (X'_2)' = X'_1 A_{12} X_2$$

which is the same as the other bilinear form. Further $X'_1 A_{12} X_2$ and $X'_2 A_{21} X_1$ are 1×1 matrices or scalars and hence they are equal.

(i) If P and Q are 1×1 matrices and if $P' = Q$ then $P = Q$.

Then, when $A = A'$,

$$u = X'_1 A_{11} X_1 + 2X'_1 A_{12} X_2 + X'_2 A_{22} X_2. \tag{2.7.3}$$

We can study the quadratic forms involving the subvectors X_1 and X_2 as well as the bilinear form involving X_1 and X_2 by using the representation in (2.7.3). As a numerical example, consider

$$u = 2x_1^2 - x_2^2 + x_3^2 + 2x_1x_2 - 4x_2x_3$$

$$= (x_1, x_2, | x_3) \begin{bmatrix} 2 & 1 & | & 0 \\ 1 & -1 & | & -2 \\ -- & -- & -- & -- \\ 0 & -2 & | & 1 \end{bmatrix} \begin{bmatrix} x_1 \\ x_2 \\ -- \\ x_3 \end{bmatrix}.$$

Consider the partitioning

$$X' = (X_1', X_2'), \quad X_1' = (x_1, x_2), \quad X_2' = x_3$$

and the corresponding conformal partitioning of A. Then

$$u = X_1'A_{11}X_1 + X_2'A_{22}X_2 + 2X_1'A_{12}X_2$$
$$= [2x_1^2 + 2x_1x_2 - x_2^2] + [x_3^2] + 2[-2x_2x_3].$$

Note that

$$-2x_2x_3 = X_1'A_{12}X_2 = X_2'A_{21}X_1.$$

2.7.3 Partitioning of bilinear forms

A bilinear form in the vectors X and Y is a homogeneous linear form in X as well as in Y. Let X be $p \times 1$ and Y be $q \times 1$. Let A be a $p \times q$ matrix of constants. Then a bilinear form can be written in the form

$$w = X'AY. \tag{2.7.4}$$

Bilinear forms have applications in studying covariances and correlations, in analysis of covariance techniques in design of experiments and in many related areas. Some theoretical aspects of bilinear forms in random variables may be seen from the book *Bilinear Forms and Zonal Polynomials* [8]. If a study of bilinear forms involving some subvectors of X and Y is undertaken then we need to partition the bilinear forms. Let

$$X = \begin{pmatrix} X_1 \\ X_2 \end{pmatrix}, \quad Y = \begin{pmatrix} Y_1 \\ Y_2 \end{pmatrix}, \quad A = \begin{pmatrix} A_{11} & A_{12} \\ A_{21} & A_{22} \end{pmatrix}$$

where X_1 is $p_1 \times 1$, X_2 is $p_2 \times 1$, $p_1 + p_2 = p$, Y_1 is $q_1 \times 1$, Y_2 is $q_2 \times 1$, $q_1 + q_2 = q$, A_{11} is $p_1 \times p_1$ and A_{22} is $q_2 \times q_2$. Then A_{12} is $p_1 \times q_2$, A_{21} is $p_2 \times q_1$ and the partitioning is conformal.

Then

$$w = X'AY$$

$$= (X_1', X_2') \begin{pmatrix} A_{11} & A_{12} \\ A_{21} & A_{22} \end{pmatrix} \begin{pmatrix} Y_1 \\ Y_2 \end{pmatrix}$$

$$= X_1' A_{11} Y_1 + X_2' A_{22} Y_2$$

$$+ X_1' A_{12} Y_2 + X_2' A_{21} Y_1. \tag{2.7.5}$$

Note that here $A_{12}' \neq A_{21}$. In fact the whole matrix A is rectangular if $p \neq q$. In (2.7.5) we get two bilinear forms involving (X_1, Y_1) and (X_2, Y_2) and two bilinear forms involving (X_1, Y_2) and (X_2, Y_1). As a numerical example, let us consider the following bilinear form in $X' = (x_1, x_2, x_3)$ and $Y' = (y_1, y_2)$:

$$w = x_1 y_1 - x_2 y_1 + x_3 y_1 + 2x_1 y_2 + x_2 y_2 + x_3 y_2$$

$$= (x_1, \mid x_2, x_3) \begin{pmatrix} 1 & \mid & 2 \\ -- & -- & -- \\ -1 & \mid & 1 \\ 1 & \mid & 1 \end{pmatrix} \begin{pmatrix} y_1 \\ -- \\ y_2 \end{pmatrix}$$

$$= (X_1', X_2') \begin{pmatrix} A_{11} & A_{12} \\ A_{21} & A_{22} \end{pmatrix} \begin{pmatrix} Y_1 \\ Y_2 \end{pmatrix},$$

$$X_1' = x_1, \quad X_2' = (x_2, x_3), \quad Y_1 = y_1, \quad Y_2 = y_2,$$

$$A_{11} = 1, \quad A_{12} = 2, \quad A_{21} = \begin{pmatrix} -1 \\ 1 \end{pmatrix}, \quad A_{22} = \begin{pmatrix} 1 \\ 1 \end{pmatrix}.$$

Note that

$$X_1' A_{11} Y_1 = x_1(1)y_1 = x_1 y_1,$$

$$X_2' A_{22} Y_2 = (x_2, x_3) \begin{pmatrix} 1 \\ 1 \end{pmatrix} y_2 = x_2 y_2 + x_3 y_2,$$

$$X_1' A_{12} Y_2 = x_1(2)y_2 = 2x_1 y_2,$$

$$X_2' A_{21} Y_1 = (x_2, x_3) \begin{pmatrix} -1 \\ 1 \end{pmatrix} y_1 = -x_2 y_1 + x_3 y_1.$$

2.7.4 Inverses of partitioned matrices

Let A be a nonsingular $n \times n$ matrix so that its regular inverse A^{-1} exists. Let us partition A and A^{-1} as follows:

$$A = \begin{bmatrix} A_{11} & A_{12} \\ A_{21} & A_{22} \end{bmatrix}, \quad A^{-1} = \begin{bmatrix} A^{11} & A^{12} \\ A^{21} & A^{22} \end{bmatrix}.$$

A convenient standard notation of the submatrices of the inverse A^{-1} is used here by writing the corresponding superscripts. Let us investigate the relationships among the submatrices. Suppose that the partitioning is conformal to take the product AA^{-1}. Suppose A_{11} and A^{11} are $r \times r$ so that the remaining are automatically defined. Let the identity matrix I_n be partitioned as

$$I_n = \begin{bmatrix} I_r & O \\ O & I_{n-r} \end{bmatrix}.$$

Then writing the equations $AA^{-1} = I$ we have

$$AA^{-1} = I \Rightarrow$$
$$\begin{bmatrix} A_{11} & A_{12} \\ A_{21} & A_{22} \end{bmatrix} \begin{bmatrix} A^{11} & A^{12} \\ A^{21} & A^{22} \end{bmatrix} = \begin{bmatrix} I_r & O \\ O & I_{n-r} \end{bmatrix} \Rightarrow$$

(1) $\quad A_{11}A^{11} + A_{12}A^{21} = I_r$
(2) $\quad A_{11}A^{12} + A_{12}A^{22} = O$
(3) $\quad A_{21}A^{11} + A_{22}A^{21} = O$
(4) $\quad A_{21}A^{12} + A_{22}A^{22} = I_{n-r}.$

Premultiplying (2) by A_{11}^{-1}, if A_{11} is nonsingular, yields

$$A^{12} = -A_{11}^{-1}A_{12}A^{22}.$$

Substitution in (4) yields

$$(-A_{21}A_{11}^{-1}A_{12} + A_{22})A^{22} = I \Rightarrow$$
$$A^{22} = (A_{22} - A_{21}A_{11}^{-1}A_{12})^{-1}.$$

From symmetry it follows that, assuming that the inverses exist,

$$(A^{11})^{-1} = A_{11} - A_{12}A_{22}^{-1}A_{21} \tag{2.7.6}$$
$$(A^{22})^{-1} = A_{22} - A_{21}A_{11}^{-1}A_{12} \tag{2.7.7}$$
$$(A_{11})^{-1} = A^{11} - A^{12}(A^{22})^{-1}A^{21} \tag{2.7.8}$$
$$(A_{22})^{-1} = A^{22} - A^{21}(A^{11})^{-1}A^{12}. \tag{2.7.9}$$

Similarly from (1), (2), (3), (4) we can solve for $A^{12}, A^{21}, A_{12}, A_{21}$ in terms of the other submatrices. The results in equations (2.7.6) to (2.7.9) are widely applicable in various types of problems.

Example 2.7.2. Write $z = X_1'A_{11}X_1 + 2X_1'A_{12}X_2$ as the sum of two quadratic forms where one of them contains X_1 and X_2 and the other contains only X_2, assuming $A_{11} = A_{11}'$ and nonsingular.

Solution 2.7.2. This is similar to completing the square when we have a term containing the square of a scalar variable and a second term which is linear in the variable. In order to see that the result can be achieved let us open up the following quadratic form, where C is a vector such that $X_1 + C$ is defined:

$$(X_1 + C)'A_{11}(X_1 + C) = X_1'A_{11}X_1 + 2X_1'A_{11}C + C'A_{11}C.$$

Comparing this with z, that is, comparing the quadratic and linear terms in X_1, we see that C corresponds to the following:

$$A_{11}C = A_{12}X_2 \;\Rightarrow\; C = A_{11}^{-1}A_{12}X_2.$$

Hence the quantity to be added and subtracted is

$$C'A_{11}C = (A_{11}^{-1}A_{12}X_2)'A_{11}(A_{11}^{-1}A_{12}X_2)$$
$$= X_2'A_{12}'A_{11}^{-1}A_{11}A_{11}^{-1}A_{12}X_2$$
$$= X_2'A_{12}'A_{11}^{-1}A_{12}X_2.$$

Therefore

$$z = X_1'A_{11}X_2 + 2X_1'A_{12}X_2$$
$$= (X_1 + C)'A_{11}(X_1 + C) - X_2'A_{12}'A_{11}^{-1}A_{12}X_2,$$
$$C = A_{11}^{-1}A_{12}X_2.$$

Example 2.7.3 (Gaussian density). In multivariate statistical analysis the most prominent density is the Gaussian density. If X is a real $p \times 1$ vector random variable then X is said to have a Gaussian density if the density of X is of the form

$$f(X) = c\, e^{-\frac{1}{2}(X-\mu)'V^{-1}(X-\mu)}$$

where μ is a constant vector, $V = V'$, V is such that the exponent of e remains negative for all values of X and μ, and c is a normalizing constant such that the integral of f over X will be unity. If X is partitioned as $X = \binom{X_1}{X_2}$ where X_1 is $r \times 1$, $r < p$ evaluate the density of X_1 (marginal density of X_1 is available from $f(X)$ by integrating out X_2, the remaining variables).

Solution 2.7.3. Consider the corresponding partitioning of μ and V, that is,

$$\mu = \begin{pmatrix} \mu_1 \\ \mu_2 \end{pmatrix}, \quad V = \begin{pmatrix} V_{11} & V_{12} \\ V_{21} & V_{22} \end{pmatrix},$$
$$V_{21} = V_{12}', \quad V^{-1} = \begin{pmatrix} V^{11} & V^{12} \\ V^{21} & V^{22} \end{pmatrix}$$

where μ_1 is $r \times 1$ and V_{11} is $r \times r$. For convenience let $Y_1 = X_1 - \mu_1$ and $Y_2 = X_2 - \mu_2$. Then

$$(X - \mu)' V^{-1}(X - \mu) = [Y_1', Y_2'] \begin{bmatrix} V^{11} & V^{12} \\ V^{21} & V^{22} \end{bmatrix} \begin{bmatrix} Y_1 \\ Y_2 \end{bmatrix}$$

$$= Y_1' V^{11} Y_1 + 2Y_1' V^{12} Y_2 + Y_2' V^{22} Y_2.$$

Writing the linear term and the quadratic term in Y_2 with the help of the result in Example 2.7.2 we have

$$2Y_1' V^{12} Y_2 + Y_2' V^{22} Y_2 = (Y_2 + C)' V^{22}(Y_2 + C)$$
$$- Y_1' V^{12}(V^{22})^{-1} V^{21} Y_1,$$
$$C = (V^{22})^{-1} V^{21} Y_1.$$

Then

$$(X - \mu)' V^{-1}(X - \mu) = Y_1' [V^{11} - V^{12}(V^{22})^{-1} V^{21}] Y_1$$
$$+ (Y_2 + C)' V^{22}(Y_2 + C).$$

Then from (2.7.7)

$$V^{11} - V^{12}(V^{22})^{-1} V^{21} = V_{11}^{-1}.$$

Therefore

$$(X - \mu)' V^{-1}(X - \mu) = (X_1 - \mu_1)' V_{11}^{-1}(X_1 - \mu_1)$$
$$+ (Y_2 + C)' V^{22}(Y_2 + C), \quad Y_2 = X_2 - \mu_2.$$

We want to integrate out X_2 from $f(X)$. That is, denoting $\int_{X_2} (\cdot) dX_2$ as the integral over X_2,

$$\int_{X_2} f(X) dX_2 = c \int_{X_2} e^{-\frac{1}{2}(X-\mu)' V^{-1}(X-\mu)} dX_2$$

$$= c\, e^{-\frac{1}{2}(X_1-\mu_1)' V_{11}^{-1}(X_1-\mu_1)}$$
$$\times \int_{X_2} e^{-\frac{1}{2}(Y_2+C)' V^{22}(Y_2+C)} dX_2.$$

But

$$dX_2 = d(X_2 - \mu_2) = dY_2 = d(Y_2 + C)$$

because all other quantities are fixed as far as the integral over X_2 is concerned. Then the integral produces only a constant so that the normalizing constant c changes to another normalizing constant c_1. Then the marginal density of X_1, denoted by $f_1(X_1)$, is given by

$$f_1(X_1) = c_1 e^{-\frac{1}{2}(X_1-\mu_1)'V_{11}^{-1}(X_1-\mu_1)},$$

exactly of the same form as $f(X)$ with p replaced by r and with the corresponding changes in μ and V. Thus all subsets of the $p \times 1$ vector X have the densities belonging to the same family as $f(X)$ since the exponent is a symmetric function in the components of $X - \mu$ and since it is proved above that a subvector has the same type of Gaussian density. If $f_2(X_2)$, which is available from $f_1(X_1)$ with the corresponding changes or from $f(X)$ directly, denotes the marginal density of X_2 then the conditional density of X_1 given X_2 (given X_2 means X_2 is assumed to be a constant vector) is given by $f(X)/f_2(X_2)$. Show that this conditional density also belongs to the same family as $f(X)$. (Exercise for the student.)

Before concluding this section a few more applications of matrices will be pointed out. More will be considered after introducing the notion of determinants in the next chapter.

2.7.5 Regression analysis

In statistics, econometrics and other areas, prediction of a variable by observing other variables or at preassigned values of other variables is an important activity. For example the world market price for wheat on the next first of January, say y, is a function of many variables such as x_1 = the current Canadian stock of wheat, x_2 = the USA stock of wheat, x_3 = the Australian stock of wheat, x_4 = the drought situation in a wheat buying country and so on. If y is to be predicted at preassigned values of $x_1, x_2, \ldots, x_k$ (say k real variables) then $y = f(x_1, \ldots, x_k)$, some scalar function of $x_1, \ldots, x_k$. If we assume f to be linear then the prediction function is

$$y = a_0 + a_1 x_1 + \cdots + a_k x_k \tag{2.7.10}$$

where $x_1, \ldots, x_k$ can be preassigned but $a_1, \ldots, a_k$ are unknown. The above model is called a linear regression model because the model assumes that the expected value of y at preassigned values of $x_1, \ldots, x_k$ ("preassigned" means that if $x_1 = 1$ million tons, if $x_2 = 3$ million tons, etc. what will be y, y = the price per bushel on next January first) is of the form in (2.7.10). [Regression of y on $x_1, \ldots, x_k$ is the conditional expectation of y at given values of $x_1, \ldots, x_k$ because it can be proved that this conditional expectation is the best predictor of y, best in the minimum mean square sense. For evaluating this regression function we need the conditional distribution of y at given values of $x_1, \ldots, x_k$. In the absence of conditional distribution, we will assume that this best predictor is of a certain form, such as a linear function.] We have the data points from previous years' readings on y as well as on $x_1, \ldots, x_k$. If the assumed model is an exact mathematical relationship then for every data point $(y_j, x_{1j}, \ldots, x_{kj})$ the equation (2.7.10) is satisfied. This is not the reality. The model is simply assumed to hold. There may or may not be such a relationship. Hence if ϵ_j denotes the error in y_j for using the model in (2.7.10) then

$$\epsilon_j = y_j - [a_0 + a_1 x_{1j} + \cdots + a_k x_{kj}], \quad j = 1, \ldots, n, \tag{2.7.11}$$

if there are n data points. Since $k + 1$ parameters $a_0, \ldots, a_k$ are to be estimated n has to be at least $k + 1$. Let, for $n \geq k + 1$,

$$\epsilon = \begin{bmatrix} \epsilon_1 \\ \vdots \\ \epsilon_n \end{bmatrix}, \quad Y = \begin{bmatrix} y_1 \\ \vdots \\ y_n \end{bmatrix}, \quad \beta = \begin{bmatrix} a_0 \\ \vdots \\ a_k \end{bmatrix},$$

$$X = \begin{bmatrix} 1 & x_{11} & \cdots & x_{k1} \\ 1 & x_{12} & \cdots & x_{k2} \\ \vdots & \vdots & \cdots & \vdots \\ 1 & x_{1n} & \cdots & x_{kn} \end{bmatrix}.$$

Then

$$Y = X\beta + \epsilon$$

and the error sum of squares is then

$$\epsilon'\epsilon = (Y - X\beta)'(Y - X\beta) \tag{2.7.12}$$

where X is a known matrix, Y is a known set of observations, β is the only unknown vector. The maximum value of $\epsilon'\epsilon$ for arbitrary β is at $+\infty$, being non-negative. If β is estimated by minimizing (2.7.12) the method is called *the method of least squares*. Recall from Chapter 1 the partial differential operator, which in the present situation is,

$$\frac{\partial}{\partial \beta} = \begin{pmatrix} \frac{\partial}{\partial a_0} \\ \vdots \\ \frac{\partial}{\partial a_k} \end{pmatrix}$$

and the effects of operating on a quadratic form and linear form are already considered earlier. Thus

$$\frac{\partial}{\partial \beta}(\epsilon'\epsilon) = 0 \implies -2X'(Y - X\beta) = 0$$

$$\implies X'X\beta = X'Y. \tag{2.7.13}$$

Since x_{ij}'s are preselected, without loss of generality we can assume $X'X$ to be non-singular (when data collected from the field are used sometimes $X'X$ can be singular or nearly singular). If $X'X$ is nonsingular then, denoting the estimated value of β by $\hat{\beta}$, we have

$$\hat{\beta} = (X'X)^{-1}X'Y. \tag{2.7.14}$$

The following properties are easy to establish (left as exercises to the student):

(ii) The least square minimum S^2, that is the right side of (2.7.12) when β is replaced by $\hat{\beta}$, is a quadratic form of the type

$$Y'[I - X(X'X)^{-1}X']Y = Y'[I - B]Y.$$

(iii) $I - B$ is idempotent of rank $n - (k+1)$, B is idempotent of rank $k+1$ and further, $I - B$ and B are orthogonal to each other. Also

$$Y'Y = Y'[I - B]Y + Y'BY.$$

When $I - B$ and B are orthogonal to each other and when ϵ has a standard multivariate Gaussian distribution it can be proved that $Y'[I - B]Y$ and $Y'BY$ are statistically independently distributed. Comparison of the sum of squares due to β, namely $Y'BY$, with the residual sum of squares, namely $Y'[I - B]Y$, is the basis in regression analysis and in a large variety of statistical inference problems, such as testing statistical hypotheses on β or on the individual components of β.

2.7.6 Design of experiments

Another prominent area of applied statistics is the topics of design of experiments and analysis of variance. Suppose that 3 different methods of teaching (say comparison of instructors) are to be studied for their effectiveness. Suppose that 3 sets of students of, say 30 each, with exactly the same background are selected and subjected to the 3 different methods, one set of 30 under method 1, another set of 30 under method 2 and a third set of 30 under method 3. Designing aspect of the experiment is to control all possible known factors, other than the methods of teaching, which may contribute towards the grade of the student. Let y_{ij} be the grade of the j-th student under method i. Here $i = 1, 2, 3$ and $j = 1, 2, \ldots, 30$. In general, we may want to compare k methods ($i = 1, \ldots, k$) and under the i-th method there may be n_i students (n_1 students under method 1, n_2 students under method 2 and so on). Then a linear, additive, fixed effect model is the following:

$$y_{ij} = \mu + \alpha_i + e_{ij} \tag{2.7.15}$$

where μ is a general effect (the student would have got some grade if she/he had studied on her/his own; no instructor or method was involved), α_i the deviation from the general effect due to the i-th method of teaching, e_{ij} the random part (sum total contribution coming from all unknown factors. Remember that all known factors which may contribute towards y_{ij} are controlled by properly designing the experiment). Note that $\mu, \alpha_1, \ldots, \alpha_k$ are all unknown. y_{ij}'s are observed. The final aim is to test statistical hypotheses on α_i's such as the hypothesis that all the methods are equally effective $(\alpha_1 = \cdots = \alpha_k)$. First step towards the analysis is to estimate $\mu, \alpha_1, \ldots, \alpha_k$. We use the method of least squares. When $\mu, \alpha_1, \ldots, \alpha_k$ are all assumed to be unknown constants we minimize the error sum of squares for estimating the parameters.

$$\sum_{i=1}^{k} \sum_{j=1}^{n_i} e_{ij}^2 = \sum \sum (y_{ij} - \mu - \alpha_i)^2. \tag{2.7.16}$$

The following results can be easily established (left as exercises to the student):

(iv) The least square estimate of α_i is given by

$$\hat{\alpha}_i = \frac{y_{i.}}{n_i} - \frac{y_{..}}{n_.}, \quad y_{i.} = \sum_{j=1}^{n_i} y_{ij}, \quad y_{..} = \sum_{i=1}^{k} \sum_{j=1}^{n_i} y_{ij}.$$

(v) The least square minimum, in this case, is given by

$$S^2 = \sum_{ij} y_{ij}^2 - \sum_{i=1}^{k} \frac{y_{i.}^2}{n_i}.$$

The problem described above is part of the analysis in a one-way classification model arising from a completely randomized design in the field of Design of Experiments.

Exercises 2.7

2.7.1. Let

$$A = \begin{bmatrix} 3 & 0 & -1 & 2 \\ 1 & -1 & 0 & 1 \\ 1 & 1 & 1 & 1 \end{bmatrix}, \quad B = \begin{bmatrix} 1 & 2 \\ -1 & 0 \\ 1 & 1 \\ 2 & 1 \end{bmatrix},$$

$$C = \begin{bmatrix} 1 & 2 & 1 \\ -1 & 0 & 1 \\ 1 & 1 & 1 \\ 2 & 1 & 1 \end{bmatrix}, \quad A = \begin{bmatrix} A_{11} & A_{12} \\ A_{21} & A_{22} \end{bmatrix}, \quad A_{11} = \begin{bmatrix} 3 & 0 \\ 1 & -1 \end{bmatrix}.$$

(a) Give all possible partitioning of B so that AB is defined. Evaluate AB by using the submatrices for each partition of B.

(b) Give all possible partitioning of C so that AC is defined. Evaluate AC in terms of the product of submatrices for each partition of C.

2.7.2. By using equations (2.7.6) to (2.7.9) and the corresponding equations for A_{12} and A_{21} evaluate the inverses of the following matrices A, B and C, assuming the inverses exist, by partitioning into convenient submatrices:

$$A = \begin{bmatrix} a & b \\ c & d \end{bmatrix}, \quad B = \begin{bmatrix} a & 0 & 0 \\ 0 & b & c \\ 0 & d & e \end{bmatrix}, \quad C = \begin{bmatrix} a_1 & a_2 & a_3 \\ b_1 & b_2 & b_3 \\ c_1 & c_2 & c_3 \end{bmatrix}.$$

2.7.3. By partitioning the matrix A and looking at the leading submatrices answer the following question: If $X'AX = 0$ for all possible vectors X is $A = O$?

2.7.4. Variance, covariance, correlations. E = expected value is an operator in Statistics. Let X be a $p \times 1$ vector of real scalar random variables. Then $\mu = E(X)$ = the mean value of X, $V = E[(X - \mu)(X - \mu)']$ is the covariance matrix of X. If $V = (v_{ij})$ then v_{ii} is the variance of x_i, the i-th component in X, and v_{ij} is the covariance between x_i and x_j. Then $\rho_{ij} = \frac{v_{ij}}{\sqrt{v_{ii}}\sqrt{v_{jj}}}$ is the correlation between x_i and x_j. Let

$$X = \begin{pmatrix} x_1 \\ X_2 \end{pmatrix}, \quad X_2' = (x_2, x_3, \dots, x_p).$$

Let V be partitioned correspondingly. That is,

$$V = \begin{pmatrix} V_{11} & V_{12} \\ V_{21} & V_{22} \end{pmatrix}$$

where v_{11} is 1×1. Let $u = a'X_2$, $a' = (a_2, \dots, a_p)$ is a constant vector. This means that we are considering a linear function of X_2. Show the following: (a) the variance of u is $a'V_{22}a$, (b) the covariance between x_1 and X_2 is $V_{12} = V_{21}'$, (c) the maximum correlation possible between x_1 and u is

$$\frac{V_{12}V_{22}^{-1}V_{21}}{v_{11}} = \rho_{1(2\dots p)}^2$$

which is called the square of the multiple correlation between x_1 and X_2.

2.7.5. Show that $\rho_{1(2.3)}^2 \geq \rho_{1(2)}^2$ where $\rho_{1(\cdot)}^2$ is defined in Exercise 2.7.4. Generalize the result to show that $\rho_{1(2\dots p)}^2$ increases with p.

2.7.6. If A is real symmetric then show that A can be written as

$$PAP' = \begin{bmatrix} I_r & O & O \\ O & -I_s & O \\ O & O & O \end{bmatrix}$$

where $r + s$ is the rank of A and P is a nonsingular matrix. [The diagonal block O may be absent.]

2.7.7. If A is $n \times m$, $n > m$ such that $A'A = I_m$ then show that there exists an $n \times (n - m)$ matrix B such that (A, B) is orthonormal.

2.7.8. Kronecker product. Let $A = (a_{ij})$ be $p \times p$ and $B = (b_{ij})$ be $q \times q$. Consider the $pq \times pq$ matrix $A \otimes B$,

$$A \otimes B = \begin{bmatrix} a_{11}B & a_{12}B & \cdots & a_{1p}B \\ a_{21}B & a_{22}B & \cdots & a_{2p}B \\ \vdots & \vdots & \cdots & \vdots \\ a_{p1}B & a_{p2}B & \cdots & a_{pp}B \end{bmatrix}.$$

Then $A \otimes B$ is known as the Kronecker product of A with B.
(1) Evaluate the Kronecker product $A \otimes B$ where

$$A = \begin{bmatrix} 2 & 1 & 5 \\ 3 & 0 & -1 \\ 1 & 4 & 2 \end{bmatrix}, \quad B = \begin{bmatrix} 2 & 0 \\ 1 & -1 \end{bmatrix}.$$

(2) Compute $B \otimes A$ and compare with the result in (1).

2.7.9. Let A be an $m \times n$ real matrix of rank r. Show that there exists a nonsingular $m \times m$ matrix B and an orthonormal matrix C such that

$$A = B \begin{bmatrix} I_r & O \\ O & O \end{bmatrix} C.$$

2.7.10. Let A be an $m \times n$ real matrix, $m \geq n$. Show that there exists an $m \times m$ orthonormal matrix B such that

$$BA = \begin{pmatrix} T \\ O \end{pmatrix}$$

where T is an $n \times n$ upper triangular matrix and O is an $(m - n) \times n$ null matrix.

2.7.11. Let X be an $n \times 1$ vector of unit length. Show that $I_n - 2XX'$ is an orthonormal matrix.

2.7.12. Let A be an $m \times n$ real matrix of rank r with $m \geq n$. Show that there exists an $m \times (m - n)$ semiorthonormal matrix Q, $Q'Q = I$ and an $n \times n$ upper triangular matrix P such that

$$A = QP.$$

2.7.13. Consider the following diagonal block matrix D where A, B, C are $n \times n$, $m \times m$ and $r \times r$ square matrices. If D is nonsingular then show that $D^{-1} = \text{diag}(A^{-1}, B^{-1}, C^{-1})$ that is,

$$D = \begin{bmatrix} A & O & O \\ O & B & O \\ O & O & C \end{bmatrix}, \quad D^{-1} = \begin{bmatrix} A^{-1} & O & O \\ O & B^{-1} & O \\ O & O & C^{-1} \end{bmatrix}.$$

2.7.14. If $A = \begin{bmatrix} B & u \\ v' & d \end{bmatrix}$ and $A^{-1} = \begin{bmatrix} C & w \\ x' & \delta \end{bmatrix}$ where u, v, w are vectors and d and δ are scalars (1×1 matrices), then show that

$$B^{-1} = C - w\delta^{-1}x', \quad C = B^{-1} + \delta B^{-1} uv' B^{-1},$$
$$w = -\delta B^{-1} u, \quad x' = -\delta v' B^{-1}.$$

2.7.15. If $A = \begin{bmatrix} I & B \\ O & C \end{bmatrix}$ and nonsingular then show that

$$A^{-1} = \begin{bmatrix} I & -BC^{-1} \\ O & C^{-1} \end{bmatrix}.$$

Additional problems on vectors and matrices

Use the 8 conditions listed after Example 2.6.7 as axioms to define a vector space, assume scalar multiplication and addition are as done in the case of matrices and assume that the scalars are real or complex numbers. Then establish the results in Exercises 2.1 to 2.4 below.

2.1. Show that the following sets are vector spaces satisfying all the conditions as mentioned above.
(i)   The collection of vectors as n-tuples for a fixed n, as we have defined in Chapter 1 where the elements are real or complex numbers;
(ii) The collection of all polynomials in a real scalar variable of degree less than n, for a given n, with coefficients real or complex numbers;
(iii) The collection of all real-valued functions of a real variable which are differentiable;
(iv) The collection of all $n \times n$ matrices, with given n, where the elements are real or complex numbers.

2.2. Construct one basis each for the vector spaces in Exercise 2.1 (i)–(iv).

2.3. Let **V** be a collection of couplets of real and positive numbers, that is, **V** = $\{(\alpha, \beta) : \alpha > 0, \beta > 0\}$. Define addition and scalar multiplication as follows:

$$(\alpha_1, \beta_1) + (\alpha_2, \beta_2) = (\alpha_1 \alpha_2, \beta_1 \beta_2) \text{ for every } (\alpha_1, \beta_1) \text{ and } (\alpha_2, \beta_2) \text{ in } \mathbf{V}.$$
$$c(\alpha, \beta) = (\alpha^c, \beta^c) \text{ for every real number } c \text{ and } (\alpha, \beta) \text{ in } \mathbf{V}.$$

Show that **V** is a vector space over the field of real numbers.

2.4. Show that the concepts of linear dependence and linear independence of vectors, basis of a vector space and dimensions of vector spaces as defined for ordered n-tuples of real numbers also go through for the general vector space defined above. [More on abstract vector spaces may be seen from more mathematically oriented books on vectors and matrices, see for example, [9]

2.5. Find matrices A for which $A'A = O$, $A \neq O$.

2.6. Show that

$$\begin{bmatrix} \cos\theta & k\sin\theta \\ -\frac{1}{k}\sin\theta & \cos\theta \end{bmatrix}^n = \begin{bmatrix} \cos n\theta & k\sin n\theta \\ -\frac{1}{k}\sin n\theta & \cos n\theta \end{bmatrix}, \quad k \neq 0, \ n = 1, 2, \ldots.$$

2.7. Kronecker product. The Kronecker product $A \otimes B$ is defined as $A \otimes B = (a_{ij}B)$. If $A = (a_{ij})$ is $p \times q$ and $B = (b_{ij})$ is $m \times n$ then $A \otimes B$ is $pm \times qn$. For example let

$$A = \begin{bmatrix} 1 & -1 & 0 \\ 2 & 3 & -2 \end{bmatrix}, \quad B = \begin{bmatrix} 4 & 5 \\ -3 & 7 \end{bmatrix} \Rightarrow$$

$$A \otimes B = \begin{bmatrix} (1)B & (-1)B & (0)B \\ (2)B & (3)B & (-2)B \end{bmatrix} = \begin{bmatrix} 4 & 5 & -4 & -5 & 0 & 0 \\ -3 & 7 & 3 & -7 & 0 & 0 \\ 8 & 10 & 12 & 15 & -8 & -10 \\ -6 & 14 & -9 & 21 & 6 & -14 \end{bmatrix}$$

and

$$B \otimes A = \begin{bmatrix} (4)A & (5)A \\ (-3)A & (7)A \end{bmatrix} = \begin{bmatrix} 4 & -4 & 0 & 5 & -5 & 0 \\ 8 & 12 & -8 & 10 & 15 & -10 \\ -3 & 3 & 0 & 7 & -7 & 0 \\ -6 & -9 & 6 & 14 & 21 & -14 \end{bmatrix} \neq A \otimes B.$$

Write down $A \otimes B$ and $B \otimes A$ if

$$A = \begin{bmatrix} 1 & -1 \\ -2 & 0 \end{bmatrix} \quad \text{and} \quad B = \begin{bmatrix} 3 & 2 \\ -3 & 4 \end{bmatrix}.$$

2.8. vec(X). Let $X = (x_{ij})$ be a $p \times q$ matrix. Let the j-th column of X be denoted by $x_{(j)}$. Consider the $pq \times 1$ vector formed by appending $x_{(1)}, x_{(2)}, \ldots, x_{(q)}$ into a long column. This is defined as vec(X). That is,

$$\text{vec}(X) = \begin{bmatrix} x_{(1)} \\ x_{(2)} \\ \vdots \\ x_{(q)} \end{bmatrix}.$$

For example, let

$$A = \begin{bmatrix} 1 & -1 & 1 \\ 2 & 0 & 5 \end{bmatrix} \Rightarrow \text{vec}(A) = \begin{bmatrix} 1 \\ 2 \\ -1 \\ 0 \\ 1 \\ 5 \end{bmatrix} \quad \text{and} \quad [\text{vec}(A)]' = (1, 2, -1, 0, 1, 5).$$

Form vec(A) for the following matrices:

(i) $A = \begin{bmatrix} 1 & 0 & -1 \\ 2 & 1 & -1 \\ 0 & 1 & 0 \end{bmatrix}$, (ii) $A = \begin{bmatrix} 1 & 1 \\ -1 & 0 \\ 2 & 5 \end{bmatrix}$, (iii) $A = I_3$.

2.9. Show that if A is $p \times q$, X is $q \times r$ and B is $r \times s$ then the $ps \times 1$ vector

$$\text{vec}(AXB) = (B' \otimes A)\text{vec}(X).$$

2.10. If $Y = AX$ where X and Y are $p \times q$ matrices of functionally independent real variables, A is a $p \times p$ nonsingular matrix of constants and if $(d\mathbf{Y})$ and $(d\mathbf{X})$ represent the matrices of differentials in Y and X respectively then show that

$$\text{vec}(d\mathbf{Y}) = (I \otimes A)\text{vec}(d\mathbf{X}).$$

2.11. If A, B, C, D are matrices for which the following sums and products are defined then establish the following results:
(i) $A \otimes B \otimes C = (A \otimes B) \otimes C = A \otimes (B \otimes C)$
(ii) $(A + B) \otimes (C + D) = (A \otimes C) + A \otimes D + D \otimes C + B \otimes D$
(iii) $(A \otimes B)(C \otimes D) = AC \otimes BD$
(iv) $\alpha \otimes A = \alpha A = A \otimes \alpha$, α a scalar
(v) $(A \otimes B)' = A' \otimes B'$
(vi) $\text{tr}(A \otimes B) = [\text{tr}(A)][\text{tr}(B)]$
(vii) $(A \otimes B)^{-1} = A^{-1} \otimes B^{-1}$
(viii) $a' \otimes b = ba' = b \otimes a'$

where a and b are two column vectors, not necessarily of the same order. Also if A is a partitioned matrix where $A_{11}, A_{12}, A_{21}, A_{22}$ are submatrices and if B is any other matrix then show that

$$A = \begin{bmatrix} A_{11} & A_{12} \\ A_{21} & A_{22} \end{bmatrix} \Rightarrow A \otimes B = \begin{bmatrix} A_{11} \otimes B & A_{12} \otimes B \\ A_{21} \otimes B & A_{22} \otimes B \end{bmatrix}.$$

2.12. For two matrices A and B show that $A \otimes B$ is nonsingular if and only if A and B are nonsingular.

2.13. For two matrices A and B show that $\text{vec}(A) = \text{vec}(B)$ does not necessarily imply that $A = B$.

2.14. For any two column vectors a and b show that

$$\text{vec}(a) = \text{vec}(a') \quad \text{and} \quad \text{vec}(ab') = b \otimes a.$$

2.15. For three matrices A, B, C for which the product ABC is defined show that

$$\text{vec}(ABC) = (C' \otimes A)\text{vec}(B) \quad \text{and} \quad \text{vec}(AB) = (B' \otimes I)\text{vec}(A).$$

2.16. If A is $m \times n$, B is $n \times r$ and α is $r \times 1$ then show that

$$\text{vec}(AB) = (B' \otimes I_m)\text{vec}(A) = (B' \otimes A)\text{vec}(I_n) = (I_q \otimes A)\text{vec}(B)$$
$$AB\alpha = (\alpha' \otimes A)\text{vec}(B) = (A \otimes \alpha')\text{vec}(B').$$

2.17. If A, B, C are $n \times n$ matrices and further, if $C = C'$ then show that

$$[\text{vec}(C)]'(A \otimes B)\text{vec}(C) = [\text{vec}(C)]'(B \otimes A)\text{vec}(C).$$

2.18. For any $m \times n$ matrix A show that

$$\text{vec}(A) = (I_n \otimes A)\text{vec}(I_n) = (A' \otimes I_m)\text{vec}(I_m).$$

2.19. For any four matrices A, B, C, D let $ABCD$ be defined such that $ABCD$ is a square matrix. Then show that

$$\text{tr}(ABCD) = [\text{vec}(D')]'(C' \otimes A)\text{vec}(B) = [\text{vec}(D)]'(A \otimes C')\text{vec}(B').$$

2.20. If $\rho(A)$ and $\rho(B)$ denote the ranks of A and B respectively then show that

$$\rho(A)\,\rho(B) = \rho(A \otimes B).$$

2.21. Let A be a square matrix of order n and let $p(x)$ be a polynomial of degree k in the scalar variable x, that is

$$p(x) = a_0 + a_1 x + \cdots + a_k x^k.$$

Let $p(x), q(x), h(x), t(x)$ be polynomials in x such that

$$p(x) + q(x) = h(x) \quad \text{and} \quad p(x)\,q(x) = t(x).$$

Define

$$p(A) = a_0 I + a_1 A + \cdots + a_k A^k.$$

Then show that

$$p(A) + q(A) = h(A) \quad \text{and} \quad p(A)\,q(A) = t(A).$$

2.22. If $\rho(\cdot)$ denotes the rank of $(\cdot)$ and if AB and $A + B$ are defined then show that

$$\rho(AB) \le \rho(A), \quad \rho(AB) \le \rho(B), \quad \rho(A + B) \le \rho(A) + \rho(B).$$

2.23. For the following matrix A take A^k, $k = 1, 2, \ldots$ by using a computer. Then show that for all $k \ge 16$ the matrix A^k remains the same (up to three decimal places) where

$$A = \begin{bmatrix} 0.8 & 0.2 & 0.1 \\ 0.1 & 0.7 & 0.3 \\ 0.1 & 0.1 & 0.6 \end{bmatrix}.$$

2.24. Take a 2×2 singly stochastic matrix A with nonzero elements and with the column sums equal to 1. Take powers of A by using a computer. Explain the behavior of A^k for large k.

2.25. Repeat the process in Exercise 2.24 for a 3×3 singly stochastic matrix A and explain the behavior for large k.

3 Determinants

3.0 Introduction

A determinant is an explicit scalar function of the elements of a square matrix. It is defined only for square matrices. A scalar function means a function which when evaluated is a 1×1 matrix. If the elements are real or complex numbers then this function will also be a real or complex number. A few scalar functions of a 2×2 matrix, $A = \begin{pmatrix} x_1 & x_2 \\ x_3 & x_4 \end{pmatrix}$, denoted by u_1, u_2, u_3, are the following:

$$u_1 = x_1^2 + x_2^2 + x_3^2 + x_4^2 = \text{tr}(AA'),$$
$$u_2 = x_1 x_4 - x_2 x_3,$$
$$u_3 = x_1 + x_4 = \text{tr}(A).$$

All the above functions are 1×1 matrices or scalars. Out of these three we will be interested in a function of the type u_2. Why are we interested in such a scalar function? It is mainly because of its applications in various fields. In Chapter 2 we have seen that matrices appear naturally in many areas. More examples from several other areas could also have been given in Chapter 2. In all such problems, where square matrices come in, determinants can enter automatically.

We will define the determinant of an $n \times n$ matrix as a scalar function satisfying certain conditions. Let $\alpha_1, \alpha_2, \ldots, \alpha_n$ denote the n rows (or columns) of an $n \times n$ matrix $A = (a_{ij})$. Then, if they are rows,

$$\alpha_i = (a_{i1}, \ldots, a_{in}), \quad i = 1, \ldots, n.$$

That is,

$$\alpha_1 = (a_{11}, a_{12}, \ldots, a_{1n}) = \text{ first row,}$$
$$\alpha_2 = (a_{21}, a_{22}, \ldots, a_{2n}) = \text{ second row, and so on}$$

3.1 Definition of the determinant of a square matrix

Let f be a scalar function (not a vector function or a matrix function) of $\alpha_1, \ldots, \alpha_n$, called the *determinant* of A, satisfying the following conditions:

$$f(\alpha_1, \ldots, c\alpha_i, \ldots, \alpha_n) = cf(\alpha_1, \ldots, \alpha_i, \ldots, \alpha_n), \qquad (\alpha)$$

where c is a scalar. This condition means that if any row (column) is multiplied by a scalar then it is equivalent to multiplying the whole determinant by that scalar. This scalar quantity can also be zero.

$$f(\alpha_1, \ldots, \alpha_i, \ldots, \alpha_i + \alpha_j, \ldots, \alpha_n) = f(\alpha_1, \ldots, \alpha_i, \ldots, \alpha_j, \ldots, \alpha_n). \qquad (\beta)$$

This condition says that if the i-th row (column) is added to the j-th row (column) the value of the determinant remains the same. The i-th row (column) remains the same but the new j-th row is $\alpha_i + \alpha_j$, that is, the original j-th row (column) plus the original i-th row (column). Combination of conditions (α) and (β) shows that the value of the determinant remains the same if a constant multiple of one row (column) is added to another row (column).

Let α_i be written as a sum of two vectors, $\alpha_i = \beta_i + \gamma_i$. Then the next condition is that

$$f(\alpha_1, \dots, \beta_i + \gamma_i, \dots, \alpha_n) = f(\alpha_1, \dots, \beta_i, \dots, \alpha_n) + f(\alpha_1, \dots, \gamma_i, \dots, \alpha_n). \tag{γ}$$

This means that if the i-th row (column) is split as the sum of two vectors, $\beta_i + \gamma_i$, then the determinant becomes sum of two determinants where in one the i-th row (column) is replaced by β_i and in the other the i-th row (column) is replaced by γ_i.

$$f(e_1, \dots, e_n) = 1 \tag{δ}$$

where $e_1, \dots, e_n$ are the basic unit vectors. This condition says that the determinant of an identity matrix is 1.

The above conditions can be called the postulates or axioms to define the determinant of a square matrix. The standard notations used to denote the determinant of a square matrix A are the following:

$$|A|, \ \det(A) = \text{determinant of } A.$$

In the first notation above, A is enclosed by vertical bars. The matrix was enclosed by ordinary or square brackets.

$$A = \begin{bmatrix} a & b \\ c & d \end{bmatrix} = \text{matrix } A, \quad |A| = \begin{vmatrix} a & b \\ c & d \end{vmatrix} = \text{determinant of } A.$$

Let us evaluate the determinant of this 2×2 matrix A by using the postulates above. Let $a \neq 0$. Then by postulate (α),

$$|A| = a \begin{vmatrix} 1 & b/a \\ c & d \end{vmatrix}.$$

We have divided the first row by a and in order to keep the value of the determinant the same we kept a outside also because dividing any row by $a \neq 0$ is equivalent to dividing the whole determinant by a. In other words we have taken a outside from the first row. Now add $(-c)$ times the first row of the determinant on the right to the second row. By postulates (α) and (β), the value of the determinant remains the same. Then

$$|A| = a \begin{vmatrix} 1 & \frac{b}{a} \\ c & \end{vmatrix} = a \begin{vmatrix} 1 & \frac{b}{a} \\ 0 & d - \frac{cb}{a} \end{vmatrix} \quad \text{(postulates } (\alpha), (\beta))$$

$$= a\left(d - \frac{cb}{a}\right)\begin{vmatrix} 1 & \frac{b}{a} \\ 0 & 1 \end{vmatrix} \quad \text{(postulate } (\alpha))$$

$$= a\left(d - \frac{cb}{a}\right)\begin{vmatrix} 1 & 0 \\ 0 & 1 \end{vmatrix} \quad \text{(postulates } (\alpha), (\beta))$$

$$= a\left(d - \frac{cb}{a}\right) \quad \text{(postulate } (\delta))$$

$$= ad - bc.$$

If $a = 0$ then by adding other rows or columns one can bring a nonzero element at the $(1,1)$-th position without altering the value of the determinant, unless all elements on the first row or first column are zeros. In such a case the determinant is zero by postulate (α) itself. Thus, in general, we have the following result:

(i) The determinant of a 2×2 matrix $A = \left[\begin{smallmatrix} a & b \\ c & d \end{smallmatrix}\right]$ is $|A| = ad - bc$.

For example,

$$A = \begin{bmatrix} 2 & 3 \\ -1 & 5 \end{bmatrix} \Rightarrow |A| = (2)(5) - (3)(-1) = 13;$$

$$B = \begin{bmatrix} 2 & 0 \\ 0 & 5 \end{bmatrix} \Rightarrow |B| = (2)(5) - (0)(0) = 10;$$

$$C = \begin{bmatrix} 2 & 3 \\ 0 & 5 \end{bmatrix} \Rightarrow |C| = (2)(5) - (3)(0) = 10.$$

3.1.1 Some general properties

A few properties follow immediately from the definition itself.

(ii) The determinant of a square null matrix is zero. The determinant of a square matrix with one or more rows or columns null is zero.
(iii) The determinant of a diagonal matrix is the product of the diagonal elements. [This means that if any diagonal element in a diagonal matrix is zero then the determinant is zero.]
(iv) The determinant of a triangular (upper or lower) matrix is the product of the diagonal elements. [This means that if any diagonal element in a triangular matrix is zero then its determinant is zero.]

For example,

$$O = \begin{bmatrix} 0 & 0 & 0 \\ 0 & 0 & 0 \\ 0 & 0 & 0 \end{bmatrix} \Rightarrow |O| = 0;$$

$$A = \begin{bmatrix} 0 & a_2 & a_3 \\ 0 & b_2 & b_3 \\ 0 & c_2 & c_3 \end{bmatrix} \Rightarrow |A| = 0;$$

$$D = \begin{bmatrix} d_1 & 0 & 0 \\ 0 & d_2 & 0 \\ 0 & 0 & d_3 \end{bmatrix} \Rightarrow |D| = d_1 d_2 d_3;$$

$$T = \begin{bmatrix} a_1 & a_2 & a_3 \\ 0 & b_2 & b_3 \\ 0 & 0 & c_3 \end{bmatrix} \Rightarrow |T| = a_1 b_2 c_3.$$

In A above, it is equivalent to multiplying the first column by the scalar $c = 0$ and hence by postulate (α) the determinant of A is zero. In the triangular case T above, by repeated application of postulate (α) and (β) the elements at the $(1,2)$-th, $(1,3)$-th and $(2,3)$-th positions in T can be made zero without affecting the value of the determinant. Then applying (iii) above the result follows:

(v) If any two rows (columns) are interchanged then the value of the determinant of the new matrix (obtained after the interchange) is (-1) times the value of the determinant of the original matrix.

For example,

$$A = \begin{bmatrix} 1 & 0 & -1 \\ 1 & 1 & 1 \\ 0 & -1 & 2 \end{bmatrix} \quad \text{and} \quad A_1 = \begin{bmatrix} 0 & -1 & 2 \\ 1 & 1 & 1 \\ 1 & 0 & -1 \end{bmatrix} \Rightarrow$$

$$|A_1| = -|A| \quad \text{or} \quad |A| = -|A_1|.$$

Here A_1 is obtained from A by interchanging the first and the third rows. Let

$$A_2 = \begin{bmatrix} 0 & 1 & -1 \\ 1 & 1 & 1 \\ -1 & 0 & 2 \end{bmatrix} \Rightarrow |A_2| = -|A|.$$

Here A_2 is obtained by interchanging the first and second columns.

Property (v) can be easily established. An outline of the proof is given here. Let $\alpha_1, \dots, \alpha_n$ denote the rows (columns) of a matrix A. The i-th and j-th positions are indicated in the following sequences of steps and the postulate used is indicated at the right of each line:

$$|A| = f(\alpha_1, \dots, \alpha_i, \dots, \alpha_j, \dots, \alpha_n)$$
$$= f(\alpha_1, \dots, \alpha_i, \dots, \alpha_i + \alpha_j, \dots, \alpha_n) \qquad (\beta)$$
$$= -f(\alpha_1, \dots, \alpha_i, \dots, -\alpha_i - \alpha_j, \dots, \alpha_n) \qquad (\alpha)$$

$$= -f(\alpha_1, \dots, -\alpha_j, \dots, -\alpha_i - \alpha_j, \dots, \alpha_n) \qquad (\beta)$$
$$= f(\alpha_1, \dots, \alpha_j, \dots, -\alpha_i - \alpha_j, \dots, \alpha_n) \qquad (\alpha)$$
$$= f(\alpha_1, \dots, \alpha_j, \dots, -\alpha_i, \dots, \alpha_n) \qquad (\beta)$$
$$= -f(\alpha_1, \dots, \alpha_j, \dots, \alpha_i, \dots, \alpha_n). \qquad (\alpha)$$

In the above steps the i-th row (column) α_i and the j-th row (column) α_j are shown to indicate the changes in these rows (columns) at the i-th and the j-th positions. For example, step 2 above says that the i-th row (column) is added to the j-th row (column) and the value of the determinant remaining the same.

Example 3.1.1. Evaluate the determinant of the following matrix:

$$A = \begin{bmatrix} 1 & 1 & -1 & 2 \\ 0 & -1 & 3 & 0 \\ 2 & 1 & -2 & 4 \\ 5 & 0 & -1 & 1 \end{bmatrix}.$$

Solution 3.1.1. We add (-2) times the first row to the third row, the value of the determinant remains the same according to postulate (β). We will indicate the steps by using our standard notations of Chapters 1 and 2.

$$|A| = \begin{vmatrix} 1 & 1 & -1 & 2 \\ 0 & -1 & 3 & 0 \\ 2 & 1 & -2 & 4 \\ 5 & 0 & -1 & 1 \end{vmatrix} = \begin{vmatrix} 1 & 1 & -1 & 2 \\ 0 & -1 & 3 & 0 \\ 0 & -1 & 0 & 0 \\ 5 & 0 & -1 & 1 \end{vmatrix}.$$

Now, starting with the determinant on the right above we add (-5) times the first row to the last row, the value of the determinant remains the same according to postulate (β). That is,

$$-5(1) + (4) \implies |A| = \begin{vmatrix} 1 & 1 & -1 & 2 \\ 0 & -1 & 3 & 0 \\ 0 & -1 & 0 & 0 \\ 0 & -5 & 4 & -9 \end{vmatrix}.$$

Now we start with the second row and apply postulate (β) repeatedly.

$$-1(2) + (3); \quad -5(2) + (4) \implies$$

$$|A| = \begin{vmatrix} 1 & 1 & -1 & 2 \\ 0 & -1 & 3 & 0 \\ 0 & 0 & -3 & 0 \\ 0 & 0 & -11 & -9 \end{vmatrix}.$$

For the next step we take out (-3) from the third row by using postulate (α) and then add 11 times the third row to the fourth row. That is,

$$|A| = (-3) \begin{vmatrix} 1 & 1 & -1 & 2 \\ 0 & -1 & 3 & 0 \\ 0 & 0 & 1 & 0 \\ 0 & 0 & -11 & -9 \end{vmatrix}$$

$$= (-3) \begin{vmatrix} 1 & 1 & -1 & 2 \\ 0 & -1 & 3 & 0 \\ 0 & 0 & 1 & 0 \\ 0 & 0 & 0 & -9 \end{vmatrix} = (-3)(1)(-1)(1)(-9) = -27.$$

The last step above is obtained by multiplying the diagonal elements because the matrix in the determinant is triangular. The above operations were in fact elementary operations. Our aim was to reduce the matrix to a triangular form (upper or lower) so that we know that the determinant would be the product of the diagonal elements there.

Example 3.1.2. Evaluate the determinant of the following matrix:

$$A = \begin{bmatrix} 0 & 2 & 1 & 5 \\ 4 & 0 & 1 & 2 \\ -7 & 1 & -1 & 0 \\ 5 & 0 & 1 & 2 \end{bmatrix}.$$

Solution 3.1.2. The element at the $(1,1)$-th position is zero. We will bring a nonzero element at the $(1,1)$-th position by interchanging rows or columns. For the above A if the first and the third columns are interchanged then we can bring a convenient number at the $(1,1)$-th position. Remember that by this interchange the original determinant is multiplied by (-1):

$$|A| = - \begin{vmatrix} 1 & 2 & 0 & 5 \\ 1 & 0 & 4 & 2 \\ -1 & 1 & -7 & 0 \\ 1 & 0 & 5 & 2 \end{vmatrix}.$$

Now we do the following operations on the rows, without altering the value of the determinant:

$$-1(1) + (2); \quad (1) + (3); \quad -1(1) + (4) \Rightarrow$$

$$|A| = - \begin{vmatrix} 1 & 2 & 0 & 5 \\ 0 & -2 & 4 & -3 \\ 0 & 3 & -7 & 5 \\ 0 & -2 & 5 & -3 \end{vmatrix}.$$

By adding (-1) times the second row to the fourth row we can get rid off the number at the $(4,2)$-th position. That is,

$$|A| = - \begin{vmatrix} 1 & 2 & 0 & 5 \\ 0 & -2 & 4 & -3 \\ 0 & 3 & -7 & 5 \\ 0 & 0 & 1 & 0 \end{vmatrix}.$$

In order to get rid off the element at the $(3,2)$-th position we can divide the second row by (-2) and then operate with the new second row. This brings in fractions and the computations become complicated. But fractions can be avoided by multiplying the second row by 3 and the third row by 2 and then operating with the second row. [When a row is multiplied by a nonzero scalar remember to keep its reciprocal outside to maintain the value of the determinant.]

$$|A| = -\frac{1}{(2)(3)} \begin{vmatrix} 1 & 2 & 0 & 5 \\ 0 & -6 & 12 & -9 \\ 0 & 6 & -14 & 10 \\ 0 & 0 & 1 & 0 \end{vmatrix}$$

$$= -\frac{1}{6} \begin{vmatrix} 1 & 2 & 0 & 5 \\ 0 & -6 & 12 & -9 \\ 0 & 0 & -2 & 1 \\ 0 & 0 & 1 & 0 \end{vmatrix}.$$

The above line is the result of the operation $[(2) + (3) \Rightarrow]$. Now we can either interchange the third and the fourth rows to avoid fractions or multiply the fourth row by 2 and then add the third row to the fourth row. Using the latter we have

$$|A| = -\frac{1}{12} \begin{vmatrix} 1 & 2 & 0 & 5 \\ 0 & -6 & 12 & -9 \\ 0 & 0 & -2 & 1 \\ 0 & 0 & 2 & 0 \end{vmatrix}$$

$$= -\frac{1}{12} \begin{vmatrix} 1 & 2 & 0 & 5 \\ 0 & -6 & 12 & -9 \\ 0 & 0 & -2 & 1 \\ 0 & 0 & 0 & 1 \end{vmatrix}.$$

Now the matrix is brought to a triangular form and hence the determinant is the product of the diagonal elements. That is,

$$|A| = -\frac{1}{12}(1)(-6)(-2)(1) = -1.$$

When evaluating a determinant the following are the steps to remember

(1) If the element at the $(1,1)$-th position is 1 or the least common multiple of the elements in the first column then start the operations. With the help of the first row

make all elements in the first column, except the first one, zeros. Adding any multiple of any row to any other row will not change the value of the determinant. The basic aim is to bring the matrix of the determinant to a triangular form.

(2) If the element at the $(1,1)$-th position is not 1 or not the least common multiple of the elements in the first column then either by adding a row (column) to the first row (column) or by interchange bring a suitable number to the $(1,1)$-th position. For each interchange keep one (-1) each outside because by one interchange of rows or columns the determinant is multiplied by (-1). If these operations do not bring a convenient number to the $(1,1)$-th position then multiply the rows (columns), as many of them as necessary, so that when suitable multiples of the resulting first row is added to the other rows fractions are avoided. When any row (column) is multiplied by a number $c \neq 0$ remember to keep $\frac{1}{c}$ outside to maintain the value of the determinant.

(3) After reducing the elements in the first column, except the first one, to zeros start with the second row. Follow through the above steps with regard to the element at the $(2,2)$-th position. Operate only with the rows and columns from the second onward. Otherwise the triangular nature of the final format will not be achieved. Try to make all elements on the second column, starting from the third element onward, to zeros. Repeat the same process with the third row, fourth row and so on.

(4) The value of the determinant is the product of the diagonal elements in the final triangular format, multiplied by the quantities which were kept outside to maintain the value of the determinant.

(5) For matrices of order $n = 2, 3$, that is $n \times n$ matrices with $n = 2$ or $n = 3$, evaluate the determinants directly. The 2×2 case is already dealt with in property (i) and a mechanical procedure for the 3×3 case will be considered in the next section. For $n \geq 4$ use the steps (1) to (4) given above.

When every element of a matrix is multiplied by a scalar quantity c we say that the matrix is multiplied by c whereas when any one particular row or column of a matrix is multiplied by c its determinant is multiplied by c. Thus we have the following result:

(vi) When an $n \times n$ matrix A is multiplied by the scalar c its determinant is multiplied by c^n. That is,

$$|cA| = c^n|A|, \quad \text{for example, } |-A| = (-1)^n|A|, \ |2A| = 2^n|A|.$$

(vii) The value of the determinant of a matrix of real numbers can be negative, positive or zero.

From our postulate (β) the value of the determinant remains the same if any multiple of any row(column) is added to any other row (column). Thus if one or more rows (columns) are linearly dependent on other rows (columns) then these dependent rows (columns) can be made null by linear operations. Then the determinant is zero. Thus we have the following result:

(viii) The determinant of a singular matrix (rows/columns are linearly dependent) is zero and that of a nonsingular matrix is nonzero or one can define singularity or nonsingularity of a matrix through this property of its determinant.

Definition 3.1.1. A square matrix A is singular iff $|A| = 0$ and nonsingular iff $|A| \neq 0$.

3.1.2 A mechanical way of evaluating a 3×3 determinant

We can evaluate an $n \times n$ determinant, by adding suitable combinations of rows (columns) to other rows (columns) and using the basic properties of determinants as shown in the illustrative examples. If the steps are applied to a 3×3 determinant the final answer can be shown to be equivalent to the expression obtained by the following mechanical procedure. Let

$$A = \begin{bmatrix} a_1 & a_2 & a_3 \\ b_1 & b_2 & b_3 \\ c_1 & c_2 & c_3 \end{bmatrix}.$$

Write down the first two columns as the fourth and the fifth columns. Then we have the configuration

$$\begin{matrix} a_1 & a_2 & a_3 & a_1 & a_2 \\ b_1 & b_2 & b_3 & b_1 & b_2 \\ c_1 & c_2 & c_3 & c_1 & c_2 \end{matrix}$$

$$\begin{matrix} a_1 & a_2 & a_3 & a_3 & a_1 & a_2 \\ \searrow & \searrow & \searrow & \swarrow & \swarrow & \swarrow \end{matrix}$$

Take the product of the leading diagonal elements starting with the $(1,1)$-th, then with the $(1,2)$-th, and then with the $(1,3)$-th elements. We have three terms from this operation. Now we look at the second diagonal, the diagonal going from the bottom left corner and up. Take the product of the elements in the second diagonal starting with the $(3,1)$-th element, then with the $(3,2)$-th element and then with the $(3,3)$-th element. We have a second set of three elements. Multiply each element in the second set by (-1). The sum of these two sets of 6 elements is the value of the determinant. The operations are shown symbolically as above. The final answer is the following:

$$|A| = \begin{vmatrix} a_1 & a_2 & a_3 \\ b_1 & b_2 & b_3 \\ c_1 & c_2 & c_3 \end{vmatrix} = a_1 b_2 c_3 + a_2 b_3 c_1 + a_3 b_1 c_2$$

$$- a_3 b_2 c_1 - a_1 b_3 c_2 - a_2 b_1 c_3. \tag{3.1.1}$$

[The student may verify it by evaluating the determinant directly.] This mechanical procedure works only for the 3×3 case.

Example 3.1.3. Evaluate the determinant of A by using the mechanical procedure in (3.1.1) and verify the result by evaluating it with the help of the postulates, where,

$$A = \begin{bmatrix} 1 & 2 & 4 \\ -1 & 3 & 2 \\ -2 & 0 & -1 \end{bmatrix}.$$

Solution 3.1.3. For getting the solution from the mechanical procedure we write the first two columns again and then from the leading and the second diagonals of the new configuration we obtain the final answer. The new configuration is the following:

$$\begin{matrix} 1 & 2 & 4 & 1 & 2 \\ -1 & 3 & 2 & -1 & 3 \\ -2 & 0 & -1 & -2 & 0 \end{matrix} \rightarrow [(1)(3)(-1) + (2)(2)(-2) + (4)(-1)(0)]$$

$$- [(4)(3)(-2) + (1)(2)(0) + (2)(-1)(-1)]$$

$$= [(-3) + (-8) + (0)] - [(-24) + (0) + (2)]$$

$$= 11.$$

In order to verify the result let us evaluate the determinant directly through elementary operations. [The operations are listed on the right of each line.]

$$|A| = \begin{vmatrix} 1 & 2 & 4 \\ -1 & 3 & 2 \\ -2 & 0 & -1 \end{vmatrix}$$

$$= \begin{vmatrix} 1 & 2 & 4 \\ 0 & 5 & 6 \\ 0 & 4 & 7 \end{vmatrix} \quad [(1) + (2); 2(1) + (3) \Rightarrow]$$

$$= \frac{1}{20} \begin{vmatrix} 1 & 2 & 4 \\ 0 & 20 & 24 \\ 0 & 20 & 35 \end{vmatrix} \quad [4(2); 5(3); \Rightarrow]$$

$$= \frac{1}{20} \begin{vmatrix} 1 & 2 & 4 \\ 0 & 20 & 24 \\ 0 & 0 & 11 \end{vmatrix} \quad [-1(2) + (3) \Rightarrow]$$

$$= \frac{1}{20} (1)(20)(11) = 11.$$

Let us evaluate a 3×3 determinant of $A = (a_{ij})$ by using the postulates (γ) and then (α) and (β). Let us open up the first row.

$$(a_{11}, a_{12}, a_{13}) = a_{11}(1, 0, 0) + a_{12}(0, 1, 0) + a_{13}(0, 0, 1).$$

Then by postulate (y)

$$|A| = \begin{vmatrix} a_{11} & a_{12} & a_{13} \\ a_{21} & a_{22} & a_{23} \\ a_{31} & a_{32} & a_{33} \end{vmatrix}$$

$$= a_{11} \begin{vmatrix} 1 & 0 & 0 \\ a_{21} & a_{22} & a_{23} \\ a_{31} & a_{32} & a_{33} \end{vmatrix} + a_{12} \begin{vmatrix} 0 & 1 & 0 \\ a_{21} & a_{22} & a_{23} \\ a_{31} & a_{32} & a_{33} \end{vmatrix} + a_{13} \begin{vmatrix} 0 & 0 & 1 \\ a_{21} & a_{22} & a_{23} \\ a_{31} & a_{32} & a_{33} \end{vmatrix}.$$

Consider the first matrix on the right. Using the first row by adding suitable multiple to the other rows (postulates (α) and (β)) we can wipe out all elements in the first column, except the first element which is 1. Similarly, by using the first row we can wipe out all elements in the second column, except the first one, in the second matrix on the right. We can wipe out all elements in the third column, except the first one, in the third matrix on the right. The result is the following:

$$|A| = a_{11} \begin{vmatrix} 1 & 0 & 0 \\ 0 & a_{22} & a_{23} \\ 0 & a_{32} & a_{33} \end{vmatrix} + a_{12} \begin{vmatrix} 0 & 1 & 0 \\ a_{21} & 0 & a_{23} \\ a_{31} & 0 & a_{23} \end{vmatrix} + a_{13} \begin{vmatrix} 0 & 0 & 1 \\ a_{21} & a_{22} & 0 \\ a_{31} & a_{32} & 0 \end{vmatrix}.$$

Then by transpositions (one transposition is one interchange of adjacent rows or columns which will result in one minus sign coming out of the determinant also) of the columns we can bring the second column in the second matrix on the right to the first column position and the third column in the third matrix on the right to the first column position. In the second matrix on the right we need one transposition and hence one minus sign will come out. In the third matrix on the right we need two transpositions and hence $(-1)^2 = 1$ will come out. The final result is the following:

$$|A| = a_{11} \begin{vmatrix} 1 & 0 & 0 \\ 0 & a_{22} & a_{23} \\ 0 & a_{32} & a_{33} \end{vmatrix} - a_{12} \begin{vmatrix} 1 & 0 & 0 \\ 0 & a_{21} & a_{23} \\ 0 & a_{31} & a_{33} \end{vmatrix} + a_{13} \begin{vmatrix} 1 & 0 & 0 \\ 0 & a_{21} & a_{22} \\ 0 & a_{31} & a_{32} \end{vmatrix}. \qquad \text{(i)}$$

Now, we will consider each determinant on the right. For example, consider the first one. Open up the second row by using postulate y. Then we have

$$\begin{vmatrix} 1 & 0 & 0 \\ 0 & a_{22} & a_{23} \\ 0 & a_{32} & a_{33} \end{vmatrix} = a_{22} \begin{vmatrix} 1 & 0 & 0 \\ 0 & 1 & 0 \\ 0 & a_{32} & a_{33} \end{vmatrix} + a_{23} \begin{vmatrix} 1 & 0 & 0 \\ 0 & 0 & 1 \\ 0 & a_{32} & a_{33} \end{vmatrix}. \qquad \text{(ii)}$$

Now by using the second row wipe out the second column elements in the second matrix and third column elements in the second matrix on the right and make one transposition in the third column to the second column position (natural order). This will bring one minus sign there. That is,

$$\begin{vmatrix} 1 & 0 & 0 \\ 0 & a_{22} & a_{23} \\ 0 & a_{32} & a_{33} \end{vmatrix} = a_{22} \begin{vmatrix} 1 & 0 & 0 \\ 0 & 1 & 0 \\ 0 & 0 & a_{33} \end{vmatrix} - a_{23} \begin{vmatrix} 1 & 0 & 0 \\ 0 & 1 & 0 \\ 0 & 0 & a_{32} \end{vmatrix}.$$

Now take out the elements from the third row in both matrices and we have

$$a_{11} \begin{vmatrix} 1 & 0 & 0 \\ 0 & a_{22} & a_{23} \\ 0 & a_{32} & a_{33} \end{vmatrix} = a_{11}a_{22}a_{33} - a_{11}a_{23}a_{32}.$$

When writing the elements keep the first subscripts in the natural order $1, 2, 3$. Note that the second term is $a_{11}a_{23}a_{32}$ and here the second subscripts are in the order $1, 3, 2$. One transposition is needed to bring $1, 3, 2$ into the natural order $1, 2, 3$ and hence the multiplicative factor outside is $(-1)^1 = -1$. Now open up the second and third terms in (ii) to get the balance of the terms in the mechanical procedure indicated earlier. This is how we obtained the terms in the mechanical procedure. Note that when it is a 3×3 determinant, when we open up, we have $3! = 6$ terms. In each term, one and only one element will come from each row and each column of A. Write these $3!$ terms with the first subscripts in the natural order $1, 2, 3$. Now look at the second subscripts. If the number of transpositions needed to bring this into the natural order $1, 2, 3$ is odd then multiply by -1, if even then multiply by $+1$. The possible terms in our case are the following:

$$a_{11}a_{22}a_{33}, \quad a_{11}a_{23}a_{32},$$
$$a_{12}a_{21}a_{33}, \quad a_{12}a_{23}a_{31},$$
$$a_{13}a_{22}a_{31}, \quad a_{13}a_{21}a_{32}.$$

In the first term the second subscripts are in the order $(1, 2, 3)$. This is in the natural order. Hence the number of transpositions needed to bring this to the natural order is 0 and hence we multiply by $(-1)^0 = +1$. The term $a_{11}a_{23}a_{32}$ has the second subscripts in the order $(1, 3, 2)$. One transposition, namely the interchange of the second and third elements, will bring this to the natural order $(1, 2, 3)$. Hence we multiply this term by $(-1)^1 = -1$. The various terms and the signs are as given below:

Term	Sign	Final term
$a_{11}a_{22}a_{33}$	$+1$	$a_{11}a_{22}a_{33}$
$a_{11}a_{23}a_{32}$	-1	$-a_{11}a_{23}a_{32}$
$a_{12}a_{21}a_{33}$	-1	$-a_{12}a_{21}a_{33}$
$a_{12}a_{23}a_{31}$	$+1$	$a_{12}a_{23}a_{31}$
$a_{13}a_{22}a_{31}$	-1	$-a_{13}a_{22}a_{31}$
$a_{13}a_{21}a_{32}$	$+1$	$a_{13}a_{21}a_{32}$

Then

$$|A| = [a_{11}a_{22}a_{33} + a_{12}a_{23}a_{31} + a_{13}a_{21}a_{32}]$$
$$- [a_{11}a_{23}a_{32} + a_{12}a_{21}a_{33} + a_{13}a_{22}a_{31}]$$

the same result obtained from the mechanical procedure in (3.1.1).

By using the same procedure as used to open up a 3×3 determinant, we can open up an $n \times n$ determinant $|A| = |(a_{ij})|$. There will be $n!$ terms. In each of these $n!$ terms there will be one and only one element coming from each row and each column of A. A typical element can be written in the following form:

$$a_{1i_1} a_{2i_2} \cdots a_{ni_n}$$

where $i_1, \dots, i_n$ represent the column numbers and the first subscripts the row numbers (taken in the natural order, $1, 2, \dots, n$) then a term will be multiplied by $(+1)$ if the number of transpositions needed (one transposition is one interchange of two adjacent columns or adjacent second subscripts) to bring $(i_1, \dots, i_n)$ into the natural order $(1, 2, \dots, n)$ is even and if this number is odd then the term is multiplied by (-1). The final value of the determinant can be written as follows, which can also be used as a definition for the determinant:

(ix) Let $A = (a_{ij})$ be an $n \times n$ matrix. Then its determinant is given by

$$|A| = \sum_{i_1} \sum_{i_2} \cdots \sum_{i_n} (-1)^{\rho(i_1, \dots, i_n)} a_{1i_1} a_{2i_2} \cdots a_{ni_n} \tag{3.1.2}$$

where $\rho(i_1, \dots, i_n)$ stands for the number of transpositions needed to bring $(i_1, \dots, i_n)$ to the natural order $(1, 2, \dots, n)$.

We may note that when n is large then (3.1.2) is not that easy to compute. Even for $n = 3$ it is found to be quite involved. Hence in a practical situation we will use the postulates mentioned in the beginning, along with the properties seen so far, to evaluate a determinant. Even though (3.1.2) is not the most efficient way of evaluating a determinant this representation has many theoretical uses.

Let us see what happens if one row of a 2×2 matrix is split into two rows. Let

$$A = \begin{bmatrix} a & b \\ c & d \end{bmatrix} = \begin{bmatrix} a_1 + a_2 & b_1 + b_2 \\ c & d \end{bmatrix}.$$

Here the first row is written as a sum to two rows, namely,

$$(a, b) = (a_1, b_1) + (a_2, b_2).$$

Then the determinant

$$|A| = ad - bc = (a_1 + a_2)d - (b_1 + b_2)c$$
$$= (a_1 d - b_1 c) + (a_2 d - b_2 c)$$
$$= \begin{vmatrix} a_1 & b_1 \\ c & d \end{vmatrix} + \begin{vmatrix} a_2 & b_2 \\ c & d \end{vmatrix}.$$

The determinant has split into a sum of two determinants. Thus it is clear that if (3.1.2) is used as a definition for the determinant then postulate (γ) can be derived from (3.1.2) itself. Let us try to verify this result with a numerical example.

Example 3.1.4. Let

$$A = \begin{bmatrix} 1 & 1 & -1 \\ 2 & 0 & 1 \\ -2 & 1 & 3 \end{bmatrix}.$$

(a) Evaluate the determinant of A; (b) Let

$$(1,1,-1) = (0,1,-1) + (1,0,0).$$

Evaluate the determinant as the sum of two determinants with the first row replaced by $(0,1,-1)$ and $(1,0,0)$ respectively and show that this sum is equal to the determinant of A evaluated in (a).

Solution 3.1.4. By elementary operations

$$|A| = \begin{vmatrix} 1 & 1 & -1 \\ 2 & 0 & 1 \\ -2 & 1 & 3 \end{vmatrix} = \begin{vmatrix} 1 & 1 & -1 \\ 0 & -2 & 3 \\ 0 & 3 & 1 \end{vmatrix}$$
$$= \frac{1}{6} \begin{vmatrix} 1 & 1 & -1 \\ 0 & -6 & 9 \\ 0 & 6 & 2 \end{vmatrix} = \frac{1}{6} \begin{vmatrix} 1 & 1 & -1 \\ 0 & -6 & 9 \\ 0 & 0 & 11 \end{vmatrix}$$
$$= \frac{1}{6}(1)(-6)(11) = -11.$$

Let A_1 and A_2 be the matrices with the first row of A replaced by the vectors $(0,1,-1)$ and $(1,0,0)$ respectively. Let us evaluate their determinants through elementary operations. In A_1 since the $(1,1)$-th element is 0 interchange the first and the second columns before starting the row operations.

$$|A_1| = \begin{vmatrix} 0 & 1 & -1 \\ 2 & 0 & 1 \\ -2 & 1 & 3 \end{vmatrix} = - \begin{vmatrix} 1 & 0 & -1 \\ 0 & 2 & 1 \\ 1 & -2 & 3 \end{vmatrix}$$

$$= - \begin{vmatrix} 1 & 0 & -1 \\ 0 & 2 & 1 \\ 0 & -2 & 4 \end{vmatrix} = - \begin{vmatrix} 1 & 0 & -1 \\ 0 & 2 & 1 \\ 0 & 0 & 5 \end{vmatrix}$$

$$= -(1)(2)(5) = -10;$$

$$|A_2| = \begin{vmatrix} 1 & 0 & 0 \\ 2 & 0 & 1 \\ -2 & 1 & 3 \end{vmatrix} = - \begin{vmatrix} 1 & 0 & 0 \\ 2 & 1 & 0 \\ -2 & 3 & 1 \end{vmatrix} \qquad \text{[columns 2, 3 interchanged]}$$

$$= -(1)(1)(1) = -1.$$

Therefore

$$|A_1| + |A_2| = (-10) + (-1) = -11 = |A|.$$

The result is verified.

Before closing this section let us examine a few more properties which follow from the observations so far. From the representation of the determinant as the sum of all possible terms consisting of one element each from each row and each column, given in property (ix), or from the postulates themselves, it follows that any matrix A and its transpose A' have the same determinant.

(x) For any $n \times n$ matrix A with A' denoting its transpose

$$|A| = |A'|.$$

For a skew symmetric matrix, $A' = -A$ which means that

$$|A| = |A'| = |-A| = (-1)^n |A|.$$

Therefore if n is odd then

$$|A| = -|A| \implies |A| = 0.$$

Therefore we have the following result:

(xi) If an $n \times n$ matrix A is skew symmetric then $|A| = 0$ if n is odd or all skew symmetric matrices of odd order are singular.

3.1.3 Diagonal and triangular block matrices

Consider the matrices

$$A = \begin{bmatrix} P & O \\ O & S \end{bmatrix}, \quad B = \begin{bmatrix} P & Q \\ O & S \end{bmatrix}, \quad C = \begin{bmatrix} P & O \\ R & S \end{bmatrix} \qquad (3.1.3)$$

where O indicates a null matrix, P is $r \times r$, S is $s \times s$. Then A is called a *diagonal block matrix* or *a block diagonal* matrix, and B and C are called *triangular block* matrices where B is an upper triangular type and C is a lower triangular block matrix. By elementary operations on the first r columns or on the last s rows we can reduce B to the form A without affecting the value of the determinant. Similarly by using the first r rows or the last s columns of C we can reduce C to the form in A without affecting the value of the determinant. Thus it follows that A, B and C have the same determinants. That is,

$$|A| = |B| = |C|. \tag{3.1.4}$$

Now, notice that by operating on the first r rows and columns in A we can reduce A to the form

$$|A| = |P| \begin{vmatrix} I_r & O \\ O & S \end{vmatrix}$$

if $|P| \neq 0$ or to zero if $|P| = 0$. Then operating on the last s rows and columns of A we can write

$$\begin{vmatrix} I_r & O \\ O & S \end{vmatrix} = |S| \begin{vmatrix} I_r & O \\ O & I_s \end{vmatrix}$$

if $|S| \neq 0$ or to 0 if $|S| = 0$. But from postulate (δ) we have

$$\begin{vmatrix} I_r & O \\ O & I_s \end{vmatrix} = |I| = 1.$$

Therefore we have the following result:

(xii) The determinant of a diagonal or triangular block matrix is the product of the determinants of the diagonal blocks.

In our illustrative example

$$|A| = |B| = |C| = |P|\,|S|. \tag{3.1.5}$$

Example 3.1.5. Evaluate the following determinant directly as well by using property (xii):

$$|A| = \begin{vmatrix} 3 & 0 & 0 \\ 2 & 1 & 3 \\ 1 & 2 & 4 \end{vmatrix}.$$

Solution 3.1.5. The matrix in this determinant can be looked upon as a triangular block matrix. That is,

$$A = \begin{bmatrix} P & Q \\ R & S \end{bmatrix}, \quad P = (3), \quad Q = (0,0),$$

$$R = \begin{bmatrix} 2 \\ 1 \end{bmatrix}, \quad S = \begin{bmatrix} 1 & 3 \\ 2 & 4 \end{bmatrix}.$$

Then by using property (xii),

$$|A| = |P|\,|S| = (3) \begin{vmatrix} 1 & 3 \\ 2 & 4 \end{vmatrix}$$

$$= (3)[(1)(4) - (3)(2)] = -6.$$

For evaluating $|A|$ directly we can add (-3) times the second column to the third column to obtain

$$|A| = \begin{vmatrix} 3 & 0 & 0 \\ 2 & 1 & 3 \\ 1 & 2 & 4 \end{vmatrix} = \begin{vmatrix} 3 & 0 & 0 \\ 2 & 1 & 0 \\ 1 & 2 & -2 \end{vmatrix}$$

$$= (3)(1)(-2) = -6.$$

One observation can be made from this example which is also a particular case of property (xii).

(xiii) If the first row (column) elements of an $n \times n$ matrix $B = (b_{ij})$ are zeros except the first element, say b_{11}, then the determinant of B is given by

$$|B| = b_{11}|B_{11}| \tag{3.1.6}$$

where B_{11} is the matrix obtained by deleting the first row and the first column of B or deleting the row and column containing b_{11}.

This property will be exploited in the next section. It is not a special property of the first row (column). If any other row (column) has all elements zeros except one element then by transpositions of rows and columns (remember to multiply the determinant by (-1) each time a transposition is done. Transpositions are done instead of direct interchange of rows (columns) in order to maintain the order of the elements in the remaining matrix) we can bring the nonzero element to the $(1,1)$-th position. For an element in the (i,j)-th position the number of column transpositions needed is j and then the number of row transpositions needed is i and hence the total number of transpositions needed is $i + j$ and then multiplicative factor is $(-1)^{i+j}$. Then we can apply property (xiii).

We can extend the above ideas to look for the determinant of a product of matrices. Let us recall the basic elementary matrices of the E and the F types from Section 2.3. Postulate (β) of the definition of a determinant says that

$$|A| = |FA| = |F|\,|A| = |A|$$

where F is a basic elementary matrix of the F type. From postulate (α) we have

$$|EA| = |E|\,|A| \quad \text{or} \quad |A| = |E^{-1}|\,|EA|$$

where E is a basic elementary matrix of the E type. Note that F type elementary operation (postulate (β)) has no effect on the determinant. Then $|AB| = |E^{-1}F^{-1}|\,|(F_1E_1A)B|$. Operating on the left of AB with elementary matrices is equivalent to operating on the left of A with elementary matrices. $EAB = (EA)B$ and $FAB = (FA)B$. Then

$$|E_k^{-1}F_r^{-1}\cdots E_1^{-1}F_1^{-1}|\,|(E_kF_r\cdots F_1E_1A)B| = |AB|. \tag{i}$$

Suppose that $E_kF_r\cdots F_1E_1A = D_1$, a diagonal matrix where $D_1 = I$ if A is nonsingular, otherwise D_1 is a diagonal matrix having at least one zero diagonal element which makes D_1B singular and $|D_1B| = 0$. When $E_kF_r\cdots F_1E_1A = I$ we have $A = E_k^{-1}F_r^{-1}\cdots E_1^{-1}$ and $|E_k^{-1}\cdots F_1^{-1}E_1^{-1}| = |A|$. Then from equation (i) above, when A is nonsingular or when $D_1 = I$, $|AB| = |E_k^{-1}\cdots F_1^{-1}E_1^{-1}| \times |IB| = |A| \times |B|$.

(xiv) For $n \times n$ matrices A and B

$$|AB| = |A|\,|B|$$

and the property can be extended to any number of $n \times n$ matrices

$$|ABC\cdots| = |A|\,|B|\,|C|\cdots.$$

Example 3.1.6. Evaluate the determinants $|AB|, |A|, |B|$ and verify that $|AB| = |A|\,|B|$ where

$$A = \begin{bmatrix} 1 & 2 \\ 1 & -1 \end{bmatrix}, \quad B = \begin{bmatrix} 0 & 1 \\ 2 & 1 \end{bmatrix}.$$

Solution 3.1.6.

$$|A| = \begin{vmatrix} 1 & 2 \\ 1 & -1 \end{vmatrix} = (1)(-1) - (1)(2) = -3;$$

$$|B| = \begin{vmatrix} 0 & 1 \\ 2 & 1 \end{vmatrix} = (0)(1) - (2)(1) = -2;$$

$$|AB| = \begin{vmatrix} \begin{bmatrix} 1 & 2 \\ 1 & -1 \end{bmatrix}\begin{bmatrix} 0 & 1 \\ 2 & 1 \end{bmatrix} \end{vmatrix} = \begin{vmatrix} 4 & 3 \\ -2 & 0 \end{vmatrix}$$

$$= (4)(0) - (-2)(3) = 6 = (-3)(-2) = |A|\,|B|.$$

By using property (xiv) we can establish many results. Note that $A^r = AA\cdots A$, product where A is multiplied by A a total of r times.

(xv)
$$|A^r| = |A|^r, \quad r = 0, 1, 2, \ldots$$
$$|A^{-r}| = |A|^{-r} = \frac{1}{|A|^r}, \quad r = 0, 1, 2, \ldots, \quad |A| \neq 0.$$

This fact follows from the property

$$AA^{-1} = I \implies 1 = |AA^{-1}| = |A| \, |A^{-1}| \implies$$
$$|A^{-1}| = \frac{1}{|A|}$$

where $A^0 = I$ by convention.

(xvi) The determinant of on orthonormal matrix A is ± 1.

$$AA' = I \implies 1 = |AA'| = |A| \, |A'| = |A|^2 \implies |A| = \pm 1.$$

(xvii) The determinant of an idempotent matrix is either 1 or 0.

$$A = A^2 \implies |A| = |A|^2 \implies |A|[1 - |A|] = 0 \implies |A| = 0, 1.$$

(xviii) The determinant of a nilpotent matrix is zero.

$$A^r = O, \quad r \geq 2 \implies 0 = |A^r| = |A|^r \implies |A| = 0.$$

Exercises 3.1

3.1.1. Evaluate the determinants of the following matrices:

$$A = \begin{bmatrix} 2 & 0 & 0 \\ 0 & -1 & 0 \\ 0 & 0 & 0 \end{bmatrix}, \quad B = \begin{bmatrix} 1 & 0 & -1 \\ 0 & 3 & 1 \\ 0 & 0 & 4 \end{bmatrix},$$

$$C = \begin{bmatrix} a_{11} & a_{12} & \cdots & a_{1n} \\ 0 & a_{22} & \cdots & a_{2n} \\ \vdots & \vdots & \cdots & \vdots \\ 0 & 0 & \cdots & a_{nn} \end{bmatrix}.$$

3.1.2. Evaluate the determinants of the following matrices:

$$A = \begin{bmatrix} 1 & 0 & -1 & 2 \\ 3 & 0 & -2 & 1 \\ 1 & 0 & 2 & 3 \\ 0 & 0 & 4 & 5 \end{bmatrix}, \quad B = \begin{bmatrix} 1 & 2 & -1 & 1 \\ 3 & 0 & -1 & 1 \\ 2 & 1 & -1 & 0 \\ 8 & 0 & -1 & 1 \end{bmatrix},$$

$$C = \begin{bmatrix} 4 & 1 & 7 & -3 \\ 3 & 0 & 8 & 1 \\ 5 & -1 & 4 & 2 \\ 2 & 1 & 2 & 5 \end{bmatrix}.$$

3.1.3. Let a and b be $n \times 1$ column vectors, $a' = (a_1, \ldots, a_n)$, $b' = (b_1, \ldots, b_n)$. Evaluate the determinants of

$$(1) \quad ab', \qquad (2) \quad a'b.$$

3.1.4. Consider the $n \times n$ matrix

$$A = \begin{bmatrix} a & b & \cdots & b \\ b & a & \cdots & b \\ \vdots & \vdots & \cdots & \vdots \\ b & b & \cdots & a \end{bmatrix}$$

(the diagonal elements are all a and the off-diagonal elements are all b). Show that
(1) $|A| = (a - b)^{n-1}[a + (n-1)b]$.
(2) Verify the result in (1) for a 4×4 matrix with $a = 2$, $b = 3$, by direct evaluation.

3.1.5. Evaluate the determinant of

$$V_3 = \begin{bmatrix} 1 & a_1 & a_1^2 \\ 1 & a_2 & a_2^2 \\ 1 & a_3 & a_3^2 \end{bmatrix}$$

and show that

$$|V_3| = (a_2 - a_1)(a_3 - a_1)(a_3 - a_2) = \prod_{i>j}(a_i - a_j).$$

3.1.6. If V_n is an $n \times n$ matrix created as in Exercise 3.1.5, such a matrix is called a *Vandermonde's matrix* and its determinant is called a *Vandermonde's determinant*, show that

$$|V_n| = \prod_{i>j}(a_i - a_j).$$

3.1.7. Show that

$$\begin{vmatrix} 2 & 5 & 7 & 0 & 0 \\ -1 & 3 & 2 & 0 & 0 \\ 5 & 4 & 1 & 0 & 0 \\ 0 & 0 & 0 & 6 & 4 \\ 0 & 0 & 0 & -1 & 1 \end{vmatrix} = \begin{vmatrix} 2 & 5 & 7 \\ -1 & 3 & 2 \\ 5 & 4 & 1 \end{vmatrix} \times \begin{vmatrix} 6 & 4 \\ -1 & 1 \end{vmatrix}.$$

3.1.8. Show that

$$\begin{vmatrix} 2 & 5 & 7 & 1 & 4 \\ -1 & 3 & 2 & 6 & 5 \\ 5 & 4 & 1 & 0 & 2 \\ 0 & 0 & 0 & 6 & 4 \\ 0 & 0 & 0 & -1 & 1 \end{vmatrix} = \begin{vmatrix} 2 & 5 & 7 \\ -1 & 3 & 2 \\ 5 & 4 & 1 \end{vmatrix} \begin{vmatrix} 6 & 4 \\ -1 & 1 \end{vmatrix}.$$

3.1.9. Let

$$A = \begin{bmatrix} a_1 & a_2 & a_3 \\ b_1 & b_2 & b_3 \\ c_1 & c_2 & c_3 \end{bmatrix}$$

and let

$$(a_1, a_2, a_3) = a_1(1, 0, 0) + a_2(0, 1, 0) + a_3(0, 0, 1)$$
$$= (a_1, 0, 0) + (0, a_2, 0) + (0, 0, a_3).$$

Show that

$$|A| = a_1|A_{11}| - a_2|A_{12}| + a_3|A_{13}|$$

where A_{ij} represents the matrix obtained by deleting the row and column containing the element a_{ij}, namely the i-th row and the j-th column. Verify the result for

$$A = \begin{bmatrix} 7 & 8 & -10 \\ 15 & 20 & 8 \\ 10 & 14 & 25 \end{bmatrix}.$$

[Such a decomposition helps to reduce the order of the determinants involved thereby such a procedure is helpful when the elements of the matrix are large numbers.]

3.1.10. Evaluate the determinants $|A|, |B|, |AB|$ and verify the result that $|AB| = |A||B|$, where

$$A = \begin{bmatrix} 1 & 0 & -1 \\ 2 & 1 & 1 \\ 1 & -1 & 1 \end{bmatrix}, \quad B = \begin{bmatrix} 2 & 1 & 1 \\ -1 & 0 & 1 \\ 1 & 1 & -1 \end{bmatrix}.$$

3.1.11. Construct (1) a 2×2 orthonormal matrix, (2) a 3×3 orthonormal matrix. Evaluate the determinants and show that the value is ±1 in each case.

3.1.12. Construct a nilpotent matrix of order 3 and evaluate its determinant and show that it is zero.

3.1.13. Let $J' = (1, \ldots, 1)$, where J is an $n \times 1$ vector of unities. Let $B = \frac{1}{n}JJ'$, $A = I_n - B$. Show that the determinants of both A and B are zeros, or both A and B are singular.

3.1.14. Let

$$A = \begin{bmatrix} 1 & 2 & 3 \\ 4 & 6 & 5 \\ 7 & 9 & 8 \end{bmatrix}.$$

First evaluate the determinant, $|A|$, by elementary operations. Then show that $|A|$ is also equal to the following:

$$|A| = (1) \begin{vmatrix} 6 & 5 \\ 9 & 8 \end{vmatrix} - (2) \begin{vmatrix} 4 & 5 \\ 7 & 8 \end{vmatrix} + (3) \begin{vmatrix} 4 & 6 \\ 7 & 9 \end{vmatrix}.$$

3.1.15. For the same A in Exercise 3.1.14 show that the determinant is also equal to the following:

$$|A| = -(4) \begin{vmatrix} 2 & 3 \\ 9 & 8 \end{vmatrix} + (6) \begin{vmatrix} 1 & 3 \\ 7 & 8 \end{vmatrix} - (5) \begin{vmatrix} 1 & 2 \\ 7 & 9 \end{vmatrix}$$

$$= (7) \begin{vmatrix} 2 & 3 \\ 6 & 5 \end{vmatrix} - (9) \begin{vmatrix} 1 & 3 \\ 4 & 5 \end{vmatrix} + (8) \begin{vmatrix} 1 & 2 \\ 4 & 6 \end{vmatrix}.$$

3.1.16. Evaluate the determinant of the following matrix A in two steps:

$$A = \begin{bmatrix} 0 & 1 & -1 & 2 & 4 & 1 & -2 & 5 \\ 1 & -1 & 1 & 2 & 5 & 8 & 9 & 7 \\ 1 & 1 & -1 & 6 & 13 & 10 & 5 & 17 \\ 80 & 40 & 20 & 11 & 20 & 15 & 2 & 8 \\ 32 & 47 & 51 & 60 & 27 & 40 & 90 & 19 \\ 41 & 42 & 40 & 38 & 30 & 22 & 28 & 15 \\ 80 & 80 & 82 & 85 & 44 & 20 & 27 & 15 \\ 79 & 70 & 72 & 90 & 95 & 92 & 94 & 60 \end{bmatrix}.$$

3.1.17. Evaluate the determinant of the following matrix A in one step:

$$A = \begin{bmatrix} 0 & 1 & 1 & -1 & 1 \\ -1 & 0 & 2 & -3 & 5 \\ -1 & -2 & 0 & 4 & 2 \\ 1 & 3 & -4 & 0 & 3 \\ -1 & -5 & -2 & -3 & 0 \end{bmatrix}.$$

3.1.18. If A, B and $A + B$ are nonsingular $n \times n$ matrices is $A^{-1} + B^{-1}$ singular or nonsingular. Prove your assertion.

3.1.19. Let A be $m \times n$ and B be $n \times m$, $m < n$. Let $C = AB$ the $m \times m$ matrix thereby $|C|$ is defined. Show that

$$
\text{(i)} \quad \begin{bmatrix} I_m & A \\ O & I_n \end{bmatrix} \begin{bmatrix} A & O \\ -I_n & B \end{bmatrix} = \begin{bmatrix} O & AB \\ -I_n & B \end{bmatrix};
$$

$$
\text{(ii)} \quad \begin{vmatrix} A & O \\ -I_n & B \end{vmatrix} = \begin{vmatrix} O & AB \\ -I_n & B \end{vmatrix} = (-1)^{n(m+1)}|AB|.
$$

3.1.20. Verify the results in (i), (ii) of Exercise 3.1.19 for

$$
A = \begin{bmatrix} 1 & 1 & -1 \\ 1 & 2 & 1 \end{bmatrix}, \quad B = \begin{bmatrix} 1 & 1 \\ 0 & 1 \\ 1 & 3 \end{bmatrix}.
$$

3.2 Cofactor expansions

A convenient way of evaluating a determinant as well to study some theoretical properties of a determinant is to expand a determinant in terms of what is known as cofactors. We define cofactors and minors of a matrix.

3.2.1 Cofactors and minors

Definition 3.2.1 (Minors). Let $A = (a_{ij})$ be an $n \times n$ matrix. Delete m rows and m columns, $m < n$. The determinant of the resulting submatrix is called a minor. If the i-th row and the j-th column (that is, the row and column where the element a_{ij} is present) are deleted then the determinant of the resulting submatrix is called the minor of a_{ij}.

For example, let

$$
A = \begin{bmatrix} 2 & 0 & -1 \\ 1 & 2 & 4 \\ 0 & 1 & 5 \end{bmatrix} = (a_{ij}).
$$

Then

$$
\begin{vmatrix} 2 & 4 \\ 1 & 5 \end{vmatrix} = \text{minor of } a_{11}, \quad \begin{vmatrix} 1 & 4 \\ 0 & 5 \end{vmatrix} = \text{minor of } a_{12},
$$

$$
\begin{vmatrix} 1 & 2 \\ 0 & 1 \end{vmatrix} = \text{minor of } a_{13}, \quad \begin{vmatrix} 2 & -1 \\ 0 & 5 \end{vmatrix} = \text{minor of } a_{22},
$$

$$\begin{vmatrix} 2 & 0 \\ 1 & 2 \end{vmatrix} = \text{minor of } a_{33}, \text{ and so on.}$$

Definition 3.2.2 (Leading minors). If the submatrices are formed by deleting the rows and columns from the 2nd onward, from the 3rd onward, and so on then the corresponding minors are called the leading minors.

For example, for the same A above the leading minors are the following determinants:

$$2, \quad \begin{vmatrix} 2 & 0 \\ 1 & 2 \end{vmatrix}, \quad \begin{vmatrix} 2 & 0 & -1 \\ 1 & 2 & 4 \\ 0 & 1 & 5 \end{vmatrix}.$$

Definition 3.2.3 (Cofactors). Let $A = (a_{ij})$ be an $n \times n$ matrix. The cofactor of a_{ij} is defined as $(-1)^{i+j}$ times the minor of a_{ij}. That is, if the cofactor and minor of a_{ij} are denoted by $|C_{ij}|$ and $|M_{ij}|$ respectively then

$$|C_{ij}| = (-1)^{i+j}|M_{ij}|.$$

For example, let

$$A = \begin{bmatrix} 2 & 0 & 5 \\ -2 & 1 & 4 \\ 2 & 3 & 7 \end{bmatrix} = (a_{ij})$$

then

$$|C_{11}| = (-1)^{1+1} \begin{vmatrix} 1 & 4 \\ 3 & 7 \end{vmatrix} = \begin{vmatrix} 1 & 4 \\ 3 & 7 \end{vmatrix} = -5 = \text{cofactor of } a_{11};$$

$$|C_{12}| = (-1)^{1+2} \begin{vmatrix} -2 & 4 \\ 2 & 7 \end{vmatrix} = - \begin{vmatrix} -2 & 4 \\ 2 & 7 \end{vmatrix} = 22 = \text{cofactor of } a_{12};$$

$$|C_{13}| = (-1)^{1+3} \begin{vmatrix} -2 & 1 \\ 2 & 3 \end{vmatrix} = \begin{vmatrix} -2 & 1 \\ 2 & 3 \end{vmatrix} = -8 = \text{cofactor of } a_{13};$$

$$|C_{23}| = (-1)^{2+3} \begin{vmatrix} 2 & 0 \\ 2 & 3 \end{vmatrix} = - \begin{vmatrix} 2 & 0 \\ 2 & 3 \end{vmatrix} = -6 = \text{cofactor of } a_{23},$$

and so on. Immediate applications of these concepts of minors and cofactors are the evaluation of the determinant of a square matrix, the evaluation of the inverse of a nonsingular matrix and many such items through what is known as a cofactor expansion. We will establish a general result regarding the evaluation of a determinant in terms of cofactors by examining a simple case.

Example 3.2.1. By using postulate (γ) expand the determinant of the following matrix in terms of the elements of the first row and their cofactors:

$$A = \begin{bmatrix} 4 & 2 & 3 \\ 0 & -1 & 1 \\ 1 & 2 & -4 \end{bmatrix}.$$

Solution 3.2.1. Write the first row as

$$(4, 2, 3) = (4, 0, 0) + (0, 2, 0) + (0, 0, 3).$$

Then from the application of postulate (γ) repeatedly we have

$$|A| = \begin{vmatrix} 4 & 0 & 0 \\ 0 & -1 & 1 \\ 1 & 2 & -4 \end{vmatrix} + \begin{vmatrix} 0 & 2 & 0 \\ 0 & -1 & 1 \\ 1 & 2 & -4 \end{vmatrix} + \begin{vmatrix} 0 & 0 & 3 \\ 0 & -1 & 1 \\ 1 & 2 & -4 \end{vmatrix}.$$

Consider

$$\begin{vmatrix} 4 & 0 & 0 \\ 0 & -1 & 1 \\ 1 & 2 & -4 \end{vmatrix} = 4 \begin{vmatrix} 1 & 0 & 0 \\ 0 & -1 & 1 \\ 1 & 2 & -4 \end{vmatrix} = 4 \begin{vmatrix} -1 & 1 \\ 2 & -4 \end{vmatrix},$$

where the first step is done by taking out 4 from the first row (postulate (α)) and the second step is done by using property (xiii) of Section 3.1. Now consider the second determinant. By transposition of columns 1 and 2 we have

$$\begin{vmatrix} 0 & 2 & 0 \\ 0 & -1 & 1 \\ 1 & 2 & -4 \end{vmatrix} = - \begin{vmatrix} 2 & 0 & 0 \\ -1 & 0 & 1 \\ 2 & 1 & -4 \end{vmatrix}$$

$$= -2 \begin{vmatrix} 0 & 1 \\ 1 & -4 \end{vmatrix} = -2[(0)(-4) - (1)(1)] = 2.$$

The last line is done by using property (xiii) of Section 3.1. Now consider the last determinant. By two transpositions we can bring the last column to the first column, then take out 3 and then apply property (xiii) of Section 3.1 to obtain

$$\begin{vmatrix} 0 & 0 & 3 \\ 0 & -1 & 1 \\ 1 & 2 & -4 \end{vmatrix} = 3 \begin{vmatrix} 0 & -1 \\ 1 & 2 \end{vmatrix} = 3[(0)(2) - (1)(-1)] = 3.$$

Thus we have the following expansion of the determinant:

$$|A| = 4 \begin{vmatrix} -1 & 1 \\ 2 & -4 \end{vmatrix} - 2 \begin{vmatrix} 0 & 1 \\ 1 & -4 \end{vmatrix} + 3 \begin{vmatrix} 0 & -1 \\ 1 & 2 \end{vmatrix}.$$

From the above example and procedure it is clear that such an expansion is possible for a general matrix. Thus we have the following general result:

(i) Let $A = (a_{ij})$ be an $n \times n$ matrix. Let $|C_{ij}|$ denote the cofactor of a_{ij} and $|M_{ij}|$ the minor of a_{ij} respectively. Then

$$
\begin{aligned}
|A| &= a_{11}|C_{11}| + a_{12}|C_{12}| + \cdots + a_{1n}|C_{1n}| \\
&= a_{11}|M_{11}| - a_{12}|M_{12}| + \cdots + (-1)^{n+1} a_{1n}|M_{1n}| \\
&= a_{i1}|C_{i1}| + a_{i2}|C_{i2}| + \cdots + a_{in}|C_{in}| \\
&= a_{i1}(-1)^{i+1}|M_{i1}| + a_{i2}(-1)^{i+2}|M_{i2}| + \cdots + (-1)^{i+n} a_{in}|M_{1n}|
\end{aligned}
$$

for $i = 1, 2, \ldots, n$.

This means that the cofactor expansion can be carried out in terms of the elements and their cofactors of any row. The same is true for columns, that is, the expansion can be carried out in terms of the elements of any column and their cofactors.

Example 3.2.2. Let

$$
A = \begin{bmatrix} 2 & 1 & -7 \\ 1 & -1 & 2 \\ 0 & 3 & 4 \end{bmatrix}.
$$

Evaluate the following: Sum of products of the elements of the first row (1) with the cofactors of the elements of the first row, (2) with the cofactors of the elements of the second row, (3) with the cofactors of the elements of the third row.

Solution 3.2.2. (1) The cofactors of the elements of the first row are the following:

$$
|C_{11}| = \begin{vmatrix} -1 & 2 \\ 3 & 4 \end{vmatrix} = [(-1)(4) - (3)(2)] = -10,
$$

$$
|C_{12}| = - \begin{vmatrix} 1 & 2 \\ 0 & 4 \end{vmatrix} = -[(1)(4) - (0)(2)] = -4,
$$

$$
|C_{13}| = \begin{vmatrix} 1 & -1 \\ 0 & 3 \end{vmatrix} = [(1)(3) - (0)(-1)] = 3.
$$

Therefore

$$
|A| = (2)[-10] + (1)[-4] + (-7)[3] = -45.
$$

(2) Now let us expand in terms of the elements of the first row with the cofactors of the elements of the second row. Let us denote this sum by p. Then

$$
p = a_{11}|C_{21}| + a_{12}|C_{22}| + a_{13}|C_{23}|
$$

$$= (2)(-1)^3 \begin{vmatrix} 1 & -7 \\ 3 & 4 \end{vmatrix} + (1)(-1)^4 \begin{vmatrix} 2 & -7 \\ 0 & 4 \end{vmatrix} + (-7)(-1)^5 \begin{vmatrix} 2 & 1 \\ 0 & 3 \end{vmatrix}$$

$$= -2(25) + (1)(8) + (7)(6) = 0.$$

(3) Now let us expand in terms of the elements of the first row with the cofactors of the third row. Let us denote this sum by q.

Then

$$q = a_{11}|C_{31}| + a_{12}|C_{32}| + a_{13}|C_{33}|$$

$$= (2)(-1)^4 \begin{vmatrix} 1 & -7 \\ -1 & 2 \end{vmatrix} + (1)(-1)^5 \begin{vmatrix} 2 & -7 \\ 1 & 2 \end{vmatrix} + (-7)(-1)^6 \begin{vmatrix} 2 & 1 \\ 1 & -1 \end{vmatrix}$$

$$= (2)(-5) + (-1)(11) + (-7)(-3) = 0.$$

We see that sums such as the ones in (2) and (3) are zeros. This in fact is a general result which can be proved by writing the sum in each case as the sum of $n \times n$ determinants by writing each cofactor as an $n \times n$ determinant rather than as an $(n-1) \times (n-1)$ determinant. Then we will see that in sums, such as the ones in (2) and (3) above, two rows will be identical and hence the determinant is zero.

(ii) Let $A = (a_{ij})$ be an $n \times n$ matrix with $|C_{ij}|$ denoting the cofactor of a_{ij}. Then, in terms of the row elements,

$$a_{i1}|C_{k1}| + a_{i2}|C_{k2}| + \cdots + a_{in}|C_{kn}| = 0$$

for all i and k, $i \neq k$. In terms of the column elements,

$$a_{1i}|C_{1k}| + a_{2i}|C_{2k}| + \cdots + a_{ni}|C_{nk}| = 0$$

for all i and k, $i \neq k$.

Consider, for example an expansion in terms of the elements of the second row and cofactors of the first row. For proving the results we use the original representation of the minors before wiping out the column elements, that is, let q be of the following form:

$$q = a_{21}|C_{11}| + \cdots + a_{2n}|C_{1n}| \quad \text{(in terms of the cofactors)}$$

$$= a_{21}|M_{11}| - a_{22}|M_{12}| + \cdots + (-1)^{1+n}a_{2n}|M_{1n}| \quad \text{(in terms of minors)}$$

$$= a_{21} \begin{vmatrix} 1 & 0 & \cdots & 0 \\ a_{21} & a_{22} & \cdots & a_{2n} \\ \vdots & \vdots & \cdots & \vdots \\ a_{n1} & a_{n2} & \cdots & a_{nn} \end{vmatrix} + \cdots + a_{2n} \begin{vmatrix} 0 & 0 & \cdots & 1 \\ a_{21} & a_{22} & \cdots & a_{2n} \\ \vdots & \vdots & \cdots & \vdots \\ a_{n1} & a_{n2} & \cdots & a_{nn} \end{vmatrix}$$

Now, writing in terms of the original determinants we have

$$q = \begin{vmatrix} a_{21} & 0 & \cdots & 0 \\ a_{21} & a_{22} & \cdots & a_{2n} \\ \vdots & \vdots & \cdots & \vdots \\ a_{n1} & a_{n2} & \cdots & a_{nn} \end{vmatrix} = \begin{vmatrix} a_{21} & a_{22} & \cdots & a_{2n} \\ a_{21} & a_{22} & \cdots & a_{2n} \\ \vdots & \vdots & \cdots & \vdots \\ a_{n1} & a_{n2} & \cdots & a_{nn} \end{vmatrix} \quad \text{(by postulate } (\gamma))$$

$$= 0$$

since two rows are identical in the matrix. Same procedure is applicable if the expansion is taken as the elements of the i-th row with the cofactors of the j-th row, $i \neq j$.

3.2.2 Inverse of a matrix in terms of the cofactor matrix

One immediate application of the results in (i) and (ii) is to obtain the inverse of a nonsingular matrix in terms of a matrix of cofactors thereby resulting in another method of evaluating the inverse of a nonsingular matrix. One method through elementary operations is already considered in Chapter 2. Let $A = (a_{ij})$ be an $n \times n$ matrix. Let $|C_{ij}|$ be the cofactor of a_{ij}. Consider a matrix created by replacing a_{ij} by its cofactor. Let this matrix of cofactors be denoted by cof(A), called the *cofactor matrix*. Then

$$\text{cof}(A) = \begin{bmatrix} |C_{11}| & |C_{12}| & \cdots & |C_{1n}| \\ |C_{21}| & |C_{22}| & \cdots & |C_{2n}| \\ \vdots & \vdots & \cdots & \vdots \\ |C_{n1}| & |C_{n2}| & \cdots & |C_{nn}| \end{bmatrix}. \tag{3.2.1}$$

Consider the transpose of this matrix and premultiply this transpose by the matrix A. That is,

$$A[\text{cof}(A)]' = \begin{bmatrix} a_{11} & a_{12} & \cdots & a_{1n} \\ \vdots & \vdots & \cdots & \vdots \\ a_{n1} & a_{n2} & \cdots & a_{nn} \end{bmatrix} \begin{bmatrix} |C_{11}| & |C_{21}| & \cdots & |C_{n1}| \\ |C_{12}| & |C_{22}| & \cdots & |C_{n2}| \\ \vdots & \vdots & \cdots & \vdots \\ |C_{1n}| & |C_{2n}| & \cdots & |C_{nn}| \end{bmatrix}.$$

Then by the results (i) and (ii) all the diagonal elements of the product are equal to $|A|$ each and all the off-diagonal elements are zeros. Thus we have

$$A[\text{cof}(A)]' = \begin{bmatrix} |A| & 0 & \cdots & 0 \\ 0 & |A| & \cdots & 0 \\ \vdots & \vdots & \cdots & \vdots \\ 0 & 0 & \cdots & |A| \end{bmatrix} = |A| \times I. \tag{3.2.2}$$

Let

$$B = \frac{[\text{cof}(A)]'}{|A|}$$

when $|A| \neq 0$. Then

$$AB = I \implies B = A^{-1}.$$

That is, B stands for the inverse of A whenever the matrix is nonsingular. We have the following result:

(iii) Let $A = (a_{ij})$ be an $n \times n$ matrix with $|A|$ denoting the determinant, $|C_{ij}|$ the cofactor of a_{ij}, cof(A) the cofactor matrix of A, $[\text{cof}(A)]'$ the transpose of the cofactor matrix, then the inverse of A, whenever it exists, is given by

$$A^{-1} = \frac{1}{|A|} [\text{cof}(A)]', \quad |A| \neq 0. \tag{3.2.3}$$

This formula (3.2.3) gives another way of computing the inverse of a nonsingular matrix.

Example 3.2.3. Evaluate the inverse of A, if it exists, by using the cofactor matrix, where

$$A = \begin{bmatrix} 1 & 0 & 1 \\ -1 & 1 & 1 \\ 1 & 1 & 1 \end{bmatrix}.$$

Solution 3.2.3. In order to apply this procedure we have to compute all the cofactors as well as the determinant. [Hence this method of evaluating the inverse is not that efficient unless the matrix is 2×2 in which case the determinant and the cofactor matrix can be read without any computation.] The determinant of A through the cofactor expansion is given by

$$|A| = (1) \begin{vmatrix} 1 & 1 \\ 1 & 1 \end{vmatrix} - (0) \begin{vmatrix} -1 & 1 \\ 1 & 1 \end{vmatrix} + (1) \begin{vmatrix} -1 & 1 \\ 1 & 1 \end{vmatrix} = -2.$$

Since $|A| \neq 0$ the inverse exists. [If $|A|$ was equal to 0 we would have stopped the process here itself.] The various cofactors are the following:

$$|C_{11}| = \text{cofactor of } a_{11} = \begin{vmatrix} 1 & 1 \\ 1 & 1 \end{vmatrix} = 0,$$

$$|C_{12}| = \text{cofactor of } a_{12} = -\begin{vmatrix} -1 & 1 \\ 1 & 1 \end{vmatrix} = 2,$$

$$|C_{13}| = \text{cofactor of } a_{13} = \begin{vmatrix} -1 & 1 \\ 1 & 1 \end{vmatrix} = -2,$$

$$|C_{21}| = \text{cofactor of } a_{21} = -\begin{vmatrix} 0 & 1 \\ 1 & 1 \end{vmatrix} = 1,$$

$$|C_{22}| = \begin{vmatrix} 1 & 1 \\ 1 & 1 \end{vmatrix} = 0, \quad |C_{23}| = -\begin{vmatrix} 1 & 0 \\ 1 & 1 \end{vmatrix} = -1, \quad |C_{31}| = \begin{vmatrix} 0 & 1 \\ 1 & 1 \end{vmatrix} = -1,$$

$$|C_{32}| = -\begin{vmatrix} 1 & 1 \\ -1 & 1 \end{vmatrix} = -2, \quad |C_{33}| = \begin{vmatrix} 1 & 0 \\ -1 & 1 \end{vmatrix} = 1.$$

The matrix of cofactors is then

$$\mathrm{cof}(A) = \begin{bmatrix} 0 & 2 & -2 \\ 1 & 0 & -1 \\ -1 & -2 & 1 \end{bmatrix}.$$

Then

$$A^{-1} = \frac{1}{|A|}[\mathrm{cof}(A)]' = -\frac{1}{2}\begin{bmatrix} 0 & 1 & -1 \\ 2 & 0 & -2 \\ -2 & -1 & 1 \end{bmatrix}$$

$$= \begin{bmatrix} 0 & -\frac{1}{2} & \frac{1}{2} \\ -1 & 0 & 1 \\ 1 & \frac{1}{2} & -\frac{1}{2} \end{bmatrix}.$$

Let us verify the result to see whether any computational error is made.

$$AA^{-1} = \begin{bmatrix} 1 & 0 & 1 \\ -1 & 1 & 1 \\ 1 & 1 & 1 \end{bmatrix}\begin{bmatrix} 0 & -\frac{1}{2} & \frac{1}{2} \\ -1 & 0 & 1 \\ 1 & \frac{1}{2} & -\frac{1}{2} \end{bmatrix}$$

$$= \begin{bmatrix} 1 & 0 & 0 \\ 0 & 1 & 0 \\ 0 & 0 & 1 \end{bmatrix} = I.$$

The result is verified. One major disadvantage of this method, compared to the method of elementary operations, is that here one has to evaluate one $n \times n$ determinant and n^2, $(n-1) \times (n-1)$ determinants.

Example 3.2.4 (Covariance matrix). In statistical theory and its applications in many fields a concept called the covariance matrix plays a vital role. Let $X = \binom{x_1}{x_2}$ be a real bivariate vector random variable. The covariance matrix for this real bivariate case is given by

$$V = \begin{bmatrix} \sigma_1^2 & \sigma_1\sigma_2\rho \\ \sigma_1\sigma_2\rho & \sigma_2^2 \end{bmatrix}$$

where $\sigma_1 \geq 0$, $\sigma_2 \geq 0$ are the standard deviations of x_1 and x_2 respectively, their squares are the variances, and ρ with $-1 \leq \rho \leq 1$ is the correlation between x_1 and x_2. When $\sigma_1 = 0$ and $\sigma_2 = 0$ the variables are degenerate. When $\rho \pm 1$ the matrix V is singular and the variables are linearly related. Evaluate the inverse of V in the nonsingular case.

Solution 3.2.4. The determinant

$$|V| = \begin{vmatrix} \sigma_1^2 & \sigma_1\sigma_2\rho \\ \sigma_1\sigma_2\rho & \sigma_2^2 \end{vmatrix} = \sigma_1^2\sigma_2^2(1-\rho^2).$$

The cofactor matrix is then

$$\mathrm{cof}(V) = \begin{bmatrix} \sigma_2^2 & -\rho\sigma_1\sigma_2 \\ -\rho\sigma_1\sigma_2 & \sigma_1^2 \end{bmatrix}.$$

Since the cofactor matrix is symmetric here, its transpose is the same as itself. Hence

$$V^{-1} = \frac{1}{|V|}[\mathrm{cof}(V)]' = \frac{1}{\sigma_1^2\sigma_2^2(1-\rho^2)} \begin{bmatrix} \sigma_2^2 & -\rho\sigma_1\sigma_2 \\ -\rho\sigma_1\sigma_2 & \sigma_1^2 \end{bmatrix}$$

$$= \frac{1}{(1-\rho^2)} \begin{bmatrix} \frac{1}{\sigma_1^2} & -\frac{\rho}{\sigma_1\sigma_2} \\ -\frac{\rho}{\sigma_1\sigma_2} & \frac{1}{\sigma_2^2} \end{bmatrix}.$$

[This is the matrix of the quadratic form appearing in the exponent of a real bivariate normal or real Gaussian density.]

3.2.3 A matrix differential operator

Let $X = (x_{ij})$ be a matrix with functionally independent real variables as its elements, that is, x_{ij}'s are functionally independent (distinct) real variables. Functionally independent means that no element is a function of other variables. Then the matrix differential operator is defined as follows:

Definition 3.2.4 (A matrix differential operator).

$$\frac{\partial}{\partial X} = \left(\frac{\partial}{\partial x_{ij}}\right) = \begin{bmatrix} \frac{\partial}{\partial x_{11}} & \frac{\partial}{\partial x_{12}} & \cdots & \frac{\partial}{\partial x_{1n}} \\ \vdots & \vdots & \cdots & \vdots \\ \frac{\partial}{\partial x_{m1}} & \frac{\partial}{\partial x_{m2}} & \cdots & \frac{\partial}{\partial x_{mn}} \end{bmatrix};$$

$$\frac{\partial f}{\partial X} = \left(\frac{\partial f}{\partial x_{ij}}\right),$$

where f is a scalar function of the $m \times n$ matrix X. It is the matrix of the corresponding partial derivatives.

Example 3.2.5. Let X be a $p \times p$ matrix of functionally independent real variables and u its trace. Evaluate $\frac{\partial u}{\partial X}$.

Solution 3.2.5. Trace is the sum of the leading diagonal elements:

$$u = \mathrm{tr}(X) = x_{11} + x_{22} + \cdots + x_{pp}.$$

Therefore

$$\frac{\partial u}{\partial x_{11}} = 1, \quad \frac{\partial u}{\partial x_{12}} = 0, \dots, \quad \frac{\partial u}{\partial x_{1p}} = 0, \dots$$

$$\frac{\partial u}{\partial x_{ii}} = 1, \quad i = 1, \dots, p, \quad \frac{\partial u}{\partial x_{ij}} = 0, \quad i \ne j.$$

Hence

$$\frac{\partial u}{\partial X} = \begin{bmatrix} 1 & 0 & \cdots & 0 \\ 0 & 1 & \cdots & 0 \\ \vdots & \vdots & \cdots & \vdots \\ 0 & 0 & \cdots & 1 \end{bmatrix} = I_p.$$

Thus we have the following result:

(iv) When X is a $p \times p$ matrix of functionally independent real variables

$$u = \mathrm{tr}(X) \implies \frac{\partial u}{\partial X} = I_p.$$

Example 3.2.6. Let X be a $p \times p$ matrix of functionally independent real variables. Let $|X|$ be the determinant of X, $|X| \ne 0$, and $\frac{\partial}{\partial X}$ the matrix differential operator. Evaluate $\frac{\partial |X|}{\partial X}$.

Solution 3.2.6. Consider the cofactor expansion of $|X|$ in terms of the elements of the i-th row and their cofactors:

$$|X| = x_{i1} |C_{i1}| + x_{i2} |C_{i2}| + \cdots + x_{in} |C_{in}| \tag{3.2.4}$$

where $|C_{ij}|$ denotes the cofactor of x_{ij}. Note that $|C_{ij}|$ does not contain x_{ij} and hence when the partial derivative of $|X|$ is taken with respect to x_{ij} we obtain $|C_{ij}|$. Thus when the matrix differential operator operates on $|X|$ we get the cofactor matrix. That is,

$$\frac{\partial}{\partial X} |X| = \mathrm{cof}(X) = \begin{bmatrix} |C_{11}| & \cdots & |C_{1p}| \\ \vdots & \cdots & \vdots \\ |C_{p1}| & \cdots & |C_{pp}| \end{bmatrix}.$$

But we have already seen that the inverse of a nonsingular matrix is the transpose of the cofactor matrix divided by the determinant. Then

$$\frac{\partial |X|}{\partial X} = |X| (X^{-1})'.$$

Thus we have the following result:

(v) When X is a nonsingular matrix of distinct functionally independent real variables

$$\frac{\partial |X|}{\partial X} = |X|(X^{-1})'. \tag{3.2.5}$$

We can modify the above result to obtain a result for a nonsingular symmetric matrix. When X is of functionally independent real variables as elements except for the property that $X = X'$, that is, X is symmetric then following through the above procedure we can derive a result analogous to the one in (3.2.5). Observe that when X is symmetric we have

$$\frac{\partial |X|}{\partial x_{ii}} = |C_{ii}| = \text{cofactor of } x_{ii} \quad \text{and}$$

$$\frac{\partial |X|}{\partial x_{ij}} = 2|C_{ij}| = 2 \text{ times the cofactor of } x_{ij}, \quad i \neq j.$$

Thus when the matrix operator $\frac{\partial}{\partial X}$ operates on $|X|$ we have the following format:

$$\frac{\partial |X|}{\partial X} = \begin{bmatrix} |C_{11}| & 2|C_{12}| & \cdots & 2|C_{1p}| \\ 2|C_{21}| & |C_{22}| & \cdots & 2|C_{2p}| \\ \vdots & \vdots & \cdots & \vdots \\ 2|C_{p1}| & 2|C_{p2}| & \cdots & |C_{pp}| \end{bmatrix}. \tag{3.2.6}$$

The diagonal elements are not multiplied by 2 whereas all the nondiagonal elements are multiplied by 2. A convenient notation for writing the right side of (3.2.6) is the following:

$$\frac{\partial |X|}{\partial X} = 2\text{cof}(X) - \text{diag}(\text{cof}(X)) \tag{3.2.7}$$

where $\text{diag}[\text{cof}(X)]$ means a diagonal matrix created with the diagonal elements of the matrix $\text{cof}(X)$. Then converting (3.2.7) in terms of the inverse of X we have the following result:

(vi) When X is a nonsingular symmetric matrix of functionally independent real variables then

$$\frac{\partial |X|}{\partial X} = |X|[2X^{-1} - \text{diag}(X^{-1})]. \tag{3.2.8}$$

This result has many applications, especially in obtaining the maximum likelihood estimators of the parameters in a multivariate Gaussian density. This is also applicable in general maxima/minima problems or optimization problems involving traces and determinants.

Example 3.2.7. Let $X = (x_{ij})$ be a $p \times p$ matrix of distinct functionally independent p^2 real scalar variables and let $\frac{\partial}{\partial X} = (\frac{\partial}{\partial x_{ij}})$ be the matrix differential operator. Let $B = (b_{ij})$ be a $p \times p$ matrix of constants, that is, b_{ij}'s are not functions of the x_{ij}'s. Let $u = \text{tr}(BX)$. Evaluate $\frac{\partial u}{\partial X}$.

Solution 3.2.7. The first diagonal element in the product BX is

$$b_{11}x_{11} + b_{12}x_{21} + \cdots + b_{1p}x_{p1}$$

or the i-th diagonal element in BX is given by

$$b_{i1}x_{1i} + b_{i2}x_{2i} + \cdots + b_{ip}x_{pi} \tag{3.2.9}$$

and then the trace is the sum of (3.2.9) over all i, $i = 1, \dots, p$. Thus the partial derivative of $u = \text{tr}(BX)$ with respect to x_{ij} is b_{ji}. This may be noted from (3.2.9). Hence when all the elements in X are distinct we have the matrix

$$\frac{\partial u}{\partial X} = B'$$

the transpose of B. If $X = X'$ then the partial derivatives of $u = \text{tr}(BX)$ with respect to x_{ij} give

$$\frac{\partial u}{\partial x_{ij}} = \begin{cases} b_{ii} & \text{for } j = i \\ b_{ji} + b_{ij} & \text{for } j \neq i. \end{cases} \tag{3.2.10}$$

Thus the diagonal elements of B come only once. Then the matrix configuration, using the notation in (vi), is

$$\frac{\partial u}{\partial X} = B + B' - \text{diag}(B)$$

where $\text{diag}(B)$ is the diagonal matrix created by using the diagonal elements of B. Thus we have the following result:

(vii) When $B = (b_{ij})$ is a $p \times p$ matrix of constants and $X = (x_{ij})$ a $p \times p$ matrix of functionally independent real scalar variables and when $u = \text{tr}(BX)$ then

$$\frac{\partial u}{\partial X} = \begin{cases} B' & \text{if all elements in } X \text{ are distinct} \\ B + B' - \text{diag}(B) & \text{if } X = X' \\ 2B - \text{diag}(B) & \text{if } X = X' \text{ and } B = B'. \end{cases} \tag{3.2.11}$$

The result above has various types of applications in statistical analysis and related areas.

3.2.4 Products and square roots

Products and integer powers of matrices and their determinants are already considered at the end of Section 3.1. From there, many more properties follow. Even if the square matrices A and B do not commute the determinants of AB and BA are the same.

(viii)
$$|AB| = |BA| = |A|\,|B| = |B|\,|A|$$
$$|I - AB| = |I - BA| \quad \text{if } B \text{ or } A \text{ is nonsingular.}$$

Now it is natural to ask the question: is $|A^{p/q}| = |A|^{p/q}$ where p and q are integers? Let us examine a simple case with $p = 1$, $q = 2$ or the square root of a matrix A. For a given matrix A can we find a matrix B such that $B^2 = A$? This B, if such a B exists, can be called a square root of A. If such a B exists, is it unique? Can there be different matrices whose squares are all equal to the same matrix? Let us take one of the simplest cases, a 2×2 identity matrix. Let

$$A = I_2 = \begin{bmatrix} 1 & 0 \\ 0 & 1 \end{bmatrix}, \quad B_1 = \begin{bmatrix} 1 & 0 \\ 0 & 1 \end{bmatrix}, \quad B_2 = \begin{bmatrix} -1 & 0 \\ 0 & -1 \end{bmatrix},$$
$$B_3 = \begin{bmatrix} 0 & 1 \\ 1 & 0 \end{bmatrix}, \quad B_4 = \begin{bmatrix} 0 & -1 \\ -1 & 0 \end{bmatrix}.$$

Note that

$$B_1^2 = A = B_2^2 = B_3^2 = B_4^2.$$

Thus $B_1, \ldots, B_4$, all qualify to be square roots of A. For a given matrix, even if such a B exists it need not be unique unless more conditions are imposed on A. Hence in our discussions to follow we will only consider integer powers (positive powers or negative powers if the inverse exists) of a square matrix and the determinants associated with such powers.

In Chapter 2 we have seen that when a matrix A is nonsingular then in most cases it can be reduced to the form

$$A = LU$$

through elementary transformations, where L is lower triangular and U is upper triangular. Then

$$|A| = |L|\,|U|$$

which is equal to the product of all the diagonal elements in L and U.

3.2.5 Cramer's rule for solving systems of linear equations

As another application of the cofactor expansion of a determinant one can examine the solutions of a nonsingular system of linear equations. Let $A = (a_{ij})$ be an $n \times n$ matrix and nonsingular and consider the system of linear equations

$$AX = b \Rightarrow X = A^{-1}b,$$
$$b' = (b_1, \ldots, b_n), \quad X' = (x_1, \ldots, x_n).$$

Writing A^{-1} in terms of the cofactor matrix we have

$$X = \frac{1}{|A|} \begin{bmatrix} |C_{11}| & |C_{21}| & \cdots & |C_{n1}| \\ |C_{12}| & |C_{22}| & \cdots & |C_{n2}| \\ \vdots & \vdots & \cdots & \vdots \\ |C_{1n}| & |C_{2n}| & \cdots & |C_{nn}| \end{bmatrix} \begin{bmatrix} b_1 \\ b_2 \\ \vdots \\ b_n \end{bmatrix}$$

where $|C_{ij}|$ is the cofactor of a_{ij}. Then the i-th element in X is given by

$$x_i = \frac{1}{|A|} \{ b_1 |C_{1i}| + b_2 |C_{2i}| + \cdots + b_n |C_{ni}| \}.$$

The numerator on the right side corresponds to the cofactor expansion of a determinant in terms of the i-th column of A with the i-th column being b and all other columns the same as those of A. Therefore

$$x_i = \frac{|A_i|}{|A|}, \quad i = 1, 2, \ldots, n \tag{3.2.12}$$

where $|A_i|$ is the determinant of A with the i-th column of A replaced by b and other columns remaining the same. [In a practical situation we will try to solve the system through elementary transformations which may work out to be much easier and faster than evaluating determinants of the type in (3.2.12).] The rule in (3.2.12) is called *Cramer's rule* and it is more of theoretical interest rather than of practical use.

Example 3.2.8. Solve the following system of equations by Cramer's rule, if applicable:

$$x_1 + x_2 + x_3 = 3$$
$$x_1 - x_3 = 0$$
$$2x_1 + x_2 + 2x_3 = 5.$$

Solution 3.2.8. Writing the system as $AX = b$ we have

$$A = \begin{bmatrix} 1 & 1 & 1 \\ 1 & 0 & -1 \\ 2 & 1 & 2 \end{bmatrix}, \quad X = \begin{bmatrix} x_1 \\ x_2 \\ x_3 \end{bmatrix}, \quad b = \begin{bmatrix} 3 \\ 0 \\ 5 \end{bmatrix}.$$

First we need to compute $|A|$. If $|A| = 0$ then the rule does not apply.

$$|A| = \begin{vmatrix} 1 & 1 & 1 \\ 1 & 0 & -1 \\ 2 & 1 & 2 \end{vmatrix}$$

$$= \begin{vmatrix} 1 & 1 & 1 \\ 0 & -1 & -2 \\ 0 & -1 & 0 \end{vmatrix} \quad [-1(1) + (2); -2(1) + (3) \Rightarrow]$$

$$= \begin{vmatrix} 1 & 1 & 1 \\ 0 & -1 & -2 \\ 0 & 0 & 2 \end{vmatrix} \quad [-1(2) + (3) \Rightarrow]$$

$$= (1)(-1)(2) = -2.$$

The rule is applicable. Replace the first column of A by b and evaluate the determinant. According to our notation,

$$|A_1| = \begin{vmatrix} 3 & 1 & 1 \\ 0 & 0 & -1 \\ 5 & 1 & 2 \end{vmatrix}.$$

Expanding in terms of the elements and their cofactors of the second row we have

$$|A_1| = -(-1) \begin{vmatrix} 3 & 1 \\ 5 & 1 \end{vmatrix} = -2.$$

Now, replace the second column by b. Then

$$|A_2| = \begin{vmatrix} 1 & 3 & 1 \\ 1 & 0 & -1 \\ 2 & 5 & 2 \end{vmatrix}.$$

Expanding in terms of the elements of the second row and their cofactors we have

$$|A_2| = -(1) \begin{vmatrix} 3 & 1 \\ 5 & 2 \end{vmatrix} + 0 - (-1) \begin{vmatrix} 1 & 3 \\ 2 & 5 \end{vmatrix} = (-1)(1) - (-1)(-1) = -2.$$

Now,

$$|A_3| = \begin{vmatrix} 1 & 1 & 3 \\ 1 & 0 & 0 \\ 2 & 1 & 5 \end{vmatrix}.$$

Expanding in terms of the elements of the second row and their cofactors we have

$$|A_3| = -(1) \begin{vmatrix} 1 & 3 \\ 1 & 5 \end{vmatrix} = -(1)(2) = -2.$$

Hence

$$x_1 = \frac{-2}{-2} = 1 = x_2 = x_3.$$

Note that when evaluating the above determinants we looked for the rows or columns containing the maximum number of zeros. Then we used a cofactor expansion in terms of the elements of that row (column). If a cofactor expansion is going to be used to evaluate a determinant then this is a rule of thumb. Note also that Cramer's rule is rather lengthy because, in general, $n + 1$ determinants of $n \times n$ matrices are to be evaluated to complete the process. In practice, the easiest way to solve a linear system is to go through elementary operations which can also determine, at the same time, whether the system is consistent, singular, nonsingular, with many solutions or with a unique solution.

The configuration of signs when using a cofactor expansion to evaluate a determinant can be remembered from the following matrix format:

$$\begin{matrix} + & - & + & - & \cdots \\ - & + & - & + & \cdots \\ + & - & + & - & \cdots \\ \vdots & \vdots & \vdots & \vdots & \ddots \end{matrix}$$

Before we conclude this section we may observe a few more minor points. Consider a product of several $n \times n$ nonsingular matrices, for example, a product of three, ABC. Then

$$(ABC)^{-1} = C^{-1}B^{-1}A^{-1} \Rightarrow$$

$$\frac{[\mathrm{cof}(ABC)]'}{|ABC|} = \frac{[\mathrm{cof}(C)]'}{|C|} \frac{[\mathrm{cof}(B)]'}{|B|} \frac{[\mathrm{cof}(A)]'}{|A|}.$$

Therefore

(ix) $$[\mathrm{cof}(ABC)]' = [\mathrm{cof}(C)]'[\mathrm{cof}(B)]'[\mathrm{cof}(A)]'.$$

Let us see what happens to the inverse of A'.

$$(A')^{-1} = \frac{[\mathrm{cof}(A')]'}{|A'|} = \frac{\mathrm{cof}(A)}{|A|} \text{ since } |A| = |A'|$$

$$= (A^{-1})'.$$

That is,

(x)
$$(A')^{-1} = (A^{-1})'$$

and

$$[(ABC)']^{-1} = (C^{-1})'(B^{-1})'(A^{-1})'.$$

Exercises 3.2

3.2.1. If the following matrix is denoted as $A = (a_{ij})$ then evaluate the cofactors and minors of $a_{12}, a_{22}, a_{31}, a_{33}$, where

$$A = \begin{bmatrix} 2 & 0 & 1 \\ -1 & 1 & 5 \\ 0 & 1 & 4 \end{bmatrix}.$$

3.2.2. For the matrix A in Exercise 3.2.1 compute the leading minors.

3.2.3. Expand $|A|$ in terms of the elements of (a) the first row and their cofactors, (b) the third row and the corresponding cofactors, (c) the second column and the corresponding cofactors, where A is the same matrix in Exercise 3.2.1.

3.2.4. Verify property (ii) by expanding $|A|$ of the matrix in Exercise 3.2.1 in terms of the elements of the (a) first row and cofactors of the elements of the third row, (b) second row and the cofactors of the elements of the third row.

3.2.5. Evaluate A^{-1}, if it exists, by computing the cofactor matrix, where

$$A = \begin{bmatrix} 1 & 1 & 1 \\ 1 & 1 & -1 \\ 2 & 1 & 4 \end{bmatrix}.$$

3.2.6. By multiplying and then taking the trace verify the result (3.2.11) if

(1) $\quad X = (x_{ij}), \quad B = \begin{bmatrix} 2 & 1 & -1 \\ 0 & 1 & 5 \\ 2 & 3 & -1 \end{bmatrix}, \quad x_{ij}$'s are distinct;

(2) $\quad X = X'$ and the same B as in (1);

(3) $\quad X = X'$ and $\quad B = \begin{bmatrix} 2 & 1 & -1 \\ 1 & 1 & 5 \\ -1 & 5 & -1 \end{bmatrix}.$

3.2.7. From the mechanical rule of Section 3.1 write down the explicit form of the determinant of a 3×3 matrix $X = (x_{ij})$. From this explicit form compute $\frac{\partial}{\partial X}|X|$ for the cases: (1) all elements of X are distinct, (2) $X = X'$. Thus verify the result in (3.2.8).

3.2.8. Consider the $n \times n$ matrices

$$B = \frac{1}{n} \begin{bmatrix} 1 & 1 & \dots & 1 \\ \vdots & \vdots & \dots & \vdots \\ 1 & 1 & \dots & 1 \end{bmatrix}$$

$$A = I_n - B.$$

Compute the following:

(1) $|B^{281}|$,   (2) $|A^{392}|$,   (3) $|A^{250} + B^{192}|^5$.

3.2.9. Are the following statements true or false. If false give two counter examples each: (1) $|-A| = -|A|$; (2) If $A' = -A$ then $|A'| = -|A|$ and since $|A'| = |A|$, $|A| = 0$; (3) If A is a matrix with real elements and if $A = A'$ then A can be written as $A = BB'$ where B is a matrix with real elements.

3.2.10. Show that

$$\begin{vmatrix} 2 & -1 & 0 \\ -1 & 2 & -1 \\ 0 & -1 & 2 \end{vmatrix} = (3+1) = 4$$

and that for an $n \times n$ matrix

$$\begin{vmatrix} 2 & -1 & 0 & 0 & \dots & 0 & 0 \\ -1 & 2 & -1 & 0 & \dots & 0 & 0 \\ \vdots & \vdots & \vdots & \vdots & \dots & \vdots & \vdots \\ 0 & 0 & 0 & 0 & \dots & 2 & -1 \\ 0 & 0 & 0 & 0 & \dots & -1 & 2 \end{vmatrix} = (n+1).$$

3.2.11. Let D_n be the $n \times n$ determinant

$$D_n = \begin{vmatrix} 1 & -1 & 0 & 0 & \dots & 0 & 0 \\ 1 & 1 & -1 & 0 & \dots & 0 & 0 \\ \vdots & \vdots & \vdots & \vdots & \dots & \vdots & \vdots \\ 0 & 0 & 0 & 0 & \dots & 1 & 1 \end{vmatrix}.$$

Show that $D_n = D_{n-1} + D_{n-2}$. [This recurrence relation produces the Fibonacci numbers $1, 2, 3, 5, 8, 13, 21, \dots$.]

3.2.12. Let $A = (a_{ij})$ be an $n \times n$ matrix where $a_{ij} = i + j$. Evaluate the determinant of A.

3.2.13. Let

$$A = \begin{bmatrix} 1 & a & a^2 \\ 1 & b & b^2 \\ 1 & c & c^2 \end{bmatrix}.$$

Evaluate the determinant of A^{250}.

3.2.14. Let

$$A = \begin{bmatrix} a & b & b & b \\ b & a & b & b \\ b & b & a & b \\ b & b & b & a \end{bmatrix}, \quad a \neq b \neq 0.$$

Evaluate the determinant of A^{100}.

3.2.15. Show that (1) A^{-1} is symmetric if A is symmetric, (2) T^{-1} is lower (upper) triangular if T is lower (upper) triangular.

3.2.16. Show that the determinant of the $n \times n$ matrix

$$A = \begin{bmatrix} 0 & 1 & 0 & \cdots & 0 & 0 \\ 0 & 0 & 1 & \cdots & 0 & 0 \\ \vdots & \vdots & \vdots & \vdots & \vdots & \vdots \\ 0 & 0 & 0 & \cdots & 0 & 1 \\ -a_0 & -a_1 & -a_2 & \cdots & -a_{n-2} & -a_{n-1} \end{bmatrix}$$

is equal to $(-1)^n a_0$.

3.2.17. Let

$$A_n = \begin{bmatrix} a_1 & b_1 & 0 & 0 & \cdots & 0 & 0 \\ c_1 & a_2 & b_2 & 0 & \cdots & 0 & 0 \\ 0 & c_2 & a_3 & b_3 & \cdots & 0 & 0 \\ \vdots & \vdots & \vdots & \vdots & \vdots & \vdots & \vdots \\ 0 & 0 & 0 & 0 & \cdots & a_{n-1} & b_{n-1} \\ 0 & 0 & 0 & 0 & \cdots & c_{n-1} & a_n \end{bmatrix}.$$

Show that its determinant, $|A_n|$, can be written as

$$|A_n| = a_n |A_{n-1}| - b_{n-1} c_{n-1} |A_{n-2}| \quad \text{for } n \geq 3.$$

3.2.18. Let the $n \times n$ matrix A be partitioned as follows, where A_1 is $p \times p$:

$$A = \begin{bmatrix} A_1 & A_2 \\ A_3 & O \end{bmatrix}.$$

Show that

$$|A| = (-1)^{(n+1)p} |A_2| \, |A_3|.$$

3.2.19. Let Cof(A) denote the matrix of cofactors of the $n \times n$ matrix A. Then show that the determinant of this cofactor matrix is given by

$$|\text{Cof}(A)| = |A|^{n-1}.$$

3.2.20. Let I, A, B be $n \times n$. Show that if $I + AB$ is nonsingular then $I + BA$ is nonsingular and that

$$(I + AB)^{-1} = I - B(I + BA)^{-1}A.$$

3.2.21. If A is $n \times n$, U is $n \times 1$, V is $n \times 1$ and if $A + UV'$ and A are nonsingular then show that

$$(A + UV')^{-1} = A^{-1} - \frac{(A^{-1}U)(V'A^{-1})}{1 + V'A^{-1}U}.$$

3.2.22. Let $A = (a_{ij}(x))$ be an $n \times n$ matrix where the elements a_{ij}'s are differentiable functions of x. Let $\mathbf{a_1}, \mathbf{a_2}, \dots, \mathbf{a_n}$ denote the columns of A so that

$$A = (\mathbf{a_1}, \dots, \mathbf{a_n}) \quad \text{and} \quad |A| = |(\mathbf{a_1}, \dots, \mathbf{a_n})|.$$

Let $\frac{d}{dx}\mathbf{a_j}$ denote the vector of derivatives of the elements in $\mathbf{a_j}$. Then show that

$$\frac{d}{dx}|A| = \left|\left(\frac{d}{dx}\mathbf{a_1}, \dots, \mathbf{a_n}\right)\right| + \left|\left(\mathbf{a_1}, \frac{d}{dx}\mathbf{a_2}, \dots, \mathbf{a_n}\right)\right|$$
$$+ \cdots + \left|\left(\mathbf{a_1}, \dots, \frac{d}{dx}\mathbf{a_n}\right)\right|.$$

3.2.23. Hadamard's inequality. Let $A = (a_{ij})$ be an $n \times n$ matrix. Show that

$$|A|^2 \leq \prod_{j=1}^{n}\left\{\sum_{i=1}^{n}|a_{ij}|^2\right\}.$$

3.2.24. For any two $m \times n$ matrices A and B show that $\operatorname{rank}(A + B) \leq \operatorname{rank}(A) + \operatorname{rank}(B)$.

3.2.25. If A and B are matrices such that AB is defined then show that

$$\operatorname{rank}(AB) \leq \min(\operatorname{rank}(A), \operatorname{rank}(B)).$$

3.2.26. If A is an $n \times n$ nonsingular matrix and if B is $n \times m$ and C is $p \times n$ then show that

$$\operatorname{rank}(AB) = \operatorname{rank}(B), \quad \operatorname{rank}(CA) = \operatorname{rank}(C).$$

3.2.27. If A is $m \times n$ and B is $n \times m$ with $m > n$ then show that $|AB| = 0$.

3.2.28. Circulant matrix. Evaluate the determinant of the circulant matrix

$$A = \begin{bmatrix} a_0 & a_1 & a_2 & \cdots & a_{n-1} \\ a_{n-1} & a_0 & a_1 & \cdots & a_{n-2} \\ \vdots & \vdots & \vdots & \ddots & a_1 \\ a_1 & a_2 & a_3 & \cdots & a_0 \end{bmatrix}.$$

3.2.29. Show that $|A| = c^{n-1}(nx + c)$ where the $n \times n$ matrix

$$A = \begin{bmatrix} x + c & x & \cdots & x \\ x & x + c & \cdots & x \\ \vdots & \vdots & \ddots & \vdots \\ x & x & \cdots & x + c \end{bmatrix}.$$

3.3 Some practical situations

The theory of determinants has applications in all sorts of practical problems as well as in theoretical developments of many other fields. A few of these will be listed in this section.

3.3.1 Cross product

A concept called *cross product* of two vectors in 3-space is found in elementary trigonometry, with applications in physics, chemistry and engineering problems. In the notation of Chapter 1 let

$$\vec{a} = a_1\vec{i} + a_2\vec{j} + a_3\vec{k},$$
$$\vec{i} = (1,0,0), \quad \vec{j} = (0,1,0), \quad \vec{k} = (0,0,1), \quad \text{and}$$
$$\vec{b} = b_1\vec{i} + b_2\vec{j} + b_3\vec{k}$$

be two vectors in 3-space. Consider the parallelogram generated by these vectors on the plane determined by $\vec{a}$ and $\vec{b}$ as shown in Figure 3.3.1.

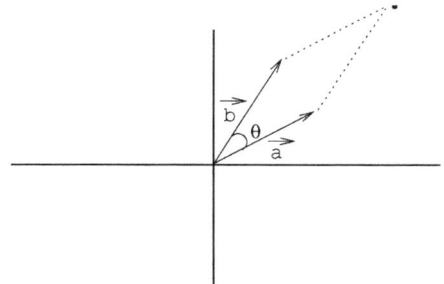

Figure 3.3.1: Parallelogram.

Let us try to construct a vector $\vec{c}$ which is orthogonal to both $\vec{a}$ and $\vec{b}$ and whose length is equal to the area of the parallelogram generated by $\vec{a}$ and $\vec{b}$. Such a vector is usually denoted by

$$\vec{c} = \vec{a} \times \vec{b}$$

the *cross product* of $\vec{a}$ with $\vec{b}$. From elementary considerations it can be shown that the vector $\vec{c}$ is obtained by opening up the following 3×3 determinant

$$\vec{c} = \vec{a} \times \vec{b} = \begin{vmatrix} \vec{i} & \vec{j} & \vec{k} \\ a_1 & a_2 & a_3 \\ b_1 & b_2 & b_3 \end{vmatrix}. \tag{3.3.1}$$

Opening up in terms of the elements of the first row and their cofactors, treating $\vec{i}, \vec{j}, \vec{k}$ as some elements of the matrix, we have

$$\vec{a} \times \vec{b} = \vec{i} \begin{vmatrix} a_2 & a_3 \\ b_2 & b_3 \end{vmatrix} - \vec{j} \begin{vmatrix} a_1 & a_3 \\ b_1 & b_3 \end{vmatrix} + \vec{k} \begin{vmatrix} a_1 & a_2 \\ b_1 & b_2 \end{vmatrix}$$

$$= (a_2 b_3 - a_3 b_2)\vec{i} - (a_1 b_3 - a_3 b_1)\vec{j} + (a_1 b_2 - a_2 b_1)\vec{k}. \tag{3.3.2}$$

Example 3.3.1. Construct the cross product of $\vec{a} = \vec{i} + \vec{j} - \vec{k}$ with $\vec{b} = 2\vec{i} + \vec{j} + \vec{k}$ by using (3.3.1) and show that this cross product vector is orthogonal to both $\vec{a}$ and $\vec{b}$ and whose length is equal to the area of the parallelogram generated by $\vec{a}$ and $\vec{b}$.

Solution 3.3.1. From (3.3.1)

$$\vec{c} = \vec{a} \times \vec{b} = \begin{vmatrix} \vec{i} & \vec{j} & \vec{k} \\ 1 & 1 & -1 \\ 2 & 1 & 1 \end{vmatrix}$$

$$= \vec{i} \begin{vmatrix} 1 & -1 \\ 1 & 1 \end{vmatrix} - \vec{j} \begin{vmatrix} 1 & -1 \\ 2 & 1 \end{vmatrix} + \vec{k} \begin{vmatrix} 1 & 1 \\ 2 & 1 \end{vmatrix}$$

$$= 2\vec{i} - 3\vec{j} - \vec{k}.$$

The dot product between $\vec{c}$ and $\vec{a}$ is then

$$\vec{c} . \vec{a} = (2)(1) + (-3)(1) + (-1)(-1) = 0.$$

The dot product between $\vec{c}$ and $\vec{b}$ is

$$\vec{c} . \vec{b} = (2)(2) + (-3)(1) + (-1)(1) = 0.$$

Thus the cross product vector of $\vec{a}$ with $\vec{b}$ is orthogonal to both $\vec{a}$ and $\vec{b}$. The length of $\vec{c}$ is given by

$$\|\vec{c}\| = \sqrt{(2)^2 + (-3)^2 + (-1)^2} = \sqrt{14}.$$

For any two vectors $\vec{U}$ and $\vec{V}$ in n-space the area of the parallelogram on the plane containing $\vec{U}$ and $\vec{V}$ is given by the following expression, see Figure 3.3.1.

$$\text{area} = \|\vec{U}\| \, \|\vec{V}\| \sin \theta.$$

[Twice the area of the triangle = 2(1/2) base times the altitude = base times the altitude $= (\|\vec{U}\|)(\|\vec{V}\|) \sin \theta$] where θ is the angle between $\vec{U}$ and $\vec{V}$. But

$$\sin \theta = \sqrt{1 - \cos^2 \theta} = \sqrt{1 - \left(\frac{\vec{U} . \vec{V}}{\|\vec{U}\| \, \|\vec{V}\|}\right)^2}$$

$$= \frac{1}{\|\vec{U}\| \, \|\vec{V}\|} \sqrt{\|\vec{U}\|^2 \|\vec{V}\|^2 - (\vec{U}.\vec{V})^2} \qquad (3.3.3)$$

holds for all $\vec{U}$ and $\vec{V}$, $\vec{U} \neq O$, $\vec{V} \neq O$. Then the area of the parallelogram is

$$\text{area} = \|\vec{U}\| \, \|\vec{V}\| \sin \theta = \sqrt{\|\vec{U}\|^2 \|\vec{V}\|^2 - (\vec{U}.\vec{V})^2}. \qquad (3.3.4)$$

Consider two vectors $\vec{a}$ and $\vec{b}$ in 3-space. Substituting in (3.3.4) we have

$$\text{area} = \sqrt{(a_1^2 + a_2^2 + a_3^2)(b_1^2 + b_2^2 + b_3^2) - [a_1 b_1 + a_2 b_2 + a_3 b_3]^2}.$$

Simplifying and rewriting we have

$$\text{area} = \sqrt{(a_2 b_3 - a_3 b_2)^2 + (a_1 b_3 - a_3 b_1)^2 + (a_1 b_2 - a_2 b_1)^2}$$
$$= \|\vec{a} \times \vec{b}\|.$$

In our illustrative example

$$[(1)(1) - (-1)(1)]^2 + [(1)(1) - (-1)(2)]^2 + [(1)(1) - (1)(2)]^2 = 14.$$

This verifies the result.

One observation is immediate. If the second and third rows of (3.3.1) are interchanged then the resulting determinant is (-1) times the original determinant which implies that

$$\vec{a} \times \vec{b} = -\vec{b} \times \vec{a}. \qquad (3.3.5)$$

3.3.2 Areas and volumes

Consider two vectors $\vec{a}$ and $\vec{b}$ in 2-space as shown in Figure 3.3.2. The lengths of $\vec{a}$ and $\vec{b}$ are

$$\|\vec{a}\| = \sqrt{a_1^2 + a_2^2}, \quad \|\vec{b}\| = \sqrt{b_1^2 + b_2^2}.$$

Let θ_1 be the angle $\vec{a}$ makes with the x-axis and θ_2 the angle $\vec{b}$ makes with the x-axis and $\theta = \theta_2 - \theta_1$, the angle between $\vec{a}$ and $\vec{b}$. One choice of $\theta_1, \theta_2, \theta$ is shown in Figure 3.3.2. [The final result will be true for all choices of $\theta_1, \theta_2, \theta$.]

$$\cos \theta = \cos(\theta_2 - \theta_1) = \cos \theta_2 \cos \theta_1 + \sin \theta_2 \sin \theta_1$$
$$= \frac{b_1}{\sqrt{b_1^2 + b_2^2}} \frac{a_1}{\sqrt{a_1^2 + a_2^2}} + \frac{b_2}{\sqrt{b_1^2 + b_2^2}} \frac{a_2}{\sqrt{a_1^2 + a_2^2}}$$
$$= \frac{\vec{a}.\vec{b}}{\|\vec{a}\| \, \|\vec{b}\|}.$$

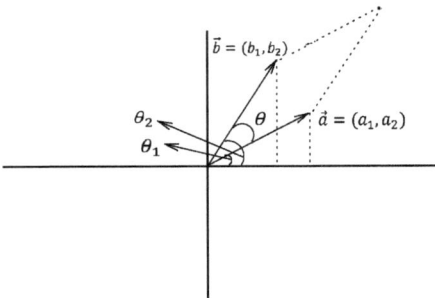

Figure 3.3.2: Area.

The area of the parallelogram is then

$$\text{area} = \|\vec{a}\| \, \|\vec{b}\| \sin\theta = \sqrt{\|\vec{a}\|^2 \|\vec{b}\|^2 - (\vec{a}.\vec{b})^2}$$
$$= \sqrt{(a_1^2 + a_2^2)(b_1^2 + b_2^2) - (a_1 b_1 + a_2 b_2)^2}$$
$$= \sqrt{(a_1 b_2 - a_2 b_1)^2} = \left\{ \begin{vmatrix} a_1 & a_2 \\ b_1 & b_2 \end{vmatrix}^2 \right\}^{1/2}$$

or the area is the absolute value of the determinant with $\vec{a}$ and $\vec{b}$ as its first and second rows.

$$\text{area} = \text{ absolute value of } \begin{vmatrix} a_1 & a_2 \\ b_1 & b_2 \end{vmatrix}.$$

Now, consider the parallelepiped generated by the three vectors $\vec{a}, \vec{b}, \vec{c}$ in a 3-space. The volume of the parallelepiped is the base area multiplied by the altitude. Consider the base area of the parallelogram generated by $\vec{b}$ and $\vec{c}$. The area is the length of the cross product, $\|\vec{b} \times \vec{c}\|$. The altitude is also equal to $\|\vec{a}\| \cos\theta$ where θ is the angle $\vec{a}$ make with the normal $\vec{N} = \vec{b} \times \vec{c}$ as shown in Figure 3.3.3.

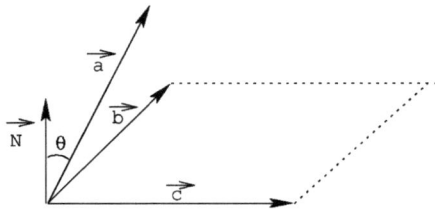

Figure 3.3.3: Volume of a parallelepiped.

Therefore the volume, denoted by v, is given by

$$v = \|\vec{b} \times \vec{c}\| \, \|\vec{a}\| \frac{[\vec{a}.(\vec{b} \times \vec{c})]}{\|\vec{a}\| \, \|\vec{b} \times \vec{c}\|} \quad \text{(substituting for } \cos\theta\text{)}$$

$$= [\vec{a}.(\vec{b} \times \vec{c})]$$
$$= a_1[b_2c_3 - b_3c_2] - a_2[b_1c_3 - b_3c_1] + a_3[b_1c_2 - b_2c_1]$$
$$= \begin{vmatrix} a_1 & a_2 & a_3 \\ b_1 & b_2 & b_3 \\ c_1 & c_2 & c_3 \end{vmatrix}.$$

Thus the volume of the parallelepiped generated by the vectors $\vec{a}, \vec{b}$ and $\vec{c}$ in a 3-space is the absolute value of the following determinant:

$$v = \text{absolute value of } \begin{vmatrix} a_1 & a_2 & a_3 \\ b_1 & b_2 & b_3 \\ c_1 & c_2 & c_3 \end{vmatrix}. \tag{3.3.6}$$

Note that $v = 0$ if any of the vectors is a linear function of the others. For example if $\vec{a}$ lies on the plane determined by $\vec{b}$ and $\vec{c}$ then $v = 0$. This formula can be generalized. Let O be the origin of a rectangular coordinate system. Let

$$X_1 = (x_{11}, x_{12}, \ldots, x_{1n}),$$
$$X_2 = (x_{21}, x_{22}, \ldots, x_{2n}),$$
$$\vdots \qquad \vdots$$
$$X_n = (x_{n1}, x_{n2}, \ldots, x_{nn})$$

be n points and consider the vectors, $\vec{OX_1}, \ldots, \vec{OX_n}$. Assuming that these are linearly independent, the volume of the parallelotope generated by $\vec{OX_1}, \ldots, \vec{OX_n}$ is given by the absolute value of the determinant

$$v_n = \text{absolute value of } |X| = |XX'|^{1/2} \tag{3.3.7}$$

where

$$X = \begin{bmatrix} x_{11} & x_{12} & \cdots & x_{1n} \\ x_{21} & x_{22} & \cdots & x_{2n} \\ \vdots & \vdots & \cdots & \vdots \\ x_{n1} & x_{n2} & \cdots & x_{nn} \end{bmatrix}.$$

Note that $|XX'| = |X||X'| = |X|^2$. But $|XX'|$ remains non-negative and hence by using this form we do not have to worry about the absolute value. This form is also useful in dealing with r points in n-space, $r < n$. If there are $n+1$ points $X_1, \ldots, X_{n+1}$ in an n-space such as 3 points in a 2-space then we can shift the origin to one of the points then the situation will be as in (3.3.7).

The volume of this parallelotope can be shown to be given by the following determinant, in absolute value:

$$v = \begin{vmatrix} x_{11} & x_{12} & \cdots & x_{1n} & 1 \\ x_{21} & x_{22} & \cdots & x_{2n} & 1 \\ \vdots & \vdots & \cdots & \vdots & \vdots \\ x_{n+11} & x_{n+12} & \cdots & x_{n+1n} & 1 \end{vmatrix}$$

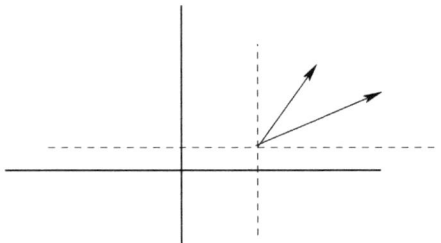

Figure 3.3.4: $n + 1$ points in n-space.

$$= \begin{vmatrix} x_{11} - x_{n+11} & \cdots & x_{1n} - x_{n+1n} & 0 \\ \vdots & \cdots & \vdots & \vdots \\ x_{n1} - x_{n+11} & \cdots & x_{nn} - x_{n+1n} & 1 \end{vmatrix}$$

$$= \begin{vmatrix} x_{11} - x_{n+11} & \cdots & x_{1n} - x_{n+1n} \\ \vdots & \cdots & \vdots \\ x_{n1} - x_{n+11} & \cdots & x_{nn} - x_{n+1n} \end{vmatrix} = \begin{vmatrix} X_1 - X_{n+1} \\ \vdots \\ X_n - X_{n+1} \end{vmatrix}. \tag{3.3.8}$$

The origin is shifted to the point X_{n+1}, as indicated in Figure 3.3.4, then the other points are $X_i - X_{n+1}, i = 1, 2, \ldots, n$ with respect to the new coordinate system. Thus (3.3.8) agrees with (3.3.7), where in (3.3.7), X_{n+1} is the origin O itself.

Example 3.3.2. Evaluate the volume (area in 2 space) of the parallelotope (parallelepiped in 3-space) created by the vectors $\vec{OX}_1, \ldots, \vec{OX}_n$ where O indicates the origin where

(a) $X_1 = (1, 1), \quad X_2 = (1, -1),$

(b) $X_1 = (2, 0, -4), \quad X_2 = (1, 1, -1), \quad X_3 = (1, 0, 1),$

(c) $X_1 = (1, 1, 1, 1), \quad X_2 = (1, 1, 1, -1),$

$\qquad X_3 = (1, 1, -1, 1), \quad X_4 = (1, -1, 1, 1).$

Solution 3.3.2. (a) In this case we have the area of the parallelogram created by $\vec{OX}_1, \vec{OX}_2$, denoted by v_2. Then

$$\begin{vmatrix} 1 & 1 \\ 1 & -1 \end{vmatrix} = -2.$$

The absolute value $= v_2 = 2$.

(b) In this case we have the volume of a parallelepiped, denoted by v_3. Consider

$$\begin{vmatrix} 2 & 0 & -4 \\ 1 & 1 & -1 \\ 1 & 0 & 1 \end{vmatrix} = - \begin{vmatrix} 1 & 0 & 1 \\ 1 & 1 & -1 \\ 2 & 0 & -4 \end{vmatrix} = - \begin{vmatrix} 1 & 0 & 1 \\ 0 & 1 & -2 \\ 0 & 0 & -6 \end{vmatrix}$$

$$= -(1)(1)(-6) = 6.$$

The absolute value is 6 and hence the volume is 6.

(c) In this case we have the volume of a parallelotope in 4-space, denoted by v_4. Consider

$$\begin{vmatrix} 1 & 1 & 1 & 1 \\ 1 & 1 & 1 & -1 \\ 1 & 1 & -1 & 1 \\ 1 & -1 & 1 & 1 \end{vmatrix} = \begin{vmatrix} 1 & 1 & 1 & 1 \\ 0 & 0 & 0 & -2 \\ 0 & 0 & -2 & 0 \\ 0 & -2 & 0 & 0 \end{vmatrix}$$

$$= - \begin{vmatrix} 1 & 1 & 1 & 1 \\ 0 & -2 & 0 & 0 \\ 0 & 0 & -2 & 0 \\ 0 & 0 & 0 & -2 \end{vmatrix} = 8.$$

The absolute value is 8 and hence $v_4 = 8$.

In a 2-space we have a parallelogram generated by $O\vec{X}_1$ and $O\vec{X}_2$, by completing the parallelogram. This parallelogram consists of two identical triangles. Then the area of one such triangle is $\frac{1}{2}v_2$ where v_2 is the area of the parallelogram. In a 3-space we have a parallelepiped. How many identical simplexes (simplices, 3-dimensional analogue of the triangle) can be packed into this parallelepiped? It can be easily seen that we can pack $6 = 3!$ such simplexes. The following are some standard notations in this area. ∇_n (nabla) and Δ_n (delta) are used to denote the volumes of the n-parallelotope and n-simplex respectively.

Notation 3.3.1.

$$\nabla_n = \text{ volume of an } n\text{-parallelotope in } n\text{-space}$$
$$\Delta_n = \text{ volume of an } n\text{-simplex in } n\text{-space}.$$

Then we have

$$\Delta_n = \frac{1}{n!} \nabla_n. \tag{3.3.9}$$

3.3.3 Jacobians of transformations

In a calculus course the instructor might have told that the Jacobian is a determinant and the curious students must have been wondering how a determinant enters into the picture. Let us see what happens if we take skew symmetric product of differentials.

Notation 3.3.2. $\wedge = $ (wedge), $dx \wedge dy$ (skew symmetric product of the differential dx with the differential dy), where x and y are real scalar independent or free variables.

Definition 3.3.1. *The skew symmetric product of the differential* of x, dx, with the differential of y, dy, is defined as

$$dx \wedge dy = -dy \wedge dx \implies dx \wedge dx = -dx \wedge dx$$
$$\implies dx \wedge dx = 0,$$

since $dx \wedge dx$ is a real scalar quantity. Let us consider some transformations. Let x and y be two real free variables and let u and v be functions of x and y. Let

$$u = f_1(x, y), \quad v = f_2(x, y). \tag{3.3.10}$$

As examples,

(a) $u = 2x - 3y, \quad v = x + y$
(b) $u = x^2 + y^2, \quad v = x - y$
(c) $u = x^2 + 2xy + y^2, \quad v = x^4 + 5.$

Taking the differentials in (3.3.10) we have, from elementary calculus,

$$du = \frac{\partial f_1}{\partial x} dx + \frac{\partial f_1}{\partial y} dy \quad \text{and} \tag{a}$$

$$dv = \frac{\partial f_2}{\partial x} dx + \frac{\partial f_2}{\partial y} dy. \tag{b}$$

Let us take the skew symmetric product of the differentials in u and v.

$$du \wedge dv = \left[\frac{\partial f_1}{\partial x} dx + \frac{\partial f_1}{\partial y} dy\right] \wedge \left[\frac{\partial f_2}{\partial x} dx + \frac{\partial f_2}{\partial y} dy\right]. \tag{c}$$

According to Definition 3.3.1 an interchange brings in a negative sign and hence when taking the product in (c) remember to keep the order and change the sign if the order is reversed. Straight multiplication of the right side in (c), keeping the order and neglecting higher orders such as $dx \wedge dx$ and $dy \wedge dy$, since they are zeros, we have

$$du \wedge dv = \frac{\partial f_1}{\partial x} \frac{\partial f_2}{\partial x} dx \wedge dx + \frac{\partial f_1}{\partial x} \frac{\partial f_2}{\partial y} dx \wedge dy$$
$$+ \frac{\partial f_1}{\partial y} \frac{\partial f_2}{\partial x} dy \wedge dx + \frac{\partial f_1}{\partial y} \frac{\partial f_2}{\partial y} dy \wedge dy$$
$$= \frac{\partial f_1}{\partial x} \frac{\partial f_2}{\partial y} dx \wedge dy + \frac{\partial f_1}{\partial y} \frac{\partial f_2}{\partial x} dy \wedge dx + 0 + 0.$$

Note that in one term we have $dx \wedge dy$ and in the other term $dy \wedge dx = -dx \wedge dy$. Therefore

$$du \wedge dv = \left[\frac{\partial f_1}{\partial x} \frac{\partial f_2}{\partial y} - \frac{\partial f_1}{\partial y} \frac{\partial f_2}{\partial x}\right] dx \wedge dy$$

$$= \begin{vmatrix} \frac{\partial f_1}{\partial x} & \frac{\partial f_1}{\partial y} \\ \frac{\partial f_2}{\partial x} & \frac{\partial f_2}{\partial y} \end{vmatrix} dx \wedge dy$$

$$= J \, dx \wedge dy.$$

The coefficient of $dx \wedge dy$ is called the Jacobian J and it is a determinant. If $J \neq 0$ then

$$dx \wedge dy = \frac{1}{J} du \wedge dv, \tag{3.3.11}$$

and the transformation $(x, y) \rightarrow (u, v)$ is one to one. Let us generalize this procedure to functions of many real variables. Let $x_1, \ldots, x_k$ be k free real scalar variables and consider k scalar functions of $x_1, \ldots, x_k$. Let

$$y_i = f_i(x_1, \ldots, x_k), \quad i = 1, 2, \ldots, k.$$

If the number of equations is not equal to the number of independent variables $x_1, \ldots, x_k$ then we cannot expect a one to one transformation. Even then for a one to one transformation we need the Jacobian to be nonzero. Only in this case one can write $dx_1 \wedge \cdots \wedge dx_k$ in terms of $dy_1 \wedge \cdots \wedge dy_k$ and vice versa. Then proceeding as before we note that

$$dy_1 \wedge \cdots \wedge dy_k = \begin{vmatrix} \frac{\partial f_1}{\partial x_1} & \cdots & \frac{\partial f_1}{\partial x_k} \\ \vdots & \cdots & \vdots \\ \frac{\partial f_k}{\partial x_1} & \cdots & \frac{\partial f_k}{\partial x_k} \end{vmatrix} dx_1 \wedge \cdots \wedge dx_k$$

where

$$J = \text{determinant of the Jacobian matrix} \left(\frac{\partial y_i}{\partial x_j} \right).$$

The (i, j)-th element in the Jacobian matrix is the partial derivative of y_i with respect to x_j. Since the transpose of this matrix also has the same determinant we could take the Jacobian matrix as $(\frac{\partial y_i}{\partial x_j})$ or $(\frac{\partial y_i}{\partial x_j})'$. Let us evaluate some Jacobians.

(a) Jacobians of linear transformations

Consider the linear transformation

$$y_i = a_{i1} x_1 + \cdots + a_{ik} x_k, \quad i = 1, \ldots, k.$$

This can be written as

$$Y = AX, \quad Y = \begin{bmatrix} y_1 \\ \vdots \\ y_k \end{bmatrix}, \quad X = \begin{bmatrix} x_1 \\ \vdots \\ x_k \end{bmatrix},$$

$$A = (a_{ij}) = \begin{bmatrix} a_{11} & \cdots & a_{1k} \\ \vdots & \cdots & \vdots \\ a_{k1} & \cdots & a_{kk} \end{bmatrix}.$$

Note that if $|A| \neq 0$ then A^{-1} exists and then

$$Y = AX \;\Rightarrow\; X = A^{-1}Y,$$

that is, Y can be written uniquely as a function of X and vice versa. In this case we have a one-to-one transformation. The coefficient matrix A in the above transformation is a matrix of constants. The Jacobian matrix in this case is

$$\left(\frac{\partial y_i}{\partial x_j}\right) = (a_{ij}) = A \;\Rightarrow\; J = |A|.$$

Thus we have an interesting result. In all the results to follow we will use the following notation.

Notation 3.3.3. When X is an $m \times n$ matrix of mn free real variables the skew symmetric product of all the differentials in X will be denoted as follows:

$$dX = \bigwedge_{i=1}^{m}\bigwedge_{j=1}^{n} dx_{ij},$$

and if $X = X'$ and $p \times p$ then

$$dX = \bigwedge_{i\geq j=1}^{p} dx_{ij}.$$

For example if

$$X = \begin{bmatrix} x_{11} & x_{12} \\ x_{21} & x_{22} \end{bmatrix} \;\Rightarrow\; dX = dx_{11} \wedge dx_{12} \wedge dx_{21} \wedge dx_{22}$$

$$= dx_{11} \wedge dx_{12} \wedge dx_{22} \quad \text{for } x_{12} = x_{21} \text{ or } X = X'$$

When taking the variables $x_{11}, x_{12}, x_{21}, x_{22}$ to form dX they can be taken in any convenient order to start with. Once they are taken in some order then that order has to be kept throughout that computation involving dX. For any transposition of differentials during the computational process the sign rule in the definition will apply.

In the symmetric case there are only $p(p+1)/2$ free scalar variables. From the above notation, if X is a $k \times 1$ vector then

$$dX = dX' = dx_1 \wedge \cdots \wedge dx_k.$$

Then we can write the result in the above linear transformation as follows:

(i) $$Y = AX, \quad |A| \neq 0, \Rightarrow dY = |A|dX,$$

Y and X are $k \times 1$ vectors of real scalar variables and $A = (a_{ij})$ is a matrix of constants.

Example 3.3.3. Evaluate the Jacobian in the following linear transformation. Is the transformation one-to-one?

$$y_1 = x_1 + x_2 + x_3$$
$$y_2 = x_1 - x_2 + x_3$$
$$y_3 = 2x_1 + x_2 - x_3.$$

Solution 3.3.3. Writing in matrix notation the transformation is

$$Y = AX, \quad Y = \begin{bmatrix} y_1 \\ y_2 \\ y_3 \end{bmatrix}, \quad X = \begin{bmatrix} x_1 \\ x_2 \\ x_3 \end{bmatrix}, \quad A = \begin{bmatrix} 1 & 1 & 1 \\ 1 & -1 & 1 \\ 2 & 1 & -1 \end{bmatrix}.$$

The Jacobian, J, is seen as the determinant

$$J = \begin{vmatrix} 1 & 1 & 1 \\ 1 & -1 & 1 \\ 2 & 1 & -1 \end{vmatrix} = \begin{vmatrix} 1 & 1 & 1 \\ 0 & -2 & 0 \\ 0 & -1 & -3 \end{vmatrix} \quad [-1(1) + 2; \; -2(1) + (3) \Rightarrow]$$

$$= \begin{vmatrix} -2 & 0 \\ -1 & -3 \end{vmatrix} = (-2)(-3) - (-1)(0) = 6.$$

Since $J \neq 0$ the transformation is one-to-one here.

(b) Linear matrix transformation

Let us consider a more general linear transformation. Let X be an $m \times n$ matrix of functionally independent mn real scalar variables, let A be an $m \times m$ nonsingular matrix of constants. Consider the following one-to-one transformation (one-to-one since $|A| \neq 0$):

$$Y = AX, \quad A = (a_{ij}) \text{ is } m \times m, \; X, Y \text{ are } m \times n.$$

What is the Jacobian in this transformation? If $X_1, \dots, X_n$ and $Y_1, \dots, Y_n$ denote the n columns of X and Y respectively then we can write this transformation in the following equivalent form:

$$[Y_1, Y_2, \dots, Y_n] = [AX_1, AX_2, \dots, AX_n].$$

Note that Y_1 does not contain $X_2, \dots, X_n$, Y_2 contains only X_2 and so on. Hence if we take the variables in Y in the order $\begin{pmatrix} Y_1 \\ \vdots \\ Y_n \end{pmatrix}$ and the variables in X in the order $(X_1', X_2', \dots, X_n')$

then by taking the partial derivatives, the Jacobian matrix has the following form:

$$
\begin{array}{ccccc}
& X_1' & X_2' & \cdots & X_n' \\[4pt]
Y_1 & A & O & \cdots & O \\
Y_2 & O & A & \cdots & O \\
\vdots & \vdots & \vdots & \cdots & \vdots \\
Y_n & O & O & \cdots & A
\end{array}
$$

observing that the partial derivative of Y_i with respect to X_j can produce A if $j = i$ and a null matrix if $j \neq i$. The determinant of the above block diagonal Jacobian matrix is the product of the determinants of the diagonal blocks. Hence the Jacobian is $|A|^n$. That is,

(ii) $\qquad\qquad Y = AX, \quad |A| \neq 0, \quad Y, X, m \times n \implies dY = |A|^n dX.$

What will be the effect if X is postmultiplied by a nonsingular $n \times n$ constant matrix B so that the transformation is one-to-one. That is,

$$Y = XB, \quad |B| \neq 0, \ Y \text{ and } X \text{ are } m \times n, \ B \text{ is } n \times n.$$

This Jacobian can be evaluated by observing the following: Look at the rows on both sides and follow through the procedure above then we have the next result.

(iii) $\qquad\qquad Y = XB, \quad |B| \neq 0, \quad Y, X, m \times n \implies dY = |B|^m dX.$

(c) Multilinear transformations

Combining the results in (ii) and (iii) above we have a general linear transformation of the type $Y = AXB$, where Y and X are $m \times n$ matrices of mn free real scalar variables and A, $m \times m$ and B, $n \times n$ are nonsingular matrices of constants. This transformation can be looked upon as $Z = AX$ and $Y = ZB$ or $U = XB$ and $Y = AU$ and then apply the above results. Then we have the following:

(iv) For X and Y, $m \times n$, $|A| \neq 0$, $|B| \neq 0$ where A is $m \times m$ and B is $n \times n$ matrices of constants,

$$Y = AXB \implies dY = |A|^n |B|^m dX.$$

Example 3.3.4. Consider the following linear transformation involving the free real variables $x_{11}, x_{12}, x_{13}, x_{21}, x_{22}, x_{23}$. Evaluate the Jacobian in this linear transformation.

$$
\begin{aligned}
y_{11} &= x_{11} + x_{21}, & y_{12} &= x_{12} + x_{22}, & y_{13} &= x_{13} + x_{23}, \\
y_{21} &= x_{11} - x_{21}, & y_{22} &= x_{12} - x_{22}, & y_{23} &= x_{13} - x_{23}.
\end{aligned}
$$

Solution 3.3.4. Writing in matrix notation

$$Y = AX, \quad Y = \begin{bmatrix} y_{11} & y_{12} & y_{13} \\ y_{21} & y_{22} & y_{23} \end{bmatrix}$$

$$X = \begin{bmatrix} x_{11} & x_{12} & x_{13} \\ x_{21} & x_{22} & x_{23} \end{bmatrix}, \quad A = \begin{bmatrix} 1 & 1 \\ 1 & -1 \end{bmatrix}.$$

Since $|A| = -2 \neq 0$ it is a one-to-one transformation. Then from property (ii) above the Jacobian is

$$J = |A|^n = \begin{vmatrix} 1 & 1 \\ 1 & -1 \end{vmatrix}^3 = (-2)^3 = -8.$$

Example 3.3.5. Consider the linear transformation involving the free real variables $x_{11}, x_{12}, x_{13}, x_{21}, x_{22}, x_{23}$.

$$y_{11} = x_{11} + x_{21} + x_{12} + x_{22} + x_{13} + x_{23}, \quad y_{12} = -(x_{13} + x_{23}),$$
$$y_{13} = 3(x_{12} + x_{22}) + 2(x_{13} + x_{23}),$$
$$y_{21} = x_{11} - x_{21} + x_{12} - x_{22} + x_{13} - x_{23}, \quad y_{22} = -(x_{13} - x_{23}),$$
$$y_{23} = 3(x_{12} - x_{22}) + 2(x_{13} - x_{23}).$$

Evaluate the Jacobian in this linear transformation. Is the transformation one-to-one?

Solution 3.3.5. Writing the transformation in matrix notation we have

$$Y = AXB, \quad A = \begin{bmatrix} 1 & 1 \\ 1 & -1 \end{bmatrix}, \quad B = \begin{bmatrix} 1 & 0 & 0 \\ 1 & 0 & 3 \\ 1 & -1 & 2 \end{bmatrix};$$

$$X = \begin{bmatrix} x_{11} & x_{12} & x_{13} \\ x_{21} & x_{22} & x_{23} \end{bmatrix}, \quad Y = \begin{bmatrix} y_{11} & y_{12} & y_{13} \\ y_{21} & y_{22} & y_{23} \end{bmatrix}.$$

Here $|A| = -2 \neq 0$, $|B| = 3 \neq 0$. Hence the transformation is one-to-one. From property (iv) the Jacobian is given by

$$J = |A|^n |B|^m = (-2)^3 (3)^2 = -72.$$

Before concluding this subsection let us consider a nonlinear transformation.

(d) Jacobian in a nonlinear transformation

Let $X = (x_{ij})$ with $x_{11} > 0$, $x_{22} > 0$, $x_{11}x_{22} - x_{12}^2 > 0$ be a 2×2 symmetric matrix and T a 2×2 lower triangular matrix with positive diagonal elements. Assume that it is possible to express $X = TT'$. [Not all symmetric matrices can be written in this form. The matrices which can be written in this form fall into the category of non-negative definite matrices. Definiteness of matrices will be discussed in the next chapter in a systematic way. When the matrix X is positive definite then the conditions stated above will be necessary.] Then our transformation is given by

$$\begin{bmatrix} x_{11} & x_{12} \\ x_{12} & x_{22} \end{bmatrix} = \begin{bmatrix} t_{11} & 0 \\ t_{21} & t_{22} \end{bmatrix} \begin{bmatrix} t_{11} & t_{21} \\ 0 & t_{22} \end{bmatrix}$$

$$= \begin{bmatrix} t_{11}^2 & t_{11}t_{21} \\ t_{21}t_{11} & t_{21}^2 + t_{22}^2 \end{bmatrix}.$$

Is this transformation one-to-one? Here $t_{11}^2 = x_{11} \Rightarrow t_{11} = \pm\sqrt{x_{11}}$. Hence we must have $t_{11} > 0$ or strictly negative to have t_{11} uniquely defined in terms of x_{ij}'s. Let $t_{11} > 0$. Then $t_{11} = \sqrt{x_{11}}$ is uniquely defined. [Note that when X is real positive definite then all diagonal elements of X must be positive. That is, $x_{jj} > 0$, $j = 1, \ldots, p$ for a $p \times p$ matrix.] $x_{12} = t_{11}t_{21} \Rightarrow t_{21} = \frac{x_{12}}{\sqrt{x_{11}}}$ or t_{21} is uniquely defined. $x_{22} = t_{21}^2 + t_{22}^2 \Rightarrow t_{22}^2 = x_{22} - t_{21}^2$. There are two possible values for t_{22}. Hence if $t_{11} > 0$ and $t_{22} > 0$ the transformation is one-to-one. This can be proved in general also for X a $p \times p$ symmetric matrix which can be written as TT' where T is $p \times p$ lower triangular with positive diagonal elements. In this transformation $p(p + 1)/2$ elements, t_{ij}'s, $i \geq j$, in T go to $p(p + 1)/2$ elements x_{ij}'s, $i \geq j$, in X. Let us evaluate the Jacobian in this transformation. Let us look at the 2×2 case, from where the general case will be obvious. We have

$$x_{11} = t_{11}^2, \quad x_{12} = t_{11}t_{21}, \quad x_{22} = t_{21}^2 + t_{22}^2.$$

Take the x_{ij}'s in the order x_{11}, x_{12}, x_{22} and the t_{ij}'s in the order t_{11}, t_{21}, t_{22} and form the matrix of partial derivatives.

$$\frac{\partial x_{11}}{\partial t_{11}} = 2t_{11}, \quad \frac{\partial x_{12}}{\partial t_{21}} = t_{11}, \quad \frac{\partial x_{22}}{\partial t_{22}} = 2t_{22},$$

$$\frac{\partial x_{11}}{\partial t_{21}} = 0, \quad \frac{\partial x_{11}}{\partial t_{22}} = 0.$$

The matrix of partial derivatives is given by

	t_{11}	t_{21}	t_{22}
x_{11}	$2t_{11}$	0	0
x_{12}	$*$	t_{11}	0
x_{22}	$*$	$*$	$2t_{22}$

Since the Jacobian matrix is in a triangular form we will not be interested in the elements marked by $*$ in the above configuration. The determinant is the product of the diagonal elements. In this case it is $2^2 t_{11}^2 t_{22}$.

Suppose we have a $p \times p$ matrix X, symmetric and positive definite, and a $p \times p$ lower triangular matrix T such that $X = TT'$ and $t_{jj} > 0$, $j = 1, \ldots, p$. Then when the Jacobian matrix is formed by taking x_{ij}'s in the order $x_{11}, x_{12}, \ldots, x_{1p}, x_{22}, \ldots, x_{2p}, \ldots, x_{pp}$ and the t_{ij}'s in the order $t_{11}, t_{21}, \ldots, t_{p1}, t_{22}, \ldots, t_{p2}, \ldots, t_{pp}$ then we have the following quantities in the diagonal of the Jacobian matrix. When $x_{11}, \ldots, x_{1p}$ are considered we have one 2 and t_{11} appearing p times. When $x_{22}, \ldots, x_{2p}$ are considered we have one 2 and t_{22} appearing $p - 1$ times, and so on. Hence the final result will be the following:

(v) If a symmetric matrix X can be written as TT' where T is lower triangular with positive diagonal elements then

$$X = TT' \implies$$
$$dX = 2^p t_{11}^p t_{22}^{p-1} \cdots t_{pp} \, dT = 2^p \left\{ \prod_{j=1}^p t_{jj}^{p+1-j} \right\} dT.$$

This Jacobian has some very interesting applications, especially in evaluating some very complicated integrals.

Example 3.3.6. Evaluate the following multiple integral

$$\int \int \int [x_{11}x_{22} - x_{12}^2]^\alpha \, e^{-(x_{11}+x_{22})} dx_{11} \wedge dx_{12} \wedge dx_{22}$$

where $x_{11} > 0$, $x_{22} > 0$, $x_{11}x_{22} - x_{12}^2 > 0$.

Solution 3.3.6. Writing in matrix and determinant notations the integral that we want to evaluate can be written as follows, observing that,

$$\begin{vmatrix} x_{11} & x_{12} \\ x_{12} & x_{22} \end{vmatrix} = x_{11}x_{22} - x_{12}^2,$$

$$X = \begin{bmatrix} x_{11} & x_{12} \\ x_{12} & x_{22} \end{bmatrix}, \quad \text{tr}(X) = x_{11} + x_{22},$$

$$\int_X |X|^\alpha \, e^{-\text{tr}(X)} dX.$$

The conditions on x_{ij}'s imply that X can be written as TT' where T is lower triangular with positive diagonal elements. Consider the transformation

$$X = TT' = \begin{bmatrix} t_{11}^2 & t_{11}t_{21} \\ t_{11}t_{21} & t_{21}^2 + t_{22}^2 \end{bmatrix}, \quad T = \begin{bmatrix} t_{11} & 0 \\ t_{21} & t_{22} \end{bmatrix}.$$

Then

$$dX = 2^2 t_{11}^2 t_{22} \, dT, \quad t_{11} > 0, \; t_{22} > 0.$$

Let the integral be denoted by y. Then

$$y = \int_X |X|^\alpha \, e^{-\operatorname{tr}(X)} dX$$
$$= \int_T |TT'|^\alpha \, e^{-\operatorname{tr}(TT')} 2^2 t_{11}^2 t_{22} \, dT.$$

Note that

$$|X| = |TT'| = t_{11}^2 t_{22}^2,$$
$$\operatorname{tr}(TT') = t_{11}^2 + t_{21}^2 + t_{22}^2.$$

Then

$$y = 2^2 \int_0^\infty (t_{11}^2)^{\alpha+1} \, e^{-t_{11}^2} dt_{11}$$
$$\times \int_0^\infty (t_{22}^2)^{\alpha+\frac{1}{2}} \, e^{-t_{22}^2} dt_{22} \int_{-\infty}^\infty e^{-t_{21}^2} dt_{21}.$$

Observe that t_{11} and t_{22} are restricted to be positive whereas t_{21} is free to vary. We need to evaluate only two types of integrals.

$$2\int_0^\infty (u^2)^\beta \, e^{-u^2} du \quad \text{and} \quad \int_{-\infty}^\infty e^{-z^2} dz,$$

for $\beta = \alpha + \frac{1}{2}, \alpha + 1$. Substituting

$$v = u^2 \;\Rightarrow\; u = v^{\frac{1}{2}} \;\Rightarrow\; du = \frac{1}{2} v^{-\frac{1}{2}} dv$$

we have

$$2\int_0^\infty (u^2)^\beta \, e^{-u^2} du = \int_0^\infty v^{\beta+\frac{1}{2}-1} \, e^{-v} dv$$
$$= \Gamma\!\left(\beta + \frac{1}{2}\right) \quad \text{for } \Re\!\left(\beta + \frac{1}{2}\right) > 0$$

where $\Gamma(\cdot)$ is a gamma function and $\Re(\cdot)$ denotes the real part of $(\cdot)$. [For the sake of those students who are unfamiliar with gamma functions a definition will be given after the discussion of this example.]

$$\int_{-\infty}^\infty e^{-z^2} dz = 2\int_0^\infty e^{-z^2} dz \quad \text{(since } e^{-z^2} \text{ is even and the integral exists)}$$
$$= \int_0^\infty w^{\frac{1}{2}-1} e^{-w} dw \quad \text{(put } z^2 = w, w > 0)$$
$$= \Gamma\!\left(\frac{1}{2}\right) = \sqrt{\pi}.$$

Hence the answer is that

$$y = \sqrt{\pi}\,\Gamma\!\left(\alpha + \frac{3}{2}\right)\Gamma(\alpha + 1).$$

Definition 3.3.2 (A gamma function $\Gamma(\alpha)$). It can be defined in many ways. $\Gamma(\alpha)$ is defined for all $\alpha \neq 0, -1, -2, \ldots$. An integral representation of $\Gamma(\alpha)$ is the following: The standard notation used is $\Gamma(z)$, (gamma z or gamma of z; it is a function of z).

$$\Gamma(z) = \int_0^\infty x^{z-1}\, e^{-x} dx \quad \text{for } \Re(z) > 0.$$

For the convergence of the integral the condition $\Re(z) > 0$ is needed. If z is real then the condition reduces to $z > 0$. This condition is needed only if we are using an integral representation. Otherwise the condition is $z \neq 0, -1, -2, \ldots$ A few properties which follow from the definition itself are the following:

$$\Gamma(\alpha) = (\alpha - 1)\Gamma(\alpha - 1) \quad \text{for } \Re(\alpha - 1) > 0. \tag{3.3.12}$$

This property is evident from the integral representation, by using integration by parts. Extending this result we have

$$\Gamma(\alpha) = (\alpha - 1)(\alpha - 2)\ldots(\alpha - r)\Gamma(\alpha - r), \quad \Re(\alpha - r) > 0. \tag{3.3.13}$$

$$\Gamma(n) = (n - 1)! \quad \text{when } n \text{ is a positive integer.} \tag{3.3.14}$$

The next result can be established by considering a double integral.

$$\Gamma\left(\frac{1}{2}\right) = \sqrt{\pi}. \tag{3.3.15}$$

3.3.4 Functions of matrix argument

Consider a $p \times p$ matrix X. We can define several scalar functions on X. For example

(a) $\quad f_1(X) = |X| = $ determinant of X

(b) $\quad f_2(X) = 2|X|^2 - 3|X| + 5$

(c) $\quad f_3(X) = \text{tr}(X) = x_{11} + \cdots + x_{pp} = $ trace of X

are all scalar functions of X. We could have also defined matrix functions. For example,

(α) $\quad g_1(X) = [I - X]^{-1}$

(β) $\quad g_2(X) = I + 3X + X^2 - 5X^3$

are matrix functions of X. We are interested in real-valued scalar functions of matrix argument in this subsection here, that is, functions of the types (a), (b), (c) above. One of the very basic functions in the theory of scalar functions of matrix argument is a matrix-variate gamma, analogous to Definition 3.3.2. Since the algebra can get very involved we will only introduce a matrix-variate gamma and stop the discussion. Consider the following integral, denoting it by

$$\Gamma_p(\alpha) = \int_X |X|^{\alpha - \frac{p+1}{2}} e^{-\text{tr}(X)} dX$$

where X is $p \times p$ such that it can be expressed in the form $X = TT'$ where T is lower triangular with positive diagonal elements. Then making the transformation $X = TT'$ we have the Jacobian from property (v). That is,

$$\Gamma_p(\alpha) = \int_X |X|^{\alpha - \frac{p+1}{2}} e^{-\text{tr}(X)} dX$$

$$= \int_T |TT'|^{\alpha - \frac{p+1}{2}} e^{-\text{tr}(TT')} 2^p \left\{ \prod_{j=1}^p t_{jj}^{p+1-j} \right\} dT$$

$$= \int_T \left(\prod_{j=1}^p t_{jj}^2 \right)^{\alpha - \frac{p+1}{2}} 2^p \left\{ \prod_{j=1}^p t_{jj}^{p+1-j} \right\}$$

$$\times e^{-(t_{11}^2 + t_{21}^2 + \cdots + t_{pp}^2)} dT$$

$$= \left\{ \prod_{j=1}^p 2 \int_0^\infty (t_{jj}^2)^{\alpha - \frac{j}{2}} e^{-t_{jj}^2} dt_{jj} \right\} \left\{ \prod_{i>j} \int_{-\infty}^\infty e^{-t_{ij}^2} dt_{ij} \right\}.$$

Evaluating with the help of the gamma integral of Definition 3.3.2 we have

$$\Gamma_p(\alpha) = \pi^{\frac{p(p-1)}{4}} \Gamma(\alpha) \Gamma\left(\alpha - \frac{1}{2}\right) \Gamma(\alpha - 1) \cdots \Gamma\left(\alpha - \frac{p-1}{2}\right)$$

for $\Re(\alpha) > \frac{p-1}{2}$.

Definition 3.3.3. *A real matrix-variate gamma: Notation* $\Gamma_p(\alpha)$ *(gamma p alpha),*

$$\Gamma_p(\alpha) = \pi^{\frac{p(p-1)}{4}} \Gamma(\alpha) \Gamma\left(\alpha - \frac{1}{2}\right) \cdots \Gamma\left(\alpha - \frac{p-1}{2}\right), \quad \Re(\alpha) > \frac{p-1}{2}$$

$$= \int_{X=X'>0} |X|^{\alpha - \frac{p+1}{2}} e^{-\text{tr}(X)} dX, \quad \Re(\alpha) > \frac{p-1}{2}.$$

Since the integral on the right gives $\Gamma_p(\alpha)$ if we divide both sides by $\Gamma_p(\alpha)$ we can create a matrix-variate statistical density out of this function, known as the real *matrix-variate gamma density*.

Definition 3.3.4. *A real matrix-variate gamma density*

$$f(X) = \frac{1}{\Gamma_p(\alpha)} |X|^{\alpha - \frac{p+1}{2}} e^{-\text{tr}(X)}, \quad X = X' > 0 \quad \Re(\alpha) > \frac{p-1}{2}$$

where the matrix is $p \times p$ symmetric and can be written in the form $X = TT'$. The notation $U = U' > 0$ means the matrix U is symmetric positive definite. This concept of definiteness will be introduced properly later on. For the time being take it as meaning that X can be expressed in the form $X = TT'$ where T is lower triangular with positive diagonal elements.

3.3.5 Partitioned determinants and multiple correlation coefficient

The idea of partitioning a matrix was introduced in Chapter 2. Let us examine the effect of partitioning on determinants. Let an $n \times n$ matrix A be partitioned as follows:

$$A = \begin{bmatrix} A_{11} & A_{12} \\ A_{21} & A_{22} \end{bmatrix}$$

where A_{11} is $r \times r$ thereby A_{12} is $r \times (n-r)$, A_{21} is $(n-r) \times r$, A_{22} is $(n-r) \times (n-r)$. Let us evaluate the determinant of A in terms of the determinants of the submatrices. Recall the steps in the actual evaluation of a determinant. We were adding suitable multiples of rows (columns) to other rows (columns) to reduce the matrix to a triangular form or to a block triangular form. Instead of adding one row (column) at a time we could have added suitable multiples of a block of rows to another block of rows. The result would have been the same. Suppose that we want to bring a null matrix at the position of A_{21}. What suitable combinations of the first block of rows, namely (A_{11}, A_{12}) to be added to the second block of rows, (A_{21}, A_{22}), so that a null matrix can be produced at the place of A_{21}? A suitable multiple is $-A_{21}A_{11}^{-1}$ times the first block (A_{11}, A_{12}) to be added to the second block (A_{21}, A_{22}). The value of the determinant remains the same. [Remember to keep the order of multiplication of the matrices involved.] Then

$$|A| = \begin{vmatrix} A_{11} & A_{12} \\ A_{21} & A_{22} \end{vmatrix} = \begin{vmatrix} A_{11} & A_{12} \\ 0 & A_{22} - A_{21}A_{11}^{-1}A_{12} \end{vmatrix}.$$

This can be done if A_{11} is nonsingular. Since the above is a triangular block matrix its determinant is the product of the determinants of the diagonal blocks. That is,

$$|A| = |A_{11}|\,|A_{22} - A_{21}A_{11}^{-1}A_{12}| \quad \text{for } |A_{11}| \neq 0. \tag{3.3.16}$$

From symmetry it follows that

$$|A| = |A_{22}|\,|A_{11} - A_{12}A_{22}^{-1}A_{12}| \quad \text{for } |A_{22}| \neq 0. \tag{3.3.17}$$

In (3.3.16) and (3.3.17) the submatrices enter into a cyclic order. If we start with A_{11} then it goes $A_{11} - A_{12}A_{22}^{-1}A_{21}$ and if we start with A_{22} then it goes $A_{22} - A_{21}A_{11}^{-1}A_{12}$. A major advantage of the formulae (3.3.16) and (3.3.17) is that the orders of the determinants on the right are reduced to r and $(n-r)$ both of which will be less than the order n on the left when $1 \leq r < n$. Thus the computations are made a little bit easier. For example if we have a 16×16 determinant the evaluation can be reduced to the evaluation of two 8×8 determinants, the latter will be considerably easier, but the penalty is that one of the matrices involves product and an inverse.

Example 3.3.7. Evaluate the following 4×4 determinant by partitioning into 2×2 blocks. Repeat the process with a partition where A_{11} is 1×1.

$$|A| = \begin{vmatrix} 1 & 0 & 1 & -1 \\ 0 & 1 & 1 & 1 \\ 1 & 1 & 0 & 0 \\ 1 & -1 & 1 & 1 \end{vmatrix}.$$

Solution 3.3.7. Let

$$A = \begin{bmatrix} A_{11} & A_{12} \\ A_{21} & A_{22} \end{bmatrix}, \quad A_{11} = \begin{bmatrix} 1 & 0 \\ 0 & 1 \end{bmatrix}, \quad A_{22} = \begin{bmatrix} 0 & 0 \\ 1 & 1 \end{bmatrix}.$$

Then $|A_{11}| \neq 0$ whereas $|A_{22}| = 0$. By using the formula (3.3.16) we have

$$|A| = |A_{11}| \, |A_{22} - A_{21} A_{11}^{-1} A_{12}|$$

$$= \begin{vmatrix} 1 & 0 \\ 0 & 1 \end{vmatrix} \left| \begin{bmatrix} 0 & 0 \\ 1 & 1 \end{bmatrix} - \begin{bmatrix} 1 & 1 \\ 1 & -1 \end{bmatrix} \begin{bmatrix} 1 & 0 \\ 0 & 1 \end{bmatrix}^{-1} \begin{bmatrix} 1 & -1 \\ 1 & 1 \end{bmatrix} \right|.$$

Since $A_{11} = I_2$, $A_{11}^{-1} = I_2$, $|A_{11}| = 1$.

$$A_{22} - A_{21} A_{11}^{-1} A_{12} = \begin{bmatrix} 0 & 0 \\ 1 & 1 \end{bmatrix} - \begin{bmatrix} 1 & 1 \\ 1 & -1 \end{bmatrix} I \begin{bmatrix} 1 & -1 \\ 1 & 1 \end{bmatrix}$$

$$= \begin{bmatrix} 0 & 0 \\ 1 & 1 \end{bmatrix} - \begin{bmatrix} 2 & 0 \\ 0 & -2 \end{bmatrix}$$

$$= \begin{bmatrix} -2 & 0 \\ 1 & 3 \end{bmatrix}.$$

$$|A_{22} - A_{21} A_{11}^{-1} A_{12}| = \begin{vmatrix} -2 & 0 \\ 1 & 3 \end{vmatrix} = -6 \Rightarrow |A| = -6.$$

When A_{11} is 1×1 we have $A_{11} = 1$, $|A_{11}| = 1$, $A_{11}^{-1} = 1$. Again using the formula (3.3.16) we have

$$A_{22} - A_{21} A_{11}^{-1} A_{12} = \begin{bmatrix} 1 & 1 & 1 \\ 1 & 0 & 0 \\ -1 & 1 & 1 \end{bmatrix} - \begin{bmatrix} 0 \\ 1 \\ 1 \end{bmatrix} [1][0, 1, -1]$$

$$= \begin{bmatrix} 1 & 1 & 1 \\ 1 & 0 & 0 \\ -1 & 1 & 1 \end{bmatrix} - \begin{bmatrix} 0 & 0 & 0 \\ 0 & 1 & -1 \\ 0 & 1 & -1 \end{bmatrix}$$

$$= \begin{bmatrix} 1 & 1 & 1 \\ 1 & -1 & 1 \\ -1 & 0 & 2 \end{bmatrix};$$

$$|A_{22} - A_{21}A_{11}^{-1}A_{12}| = \begin{vmatrix} 1 & 1 & 1 \\ 1 & -1 & 1 \\ -1 & 0 & 2 \end{vmatrix}$$

$$= \begin{vmatrix} 1 & 1 & 1 \\ 0 & -2 & 0 \\ 0 & 1 & 3 \end{vmatrix} \quad [-1(1) + (2); (1) + (3) \Rightarrow]$$

$$= \begin{vmatrix} -2 & 0 \\ 1 & 3 \end{vmatrix} = -6 \Rightarrow |A| = -6.$$

Note 3.1. If partitioning technique is used to evaluate a determinant then select the submatrices appropriately so that the computations can be minimized.

We can obtain a very interesting result when A_{11} or A_{22} is 1×1. Let the $p \times p$ matrix $V = (v_{ij})$ be partitioned as follows:

$$V = \begin{bmatrix} V_{11} & V_{12} \\ V_{21} & V_{22} \end{bmatrix}$$

where $V_{11} = v_{11}$ is 1×1 thereby V_{12} is $1 \times (p-1)$, V_{21} is $(p-1) \times 1$ and V_{22} is $(p-1) \times (p-1)$. Let $v_{11} \neq 0$, $|V_{22}| \neq 0$. Then from (3.3.16) and (3.3.17) we have

$$|V| = v_{11}\left|V_{22} - \frac{V_{21}V_{12}}{v_{11}}\right| \tag{a}$$

$$= |V_{22}|\,[v_{11} - V_{12}V_{22}^{-1}V_{21}]. \tag{b}$$

That is, the scalar quantity,

$$v_{11} - V_{12}V_{22}^{-1}V_{21} = \frac{|V|}{|V_{22}|} = \frac{v_{11}}{|V_{22}|}\left|V_{22} - \frac{V_{21}V_{12}}{v_{11}}\right|. \tag{3.3.18}$$

But

$$\left|V_{22} - \frac{V_{21}V_{12}}{v_{11}}\right| = |V_{22}|\left|\frac{v_{11}I - V_{22}^{-1}V_{21}V_{12}}{v_{11}}\right|.$$

Comparing with (a) and (b) above we have

$$v_{11} - V_{12}V_{22}^{-1}V_{21} = v_{11}\left|I - \frac{V_{22}^{-1}V_{21}V_{12}}{v_{11}}\right|. \tag{3.3.19}$$

The beauty of the relationship is that on one side we have a scalar quantity whereas on the other side we have a $(p-1) \times (p-1)$ determinant. This formula is often used in statistical and other problems to reduce a $(p-1) \times (p-1)$ determinant to a scalar quantity. Also from (a) and (b) above we have

$$1 - \frac{V_{12}V_{22}^{-1}V_{21}}{v_{11}} = \frac{|V|}{v_{11}|V_{22}|} \Rightarrow \tag{c}$$

$$\frac{V_{12}V_{22}^{-1}V_{21}}{v_{11}} = 1 - \frac{|V|}{v_{11}|V_{22}|}.$$

When V is a variance–covariance matrix associated with a real vector random variable

$$X = \begin{pmatrix} x_1 \\ \vdots \\ x_p \end{pmatrix} = \begin{pmatrix} x_1 \\ X_2 \end{pmatrix}, \quad X_2 = \begin{pmatrix} x_2 \\ \vdots \\ x_p \end{pmatrix}$$

then $\frac{V_{12}V_{22}^{-1}V_{21}}{v_{11}}$ is called the square of the *multiple correlation coefficient* of x_1 on $X_2 = \begin{pmatrix} x_2 \\ \vdots \\ x_p \end{pmatrix}$ and it is usually denoted by

$$\rho^2_{1(2...p)} = \frac{V_{12}V_{22}^{-1}V_{21}}{v_{11}}$$

$$= v_{11}^{-\frac{1}{2}} V_{12} V_{22}^{-1} V_{21} v_{11}^{-\frac{1}{2}}. \tag{3.3.20}$$

One can show $\rho_{1(2...p)}$ to be the maximum correlation between x_1 and an arbitrary linear function of $x_2, \ldots, x_p$. In prediction problems when a variable such as x_1 is predicted by using a linear function of other variables such as $x_2, \ldots, x_p$ then $\rho_{1(2...p)}$ is often used to measure how good is the predictor in the sense, larger the value of $\rho_{1(2...p)}$ better the predictor.

If V_{11} is $p_1 \times p_1$, V_{22} is $p_2 \times p_2$ such that $p_1 + p_2 = p$ then (3.3.20) is no longer a scalar quantity. If we write the last expression in (3.3.20) with V_{11} a $p_1 \times p_1$ matrix, that is,

$$P = V_{11}^{-\frac{1}{2}} V_{12} V_{22}^{-1} V_{21} V_{11}^{-\frac{1}{2}} \tag{3.3.21}$$

where $V_{11}^{\frac{1}{2}}$ indicates a positive definite square root of V_{11} then P in (3.3.21) is known as the *canonical correlation matrix* which plays a vital role in canonical correlation analysis. This field is also mainly concerned about prediction problems, predicting a set of variables with the help of another set of variables, a generalization of the first situation where one scalar variable is predicted by using a set of other variables. These areas are very rich in real-life situations where matrices and determinants play very important roles.

Other concepts associated with the concept of multiple correlation and canonical correlations are the concepts of partial correlations, correlation ratios and partial correlation matrices which come into regression problems, residual analysis, model building and other prediction and estimation problems. These quantities can be written up in terms of partitioned matrices and the corresponding determinants.

Example 3.3.8. Compute the multiple correlation coefficient of x_1 on (x_2, x_3) from the following variance–covariance matrix of X, where $X' = (x_1, x_2, x_3)$:

$$V = \begin{bmatrix} 2 & 1 & 1 \\ 1 & 3 & 2 \\ 1 & 2 & 5 \end{bmatrix}.$$

Solution 3.3.8. Let $v_{11} = 2$. Taking the expression from (c) above

$$\rho^2_{1(2.3)} = \frac{V_{12}V_{22}^{-1}V_{21}}{v_{11}} = 1 - \frac{|V|}{v_{11}|V_{22}|};$$

$$|V_{22}| = \begin{vmatrix} 3 & 2 \\ 2 & 5 \end{vmatrix} = 11;$$

$$|V| = \begin{vmatrix} 2 & 1 & 1 \\ 1 & 3 & 2 \\ 1 & 2 & 5 \end{vmatrix} = 2\begin{vmatrix} 3 & 2 \\ 2 & 5 \end{vmatrix} - \begin{vmatrix} 1 & 2 \\ 1 & 5 \end{vmatrix} + \begin{vmatrix} 1 & 3 \\ 1 & 2 \end{vmatrix}$$

$$= (2)(11) - 3 + (-1) = 18.$$

Hence

$$\frac{|V|}{v_{11}|V_{22}|} = \frac{18}{(2)(11)} = \frac{9}{11} \Rightarrow$$

$$\rho^2_{1(2.3)} = 1 - \frac{9}{11} = \frac{2}{11}.$$

3.3.6 Maxima/minima problems

One of the problems in multivariable calculus is to look for maxima/minima of a function of many variables. Let $f(X)$ be a scalar function of the $p \times 1$ vector X of real variables. In Chapters 1 and 2 we have defined the differential operators

$$\frac{\partial}{\partial X} = \begin{bmatrix} \frac{\partial}{\partial x_1} \\ \vdots \\ \frac{\partial}{\partial x_p} \end{bmatrix}, \quad \frac{\partial}{\partial X'} = \left(\frac{\partial}{\partial x_1}, \dots, \frac{\partial}{\partial x_p} \right),$$

$$\frac{\partial}{\partial X} \frac{\partial}{\partial X'} = \begin{bmatrix} \frac{\partial^2}{\partial x_1^2} & \cdots & \frac{\partial^2}{\partial x_1 \partial x_p} \\ \vdots & \cdots & \vdots \\ \frac{\partial^2}{\partial x_p \partial x_1} & \cdots & \frac{\partial^2}{\partial x_p^2} \end{bmatrix}.$$

Then $\frac{\partial}{\partial X}$ operating on a function f equated to a null vector gives the critical points and $\frac{\partial}{\partial X} \frac{\partial f}{\partial X'}$ at these critical points will decide on the critical points being corresponding to a local maximum or a local minimum or something else.

Example 3.3.9. Check for maxima/minima in the following function

$$f = x_1^2 + 2x_2^2 + x_3^2 - 2x_1x_2 - x_2x_3 + 2x_1 + 4x_2 + x_3 + 8.$$

Solution 3.3.9. Consider

$$\frac{\partial f}{\partial X} = \begin{bmatrix} 2x_1 - 2x_2 + 2 \\ -2x_1 + 4x_2 - x_3 + 4 \\ -x_2 + 2x_3 + 1 \end{bmatrix} = \begin{bmatrix} 0 \\ 0 \\ 0 \end{bmatrix}.$$

Let us solve the equations by looking at the coefficient matrix and performing elementary operations:

$$
\begin{array}{ccc|c}
1 & -1 & 0 & -1 \\
-2 & 4 & -1 & -4 \\
0 & -1 & 2 & -1
\end{array}
\Rightarrow
\begin{array}{ccc|c}
1 & -1 & 0 & -1 \\
0 & 2 & -1 & -6 \\
-1 & 0 & -1 & -1
\end{array}
$$

$$
\Rightarrow
\begin{array}{ccc|c}
1 & -1 & 0 & -1 \\
0 & -1 & 2 & -1 \\
0 & 0 & 3 & -8
\end{array}
$$

$$
\Rightarrow x_3 = -\frac{8}{3}, \quad x_2 = -\frac{13}{3}, \quad x_1 = -\frac{16}{3}.
$$

There is only one critical point $(x_1, x_2, x_3) = (-\frac{16}{3}, -\frac{13}{3}, -\frac{8}{3})$. This point may correspond to a local maximum or a local minimum or neither. Consider the matrix of second order partial derivatives operating on f.

$$
\frac{\partial f}{\partial X'} = [2x_1 - 2x_2 + 2, -2x_1 + 4x_2 - x_3 + 4, -x_2 + 2x_3 + 1]
$$

$$
\frac{\partial}{\partial X} \frac{\partial f}{\partial X'} =
\begin{bmatrix}
2 & -2 & 0 \\
-2 & 4 & -1 \\
0 & -1 & 2
\end{bmatrix}.
$$

For a minimum this matrix at the critical point must be positive definite and for a maximum it should be negative definite. We will define definiteness of matrices in terms of determinants next. Definiteness can also be defined equivalently in terms of other quantities.

Definition 3.3.5 (Definiteness of an $n \times n$ real symmetric matrix A). It is defined only for non-null square symmetric matrices when real. Consider all the leading minors of A. Let the leading minors be denoted by $|M_1|, \ldots, |M_n|$. Then A is positive definite if $|M_j| > 0$, $j = 1, \ldots, n$ (positive semi-definite if the minors can be zero also); negative definite if $|M_1| < 0$, $|M_2| > 0$, $|M_3| < 0$, $\ldots$ (negative semi-definite if the minors can be zero also); and A is indefinite if some of the minors are negative and some positive, at least one each set, and not belonging to the above types.

For our Example 3.3.9 let us look for the definiteness of our matrix of second order partial derivatives, evaluated at the critical points. Since our matrix is free of the variables the matrix evaluated at the critical point is itself. Let us look at the leading or *principal* minors.

$$
|M_1| = 2 > 0,
$$

$$
|M_2| =
\begin{vmatrix}
2 & -2 \\
-2 & 4
\end{vmatrix}
= 4 > 0,
$$

$$|M_3| = \begin{vmatrix} 2 & -2 & 0 \\ -2 & 4 & -1 \\ 0 & -1 & 2 \end{vmatrix} = 2 \begin{vmatrix} 4 & -1 \\ -1 & 2 \end{vmatrix} - (-2) \begin{vmatrix} -2 & -1 \\ 0 & 2 \end{vmatrix} + 0$$

$$= (2)(7) - 8 = 6 > 0.$$

Hence the matrix is positive definite. Therefore the critical point corresponds to a minimum.

Another definition of definiteness of matrices in terms of eigenvalues will be introduced in the next chapter. Another one in terms of quadratic forms will be given next.

Definition 3.3.6 (Definiteness of an $n \times n$ non-null real symmetric matrix A). Consider a quadratic form $u = X'AX$, $A = A'$ where X is an $n \times 1$ non-null vector and A is the real symmetric matrix under consideration. If $u > 0$ for all possible non-null X then A is positive definite ($u \geq 0$ means positive semidefinite). If $u < 0$ for all possible non-null X then A is negative definite ($u \leq 0$ means negative semidefinite). If $u > 0$ for some values of X and $u < 0$ for some other values of X then A is indefinite.

Note 3.2. In order to avoid confusion and possible misinterpretation, one should take Definition 3.3.6 as the definition for definiteness in the real case and all other properties are to be treated as consequences.

Exercises 3.3

3.3.1. Evaluate the cross product $\vec{a} \times \vec{b}$ for the following cases: (1) $\vec{a} = \vec{i} - \vec{j} - \vec{k}$, $\vec{b} = 2\vec{i} + 3\vec{j} - \vec{k}$; (2) $\vec{a} = \vec{i}$, $\vec{b} = \vec{j}$; (3) $\vec{a} = \vec{i}$, $\vec{b} = \vec{k}$.

3.3.2. Construct a vector parallel to the line of intersection of the planes

$$x + y + z = 2, \quad 2x - 3y = z = 5.$$

3.3.3. Evaluate the volume of the parallelepiped generated by the following points with the origin:

(1) $(1, 1, -1), (1, 2, 5), (3, 2, -1),$ (2) $(2, 1, -1), (1, 1, 2), (3, 2, 1).$

3.3.4. Evaluate the volume of (a) the parallelotope, (b) the simplex generated by the following points with the origin.

(1) $(1, 1, 1, 1, 1),$ $(1, -1, 1, 2, 1),$ $(1, 1, -1, 1, -1),$

$(2, 1, 3, -1, 1),$ $(1, 0, 0, 0, 2)$

(2) $(1, 1, 1, 1),$ $(1, 1, -1, -1),$ $(1, -1, 1, -1),$ $(1, -1, -1, -1).$

3.3.5. Evaluate the volume of (a) the parallelotope, (b) the simplex, determined by the points

$$(1,2,1,1), \quad (1,3,1,2), \quad (2,1,1,1), \quad (2,3,4,1), \quad (1,-1,0,1).$$

3.3.6. Evaluate the Jacobians in the following linear transformations:

(a) $y_1 = 2x_1 - x_2 + x_3, \quad y_2 = x_1 - x_2 + 2x_3,$

  $y_3 = 2x_1 - x_2 - x_3;$

(b) $y_{11} = 2x_{11} + x_{21}, \quad y_{12} = 2x_{12} + x_{22},$

  $y_{21} = x_{11} + 3x_{21}, \quad y_{22} = x_{12} + 3x_{22};$

(c) $y_{11} = 2x_{11} + x_{21} + 2x_{12} + x_{22}, \quad y_{12} = 2x_{11} + x_{21} - 2x_{12} - x_{22},$

  $y_{21} = x_{11} + 3x_{21} + x_{12} + 3x_{22}, \quad y_{22} = x_{11} + 3x_{21} - x_{12} - 3x_{22}.$

3.3.7. Let $X = X'$ be a $p \times p$ symmetric matrix of $p(p+1)/2$ real variables. Let E and F be two basic elementary matrices of the E and F types (see Chapter 2). Evaluate the Jacobians in the following transformations:

(a) $Y = EXE',$ (b) $Y = FXF'.$

3.3.8. Let A be a $p \times p$ nonsingular matrix of constants, X a $p \times p$ symmetric matrix of $p(p+1)/2$ real variables. Evaluate the Jacobian in the linear transformation $Y = AXA'$ and show that $dY = |A|^{p+1}dX$, ignoring the sign. [Hint: A nonsingular matrix of the type A is a product of the elementary matrices of E and F types.]

3.3.9. Evaluate the Jacobian in the following nonlinear transformation. Let $x_j > 0$, $j = 1, \ldots, p$. Let $y_1 = x_1 + \cdots + x_p$, $y_2 = x_1 x_2 + x_1 x_3 + \cdots + x_{p-1} x_p$, ..., $y_p = x_1 \ldots x_p$. [These are the basic elementary symmetric functions].

3.3.10. Evaluate the Jacobian in the following generalized polar coordinate transformation:

$$x_1 = r \sin \theta_1 \sin \theta_2 \cdots \sin \theta_{k-2} \sin \theta_{k-1}$$
$$x_2 = r \sin \theta_1 \sin \theta_2 \cdots \sin \theta_{k-2} \cos \theta_{k-1}$$
$$x_3 = r \sin \theta_1 \sin \theta_2 \cdots \cos \theta_{k-2}$$
$$\vdots$$
$$x_{k-1} = r \sin \theta_1 \cos \theta_2$$
$$x_k = r \cos \theta_1$$

where $0 < \theta_j \le \pi$, $j = 1, 2, \ldots, k-2$, $0 < \theta_{k-1} \le 2\pi$, $0 < r < \infty$.

3.3.11. Evaluate the integral $\int_X e^{-\text{tr}(X)} dX$ where X is $p \times p$, $X = X'$ and X can be written as $X = TT'$ where T is lower triangular with positive diagonal elements. $\int_X$ means the integral over all such X.

3.3.12. Evaluate the integral $\int_{-\infty}^{\infty} e^{-(ax^2+bx)} dx$, $a > 0$, $b \neq 0$ and x is real scalar.

3.3.13. Let X be a $p \times 1$ vector of real scalar variables. Let A be a $p \times p$ constant matrix such that A can be written as $A = BB'$ with $|B| \neq 0$. Evaluate the following integrals:

$$\text{(a)} \quad \int_X e^{-X'AX} dX, \qquad \text{(b)} \quad \int_X e^{-(X-\mu)'A(X-\mu)} dX$$

where μ is a constant vector. [Hint: Use the transformation of the type $Y = B'X$.]

3.3.14. Let $X = \binom{X_1}{X_2}$, X_1 is $r \times 1$, X is $p \times 1$ and let $A = BB'$, with $|B| \neq 0$ a $p \times p$ matrix of constants. Evaluate the integral $\int_{X_2} e^{-X'AX} dX_2$, that is, integrate out the variables in X_2.

3.3.15. Multivariate Gaussian density. The most popular density in multivariate statistical analysis is the multivariate Gaussian density. Let X be a $p \times 1$ vector of real scalar random variables, μ a $p \times 1$ vector of constants, V a $p \times p$ real symmetric positive definite matrix, that is, which can be written as $V = BB'$, $|B| \neq 0$. Then the p-variate Gaussian density is given by

$$f(X) = \frac{1}{(2\pi)^{p/2}|V|^{1/2}} e^{-\frac{1}{2}(X-\mu)'V^{-1}(X-\mu)},$$

for $-\infty < x_i < \infty$, $-\infty < \mu_i < \infty$, with $X' = (x_1, \ldots, x_p)$, $\mu' = (\mu_1, \ldots, \mu_p)$, $V = (v_{ij})$. Show the following: (a) $f(X)$ is a density, that is to say that $f(X) \geq 0$ for all X and $\int_X f(X) dX = 1$; (b) $E(X) = \mu$, that is to say that $\int_X Xf(X) dX = \mu$; (c) Covariance matrix of X is V, that is to say that

$$V = E[(X - \mu)(X - \mu)'] = \int_X (X - \mu)(X - \mu)' f(X) dX.$$

3.3.16. Canonical form for a quadratic form. Let $u = X'AX$ be a quadratic form. Without loss of generality $A = A'$. From elementary transformations it was seen in Chapter 2 that A can be written as $A = QDQ'$ where Q is nonsingular and D is diagonal. Then if $Y = Q'X$ the quadratic form reduces to its canonical form, a linear function of squares of the form $u = \lambda_1 y_1^2 + \cdots + \lambda_p y_p^2$ when A is $p \times p$. Reduce the following quadratic forms to their canonical forms:
(a) $u = 2x_1^2 + 3x_2^2 + 2x_3^2 + 2x_1x_2 - 2x_1x_3$;
(b) $u = x_1^2 + 2x_1x_2 + 2x_2^2 + 2x_2x_3 - 2x_3^2$.

3.3.17. Show that $u = 1$ in Exercise 2.3.16(a) can be reduced to an ellipsoid in the standard form.

3.3.18. Show that for $u = 1$ in Exercise 2.3.16(b) is not an ellipsoid.

3.3.19. Write the following bilinear forms in matrix notation as $X'AY$:
(a) $u_1 = x_1y_1 + 2x_2y_1 + 2x_3y_1 - x_1y_2 + x_3y_2$;
(b) $u_2 = x_1y_1 + 2x_2y_1 + 2x_3y_1 - x_1y_2 + 2x_1y_3 + x_2y_3 - x_3y_3$.

3.3.20. Through elementary transformations an $m \times n$ matrix A can be written as $A = PDQ$ where P and Q are nonsingular matrices and D is a diagonal type matrix. By using this property reduce the bilinear forms in (a) and (b) in Exercise 3.3.19 to the form

$$\lambda_1 z_1 t_1 + \cdots + \lambda_p z_p t_p.$$

3.3.21. Show that Definitions 3.3.5 and 3.3.6 are equivalent for real symmetric matrices.

3.3.22. By using Definitions 3.3.5 and 3.3.6 show the following:
(a) If a real square matrix A is positive definite then it can always be written as $A = BB'$, $|B| \neq 0$.
(b) If B is a real rectangular matrix $m \times n$ then $A = BB'$ is either positive definite or positive semidefinite.
(c) What should be the condition on B in (b) above so that A is strictly positive definite?
(d) Show that a negative definite or indefinite matrix A cannot be written in the form $A = CC'$ for some matrix C.

3.3.23. Evaluate A^{20}, B^{30}, C^{-5} and their determinants, where

$$A = \begin{bmatrix} 2 & 5 \\ 0 & 0 \end{bmatrix}, \quad B = \begin{bmatrix} 2 & 5 \\ 0 & 1 \end{bmatrix}, \quad C = \begin{bmatrix} 2 & 5 \\ 0 & 1 \end{bmatrix}^{-1}.$$

3.3.24. Let A and B be $n \times n$ matrices. If A and B differ only in their j-th column then show that

$$2^{1-n}|A + B| = |A| + |B|.$$

3.3.25. Let $A = \begin{bmatrix} 1 & 2 & 0 \\ 0 & 1 & 4 \\ 0 & 0 & 1 \end{bmatrix}$ and $B = A^{10}$. Compute B.

3.3.26. Let $J' = (1, 1, \dots, n)$ and let $A = I - B$, $B = \frac{1}{n}JJ'$ Compute $|(AB)^5|$.

3.3.27. Let A be a 3×3 matrix with all principal minors positive. Show that $|A|$ can be written as $|A| = a^2 b^2 c^2$ for some a, b, c.

3.3.28. By using Definition 3.3.6, or otherwise, show that if $A = (a_{ij})$, $A = A'$, $n \times n$ positive definite then $a_{jj} > 0$, $j = 1, \dots, n$ and if A is negative definite then $a_{jj} < 0$, $j = 1, \dots, n$. Note that these are necessary properties but not sufficient to talk about positive definiteness or negative definiteness.

3.3.29. If $A = A'$ is $n \times n$ then prove that A is indefinite if at least one diagonal elements is negative and at least one diagonal element is positive.

3.3.30. Let $A = (a_{ij})$ be $n \times n$. Consider the determinant $A - xI|$ where x is a scalar quantity and I is an $n \times n$ identity matrix. That is, x is subtracted from the diagonal elements $a_{jj}, j = 1, \ldots, n$. Then prove that when $|A - xI|$ is opened up with $n!$ terms using (3.1.2) then show that x^{n-1} can come only from one term, namely, $(a_{11} - x)(a_{22} - x) \cdots (a_{pp} - x)$ and there is no other term of degree $n - 1$ in x out of the $n!$ terms.

4 Eigenvalues and eigenvectors

4.0 Introduction

Among all the concepts introduced so far none may have as many applications as the concept of eigenvalues. Before introducing the possible fields of applications we will define the concept, study some of the main properties and then we will have a sufficient set of properties and tools to tackle practical problems where these are applicable. Fields of applications include theoretical developments of many branches of mathematics, physics, statistics, econometrics and many other areas, and real-life problems.

Definition 4.0.1 (Eigenvalues). Eigenvalues are defined only for square matrices. Let A be an $n \times n$ matrix. Consider the equation

$$AX = \lambda X \qquad (4.0.1)$$

where λ is a scalar and X is a non-null $n \times 1$ vector. This equation is evidently satisfied by a null vector X. If the equation has a solution for a λ and for a non-null X then that λ is called an *eigenvalue* or *characteristic root* or *latent root* of A and the non-null X satisfying (4.0.1) for that particular λ is called the *eigenvector* or *characteristic vector* or *latent vector* corresponding to that eigenvalue λ.

Let us examine the equation a bit more closely.

$$AX = \lambda X \implies (A - \lambda I)X = O. \qquad (4.0.2)$$

If this homogeneous linear system $(A - \lambda I)X = O$ has to have a non-null solution X then $A - \lambda I$ must be singular. If $A - \lambda I$ is nonsingular then its regular inverse exists and then multiplying both sides by $(A - \lambda I)^{-1}$ we have the only solution as $X = O$. If $A - \lambda I$ is singular then its determinant must be zero. That is,

$$|A - \lambda I| = 0. \qquad (4.0.3)$$

The eigenvalues can also be defined as the roots of the determinantal equation in (4.0.3).

Definition 4.0.2 (Eigenvalues of an $n \times n$ matrix A). They are the n roots of the determinantal equation (4.0.3).

4.1 Eigenvalues of special matrices

Let us evaluate the eigenvalues of some special matrices.

Example 4.1.1. Determine the eigenvalues of (1) a diagonal matrix, (2) a triangular matrix.

Solution 4.1.1. Let $D = \text{diag}(d_1, \ldots, d_n)$ be a diagonal matrix. Consider the equation

$$|D - \lambda I| = 0 \implies \begin{vmatrix} d_1 - \lambda & 0 & \cdots & 0 \\ 0 & d_2 - \lambda & \cdots & 0 \\ \vdots & \vdots & \cdots & \vdots \\ 0 & 0 & \cdots & d_n - \lambda \end{vmatrix} = 0.$$

This determinantal equation is nothing but

$$(d_1 - \lambda)(d_2 - \lambda) \cdots (d_n - \lambda) = 0.$$

That is, the n roots are $\lambda_1 = d_1, \lambda_2 = d_2, \ldots, \lambda_n = d_n$ and these are the eigenvalues of D. Now consider a triangular matrix, for example, a lower triangular matrix T. Then

$$|T - \lambda I| = \begin{vmatrix} t_{11} - \lambda & 0 & \cdots & 0 \\ t_{21} & t_{22} - \lambda & \cdots & 0 \\ \vdots & \vdots & \cdots & \vdots \\ t_{n1} & t_{n2} & \cdots & t_{nn} - \lambda \end{vmatrix} = 0.$$

Since the determinant of a triangular matrix (lower or upper) is the product of the diagonal elements, the determinantal equation reduces to

$$(t_{11} - \lambda) \cdots (t_{nn} - \lambda) = 0.$$

Hence the n eigenvalues in this case are $\lambda_1 = t_{11}, \ldots, \lambda_n = t_{nn}$.

(i) The eigenvalues of a diagonal matrix are its diagonal elements.
(ii) The eigenvalues of a triangular (upper or lower) matrix are its diagonal elements.
(iii) The eigenvalues of a scalar matrix with the diagonal elements c each are c repeated n times. The eigenvalues of an identity matrix I_n are 1 repeated n times.

Example 4.1.2. Evaluate the eigenvalues of

$$A = \begin{bmatrix} 1 & 3 \\ 0 & 1 \end{bmatrix}, \quad B = \begin{bmatrix} 2 & 2 \\ 1 & 3 \end{bmatrix}.$$

Solution 4.1.2. Consider

$$|A - \lambda I| = 0 \implies \begin{vmatrix} 1 - \lambda & 3 \\ 0 & 1 - \lambda \end{vmatrix} = 0$$

$$\implies (1 - \lambda)(1 - \lambda) = 0$$

$$\implies \lambda_1 = 1, \quad \lambda_2 = 1$$

are the eigenvalues. It follows directly from property (ii) also since A is triangular.

(iv) If the eigenvalues of a matrix are 1 repeated n times this does not mean that the matrix is an identity matrix.

Now consider

$$|B - \lambda I| = 0 \implies \begin{vmatrix} 2 - \lambda & 2 \\ 1 & 3 - \lambda \end{vmatrix} = 0$$

$$\implies (2 - \lambda)(3 - \lambda) - 2 = 0.$$

That is,

$$\lambda^2 - 5\lambda + 4 = 0 \implies (\lambda - 4)(\lambda - 1) = 0$$

$$\implies \lambda_1 = 4, \quad \lambda_2 = 1$$

are the eigenvalues.

Let us see what happens if we have an idempotent matrix. Idempotent means $A = A^2$. Consider the equation

$$AX = \lambda X.$$

Premultiply both sides by A. Then we have

$$A^2 X = \lambda A X = \lambda(\lambda X) = \lambda^2 X.$$

But $A^2 = A$ and then

$$AX = \lambda X = \lambda^2 X \implies (\lambda - \lambda^2)X = O.$$

But by definition $X \neq O$ and λ is a scalar. Hence $\lambda - \lambda^2 = 0$ which means the roots are 0 or 1.

(v) The eigenvalues of an idempotent matrix are 0's and 1's.

Identity matrix is an idempotent matrix with all eigenvalues 1 each. We can have a triangular matrix with the diagonal elements 0's and 1's which means that for such a triangular matrix also the eigenvalues are 0's and 1's.

(vi) If the eigenvalues of a matrix are 0's and 1's that does not necessarily mean that the matrix is idempotent.

Example 4.1.3. Evaluate the eigenvalues of the matrix

$$A = \frac{1}{n} \begin{bmatrix} 1 & 1 & \cdots & 1 \\ 1 & 1 & \cdots & 1 \\ \vdots & \vdots & \cdots & \vdots \\ 1 & 1 & \cdots & 1 \end{bmatrix}.$$

[This is the matrix of the quadratic form $(\bar{x})^2$, $\bar{x} = (x_1 + \cdots + x_n)/n$ the average of the n quantities $x_1, \ldots, x_n$.]

Solution 4.1.3.

$$|A - \lambda I| = 0 \ \Rightarrow \ \begin{vmatrix} \frac{1}{n} - \lambda & \frac{1}{n} & \cdots & \frac{1}{n} \\ \frac{1}{n} & \frac{1}{n} - \lambda & \cdots & \frac{1}{n} \\ \vdots & \vdots & \cdots & \vdots \\ \frac{1}{n} & \frac{1}{n} & \cdots & \frac{1}{n} - \lambda \end{vmatrix} = 0.$$

Add all the rows to the first row, the value of the determinant remains the same, take out $(1 - \lambda)$ from the first row and then add $(-\frac{1}{n})$ times the first row to all other rows to obtain

$$|A - \lambda I| = (1 - \lambda) \begin{vmatrix} 1 & 1 & \cdots & 1 \\ 0 & -\lambda & \cdots & 0 \\ \vdots & \vdots & \cdots & \vdots \\ 0 & 0 & \cdots & -\lambda \end{vmatrix} = (1 - \lambda)(-\lambda)^{n-1}.$$

Hence the roots are $\lambda_1 = 1$, $\lambda_2 = 0 = \cdots = \lambda_n$. That is, one root is 1 and all other roots are 0 each. In this example we can show that our matrix A is idempotent by showing $A^2 = A$ also.

Let us examine the determinantal equation $|A - \lambda I| = 0$. If $\lambda_1, \ldots, \lambda_n$ are the n roots of this equation then

$$|A - \lambda I| = (\lambda_1 - \lambda)(\lambda_2 - \lambda) \cdots (\lambda_n - \lambda). \tag{4.1.1}$$

The right side of (4.1.1), when opened up, is a polynomial in λ of degree n. That is,

$$\begin{aligned} |A - \lambda I| = &(-1)^n \lambda^n + (-1)^{n-1}[\lambda_1 + \cdots + \lambda_n]\lambda^{n-1} \\ &+ (-1)^{n-2}[\text{sum of products of roots taken two at a time}] \\ &+ \cdots + (\lambda_1 \cdots \lambda_n). \end{aligned} \tag{4.1.2}$$

Definition 4.1.1 (The characteristic polynomial of an $n \times n$ matrix A). The polynomial on the right of (4.1.2) or when $|A - \lambda I|$ is written as a polynomial in λ then this polynomial is called the characteristic polynomial of A.

Treating (4.1.1) as a polynomial in λ and evaluating it at $\lambda = 0$ we have the following result:

(vii) The determinant of an $n \times n$ matrix A is the product of the eigenvalues of A.

Some immediate consequences of this property are the following:

(viii) If at least one of the eigenvalues of A is zero then $|A| = 0$ or A is singular, and if A is nonsingular then all the eigenvalues of A are nonzeros (may be positive or negative but none zero).

From (4.1.2) we note that the coefficient of $(-\lambda)^{n-1}$ in the expansion of $|A - \lambda I|$ is the sum of the eigenvalues. We will derive a very important result connecting the sum of the eigenvalues to the trace of the matrix. To this end let us evaluate $|A - \lambda I|$. Let $B = (b_{ij})$ be an $n \times n$ matrix. Then we had seen from Chapter 3 that

$$|B| = \sum_{i_1} \cdots \sum_{i_n} (-1)^{\rho(i_1,\ldots,i_n)} b_{1i_1} b_{2i_2} \cdots b_{ni_n}. \tag{4.1.3}$$

That is, when $|B|$ is written as an explicit sum, each term in that sum contains one and only one element from each row and each column. Now, look at the determinant

$$|A - \lambda I| = \begin{vmatrix} a_{11} - \lambda & a_{12} & \cdots & a_{1n} \\ a_{21} & a_{22} - \lambda & \cdots & a_{2n} \\ \vdots & \vdots & \cdots & \vdots \\ a_{n1} & a_{n2} & \cdots & a_{nn} - \lambda \end{vmatrix}.$$

One term in this determinant, when the determinant is written as a sum of the type in (4.1.3), is

$$(a_{11} - \lambda)(a_{22} - \lambda) \cdots (a_{nn} - \lambda).$$

This, when opened up gives a term containing λ^n, a term containing λ^{n-1} and so on. The coefficient of $(-\lambda)^{n-1}$ here is

$$a_{11} + \cdots + a_{nn} = \text{tr}(A).$$

What is the nature of any other term in the sum when $|A - \lambda I|$ is written as a sum? One element, other than $a_{11} - \lambda$ has to come from the first row. Let this be the j-th element a_{1j}. Then this rules out the presence of $a_{jj} - \lambda$ the term in the j-th column containing λ. In other words, in all other terms the exponent of λ can be only up to λ^{n-2}. Now equating the coefficient of $(-\lambda)^{n-1}$ on both sides of (4.1.2) we see that the trace of A is the sum of the eigenvalues of A also. Thus we have the following important result:

(ix) For any $n \times n$ matrix $A = (a_{ij})$

$$\text{tr}(A) = a_{11} + \cdots + a_{nn} = \lambda_1 + \cdots + \lambda_n$$

where $\lambda_1, \ldots, \lambda_n$ are the eigenvalues of A.

This does not mean that $\lambda_1 = a_{11}$, etc. Thus the determinant of A is the product of the eigenvalues and the trace is the sum of the eigenvalues of A. For example,

For the eigenvalues	the matrix is	its trace is	its determinant is
$1, -1, 2$	nonsingular	2	-2
$0, 1, 3$	singular	4	0
$1, 1, 1, 1$	nonsingular	4	1
$1, -1, 1, -1$	nonsingular	0	1
$1, 0, 1, -1$	singular	1	0

Example 4.1.4. Verify the results (vii) and (ix) from Examples 4.1.1 to 4.1.3.

Solution 4.1.4. For the diagonal and triangular matrices the results are obvious since the diagonal elements themselves are the eigenvalues. In Example 4.1.2 the eigenvalues are $1, 1$, $\text{tr}(A) = 1 + 1 = 2$, $|A| = 1$. In B, $\text{tr}(B) = 2 + 3 = 5$, the sum of the eigenvalues is $4 + 1 = 5$ and $|B| = (2)(3) - (1)(2) = 4$. The product of the eigenvalues is $(4)(1) = 4$. In Example 4.1.3, $\text{tr}(A) = \frac{1}{n} + \cdots + \frac{1}{n} = 1$. The sum of the eigenvalues is $1 + 0 + \cdots + 0 = 1$. Since A is singular $|A| = 0$. The product of the eigenvalues is $(1)(0) \cdots (0) = 0$.

Since A and A' have the same determinant we have

$$|A - \lambda I| = |A' - \lambda I|.$$

Hence we have the following result:

(x) The matrices A and A' have the same eigenvalues.

What are the eigenvalues of an orthonormal matrix? An orthonormal matrix P is such that $PP' = I$, $P'P = I$. Let λ be an eigenvalue of P and X the corresponding eigenvector. Then

$$PX = \lambda X.$$

Premultiplying by P' we have, since $P'P = I$, $IX = X$,

$$X = \lambda P'X = \lambda^2 X$$

since P and P' have the same eigenvalues. This means, $(\lambda^2 - 1)X = O$ where $X \neq O$ and λ is a scalar. Therefore $\lambda = \pm 1$.

(xi) The eigenvalues of an orthonormal matrix are ± 1.
(xii) For $n \times n$ matrices A and B, $|AB| =$ product of the eigenvalues of A and B.
(xiii) $\text{tr}(A + B) = \text{tr}(A) + \text{tr}(B) =$ sum of the eigenvalues of A and B, when $A + B$ is defined.
(xiv) The eigenvalues of a null matrix are zeros.

(xv) If λ is an eigenvalue of A then $k\lambda$ is an eigenvalue of kA where k is a scalar quantity.

(xvi) If λ is an eigenvalue of A then $1 \pm k\lambda$ is an eigenvalue of $I \pm kA$ where k is a scalar quantity.

(xvii) If $A = QBQ^{-1}$ then

$$|A - \lambda I| = |QBQ^{-1} - \lambda I| = |B - \lambda I|$$

and therefore A and B have the same eigenvalues.

(xviii) If $A = A'$ and if $A = PDP'$, where P is orthonormal and D is diagonal then the eigenvalues of A are the diagonal elements in D.

Exercises 4.1

4.1.1. Evaluate the eigenvalues of the following matrices:

$$A = \begin{bmatrix} 1 & 1 \\ 1 & 2 \end{bmatrix}, \quad B = \begin{bmatrix} 2 & 1 \\ -1 & 0 \end{bmatrix}, \quad C = \begin{bmatrix} 1 & 1 \\ 2 & 2 \end{bmatrix}.$$

4.1.2. Evaluate the eigenvalues of the following matrices:

$$A = \begin{bmatrix} 1 & 1 & 1 \\ 1 & -1 & 1 \\ -1 & 0 & 1 \end{bmatrix}, \quad B = \begin{bmatrix} 0 & 1 & -1 \\ 2 & 0 & 1 \\ 1 & 1 & -1 \end{bmatrix}, \quad C = \begin{bmatrix} 2 & 1 & 1 \\ 1 & 3 & 0 \\ 1 & 0 & 1 \end{bmatrix},$$

$$D = \begin{bmatrix} 0 & 1 & 1 \\ -1 & 0 & 2 \\ -1 & -2 & 0 \end{bmatrix}.$$

4.1.3. Let U and V be $n \times 1$ non-null, non-orthogonal vectors. Show that at least one eigenvalue of UV' is zero and no eigenvalue of $U'V$ is zero. What are the eigenvalues in each?

4.1.4. Show that the eigenvalues of the $n \times n$ matrix A, where

$$A = \begin{bmatrix} 1 - \frac{1}{n} & -\frac{1}{n} & \cdots & -\frac{1}{n} \\ -\frac{1}{n} & 1 - \frac{1}{n} & \cdots & -\frac{1}{n} \\ \vdots & \vdots & \cdots & \vdots \\ -\frac{1}{n} & -\frac{1}{n} & \cdots & 1 - \frac{1}{n} \end{bmatrix} = I - B$$

are one zero and $n - 1$ unities, the eigenvalues of B are one unity and $n - 1$ zeros and the eigenvalues of $I - B$ are of the form $1 -$ eigenvalues of B.

4.1.5. Construct a 2×2 matrix with real elements but whose eigenvalues are irrational.

4.1.6. Construct a 2×2 matrix with real elements but whose eigenvalues are complex quantities. Can there be only one or an odd number of complex roots for any given matrix with complex roots?

4.1.7. If $\text{tr}(A) = 0$ is A null? If not give two counter examples.

4.1.8. If $AB = O$ is $\text{tr}(AB) = 0$? If $\text{tr}(AB) = 0$ is AB null? If not give two counter examples.

4.1.9. If $A + B = I_n$ is the sum of the eigenvalues of A and B equal to n. If not give two counter examples.

4.1.10. If $A = PDQ$ where P and Q are nonsingular and D is diagonal, are the eigenvalues of A the diagonal elements in D? If not give two counter examples in the 2×2 case.

4.2 Eigenvectors

An eigenvector is defined along with an eigenvalue in Section 4.1. Here we will redefine it for the sake of completeness.

4.2.1 Some definitions and examples

Definition 4.2.1. If λ is an eigenvalue of A then any non-null vector X satisfying the equation

$$AX = \lambda X$$

is an *eigenvector of A* corresponding to the eigenvalue λ.

Some properties are immediate from this definition itself.

(i) If X_1 is an eigenvector corresponding to the eigenvalue λ_1 then cX_1, c a nonzero scalar, is also an eigenvector corresponding to λ_1. If X_1 and X_2 are two eigenvectors corresponding to the same eigenvalue λ_1 then $c_1X_1 + c_2X_2$ is also an eigenvector corresponding to λ_1 where c_1 and c_2 are nonzero scalars.
(ii) If $\lambda = 0$ is an eigenvalue of A then the eigenvector corresponding to $\lambda = 0$ is a vector in the null space of A. There are $n - r$ such linearly independent eigenvectors if the rank of A is r and if A is $n \times n$.

Example 4.2.1. Compute the eigenvalues and the eigenvectors of the matrix

$$B = \begin{bmatrix} 2 & 2 \\ 1 & 3 \end{bmatrix}.$$

Solution 4.2.1. The eigenvalues of this matrix are 4 and 1, evaluated in Example 4.1.2. Take $\lambda_1 = 1$ and consider the equation

$$BX = \lambda_1 X \implies (B - \lambda_1 I)X = O$$

$$\implies \begin{bmatrix} 2-1 & 2 \\ 1 & 3-1 \end{bmatrix} \begin{bmatrix} x_1 \\ x_2 \end{bmatrix} = \begin{bmatrix} 0 \\ 0 \end{bmatrix}$$

$$\implies x_1 + 2x_2 = 0.$$

Both rows give the same equation. Since by definition $B - \lambda_1 I$ is singular we cannot get two linearly independent equations here:

$$x_1 + 2x_2 = 0 \implies X_1 = \begin{pmatrix} x_1 \\ x_2 \end{pmatrix} = \begin{pmatrix} -2 \\ 1 \end{pmatrix}.$$

One solution is this. Any nonzero scalar multiple of X_1 is also a solution. There are plenty of solutions. Hence one eigenvector corresponding to $\lambda_1 = 1$, denoted by X_1, is

$$X_1 = \begin{pmatrix} -2 \\ 1 \end{pmatrix}.$$

If we want a normalized eigenvector corresponding to $\lambda_1 = 1$ then

$$Y_1 = \frac{1}{\sqrt{5}} \begin{pmatrix} -2 \\ 1 \end{pmatrix} = \begin{pmatrix} -\frac{2}{\sqrt{5}} \\ \frac{1}{\sqrt{5}} \end{pmatrix}$$

is that eigenvector. For $\lambda_2 = 4$

$$(B - \lambda_2 I)X = O \implies \begin{bmatrix} 2-4 & 2 \\ 1 & 3-4 \end{bmatrix} \begin{bmatrix} x_1 \\ x_2 \end{bmatrix} = \begin{bmatrix} 0 \\ 0 \end{bmatrix}$$

$$\implies -2x_1 + 2x_2 = 0 \quad \text{and} \quad x_1 - x_2 = 0$$

$$\implies x_1 = x_2.$$

For example, for $x_2 = 1$ we have $x_1 = 1$ and $X_2 = \begin{pmatrix} 1 \\ 1 \end{pmatrix}$ is an eigenvector corresponding to $\lambda_2 = 4$. A normalized eigenvector corresponding to $\lambda_2 = 4$ is

$$Y_2 = \frac{1}{\sqrt{2}} \begin{pmatrix} 1 \\ 1 \end{pmatrix} = \begin{pmatrix} \frac{1}{\sqrt{2}} \\ \frac{1}{\sqrt{2}} \end{pmatrix}.$$

Let us create a matrix of eigenvectors and see what happens:

$$BX_1 = \lambda_1 \quad \text{and} \quad BX_2 = \lambda_2 X_2 \implies$$

$$\begin{bmatrix} 2 & 2 \\ 1 & 3 \end{bmatrix} \begin{bmatrix} -2 \\ 1 \end{bmatrix} = 1 \begin{bmatrix} -2 \\ 1 \end{bmatrix} \quad \text{and}$$

$$\begin{bmatrix} 2 & 2 \\ 1 & 3 \end{bmatrix} \begin{bmatrix} 1 \\ 1 \end{bmatrix} = 4 \begin{bmatrix} 1 \\ 1 \end{bmatrix}.$$

Putting the two equations together we have,

$$B(X_1, X_2) = (X_1, X_2) \begin{bmatrix} \lambda_1 & 0 \\ 0 & \lambda_2 \end{bmatrix} \Rightarrow$$

$$\begin{bmatrix} 2 & 2 \\ 1 & 3 \end{bmatrix} \begin{bmatrix} -2 & 1 \\ 1 & 1 \end{bmatrix} = \begin{bmatrix} -2 & 1 \\ 1 & 1 \end{bmatrix} \begin{bmatrix} 1 & 0 \\ 0 & 4 \end{bmatrix}.$$

Note that X_1 is multiplied by λ_1 and X_2 is multiplied by λ_2 which means that the matrix (X_1, X_2) is postmultiplied by a diagonal matrix then the columns will be multiplied by the corresponding diagonal elements. From the above equation we have the following:

$$B(X_1, X_2) = (X_1, X_2) \begin{bmatrix} \lambda_1 & 0 \\ 0 & \lambda_2 \end{bmatrix} \Rightarrow$$

$$B = (X_1, X_2) \begin{bmatrix} \lambda_1 & 0 \\ 0 & \lambda_2 \end{bmatrix} (X_1, X_2)^{-1} \quad \text{if the inverse exists} \Rightarrow$$

$$\begin{bmatrix} 2 & 2 \\ 1 & 3 \end{bmatrix} = \begin{bmatrix} -2 & 1 \\ 1 & 1 \end{bmatrix} \begin{bmatrix} 1 & 0 \\ 0 & 4 \end{bmatrix} \begin{bmatrix} -2 & 1 \\ 1 & 1 \end{bmatrix}^{-1}.$$

Let us verify this:

$$\begin{bmatrix} -2 & 1 \\ 1 & 1 \end{bmatrix}^{-1} = \frac{1}{-3} \begin{bmatrix} 1 & -1 \\ -1 & -2 \end{bmatrix} = \frac{1}{3} \begin{bmatrix} -1 & 1 \\ 1 & 2 \end{bmatrix}.$$

Then

$$\begin{bmatrix} -2 & 1 \\ 1 & 1 \end{bmatrix} \begin{bmatrix} 1 & 0 \\ 0 & 4 \end{bmatrix} \begin{bmatrix} -2 & 1 \\ 1 & 1 \end{bmatrix}^{-1} = \begin{bmatrix} -2 & 1 \\ 1 & 1 \end{bmatrix} \begin{bmatrix} 1 & 0 \\ 0 & 4 \end{bmatrix} \left(\frac{1}{3}\right) \begin{bmatrix} -1 & 1 \\ 1 & 2 \end{bmatrix}$$

$$= \frac{1}{3} \begin{bmatrix} 6 & 6 \\ 3 & 9 \end{bmatrix} = \begin{bmatrix} 2 & 2 \\ 1 & 3 \end{bmatrix} = B.$$

It is verified. What we have seen is the following: The eigenvectors in this case are linearly independent and if we denote the matrix of eigenvectors by Q, $Q = (X_1, X_2)$ above, which is nonsingular here, then

$$BQ = QD,$$

D is a diagonal matrix with the eigenvalues being the diagonal elements. Therefore

$$B = QDQ^{-1}. \tag{4.2.1}$$

We will investigate this aspect a little further later on.

Example 4.2.2. Evaluate the eigenvalues and eigenvectors of the matrix

$$A = \begin{bmatrix} 2 & 1 \\ 2 & 2 \end{bmatrix}.$$

Solution 4.2.2.

$$|A - \lambda I| = 0 \implies \begin{vmatrix} 2 - \lambda & 1 \\ 2 & 2 - \lambda \end{vmatrix} = 0$$

$$\implies (2 - \lambda)^2 - 2 = 0$$

$$\implies \lambda^2 - 4\lambda + 2 = 0.$$

The roots of this quadratic equation are available as $\lambda_1 = 2 + \sqrt{2}$, $\lambda_2 = 2 - \sqrt{2}$.

$$\left[ax^2 + bx + c = 0 \implies x = \frac{-b \pm \sqrt{b^2 - 4ac}}{2a} \right].$$

The eigenvalues are irrational here. In order to compute one eigenvector corresponding to the eigenvalue $\lambda_1 = 2 + \sqrt{2}$ we consider the equation

$$[A - \lambda_1 I] \begin{bmatrix} x_1 \\ x_2 \end{bmatrix} = \begin{bmatrix} 0 \\ 0 \end{bmatrix} \implies \begin{bmatrix} 2 - (2 + \sqrt{2}) & 1 \\ 2 & 2 - (2 + \sqrt{2}) \end{bmatrix} \begin{bmatrix} x_1 \\ x_2 \end{bmatrix} = \begin{bmatrix} 0 \\ 0 \end{bmatrix}$$

$$\implies \begin{bmatrix} -\sqrt{2} & 1 \\ 2 & -\sqrt{2} \end{bmatrix} \begin{bmatrix} x_1 \\ x_2 \end{bmatrix} = \begin{bmatrix} 0 \\ 0 \end{bmatrix}.$$

Note that if we multiply the first row by $-\sqrt{2}$ we get the second row. Hence we need to consider only any one of the two equations. [The rows have to be dependent because the matrix $A - \lambda_1 I$ is singular.] Take the first equation:

$$-\sqrt{2}x_1 + x_2 = 0 \implies x_2 = \sqrt{2}x_1.$$

One solution is $x_1 = 1$, $x_2 = \sqrt{2}$ or

$$X_1 = \begin{pmatrix} x_1 \\ x_2 \end{pmatrix} = \begin{pmatrix} 1 \\ \sqrt{2} \end{pmatrix}.$$

The normalized X_1 is

$$Y_1 = \frac{1}{\sqrt{3}} \begin{pmatrix} 1 \\ \sqrt{2} \end{pmatrix} = \begin{pmatrix} \frac{1}{\sqrt{3}} \\ \frac{\sqrt{2}}{\sqrt{3}} \end{pmatrix}.$$

Now consider the equation corresponding to the second eigenvalue $\lambda_2 = 2 - \sqrt{2}$.

$$[A - \lambda_2 I]X = O \implies \begin{bmatrix} 2 - (2 - \sqrt{2}) & 1 \\ 2 & 2 - (2 - \sqrt{2}) \end{bmatrix} \begin{bmatrix} x_1 \\ x_2 \end{bmatrix} = \begin{bmatrix} 0 \\ 0 \end{bmatrix}$$

$$\implies \begin{bmatrix} \sqrt{2} & 1 \\ 2 & \sqrt{2} \end{bmatrix} \begin{bmatrix} x_1 \\ x_2 \end{bmatrix} = \begin{bmatrix} 0 \\ 0 \end{bmatrix}.$$

As before, the second equation is a multiple of the first equation. Taking the first equation we have $\sqrt{2}x_1 + x_2 = 0$ which gives $x_2 = -\sqrt{2}x_1$. For example $x_1 = 1$, $x_2 = -\sqrt{2}$ is a solution:

$$X_2 = \begin{pmatrix} 1 \\ -\sqrt{2} \end{pmatrix}. \quad \text{The normalized } X_2 \text{ is} \quad Y_2 = \begin{pmatrix} \frac{1}{\sqrt{3}} \\ -\frac{\sqrt{2}}{\sqrt{3}} \end{pmatrix}.$$

The matrix of eigenvectors X_1 and X_2 is then

$$Q = (X_1, X_2) = \begin{pmatrix} 1 & 1 \\ \sqrt{2} & -\sqrt{2} \end{pmatrix}.$$

Then we have the equation

$$AQ = Q\begin{bmatrix} \lambda_1 & 0 \\ 0 & \lambda_2 \end{bmatrix} = \begin{bmatrix} 1 & 1 \\ \sqrt{2} & -\sqrt{2} \end{bmatrix}\begin{bmatrix} 2+\sqrt{2} & 0 \\ 0 & 2-\sqrt{2} \end{bmatrix}.$$

This means that A, in this case, can be written as

$$A = QDQ^{-1} = \begin{bmatrix} 1 & 1 \\ \sqrt{2} & -\sqrt{2} \end{bmatrix}\begin{bmatrix} 2+\sqrt{2} & 0 \\ 0 & 2-\sqrt{2} \end{bmatrix}\begin{bmatrix} 1 & 1 \\ \sqrt{2} & -\sqrt{2} \end{bmatrix}^{-1}$$

$$= \begin{bmatrix} 1 & 1 \\ \sqrt{2} & -\sqrt{2} \end{bmatrix}\begin{bmatrix} 2+\sqrt{2} & 0 \\ 0 & 2-\sqrt{2} \end{bmatrix}\begin{bmatrix} \frac{1}{2} & \frac{1}{2\sqrt{2}} \\ \frac{1}{2} & -\frac{1}{2\sqrt{2}} \end{bmatrix}.$$

Example 4.2.3. Evaluate the eigenvalues and eigenvectors of the matrix

$$A = \begin{bmatrix} 1 & -2 \\ 1 & 2 \end{bmatrix}.$$

Solution 4.2.3. Consider the equation

$$|A - \lambda I| = 0 \;\Rightarrow\; \begin{vmatrix} 1-\lambda & -2 \\ 1 & 2-\lambda \end{vmatrix} = 0$$

$$\Rightarrow (1-\lambda)(2-\lambda) + 2 = 0$$

$$\Rightarrow \lambda^2 - 3\lambda + 4 = 0.$$

The roots are

$$\lambda = \frac{3 \pm \sqrt{3^2 - 4(4)}}{2} = \frac{3}{2} \pm i\frac{\sqrt{7}}{2}, \quad i = \sqrt{-1}.$$

Both the roots are complex here. [Complex roots and irrational roots can only come in pairs.] Let

$$\lambda_1 = \frac{3}{2} + i\frac{\sqrt{7}}{2}, \quad \lambda_2 = \frac{3}{2} - i\frac{\sqrt{7}}{2}.$$

The eigenvector corresponding to λ_1 is available by solving

$$[A - \lambda_1 I]X = O \Rightarrow$$

$$\begin{bmatrix} 1 - (\frac{3}{2} + i\frac{\sqrt{7}}{2}) & -2 \\ 1 & 2 - (\frac{3}{2} + i\frac{\sqrt{7}}{2}) \end{bmatrix} \begin{bmatrix} x_1 \\ x_2 \end{bmatrix} = \begin{bmatrix} 0 \\ 0 \end{bmatrix} \Rightarrow$$

$$\begin{bmatrix} -\frac{1}{2} - i\frac{\sqrt{7}}{2} & -2 \\ 1 & \frac{1}{2} - i\frac{\sqrt{7}}{2} \end{bmatrix} \begin{bmatrix} x_1 \\ x_2 \end{bmatrix} = \begin{bmatrix} 0 \\ 0 \end{bmatrix}.$$

It is not obvious whether the second equation is a scalar multiple of the first equation or not. Let us multiply the first equation by $-\frac{1}{4}(1 - i\sqrt{7})$. Then we get $[1, \frac{1}{2} - i\frac{\sqrt{7}}{2}]$. (Exercise for the student.) Consider the second equation

$$x_1 + \left(\frac{1}{2} - i\frac{\sqrt{7}}{2} \right)x_2 = 0.$$

For example, for $x_2 = 1, x_1 = -\frac{1}{2} + i\frac{\sqrt{7}}{2}$. One eigenvector corresponding to the eigenvalue $\lambda_1 = \frac{3}{2} + i\frac{\sqrt{7}}{2}$ is then

$$X_1 = \begin{pmatrix} -\frac{1}{2} + i\frac{\sqrt{7}}{2} \\ 1 \end{pmatrix}.$$

Now consider $\lambda_2 = \frac{3}{2} - i\frac{\sqrt{7}}{2}$ and the equation

$$[A - \lambda_2 I]X = O.$$

Proceeding as above one eigenvector is easily seen to be

$$X_2 = \begin{pmatrix} -\frac{1}{2} - i\frac{\sqrt{7}}{2} \\ 1 \end{pmatrix}.$$

Can we have a diagonalization of A by using $Q = (X_1, X_2)$? Here

$$Q = (X_1, X_2) = \begin{bmatrix} -\frac{1}{2}(1 - i\sqrt{7}) & -\frac{1}{2}(1 + i\sqrt{7}) \\ 1 & 1 \end{bmatrix},$$

$$Q^{-1} = \frac{1}{i\sqrt{7}} \begin{bmatrix} 1 & \frac{1}{2}(1 + i\sqrt{7}) \\ -1 & -\frac{1}{2}(1 - i\sqrt{7}) \end{bmatrix},$$

$$D = \begin{bmatrix} \frac{1}{2}(3 + i\sqrt{7}) & 0 \\ 0 & \frac{1}{2}(3 - i\sqrt{7}) \end{bmatrix}.$$

It is easily seen by straight multiplication (the multiplication is left to the student) that, in fact,

$$QDQ^{-1} = A. \tag{4.2.2}$$

(iii) Even if all the elements of a matrix A are real the eigenvalues could be real, rational, irrational or complex quantities. If the elements of A are real and if an eigenvalue is real then the corresponding eigenvector is real and if an eigenvalue is complex the corresponding eigenvector is complex.

(iv) Q in the representation of A in (4.2.2) is not unique. Multiply Q by a scalar $k \neq 0$ then Q^{-1} produces $\frac{1}{k}$ and A remains the same whereas Q is changed.

Example 4.2.4. Evaluate the eigenvalues and the eigenvectors of the matrix

$$A = \begin{bmatrix} 2 & 1 \\ 0 & 2 \end{bmatrix}.$$

Solution 4.2.4. Since A is triangular with the diagonal elements 2 and 2 the eigenvalues are $\lambda_1 = 2$, $\lambda_2 = 2$. An eigenvector corresponding to $\lambda_1 = 2$ is available from

$$(A - \lambda_1 I)X = O \Rightarrow$$

$$\begin{bmatrix} 2-2 & 1 \\ 0 & 2-2 \end{bmatrix} \begin{bmatrix} x_1 \\ x_2 \end{bmatrix} = \begin{bmatrix} 0 \\ 0 \end{bmatrix} \Rightarrow \begin{bmatrix} 0 & 1 \\ 0 & 0 \end{bmatrix} \begin{bmatrix} x_1 \\ x_2 \end{bmatrix} = \begin{bmatrix} 0 \\ 0 \end{bmatrix}.$$

Thus a general solution is $x_1 = a$, $x_2 = 0$, $a \neq 0$. Since the rank of the matrix $\left(\begin{smallmatrix} 0 & 1 \\ 0 & 0 \end{smallmatrix} \right)$ is 1 its null space has only one linearly independent vector or the space consists of scalar multiples of $\left(\begin{smallmatrix} 1 \\ 0 \end{smallmatrix} \right)$. We cannot find two linearly independent eigenvectors corresponding to the two roots $\lambda_1 = 2$, $\lambda_2 = 2$. Thus if Q denotes the matrix of eigenvectors then Q is singular. Therefore this matrix A does not admit a decomposition $A = QDQ^{-1}$ where D is the diagonal matrix with the diagonal elements being the eigenvalues of A and Q is the matrix of eigenvectors. We will see later that Q is singular here not because all the eigenvalues are equal or repeated but because of the special nature of the matrix involved.

Example 4.2.5. Construct a 2×2 matrix B whose eigenvalues are $\lambda_1 = 2$, $\lambda_2 = 2$ and which can be written as HDH^{-1} for some H, $|H| \neq 0$, $D = \text{diag}(2, 2)$.

Solution 4.2.5. Take any 2×2 nonsingular matrix H and consider HDH^{-1}. Since D is 2 times an identity matrix

$$HDH^{-1} = 2IHH^{-1} = 2I = D.$$

When D is a scalar matrix such as the one here, $D = cI$, $c = 2$ here, then any n-vector, $n = 2$ here, is an eigenvector and n such linearly independent vectors can be constructed and H consists of the eigenvectors. If a given $n \times n$ matrix has eigenvalues λ_1 repeated n times can it be written in the form QDQ^{-1} where $D = \text{diag}(\lambda_1, \dots, \lambda_1)$. In some cases it is possible and in some cases it is not possible.

If a matrix has distinct eigenvalues, may be one of them is zero, what can we say about the corresponding eigenvectors? Are they linearly independent or dependent? Let λ_1 and λ_2, $\lambda_1 \neq \lambda_2$ be two distinct eigenvalues of a matrix A and let X_1 and X_2 be two eigenvectors corresponding to these eigenvalues. If X_1 and X_2 are linearly dependent then there exists a non-null vector (c_1, c_2) such that $c_1 X_1 + c_2 X_2 = O$.

$$AX_1 = \lambda_1 X_1 \Rightarrow Ac_1 X_1 = \lambda_1 c_1 X_1$$
$$AX_2 = \lambda_2 X_2 \Rightarrow Ac_2 X_2 = \lambda_2 c_2 X_2$$
$$\Rightarrow A(c_1 X_1 + c_2 X_2) = (\lambda_1 c_1 X_1 + \lambda_2 c_2 X_2).$$

That is,

$$\lambda_1 c_1 X_1 + \lambda_2 c_2 X_2 = O \tag{a}$$

as well as $c_1 X_1 + c_2 X_2 = O$ by assumption. From this

$$c_1 \lambda_2 X_1 + c_2 \lambda_2 X_2 = O. \tag{b}$$

Substituting (b) in (a) we get

$$(\lambda_1 - \lambda_2) c_1 X_1 = O \Rightarrow (\lambda_1 - \lambda_2) c_1 = 0.$$

But $\lambda_1 \neq \lambda_2$. Then $c_1 = 0$. Similarly $c_2 = 0$ which means that X_1 and X_2 are linearly independent by the definition of linear independence. Note that the above proof does not depend on whether λ_1 and λ_2 are real, including one of them zero, or in the complex field and the elements of the eigenvectors could be real or in the complex field. The above method is applicable in the general situation. But if we are only concerned about just two eigenvalues then if the eigenvectors are dependent, one is a scalar multiple of the other. Then the result follows in two steps by using this property.

(v) If the eigenvalues of an $n \times n$ matrix are all distinct, some may be in the complex field, one may be zero also, then there are n linearly independent eigenvectors. Further, eigenvectors corresponding to distinct eigenvalues are linearly independent.

Thus we are guaranteed that the matrix of eigenvectors will be nonsingular if all the eigenvalues are distinct. This does not mean that if some eigenvalues are repeated then the matrix of eigenvectors is singular. Example 4.2.3 gives a counter example. Hence the situation is that in some cases when some eigenvalues are repeated the matrix of eigenvectors can become singular.

4.2.2 Eigenvalues of powers of a matrix

Let X_1 be an eigenvector corresponding to the eigenvalue λ_1 of a matrix A. Then

$$AX_1 = \lambda_1 X_1.$$

Premultiplying by A we have

$$A^2X_1 = \lambda_1 AX_1 = \lambda_1^2 X_1$$

which means that λ_1^2 is an eigenvalue of A^2 with the same eigenvector X_1. Extending this result we have the following:

(vi) If λ is an eigenvalue of A with the eigenvector X then λ^k is an eigenvalue of A^k with the same eigenvector X of A, for $k = 0, 1, 2, \ldots$.

If A^{-1} exists what about the eigenvalues of A^{-1} in terms of the eigenvalues of A? Let λ_1 be an eigenvalue of A and X_1 an eigenvector corresponding to λ_1. [If A^{-1} exists then all eigenvalues are nonzero.]

$$AX_1 = \lambda_1 X_1.$$

If A^{-1} exists then premultiply by A^{-1} to obtain

$$A^{-1}AX_1 = \lambda_1 A^{-1}X_1 \ \Rightarrow \ A^{-1}X_1 = \frac{1}{\lambda_1}X_1$$

which means that $\frac{1}{\lambda_1}$ is an eigenvalue of A^{-1}. Extending this result we have the following:

(vii) If all eigenvalues $\lambda_1, \ldots, \lambda_n$ of A are nonzero then A^{-k} has the eigenvalues $\lambda_1^{-k}, \ldots, \lambda_n^{-k}$, for $k = 0, 1, 2, \ldots$ with the same eigenvectors as those of A.
(viii) If the matrix of eigenvectors Q is nonsingular then we have

$$A^k = AA \cdots A = QDQ^{-1}QDQ^{-1} \cdots QDQ^{-1}$$
$$= QD^kQ^{-1}, \quad D = \text{diag}(\lambda_1, \ldots, \lambda_n)$$

which also means

$$Q^{-1}A^kQ = D^k$$

where $\lambda_1, \ldots, \lambda_n$ are the eigenvalues of A, for $k = 0, 1, 2, \ldots$, and if A is nonsingular then for $k = 0, -1, -2, \ldots$ also.
(ix) A being nonsingular means no eigenvalue of A is zero, Q being nonsingular means there is a set of n linearly independent eigenvectors for the $n \times n$ matrix A.

Definition 4.2.2 (Definiteness of real symmetric matrices). Definiteness is defined only for symmetric matrices, when real, and Hermitian matrices, when in the complex domain. If all the eigenvalues of an $n \times n$ real symmetric matrix A are strictly positive (eigenvalues of a real symmetric as well as Hermitian matrix are always real) then A is called *positive definite*, (if the eigenvalues are ≥ 0 then A is *positive semidefinite*), strictly negative then A is *negative definite* (if ≤ 0 then *negative semidefinite*) and if some eigenvalues are negative and some positive, at least one in each set, then the matrix A is called *indefinite*.

For example, we have the following situations when the matrix is real symmetric:

Eigenvalues	the matrix is (singularity)	the matrix is (definiteness)
$1, 1, 5$	nonsingular	positive definite
$0, 1, 4$	singular	positive semidefinite
$-1, -3, -8$	nonsingular	negative definite
$0, -3, -1$	singular	negative semidefinite
$1, 5, -2$	nonsingular	indefinite
$0, 1, -3$	singular	indefinite

4.2.3 Eigenvalues and eigenvectors of real symmetric matrices

Let $A = A'$, a real symmetric $n \times n$ matrix. Let λ_1 and λ_2 be two eigenvalues of A and X_1 and X_2 the corresponding eigenvectors. Then

$$AX_1 = \lambda_1 X_1 \implies X_2' A X_1 = \lambda_1 X_2' X_1 \tag{a}$$
$$AX_2 = \lambda_2 X_2 \implies X_1' A X_2 = \lambda_2 X_1' X_2. \tag{b}$$

But when $A = A'$ we have

$$(X_2' A X_1)' = X_1' A' X_2 = X_1' A X_2.$$

Hence the left sides of (a) and (b) are equal, because both quantities are 1×1 matrices and one is the transpose of the other. Similarly $X_1' X_2 = X_2' X_1$. Then from (a) and (b) we have

$$(\lambda_1 - \lambda_2) X_1' X_2 = 0.$$

This can happen when either $\lambda_1 = \lambda_2$ or $X_1' X_2 = 0$ or both hold. If λ_1 and λ_2 are distinct then $X_1' X_2 = 0$ which means that the eigenvectors are orthogonal to each other.

(x) When the matrix A is real symmetric the eigenvectors corresponding to distinct eigenvalues are orthogonal to each other.

This leads to some very interesting results. A scalar multiple of an eigenvector is also an eigenvector. If A is symmetric and if we have n linearly independent eigenvectors then the matrix of eigenvectors can be made into an orthonormal matrix, say P. For orthonormal matrices we know that the transposes are the inverses. Then we have a representation for $A = A'$.

$$A = PDP', \quad A = A', \quad PP' = I, \quad P'P = I$$

where D is a diagonal matrix with the diagonal elements being the eigenvalues of A, and P is the matrix of normalized eigenvectors.

Example 4.2.6. Compute the eigenvalues and the eigenvectors of

$$A = \begin{bmatrix} 1 & 2 \\ 2 & 4 \end{bmatrix}.$$

Solution 4.2.6.

$$|A - \lambda I| = 0 \implies \begin{vmatrix} 1 - \lambda & 2 \\ 2 & 4 - \lambda \end{vmatrix} = 0$$

$$\implies (1 - \lambda)(4 - \lambda) - 4 = 0$$

$$\implies \lambda(\lambda - 5) = 0.$$

Therefore $\lambda_1 = 0$, $\lambda_2 = 5$ are the eigenvalues of A. [Since the eigenvalues are positive or zero this symmetric matrix A is positive semi-definite.] Let us compute the eigenvectors. For $\lambda_1 = 0$,

$$(A - \lambda_1 I)X = O \implies \begin{bmatrix} 1 & 2 \\ 2 & 4 \end{bmatrix} \begin{bmatrix} x_1 \\ x_2 \end{bmatrix} = \begin{bmatrix} 0 \\ 0 \end{bmatrix}$$

$$\implies x_1 + 2x_2 = 0$$

$$\implies \begin{pmatrix} x_1 \\ x_2 \end{pmatrix} = \begin{pmatrix} -2 \\ 1 \end{pmatrix} = X_1$$

is one eigenvector. Let us normalize X_1. Then

$$Y_1 = \begin{pmatrix} -\frac{2}{\sqrt{5}} \\ \frac{1}{\sqrt{5}} \end{pmatrix}$$

is the normalized eigenvector corresponding to $\lambda_1 = 0$. For $\lambda_2 = 5$,

$$(A - \lambda_2 I)X = O \implies \begin{bmatrix} -4 & 2 \\ 2 & -1 \end{bmatrix} \begin{bmatrix} x_1 \\ x_2 \end{bmatrix} = \begin{bmatrix} 0 \\ 0 \end{bmatrix}$$

$$\implies X_2 = \begin{pmatrix} 1 \\ 2 \end{pmatrix}$$

is an eigenvector. A normalized eigenvector is

$$Y_2 = \begin{pmatrix} \frac{1}{\sqrt{5}} \\ \frac{2}{\sqrt{5}} \end{pmatrix}.$$

Can there be different normalized eigenvectors for the eigenvalue $\lambda_2 = 5$? The answer is in the affirmative:

$$\begin{pmatrix} -\frac{1}{\sqrt{5}} \\ -\frac{2}{\sqrt{5}} \end{pmatrix}$$

is another one, normalized as well as orthogonal to Y_1. Then the matrix of normalized eigenvectors, denoted by P, is given by

$$P = \begin{bmatrix} -\frac{2}{\sqrt{5}} & \frac{1}{\sqrt{5}} \\ \frac{1}{\sqrt{5}} & \frac{2}{\sqrt{5}} \end{bmatrix}.$$

Note the following properties for P. The length of each row vector is 1. The length of each column vector is 1. The row (column) vectors are orthogonal to each other. That is, P is an orthonormal matrix. Then we have

$$XP = P\Lambda \Rightarrow$$

$$\begin{bmatrix} 1 & 2 \\ 2 & 4 \end{bmatrix} \begin{bmatrix} -\frac{2}{\sqrt{5}} & \frac{1}{\sqrt{5}} \\ \frac{1}{\sqrt{5}} & \frac{2}{\sqrt{5}} \end{bmatrix} = \begin{bmatrix} -\frac{2}{\sqrt{5}} & \frac{1}{\sqrt{5}} \\ \frac{1}{\sqrt{5}} & \frac{2}{\sqrt{5}} \end{bmatrix} \begin{bmatrix} 0 & 0 \\ 0 & 5 \end{bmatrix}$$

where Λ is the diagonal matrix of eigenvalues. Since P' is the inverse of P when P is orthonormal we have

$$A = P\Lambda P'. \tag{4.2.3}$$

This is a very important representation for symmetric matrices A, in general.

Example 4.2.7. Compute the eigenvalues and eigenvectors of

$$A = \begin{bmatrix} 1 & 0 & -1 \\ 0 & 3 & 0 \\ -1 & 0 & 1 \end{bmatrix}.$$

Solution 4.2.7. Consider the equation

$$|A - \lambda I| = 0 \Rightarrow \begin{vmatrix} 1-\lambda & 0 & -1 \\ 0 & 3-\lambda & 0 \\ -1 & 0 & 1-\lambda \end{vmatrix} = 0$$

$$\Rightarrow (3-\lambda)(-\lambda)(2-\lambda) = 0.$$

Therefore $\lambda_1 = 0$, $\lambda_2 = 2$, $\lambda_3 = 3$ are the eigenvalues. [The matrix A is positive semidefinite.] Let us compute the eigenvectors. For $\lambda_1 = 0$,

$$(A - \lambda_1 I)X = O \Rightarrow \begin{bmatrix} 1 & 0 & -1 \\ 0 & 3 & 0 \\ -1 & 0 & 1 \end{bmatrix} \begin{bmatrix} x_1 \\ x_2 \\ x_3 \end{bmatrix} = \begin{bmatrix} 0 \\ 0 \\ 0 \end{bmatrix}$$

$$\Rightarrow X_1 = \begin{bmatrix} 1 \\ 0 \\ 1 \end{bmatrix}$$

is one such eigenvector. The corresponding normalized vector is

$$Y_1 = \begin{pmatrix} \frac{1}{\sqrt{2}} \\ 0 \\ \frac{1}{\sqrt{2}} \end{pmatrix}.$$

For $\lambda_2 = 2$,

$$(A - \lambda_2 I)X = O \Rightarrow \begin{bmatrix} -1 & 0 & -1 \\ 0 & 1 & 0 \\ -1 & 0 & -1 \end{bmatrix} \begin{bmatrix} x_1 \\ x_2 \\ x_3 \end{bmatrix} = \begin{bmatrix} 0 \\ 0 \\ 0 \end{bmatrix}$$

$$\Rightarrow X_2 = \begin{bmatrix} 1 \\ 0 \\ -1 \end{bmatrix} \quad \text{or} \quad Y_2 = \begin{bmatrix} \frac{1}{\sqrt{2}} \\ 0 \\ -\frac{1}{\sqrt{2}} \end{bmatrix}$$

is a solution. For $\lambda_3 = 3$,

$$(A - \lambda_3 I)X = O \Rightarrow \begin{bmatrix} -2 & 0 & -1 \\ 0 & 0 & 0 \\ -1 & 0 & -2 \end{bmatrix} \begin{bmatrix} x_1 \\ x_2 \\ x_3 \end{bmatrix} = \begin{bmatrix} 0 \\ 0 \\ 0 \end{bmatrix}$$

$$\Rightarrow X_3 = \begin{bmatrix} 0 \\ 1 \\ 0 \end{bmatrix} \Rightarrow Y_3 = \begin{bmatrix} 0 \\ 1 \\ 0 \end{bmatrix}.$$

Note that the matrix $(A - \lambda_i I)$, in each case $i = 1, 2, 3$, is singular with rank 2. Hence in the class of linearly independent vectors X there can be only one, $3 - 2 = 1$, vector for a given eigenvalue. [This is a general property when the eigenvalues are distinct for any matrix, need not be symmetric.] Let us consider the matrix of normalized eigenvectors:

$$P = \begin{bmatrix} \frac{1}{\sqrt{2}} & \frac{1}{\sqrt{2}} & 0 \\ 0 & 0 & 1 \\ \frac{1}{\sqrt{2}} & -\frac{1}{\sqrt{2}} & 0 \end{bmatrix}, \quad PP' = I, \quad P'P = I.$$

We have the representation

$$A = P\Lambda P'$$

where

$$\Lambda = \begin{bmatrix} 0 & 0 & 0 \\ 0 & 2 & 0 \\ 0 & 0 & 3 \end{bmatrix}$$

the diagonal matrix of eigenvalues of A.

We may also observe one interesting property from the representation $A = PDP'$. We can always write

$$\Lambda = \begin{bmatrix} \lambda_1 & 0 & 0 & \cdots & 0 \\ 0 & \lambda_2 & \cdots & & 0 \\ \vdots & \vdots & \cdots & & \vdots \\ 0 & 0 & \cdots & & \lambda_n \end{bmatrix}$$

$$= \begin{bmatrix} \lambda_1 & 0 & \cdots & 0 \\ 0 & 0 & \cdots & 0 \\ \vdots & \vdots & \cdots & \vdots \\ 0 & 0 & \cdots & . \end{bmatrix}$$

$$+ \cdots + \begin{bmatrix} 0 & 0 & \cdots & 0 \\ 0 & 0 & \cdots & 0 \\ \vdots & \vdots & \cdots & \vdots \\ 0 & 0 & \cdots & \lambda_n \end{bmatrix}$$

$$= \Lambda_1 + \Lambda_2 + \cdots + \Lambda_n.$$

That is, Λ is written as a sum of diagonal matrices of which the j-th one, namely Λ_j, has λ_j as the j-th diagonal element and all other elements zeros for $j = 1, \dots, n$. Then

$$P\Lambda P' = P\Lambda_1 P' + P\Lambda_2 P' + \cdots + P\Lambda_n P'.$$

Note that P postmultiplied by Λ_j gives the j-th column of P multiplied by λ_j and all other columns multiplied by zeros and when this is postmultiplied by P' we get only the transpose of the j-th column and all other elements zeros. Thus if $P_1, \dots, P_n$ denote the columns of P then we have the following result:

(xi) $$A = P\Lambda P' = \lambda_1 P_1 P_1' + \cdots + \lambda_n P_n P_n'. \qquad (4.2.4)$$

In Example 4.2.7 if we write by using this representation then we have

$$\begin{bmatrix} 1 & 0 & -1 \\ 0 & 3 & 0 \\ -1 & 0 & 1 \end{bmatrix} = 0 \begin{bmatrix} \frac{1}{\sqrt{2}} \\ 0 \\ \frac{1}{\sqrt{2}} \end{bmatrix} \left[\frac{1}{\sqrt{2}}, 0, \frac{1}{\sqrt{2}} \right]$$

$$+ 2 \begin{bmatrix} \frac{1}{\sqrt{2}} \\ 0 \\ -\frac{1}{\sqrt{2}} \end{bmatrix} \left[\frac{1}{\sqrt{2}}, 0, -\frac{1}{\sqrt{2}} \right] + 3 \begin{bmatrix} 0 \\ 1 \\ 0 \end{bmatrix} [0,1,0]$$

$$= 0 + 2 \begin{bmatrix} \frac{1}{2} & 0 & -\frac{1}{2} \\ 0 & 0 & 0 \\ -\frac{1}{2} & 0 & \frac{1}{2} \end{bmatrix} + 3 \begin{bmatrix} 0 & 0 & 0 \\ 0 & 1 & 0 \\ 0 & 0 & 0 \end{bmatrix}.$$

Both sides are equal. In the general case when we have the representation,

$$A = Q \Lambda Q^{-1}$$

where A need not be symmetric, note that the columns of Q^{-1} or the rows of Q^{-1} are not directly available from the corresponding rows or columns of Q as in the orthonormal case. Hence, first Q^{-1} has to be evaluated. Then look at the rows of Q^{-1}. Let the columns of Q be $Q_1, \dots, Q_n$ and the rows of Q^{-1} be $R_1, \dots, R_n$. Then the representation will be the following:

(xii) $$A = \lambda_1 Q_1 R_1 + \cdots + \lambda_n Q_n R_n. \qquad (4.2.5)$$

Let us verify this for the A in Example 4.2.1. There the various quantities are already evaluated.

$$A = \begin{bmatrix} 2 & 2 \\ 1 & 3 \end{bmatrix}, \quad Q = \begin{bmatrix} -2 & 1 \\ 1 & 1 \end{bmatrix},$$

$$Q^{-1} = \begin{bmatrix} -\frac{1}{3} & \frac{1}{3} \\ \frac{1}{3} & \frac{2}{3} \end{bmatrix}, \quad \Lambda = \begin{bmatrix} 1 & 0 \\ 0 & 4 \end{bmatrix}.$$

Then, according to the notation in (4.2.5)

$$Q_1 = \begin{bmatrix} -2 \\ 1 \end{bmatrix}, \quad Q_2 = \begin{bmatrix} 1 \\ 1 \end{bmatrix}, \quad R_1 = \left[-\frac{1}{3}, \frac{1}{3} \right], \quad R_2 = \left[\frac{1}{3}, \frac{2}{3} \right],$$

$\lambda_1 = 1, \lambda_2 = 4$. Then

$$\lambda_1 Q_1 R_1 + \lambda_2 Q_2 R_2 = 1 \begin{bmatrix} -2 \\ 1 \end{bmatrix} \left[-\frac{1}{3}, \frac{1}{3} \right] + 4 \begin{bmatrix} 1 \\ 1 \end{bmatrix} \left[\frac{1}{3}, \frac{2}{3} \right]$$

$$= \begin{bmatrix} \frac{2}{3} & -\frac{2}{3} \\ -\frac{1}{3} & \frac{1}{3} \end{bmatrix} + 4 \begin{bmatrix} \frac{1}{3} & \frac{2}{3} \\ \frac{1}{3} & \frac{2}{3} \end{bmatrix}$$

$$= \begin{bmatrix} 2 & 2 \\ 1 & 3 \end{bmatrix} = A.$$

The result is verified.

Example 4.2.8. Evaluate the eigenvalues and the eigenvectors of A and represent, if possible, A in the form $A = P\Lambda P'$ where

$$A = \frac{1}{3}\begin{bmatrix} 1 & 1 & 1 \\ 1 & 1 & 1 \\ 1 & 1 & 1 \end{bmatrix}.$$

Solution 4.2.8. The eigenvalues of such an $n \times n$ matrix are already evaluated in Example 4.1.3. The eigenvalues are 1 and the rest zeros. In our case the eigenvalues are $\lambda_1 = 1$, $\lambda_2 = 0$, $\lambda_3 = 0$. The eigenvectors corresponding to $\lambda_1 = 1$ are given by

$$(A - \lambda_1 I)X = O \implies \begin{bmatrix} -\frac{2}{3} & \frac{1}{3} & \frac{1}{3} \\ \frac{1}{3} & -\frac{2}{3} & \frac{1}{3} \\ \frac{1}{3} & \frac{1}{3} & -\frac{2}{3} \end{bmatrix}\begin{bmatrix} x_1 \\ x_2 \\ x_3 \end{bmatrix} = \begin{bmatrix} 0 \\ 0 \\ 0 \end{bmatrix}$$

$$\implies X_1 = \begin{bmatrix} 1 \\ 1 \\ 1 \end{bmatrix}$$

is an eigenvector. The normalized X_1 is

$$Y_1 = \begin{pmatrix} \frac{1}{\sqrt{3}} \\ \frac{1}{\sqrt{3}} \\ \frac{1}{\sqrt{3}} \end{pmatrix}.$$

Now consider

$$(A - \lambda_2 I)X = O \implies \frac{1}{3}\begin{bmatrix} 1 & 1 & 1 \\ 1 & 1 & 1 \\ 1 & 1 & 1 \end{bmatrix}\begin{bmatrix} x_1 \\ x_2 \\ x_3 \end{bmatrix} = \begin{bmatrix} 0 \\ 0 \\ 0 \end{bmatrix}.$$

We can obtain 2 linearly independent X since the rank of A is 1. For example

$$X_2 = \begin{pmatrix} 1 \\ 0 \\ -1 \end{pmatrix} \quad \text{and} \quad X_3 = \begin{pmatrix} 1 \\ -2 \\ 1 \end{pmatrix}$$

are two such vectors. The normalized vectors corresponding to these are

$$Y_2 = \begin{pmatrix} \frac{1}{\sqrt{2}} \\ 0 \\ -\frac{1}{\sqrt{2}} \end{pmatrix} \quad \text{and} \quad Y_3 = \begin{pmatrix} \frac{1}{\sqrt{6}} \\ -\frac{2}{\sqrt{6}} \\ \frac{1}{\sqrt{6}} \end{pmatrix}.$$

Hence the matrix of normalized eigenvectors is

$$P = \begin{bmatrix} \frac{1}{\sqrt{3}} & \frac{1}{\sqrt{2}} & \frac{1}{\sqrt{6}} \\ \frac{1}{\sqrt{3}} & 0 & -\frac{2}{\sqrt{6}} \\ \frac{1}{\sqrt{3}} & -\frac{1}{\sqrt{2}} & \frac{1}{\sqrt{6}} \end{bmatrix}.$$

It is evident that P is an orthonormal matrix, $PP' = I$, $P'P = I$. Hence

$$A = P\Lambda P', \quad \Lambda = \begin{bmatrix} 1 & 0 & 0 \\ 0 & 0 & 0 \\ 0 & 0 & 0 \end{bmatrix}.$$

Can we write this A in the form (4.2.4)?

$$A = \lambda_1 P_1 P_1' + \lambda_2 P_2 P_2' + \lambda_3 P_3 P_3'$$

$$= 1 \begin{bmatrix} \frac{1}{\sqrt{3}} \\ \frac{1}{\sqrt{3}} \\ \frac{1}{\sqrt{3}} \end{bmatrix} \begin{bmatrix} \frac{1}{\sqrt{3}}, \frac{1}{\sqrt{3}}, \frac{1}{\sqrt{3}} \end{bmatrix} + 0 + 0$$

$$= \begin{bmatrix} \frac{1}{3} & \frac{1}{3} & \frac{1}{3} \\ \frac{1}{3} & \frac{1}{3} & \frac{1}{3} \\ \frac{1}{3} & \frac{1}{3} & \frac{1}{3} \end{bmatrix}.$$

The result is verified. Here in our example two eigenvalues were equal but still we could get 3 eigenvectors which were orthonormal. This, in fact, is a general result for symmetric matrices. It need not hold for nonsymmetric matrices. In nonsymmetric cases in some situations it is possible to obtain a complete set of orthonormal vectors and some cases it is not possible. In Chapter 2 it was illustrated that any symmetric matrix A can always be written in the form

$$A = QDQ'$$

through elementary operations on the left and on the right, where D is diagonal and Q is nonsingular. Now the question remains: can we select a D and a Q for any given symmetric matrix A such that D contains all the eigenvalues of A and Q is orthonormal? The answer to this is in the affirmative. We will establish this general result after introducing some aspects of complex numbers, and matrices whose elements are complex numbers, in the next section.

Before concluding this section observe the following points: Computations of eigenvalues and eigenvectors are, in general, difficult problems. Even for a 3×3 matrix the characteristic equation, $|A - \lambda I| = 0$, is a cubic equation. Often one may have to use a computer to obtain the roots or use the complicated formula for the three roots of a cubic equation. When the degree of the characteristic polynomial is large the only way to solve may be to use a computer. Also the characteristic equation can produce irrational or complex roots.

Exercises 4.2

4.2.1. Compute the eigenvalues and eigenvectors of A, and if possible, represent A in the form $A = QDQ^{-1}$, where

(a) $A = \begin{bmatrix} 0 & 1 \\ 1 & 0 \end{bmatrix}$,   (b) $A = \begin{bmatrix} 0 & -1 \\ 1 & 0 \end{bmatrix}$,   (c) $A = \begin{bmatrix} 0 & 1 \\ -1 & 0 \end{bmatrix}$,

(d) $A = \begin{bmatrix} 0 & 1 \\ 0 & 0 \end{bmatrix}$,   (e) $A = \frac{1}{2} \begin{bmatrix} 1 & 1 \\ 1 & 1 \end{bmatrix}$.

4.2.2. Repeat Exercise 4.2.1 for

(a) $A = \begin{bmatrix} 1 & 3 \\ 2 & 5 \end{bmatrix}$,   (b) $A = \begin{bmatrix} 1 & 1 & 0 \\ 1 & -1 & 2 \\ 3 & 0 & 4 \end{bmatrix}$,   (c) $A = \begin{bmatrix} 1 & 2 & 1 \\ 2 & 4 & 2 \\ 3 & 6 & 3 \end{bmatrix}$.

4.2.3. Repeat Exercise 4.2.1 for

(a) $A = \frac{1}{4} \begin{bmatrix} 1 & 1 & 1 & 1 \\ 1 & 1 & 1 & 1 \\ 1 & 1 & 1 & 1 \\ 1 & 1 & 1 & 1 \end{bmatrix}$,   (b) $A = \begin{bmatrix} 1 & 1 & 1 & 1 \\ 1 & 1 & 1 & 1 \\ 1 & 1 & 1 & 1 \\ 1 & 1 & 1 & 1 \end{bmatrix}$,

(c) $A = \begin{bmatrix} 1 & 1 & 1 & 1 \\ 1 & -1 & 1 & -1 \\ 1 & -1 & -1 & 1 \\ 1 & 1 & -1 & -1 \end{bmatrix}$.

4.2.4. Construct a 3×3 matrix whose eigenvalues are $\lambda_1 = 1$, $\lambda_2 = 2$, $\lambda_3 = 3$ and whose eigenvectors are

$$X_1 = \begin{pmatrix} 1 \\ 1 \\ 1 \end{pmatrix}, \quad X_2 = \begin{pmatrix} 2 \\ 1 \\ 2 \end{pmatrix}, \quad X_3 = \begin{pmatrix} 3 \\ 2 \\ 2 \end{pmatrix}.$$

Is this matrix unique or can you find one more matrix with the same eigenvalues and eigenvectors?

4.2.5. Find two different matrices Q_1 and Q_2 such that the matrix in Exercise 4.2.3 (b) can be written in the form

$$A = Q_1 \Lambda Q_1^{-1} = Q_2 \Lambda Q_2^{-1}$$

where Λ is the diagonal matrix with the eigenvalues of A as the diagonal elements.

4.2.6. Evaluate the eigenvalues and eigenvectors of the following matrices. Let Q denote the matrix of eigenvectors for each case. Is Q nonsingular?

$$A = \begin{bmatrix} 2 & 0 & 0 \\ 0 & 2 & 0 \\ 0 & 0 & 2 \end{bmatrix}, \quad B = \begin{bmatrix} 1 & 2 & 3 \\ 0 & 4 & -1 \\ 0 & 0 & 0 \end{bmatrix}, \quad C = \begin{bmatrix} 0 & 0 & 0 \\ 2 & 0 & 0 \\ 1 & -1 & 1 \end{bmatrix}.$$

4.2.7. From the results obtained in Exercise 4.2.3 compute the determinant of A without any additional computation.

4.2.8. Compute the eigenvalues and eigenvectors of A^{10} for each case of (1) Exercise 4.2.1, (2) Exercise 4.2.2. List the cases where A^{-25} is defined.

4.2.9. Compute A^{100} for each case of A in Exercise 4.2.1.

4.2.10. Let $A = (a_{ij})$ and $B = (b_{ij})$ be $n \times n$ matrices. By straight multiplication and then taking the traces show that

$$\text{(a)} \quad \text{tr}(AB) = \text{tr}(BA), \qquad \text{(b)} \quad AB - BA \neq I.$$

4.2.11. If a matrix A has eigenvalues $1, -1, 2$ is the matrix Q of eigenvectors (a) singular? (b) orthonormal?

4.2.12. Repeat Exercise 4.2.11 if the eigenvalues are

$$\text{(a)} \quad 1, 1, 2, \qquad \text{(b)} \quad 1, 0, 0.$$

4.2.13. Let

$$A = \begin{bmatrix} 1 & 2 \\ 1 & 1 \end{bmatrix}.$$

Compute (a) $\text{tr}(A^{20})$, (b) $|A^{-20}|$ (the determinant of A^{-20}).

4.2.14. If a matrix A can be written as $A = QDQ^{-1}$ show that

$$\text{tr}(A^{10}) = d_1^{10} + \cdots + d_n^{10}$$

where $d_1, \ldots, d_n$ are the diagonal elements in D.

4.2.15. If A is a 3×3 symmetric matrix with the eigenvalues $0, 1, 2$ construct 4 different matrices B such that $B^2 = A$. [Here B is a square root of A.]

4.2.16. For the problem in Exercise 4.2.3 (a) represent A in the form of equation (4.2.4).

4.2.17. Represent A in the form of equation (4.2.5) where

$$A = \begin{bmatrix} 1 & 1 & 0 \\ 1 & -2 & 2 \\ 3 & 0 & 4 \end{bmatrix}.$$

4.2.18. Evaluate the eigenvalues of the following matrices:

$$A = \begin{bmatrix} a & b & \cdots & b \\ b & a & \cdots & b \\ \vdots & \vdots & \cdots & \vdots \\ b & b & \cdots & a \end{bmatrix} \quad \text{where } A \text{ is } n \times n$$

$$B = \begin{bmatrix} 0 & 0 & 0 & 1 & 1 & 1 & 1 & 1 & 1 \\ 0 & 0 & 0 & 1 & 1 & 1 & 1 & 1 & 1 \\ 0 & 0 & 0 & 1 & 1 & 1 & 1 & 1 & 1 \\ 1 & 1 & 1 & 0 & 0 & 0 & 1 & 1 & 1 \\ 1 & 1 & 1 & 0 & 0 & 0 & 1 & 1 & 1 \\ 1 & 1 & 1 & 0 & 0 & 0 & 1 & 1 & 1 \\ 1 & 1 & 1 & 1 & 1 & 1 & 0 & 0 & 0 \\ 1 & 1 & 1 & 1 & 1 & 1 & 0 & 0 & 0 \\ 1 & 1 & 1 & 1 & 1 & 1 & 0 & 0 & 0 \end{bmatrix}.$$

4.2.19. Companion matrix. Show that the $n \times n$ companion matrix A has the characteristic polynomial

$$P(\lambda) = \lambda^n + \sum_{i=0}^{n-1} c_i \lambda^i$$

where

$$A = \begin{bmatrix} 0 & 1 & 0 & \cdots & 0 \\ 0 & 0 & 1 & \cdots & 0 \\ \vdots & \vdots & \vdots & \vdots & \vdots \\ 0 & 0 & 0 & \cdots & 1 \\ -c_0 & -c_1 & -c_2 & \cdots & -c_{n-1} \end{bmatrix}.$$

4.2.20. Let A be an $n \times n$ matrix with the characteristic polynomial $P(\lambda) = \sum_{i=0}^{n} c_i \lambda^i$. Show that the scalar c_r, $0 \le r < n$, is equal to the sum of all principal minors of order $n - r$ of A multiplied by $(-1)^{n-r}$.

4.2.21. For the $n \times n$ real matrix A let $\lambda^n - a_1 \lambda^{n-1} + a_2 \lambda^{n-2} - \cdots \pm a_n = 0$ be the characteristic polynomial. Then show that

$$A^n - a_1 A^{n-1} + \cdots \pm a_n I = O.$$

4.2.22. For a nonsingular 2×2 matrix A let $\lambda^2 - a_1 \lambda + a_2 = 0$ be the characteristic equation. Show that

$$a_2 A^{-1} = a_1 I - A.$$

4.3 Some properties of complex numbers and matrices in the complex fields

In all the discussions so far the elements in our vectors and matrices were real numbers or real quantities. Now we will extend our discussion to complex numbers and matrices with the elements in the complex field.

4.3.1 Complex numbers

The very basic quantity in the field of complex numbers is denoted by i which is the positive square root of -1, that is, $i = \sqrt{-1}$. It can arise, for example, when we try to solve the equation

$$x^2 + 1 = 0 \implies x^2 = -1 \implies x = \pm\sqrt{-1} \quad \text{or} \quad x = \pm i.$$

A complex number can be written as $z = a + ib$ where a and b are real numbers and $i = \sqrt{-1}$. Then we say that a is the *real part* of z, written as $a = \Re(z)$ and b is the *imaginary part* of z, written as $b = \Im(z)$. For example,

complex number	real part	imaginary part
$2 + 3i$	2	3
$2 - 3i$	2	-3
$3i$	0	3
2	2	0

(i) A complex number with the imaginary part zero is a real number. A complex number with the real part zero is a purely imaginary number.

Definition 4.3.1 (A complex number). Any number of the form $a + ib$ where a and b are real and $i = \sqrt{-1}$ is called a complex number.

Definition 4.3.2 (Complex conjugate). The complex conjugate of $a + ib$ is defined as $a - ib$, (with i replaced by $-i$) so that

$$(a + ib)(a - ib) = a^2 + b^2$$

is always real.

Notation 4.3.1. The complex conjugate of z is usually denoted by z^c or $\bar{z}$. Since $\bar{z}$ is already used for the average we will denote the conjugate of z by z^c.

Definition 4.3.3 (The absolute value of a complex number). If $z = a + ib$ is a complex number its absolute value, denoted by $|z|$, is the positive square root of z multiplied by its conjugate z^c. That is,

$$|z| = \sqrt{zz^c} = \sqrt{(a + ib)(a - ib)} = \sqrt{a^2 + b^2} = |a + ib| = |a - ib|.$$

For example,

complex number	its conjugate	its absolute value
$3 - 4i$	$3 + 4i$	$+\sqrt{(3)^2 + (-4)^2} = 5$
$1 + 2i$	$1 - 2i$	$+\sqrt{(1)^2 + (2)^2} = \sqrt{5}$
$7i$	$-7i$	$+\sqrt{(0)^2 + (-7)^2} = 7$
-2	-2	$+\sqrt{(-2)^2 + (0)^2} = 2$

This definition of the absolute value is in agreement with the concept of the absolute value of a real number. Also real numbers can be taken as particular cases of complex numbers where the imaginary parts are zeros.

4.3.2 Geometry of complex numbers

Since any complex number is of the form $z = x + iy$ where x and y are real and $i = \sqrt{-1}$ the pair of real quantities (x, y) uniquely determine the complex number z. If we take a rectangular coordinate system and call the x-axis the real axis, and the y-axis the imaginary axis then $x + iy$ is the point (x, y) in this complex plane. Some complex numbers are marked in Figure 4.3.1.

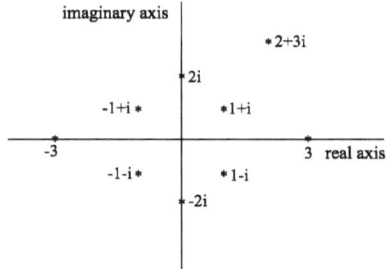

Figure 4.3.1: A complex plane.

Let us take an arbitrary point $z = x + iy$. Call the origin of the above rectangular coordinate system O and let $P = (x, y)$ be the point z in this complex plane. Let θ be the angle OP makes with the x-axis and r the length of OP. Then as shown in Figure 4.3.2

$$x = r \cos\theta, \quad y = r \sin\theta, \quad z = r \cos\theta + i r \sin\theta.$$

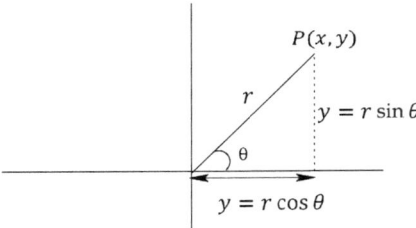

Figure 4.3.2: Geometry of complex numbers.

What about powers of z, such as $z^2, z^3, \ldots$ or z^k for some k, positive or negative. Trying to take the powers directly as

$$z^k = (r\cos\theta + ir\sin\theta)^k = r^k(\cos\theta + i\sin\theta)^k$$

is not an easy process. But one thing is certain. When $(\cos\theta + i\sin\theta)^k$ is expanded it can give terms containing $i, i^2, i^3, \ldots, i^k$. These powers of i can be reduced by using the results $i^2 = -1$, $i^3 = -i$, $i^4 = 1$, $i^5 = i$ and so on. Thus evidently, $(\cos\theta + i\sin\theta)^k$ gives rise to a quantity of the form $a + ib$ where a and b are real. Hence all powers of z^k are also complex numbers. In order to evaluate the powers in a much simpler way we will look at another representation of a complex number. To this end, consider the expansion

$$e^{i\theta} = 1 + (i\theta) + \frac{(i\theta)^2}{2!} + \frac{(i\theta)^3}{3!} + \cdots$$

$$= \left[1 - \frac{\theta^2}{2!} + \frac{\theta^4}{4!} - \cdots\right] + i\left[\theta - \frac{\theta^3}{3!} + \frac{\theta^5}{5!} - \cdots\right].$$

These two seres (series) are known from calculus or trigonometry as

$$\cos\theta = 1 - \frac{\theta^2}{2!} + \frac{\theta^4}{4!} + \cdots$$

and

$$\sin\theta = \theta - \frac{\theta^3}{3!} + \frac{\theta^5}{5!} - \cdots.$$

Hence

$$e^{i\theta} = \cos\theta + i\sin\theta \implies \tag{4.3.1}$$

$$z = r(\cos\theta + i\sin\theta) = re^{i\theta}. \tag{4.3.2}$$

This is a very important formula to deal with complex numbers. For example,

$$z^2 = \left[re^{i\theta}\right]^2 = r^2 e^{2i\theta} = r^2[\cos 2\theta + i\sin 2\theta];$$

$$z^{-1} = \left[re^{i\theta}\right]^{-1} = \frac{1}{r}\left[e^{-i\theta}\right] = \frac{1}{r}[\cos(-\theta) + i\sin(-\theta)]$$

$$= \frac{1}{r}[\cos\theta - i\sin\theta], \quad r \neq 0;$$

$$z^k = [re^{i\theta}]^k = r^k e^{ik\theta} = r^k[\cos k\theta + i\sin k\theta],$$

$$k = 0, 1, 2, \ldots, -1, -2, \ldots, \quad r \neq 0. \tag{4.3.3}$$

Could we have obtained the same result by powers of $z = r(\cos\theta + i\sin\theta)$ directly? Yes, we could have arrived at the same results but the process would have involved invoking many results from trigonometry. For example,

$$z^2 = [r(\cos\theta + i\sin\theta)]^2 = r^2[\cos^2\theta + (i\sin\theta)^2 + 2i\cos\theta\sin\theta]$$

$$= r^2[\cos^2\theta - \sin^2\theta + i(2\sin\theta\cos\theta)]$$

$$= r^2[\cos 2\theta + i\sin 2\theta]$$

since $\cos 2\theta = \cos^2\theta - \sin^2\theta$ and $\sin 2\theta = 2\sin\theta\cos\theta$. What about the exponents if we take products and ratios? For example,

$$zz^{-1} = (re^{i\theta})(re^{i\theta})^{-1} = re^{i\theta}\frac{1}{r}e^{-i\theta} = 1, \quad r \neq 0;$$

$$z^2z^{-1} = \left(r^2e^{2i\theta}\frac{1}{r}e^{-i\theta}\right) = re^{i\theta} = z, \quad r \neq 0;$$

$$z^2z^3 = (r^2e^{i2\theta})(r^3e^{i3\theta}) = r^5e^{i5\theta} = z^5;$$

$$z^2z^{-3} = \left(r^2e^{i2\theta}\frac{1}{r^3}e^{-i3\theta}\right) = \frac{1}{r}e^{-i\theta} = z^{-1}, \quad r \neq 0;$$

$$z^{\frac{1}{2}}z^{\frac{1}{2}} = (re^{i\theta})^{\frac{1}{2}}(re^{i\theta})^{\frac{1}{2}} = re^{i\theta} = z;$$

$$z^mz^n = z^{m+n}.$$

The rules of multiplication and division can be carried through exactly as in the real case. What about the complex conjugates? We have defined the conjugate of $z = x + iy$ as $z^c = x - iy$. The absolute values remain the same.

$$|z| = \sqrt{x^2 + y^2} = \sqrt{x^2 + (-y)^2} = |z^c|.$$

Writing $x - iy$ as $x + (-i)y$ we have

$$z = re^{i\theta} \implies z^c = re^{-i\theta}$$

$$\implies zz^c = (re^{i\theta})(re^{-i\theta}) = r^2,$$

which is also seen from direct multiplication, $(x + iy)(x - iy) = x^2 + y^2 = r^2$.

4.3.3 Algebra of complex numbers

What about the sums and products of complex numbers? Let

$$z_1 = x_1 + iy_1 \quad \text{and} \quad z_2 = x_2 + iy_2$$

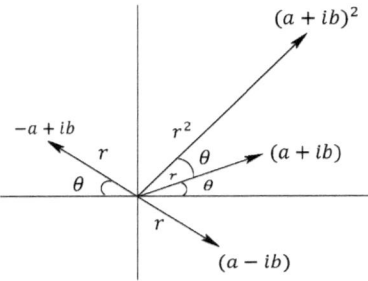

Figure 4.3.3: Powers of complex numbers.

where x_1, x_2, y_1, y_2 real. If $(x_1, y_1) \neq (x_2, y_2)$ then we have two complex numbers. A complex number, its conjugate and powers are shown in Figure 4.3.3. Writing in exponential form we have

$$z_1 = r_1 e^{i\theta_1}, \quad z_2 = r_2 e^{i\theta_2}$$

where

$$r_1 = \sqrt{x_1^2 + y_1^2}, \quad r_2 = \sqrt{x_2^2 + y_2^2},$$

$$\theta_1 = \tan^{-1}\left(\frac{y_1}{x_1}\right), \quad \theta_2 = \tan^{-1}\left(\frac{y_2}{x_2}\right), \quad x_1 \neq 0, \quad x_2 \neq 0.$$

$$z_1 + z_2 = (x_1 + iy_1) + (x_2 + iy_2) = (x_1 + y_1) + i(x_2 + y_2)$$
$$= r_1 e^{i\theta_1} + r_2 e^{i\theta_2}$$
$$= r_1[\cos\theta_1 + i\sin\theta_1] + r_2[\cos\theta_2 + i\sin\theta_2]$$
$$= (r_1\cos\theta_1 + r_2\cos\theta_2) + i(r_1\sin\theta_1 + r_2\sin\theta_2). \tag{4.3.4}$$

All these are equivalent representations for $z_1 + z_2$. Thus $z_1 + z_2$ is again a complex number.

$$z_1 z_2 = [x_1 + iy_1][x_2 + iy_2] = (x_1 x_2 - y_1 y_2) + i(x_1 y_2 + x_2 y_1) \tag{4.3.5}$$
$$= (r_1 e^{i\theta_1})(r_2 e^{i\theta_2}) = r_1 r_2 e^{i(\theta_1 + \theta_2)}$$
$$= r_1 r_2[\cos(\theta_1 + \theta_2) + i\sin(\theta_1 + \theta_2)].$$
$$zz^c = (x + iy)(x - iy) = x^2 + y^2 = (re^{i\theta})(re^{-i\theta}) = r^2$$
$$= [r(\cos\theta + i\sin\theta)][r(\cos\theta - i\sin\theta)] = r^2[\cos^2\theta + \sin^2\theta]$$
$$= r^2. \tag{4.3.6}$$

What about the conjugate of the product $z_1 z_2$?

$$(z_1 z_2)^c = [(x_1 x_2 - y_1 y_2) + i(x_1 y_2 + x_2 y_1)]^c$$
$$= (x_1 x_2 - y_1 y_2) - i(x_1 y_2 + x_2 y_1)$$
$$z_1^c z_2^c = [x_1 - iy_1][x_2 - iy_2] = (x_1 x_2 - y_1 y_2) - i(x_1 y_2 + x_2 y_1)$$
$$= (z_1 z_2)^c \implies (z_1 z_2)^c = z_1^c z_2^c. \tag{4.3.7}$$

Taking the conjugate of a product or the product of a conjugate give rise to the same result. Some numerical examples are the following:

$$z = 2 + 3i, \quad z^C = 2 - 3i, \quad |z| = \sqrt{2^2 + 3^2} = \sqrt{13},$$

$$z = \sqrt{13}e^{i\theta}, \quad \tan\theta = \frac{3}{2};$$

$$z_1 = 3i, \quad z_1^C = -3i, \quad |z_1| = \sqrt{(0)^2 + (3)^2} = 3,$$

$$\cos\theta = 0, \quad \sin\theta = 1, \quad \theta = \frac{\pi}{2}$$

$$z_1 = 3e^{i\frac{\pi}{2}} = 3e^{i(\frac{\pi}{2}+2m\pi)}, \quad m = 0,1,2,\dots;$$

$$z_2 = 2, \quad z_2^C = 2, \quad |z_2| = \sqrt{(2)^2 + (0)^2} = 2, \quad \cos\theta = 1, \quad \sin\theta = 0, \quad \theta = 0$$

$$z_2 = 2e^{i(0+2m\pi)} = 2 \quad \text{(in agreement with the rules for real numbers);}$$

$$z_3 = 1 + 2i, \quad z_4 = 3 + 4i \implies z_3 + z_4 = (1 + 2i) + (3 + 4i) = 4 + 6i;$$

$$z_3 - z_4 = (1 + 2i) - (3 + 4i) = -2 - 2i;$$

$$z_3 + z_4^C = (1 + 2i) + (3 - 4i) = 4 - 2i = z_4^C + z_3;$$

$$z_3 - z_4^C = (1 + 2i) - (3 - 4i) = -2 + 6i;$$

$$2z_3 + 5z_4 = 2(1 + 2i) + 5(3 + 4i) = (2 + 4i) + (15 + 20i) = 17 + 24i;$$

$$z_3 z_4 = (1 + 2i)(3 + 4i) = [(1)(3) + i^2(2)(4)] + i[(1)(4) + (2)(3)]$$
$$= 3 - 8 + 10i = -5 + 10i;$$

$$z_3^2 = (1 + 2i)^2 = (1)^2 + (2i)^2 + 2(1)(2i) = 1 - 4 + 4i = -3 + 4i.$$

In the real case the square is a non-negative number. In the complex case we cannot talk about non-negativity since it is again a complex number.

$$z_3 z_4^C = (1 + 2i)(3 - 4i) = [(1)(3) + (2)(-4)i^2] + [(1)(-4) + (2)(3)]i$$
$$= 11 + 2i;$$

$$(z_3 z_4)^C = (-5 + 10i)^C = -5 - 10i;$$

$$z_3^C z_4^C = (1 - 2i)(3 - 4i) = -5 - 10i = (z_3 z_4)^C;$$

$$|z_3 z_4| = |(-5 - 10i)| = \sqrt{(-5)^2 + (-10)^2} = \sqrt{125} = |(z_3 z_4)^C|;$$

$$|z_3||z_4| = |(1 + 2i)||(3 + 4i)| = \sqrt{(1)^2 + (2)^2}\sqrt{(3)^2 + (4)^2}$$
$$= \sqrt{5}\sqrt{25} = \sqrt{125} = |z_3 z_4|.$$

This is a general result. If $z_1,\dots,x_k$ are complex numbers then the absolute value of the product or the absolute value of the product of their conjugates is the product of the absolute values.

(ii)
$$|z_1 z_2 \cdots z_k| = |z_1| \, |z_2| \cdots |z_k|$$
$$= |z_1^c| \, |z_2^c| \cdots |z_k^c| = |(z_1 \cdots z_k)^c|. \tag{4.3.8}$$

How do we compute the ratios of complex numbers? Since a complex number multiplied by its conjugate is a real number we may multiply the complex numbers appearing in the denominator by the conjugates to make the denominator real. This is a convenient technique that can be employed when evaluating a ratio. Let $z_1 = x_1 + iy_1$ and $z_2 = x_2 + iy_2$. Then

$$
\begin{aligned}
\frac{z_1}{z_2} &= \frac{x_1 + iy_1}{x_2 + iy_2} = \frac{(x_1 + iy_1)(x_2 - iy_2)}{(x_2 + iy_2)(x_2 - iy_2)} \\
&= \frac{[x_1 x_2 + y_1 y_2] + i[-x_1 y_2 + x_2 y_1]}{x_2^2 + y_2^2} \\
&= \frac{(x_1 x_2 + y_1 y_2)}{x_2^2 + y_2^2} + i \frac{(-x_1 y_2 + x_2 y_1)}{x_2^2 + y_2^2}.
\end{aligned}
$$

As numerical examples we have

$$
\begin{aligned}
\frac{(1 + 3i)}{(2 - i)} &= \frac{(1 + 3i)(2 + i)}{(2 - i)(2 + i)} = \frac{[(1)(2) - (3)(1)] + i[(1)(1) + (3)(2)]}{2^2 + 1^2} \\
&= -\frac{1}{5} + \frac{7}{5}i; \\
\frac{1 + 3i}{5i} &= \frac{(1 + 3i)(-5i)}{(5i)(-5i)} = \frac{3}{5} - \frac{1}{5}i.
\end{aligned}
$$

Instead of $-5i$ we could have kept 5 outside and multiplied both numerator and the denominator by $-i$.

$$
\begin{aligned}
\frac{1 + 3i}{2} &= \frac{1}{2} + \frac{3}{2}i; \\
\frac{(1 - 3i)}{(1 + i)(2 + i)} &= \frac{(1 - 3i)(1 - i)(2 - i)}{(1 + i)(1 - i)(2 + i)(2 - i)} \\
&= \frac{(-2 - 4i)(2 - i)}{(1^2 + 1^2)(2^2 + 1^2)} = -\frac{4}{5} - \frac{3}{5}i; \\
\frac{(2 + i)}{5i(1 - i)(2 - i)} &= \frac{(2 + i)}{5i(1 - i)(2 - i)} \frac{(-i)(1 + i)(2 + i)}{(-i)(1 + i)(2 + i)} \\
&= -\frac{(2 + i)(-1 + i)(2 + i)}{5(1^2 + 1^2)(2^2 + 1^2)} = \frac{7}{50} + \frac{1}{50}i.
\end{aligned}
$$

4.3.4 n-th roots of unity

The square roots of 1 are available by solving the equation $x^2 = 1 \Rightarrow x = \pm 1$. If we take the fourth root of 1 there must be 4 roots. What are they? $x^4 = 1 \Rightarrow x^2 = \pm 1$. Then

$x^2 = 1 \Rightarrow x = \pm 1$ and $x^2 = -1 \Rightarrow x = \pm i$. Hence the four roots of 1 are $1, -1, i, -i$. What about the n-th roots of 1 for any given n? There will be n roots. We must have a general way of computing all the roots. This can be done through the representation of a complex number in (4.3.2):

$$z = re^{i\theta} = r[\cos\theta + \sin\theta]$$
$$= r[\cos(\theta + 2m\pi) + i\sin(\theta + 2m\pi)] = re^{i(\theta + 2m\pi)}$$

for $m = 0, 1, 2, \ldots$ since any multiple of 2π will bring back to the original position. Therefore the n roots of z are given by

$$z^{\frac{1}{n}} = r^{\frac{1}{n}} e^{\frac{1}{n}(\theta + 2m\pi)}, \quad m = 0, 1, \ldots, n-1$$

where the positive n-th root of r is taken in $r^{\frac{1}{n}}$ and the remaining part is in the exponential factor for various values of m. Note that for $m = 0, 1, \ldots, n-1$ one set of n roots are available. Then for $m = n, n+1, \ldots$ the same set of roots are repeated. Then writing

$$1 = e^{i(2m\pi)} \Rightarrow 1^{\frac{1}{n}} = e^{i(\frac{2m\pi}{n})}, \quad m = 1, 2, \ldots, n.$$

But

$$e^{i(2m\pi/n)} = \cos\frac{2m\pi}{n} + i\sin\frac{2m\pi}{n}, \quad m = 1, 2, \ldots, n.$$

The n roots of 1 are then available from the above by substituting for various values of m, $m = 1, 2, \ldots, n$. What about the n-th root of -1? Note that

$$-1 = \cos\pi + i\sin\pi = \cos(\pi + 2m\pi) + i\sin(\pi + 2m\pi)$$
$$= e^{i(2m+1)\pi}.$$

Therefore

$$(-1)^{\frac{1}{n}} = e^{(2m+1)\pi/n} = \cos\left(\frac{2m+1}{n}\right)\pi + i\sin\left(\frac{2m+1}{n}\right)\pi$$

for $m = 1, \ldots, n$. For example, the 4-th roots of 1 are

$$\cos\frac{2m}{4}\pi + i\sin\frac{2m}{4}\pi, \quad m = 1, 2, 3, 4 \Rightarrow \cos\frac{\pi}{2} + i\sin\frac{\pi}{2}, \cos\pi + i\sin\pi,$$
$$\cos\left(\pi + \frac{\pi}{2}\right) + i\sin\left(\pi + \frac{\pi}{2}\right), \cos 2\pi + i\sin 2\pi$$
$$\Rightarrow 0 + i, -1 + (0)i, 0 - i, 1$$
$$\Rightarrow 1, -1, i, -i.$$

What about the square root of -1 from the formula?

$$\cos(2m+1)\frac{\pi}{2} + i\sin(2m+1)\frac{\pi}{2}, \quad m = 1, 2$$

$$\Rightarrow \cos\frac{3\pi}{2} + i\sin\frac{3\pi}{2}, \ \cos\frac{5\pi}{2} + i\sin\frac{5\pi}{2},$$

$$\Rightarrow \cos\frac{\pi}{2} - i\sin\frac{\pi}{2}, \ \cos\frac{\pi}{2} + i\sin\frac{\pi}{2}$$

$$\Rightarrow 0 - i, 0 + i \Rightarrow i, -i$$

are the roots. What about the n-th roots of any complex number $a + ib$? Take $r = \sqrt{a^2 + b^2}$ and θ such that $\tan\theta = \frac{b}{a}$ for $a \neq 0$. If $b = 0$ then it is a real number. Then take it as a multiplied by 1 if $a > 0$ or $|a|$ multiplied by -1 if $a < 0$. Then take the positive n-th root of $|a|$ multiplied by all the n roots of 1 or -1 as the case may be. If $a = 0$ then take it as ib and proceed as above taking the n-th positive root of $|b|$ and all the n roots of i or $-i$ as the case may be.

Example 4.3.1. Compute the 4-th roots of $1 + i$.

Solution 4.3.1. We can write

$$1 + i = re^{i(\theta + 2m\pi)}, \quad \tan\theta = \frac{1}{1} = 1 \Rightarrow$$

$$\theta = \frac{\pi}{4} + 2m\pi, \quad r = \sqrt{1^2 + 1^2} = \sqrt{2}.$$

Hence

$$[1+i]^{\frac{1}{4}} = 2^{\frac{1}{8}}\left[\cos\left(\frac{\pi}{16} + \frac{2m\pi}{4}\right) + i\sin\left(\frac{\pi}{16} + \frac{2m\pi}{4}\right), \ m = 1, 2, 3, 4\right].$$

The 4 roots are

$$2^{\frac{1}{8}}\left[\cos\left(\frac{\pi}{2} + \frac{\pi}{16}\right) + i\sin\left(\frac{\pi}{2} + \frac{\pi}{16}\right), \cos\left(\pi + \frac{\pi}{16}\right) + i\sin\left(\pi + \frac{\pi}{16}\right),\right.$$

$$\left.\cos\left(\frac{3\pi}{2} + \frac{\pi}{16}\right) + i\sin\left(\frac{3\pi}{2} + \frac{\pi}{16}\right), \cos\left(2\pi + \frac{\pi}{16}\right) + i\sin\left(2\pi + \frac{\pi}{16}\right)\right].$$

That is,

$$2^{\frac{1}{8}}\left[-\sin\frac{\pi}{16} + i\cos\frac{\pi}{16}, -\cos\frac{\pi}{16} - i\sin\frac{\pi}{16}, \sin\frac{\pi}{16} - i\cos\frac{\pi}{16},\right.$$

$$\left.\cos\frac{\pi}{16} + \sin\frac{\pi}{16}\right].$$

What about the 4-th power of $1 + i$?

$$1 + i = \sqrt{2}e^{i(\frac{\pi}{4})} \Rightarrow$$

$$(1+i)^4 = 4e^{4(i\pi/4)} = 4e^{i\pi} = 4[\cos\pi + i\sin\pi] = -4.$$

Let us verify it by direct multiplication.

$$(1 + i)^2 = 1 + 2i + i^2 = 1 + 2i - 1 = 2i,$$
$$(1 + i)^4 = (2i)^2 = 4(-1) = -4.$$

Thus the representation of a complex number in terms of its length or absolute value r, and θ, such that $\tan\theta = \frac{b}{a}$, $a \neq 0$, gives a convenient way of finding the powers or roots of any complex or real number. The representations for real as well as purely imaginary numbers are the following:

$$1 = e^{i(2m\pi)}, \quad m = 0, 1, 2, 3, \ldots$$
$$-1 = e^{i(2m+1)\pi}, \quad m = 0, 1, 2, \ldots$$
$$i = e^{i(\frac{\pi}{2} + 2m\pi)}, \quad m = 0, 1, \ldots$$
$$-i = e^{i(\frac{3\pi}{2} + 2m\pi)}, \quad m = 0, 1, \ldots.$$

4.3.5 Vectors with complex elements

Let us start with the concept of vectors as n-tuples or as ordered set of n elements. If some or all elements are complex what will happen to the various operations with the vectors. Let U and V, $U' = (u_1, \ldots, u_n)$, $V' = (v_1, \ldots, u_n)$, be two $n \times 1$ vectors. The definition of a scalar multiple of a vector remains the same as in the real case:

$$cU' = (cu_1, \ldots, cu_n)$$

where c is a scalar in the real or complex field. Addition remains the same:

$$cU' + dV' = (cu_1 + dv_1, \ldots, cu_n + dv_n)$$

where c and d are scalars. The definition of linear independence remains the same. Let $U_1, \ldots, U_k$ be n-vectors defined in the complex space C^n. [We had denoted real Euclidean n-space by R^n. The complex n-space will be denoted by C^n. Since each complex variable represents a pair of real variables one can look upon C^n as corresponding to R^{2n}.] Let $c_1, \ldots, c_k$ be scalars. If the equation

$$c_1 U_1 + \cdots + c_k U_k = O$$

holds only when $c_1 = 0 = \cdots = c_k$, where O denotes the null vector, then $U_1, \ldots, U_k$ are linearly independent. Otherwise they are linearly dependent, and when they are linearly dependent then at least one of them can be written as a linear function of the others. The scalars $c_1, \ldots, c_k$ could be real or complex.

The definitions of vector subspaces, their bases and dimensions remain the same as in the real case. Since the length or absolute value of a complex number is an extension of the length in the real case the length of a vector has to be redefined when the elements are in the complex field.

Definition 4.3.4 (The length of a vector). Let

$$Z = \begin{pmatrix} z_1 \\ \vdots \\ z_n \end{pmatrix} = \begin{pmatrix} x_1 + iy_1 \\ \vdots \\ x_n + iy_n \end{pmatrix}, \quad X = \begin{pmatrix} x_1 \\ \vdots \\ x_n \end{pmatrix}, \quad Y = \begin{pmatrix} y_1 \\ \vdots \\ y_n \end{pmatrix}$$

where $x_1, \ldots, x_n, y_1, \ldots, y_n$ are all real and $i = \sqrt{-1}$. Then the length of Z, denoted by $\|Z\|$, is defined as

$$\|Z\| = \sqrt{|z_1|^2 + \cdots + |z_n|^2} = \sqrt{(x_1^2 + y_1^2) + \cdots + (x_n^2 + y_n^2)} \tag{4.3.9}$$

or $\|Z\|$ satisfies the relation

(iii) $$\|Z\|^2 = \|X\|^2 + \|Y\|^2. \tag{4.3.10}$$

For example,

$$Z = \begin{bmatrix} 2 \\ 3 - i \end{bmatrix} \Rightarrow \|Z\|^2 = [2^2] + [(3)^2 + (-1)^2] = 14;$$

$$Z = (1 + i, 2 - 3i) \Rightarrow \|Z\|^2 = [(1)^2 + (1)^2] + [(2)^2 + (-3)^2] = 15;$$

$$Z = (2, 1 - i, 4i) \Rightarrow \|Z\|^2 = [(2)^2 + (0)^2] + [(1)^2 + (-1)^2] + [(0)^2 + (4)^2]$$

$$= 22.$$

If X is an $n \times 1$ real vector then the dot product of X with X is

$$X.X = X'X = x_1^2 + \cdots + x_n^2.$$

Then length of X, denoted by $\|X\|$ is given by

$$X'X = \|X\|^2. \tag{4.3.11}$$

If this relationship is to be preserved then we should redefine dot product of two vectors in the complex field slightly differently. Then when the vectors are in the complex field the dot product will be different from that in the real case.

Definition 4.3.5 (The dot product in the complex case). Let U and V be vectors in the complex space C^n. Let U^* be the conjugate transpose of U (either take the transpose first and then the complex conjugates of every element or take the complex conjugates of every element first and then transpose the vector). The dot product of U with V is defined as

$$U.V = U^*V = u_1^c v_1 + u_2^c v_2 + \cdots + u_n^c v_n. \tag{4.3.12}$$

Then the dot product of V with U will be

$$V.U = v_1^c u_1 + \cdots + v_n^c u_n.$$

Note that $U.V$ need not be equal to $V.U$ but if all the elements of U and V are real then $U.V = V.U$. If $U = V$ then

$$U.V = U.U = u_1^c u_1 + \cdots + u_n^c u_n$$

and

$$U^* U = |u_1|^2 + \cdots + |u_n|^2 = \|U\|^2 \tag{4.3.13}$$

which is consistent with (4.3.11). For example,

(1) $U = \begin{bmatrix} 2 \\ 1-i \\ 3i \end{bmatrix}, \quad V = \begin{bmatrix} 1+i \\ 2i \\ 1-3i \end{bmatrix},$

$U^* V = 2(1+i) + (1+i)(2i) + (-3i)(1-3i) = -9+i,$

$V^* U = (1-i)(2) + (-2i)(1-i) + (1+3i)(3i) = -9-i;$

(2) $U = \begin{bmatrix} 1+2i \\ i \\ 1-i \end{bmatrix}, \quad V = \begin{bmatrix} -2+i \\ 1-i \\ 1-i \end{bmatrix},$

$U^* V = (1-2i)(-2+i) + (-i)(1-i) + (1+i)(1-i) = 1+4i,$

$V^* U = (-2-i)(1+2i) + (1+i)(i) + (1+i)(1-i) = 1-4i.$

From the above examples it is evident that

$$(U^* V)^* = V^* (U^*)^* = V^* U. \tag{4.3.14}$$

If U is changed to $c_1 U$ and V is changed to $c_2 V$, where c_1 and c_2 are scalars, then the dot product of $(c_1 U)$ with $c_2 V$ is given by

$$(c_1 U)^* (c_2 V) = c_1^c c_2 U^* V.$$

That is, c_1 is changed to its complex conjugate whereas c_2 remains as it is.

4.3.6 Matrices with complex elements

Let Z be an $m \times n$ matrix where the elements are in the complex field. Then Z will be of the following form:

$$Z = \begin{bmatrix} x_{11} + i\, y_{11} & \cdots & x_{1n} + i\, y_{1n} \\ \vdots & \cdots & \vdots \\ x_{m1} + i\, y_{m1} & \cdots & x_{mn} + i\, y_{mn} \end{bmatrix}$$

$$= X + i\, Y, \quad X = (x_{ij}), \quad Y = (y_{ij})$$

where x_{ij}'s and y_{ij}'s are real. If any of the y_{ij} is zero then the corresponding element is a real quantity. If any of the x_{ij} is zero then the corresponding element is a purely imaginary quantity.

Definition 4.3.6 (Complex conjugate of a matrix). Let $Z = X + iY$ be an $m \times n$ matrix in the complex field. Then $Z^c = X - iY$ is called its complex conjugate. That is, the complex conjugate of every element in Z is taken or if $Z = (z_{ij})$ then $Z^c = (z_{ij}^c)$.

Definition 4.3.7 (The conjugate transpose of a matrix). The transpose of the conjugate matrix is called the conjugate transpose of the matrix. If $Z = X + iY$ then its conjugate transpose, denoted by Z^*, is given by $Z^* = X' - iY'$ where the primes denote the transposes. That is,

$$Z^* = \left[(X + iY)^c\right]' = [X - iY]' = X' - iY' = \left[(X + iY)'\right]^c.$$

For example,

$$
\begin{array}{cc}
\text{a matrix in the complex field} & \text{its conjugate} \\[4pt]
\begin{bmatrix} 1 & 0 & 2-i \\ 1+i & 3 & 1-i \\ 2 & 3+i & 4-i \end{bmatrix}, & \begin{bmatrix} 1 & 0 & 2+i \\ 1-i & 3 & 1+i \\ 2 & 3-i & 4+i \end{bmatrix}, \\[30pt]
(2+i, 1, 1-i) & (2-i, 1, 1+i) \\[8pt]
\text{its conjugate transpose} & \text{its conjugate transpose} \\[4pt]
\begin{bmatrix} 1 & 1-i & 2 \\ 0 & 3 & 3-i \\ 2+i & 1+i & 4+i \end{bmatrix} & \begin{pmatrix} 2-i \\ 1 \\ 1+i \end{pmatrix}
\end{array}
$$

Let us see what happens if a square matrix in the complex field is equal to its conjugate transpose. Let

$$Z = X + iY \implies Z^* = X' - iY'.$$

If $Z = Z^*$ then $X = X'$ and $Y = -Y'$. That is, the real part of the matrix has to be symmetric and the imaginary part of the matrix has to be skew symmetric. Such matrices are called Hermitian matrices.

Definition 4.3.8 (A Hermitian matrix). If Z is equal to its conjugate transpose Z^* then Z is called a Hermitian matrix.

Definition 4.3.9 (A skew Hermitian matrix). If Z is equal to (-1) times its conjugate transpose, that is $Z = -Z^*$, then Z is called skew Hermitian.

Thus if Z is skew Hermitian then the real part is skew symmetric and the imaginary part is symmetric.

(iv) The diagonal elements of a Hermitian matrix are real. The diagonal elements of a skew Hermitian matrix are purely imaginary or zero. The elements above the leading diagonal of a Hermitian matrix are the complex conjugates of the corresponding elements below the leading diagonal. The elements above the leading diagonal of a skew Hermitian matrix are minus one times the complex conjugates of the corresponding elements below the leading diagonal.

Let

$$Z = X + iY, \quad X = (x_{ij}), \quad Y = (y_{ij})$$

$$Z = Z^* \ \Rightarrow \ x_{ij} = x_{ji}, \, y_{ij} = -y_{ji}, \, y_{ii} = 0, \quad \text{for all } i \text{ and } j;$$

$$Z = -Z^* \ \Rightarrow \ x_{ij} = -x_{ji}, \, y_{ij} = y_{ji}, \, x_{ii} = 0, \quad \text{for all } i \text{ and } j.$$

For example,

(a) $\quad Z = \begin{bmatrix} 2 & 1+i & 2-i \\ 1-i & 5 & 3+2i \\ 2+i & 3-2i & 6 \end{bmatrix}, \quad Z$ is Hermitian, $\quad Z^* = Z$;

(b) $\quad Z = \begin{bmatrix} 1 & 5+4i \\ 5-4i & 7 \end{bmatrix}, \quad Z$ is Hermitian, $\quad Z^* = Z$;

(c) $\quad Z = \begin{bmatrix} 3i & 2-i \\ -2-i & -5i \end{bmatrix}, \quad Z$ is skew Hermitian, $\quad Z^* = -Z$;

(d) $\quad Z = \begin{bmatrix} 0 & 1+i & 2+i \\ -1+i & 2i & 1-i \\ -2+i & -1-i & -5i \end{bmatrix}, \quad Z$ is skew Hermitian, $\quad Z^* = -Z$.

Some properties are immediate for conjugate transposes, denoted by $*$.

$$(A + B)^* = A^* + B^*;$$

$$(AB)^* = B^* A^*;$$

$$(c_1 A + c_2 B)^* = c_1^c A^* + c_2^c B^*, \quad c_1, c_2 \text{ scalars.}$$

For $n \times 1$ vectors U and V

$$U^* U = \|U\|^2 = |u_1|^2 + \cdots + |u_n|^2 \quad \text{(square of the length)};$$

$$U^* V \neq V^* U, \quad (U^* V)^* = V^* U;$$

$$U^* V = 0 \ \Rightarrow \ U \text{ is orthogonal to } V;$$

$$U^* U = 1 \ \Rightarrow \ \text{is a normal vector};$$

$$U^* V = V^* U \text{ if } U^* V \text{ is real.}$$

Definition 4.3.10 (Orthogonality). Let U and V be two vectors in the complex space C^n. Then U is said to be orthogonal to V if $U^*V = 0$.

Note that $(U^*V)^* = V^*U = 0^* = 0$. Hence the condition $U^*V = 0$ also implies $V^*U = 0$. That is, if U is orthogonal to V then V is orthogonal to U or they are orthogonal to each other.

Definition 4.3.11 (Orthonormal system). Let $U_1, \ldots, U_k$ be k vectors in C^n. They form an orthonormal system if $U_i^*U_j = 0$ for all i and j, $i \neq j$ and $\|U_i\| = 1$ for $i = 1, 2, \ldots, k$.

Definition 4.3.12 (A unitary matrix). Consider an $n \times n$ matrix Q whose columns are orthonormal. Then Q is called a unitary matrix, $Q^*Q = I$ or

$$Q^*Q = I \implies Q^{-1} = Q^* \implies I = QQ^*.$$

When the columns are orthonormal the rows are also orthonormal for an $n \times n$ unitary matrix. For example,

(a) $\quad Q = \begin{bmatrix} \frac{1+i}{2} & \frac{1-i}{2} \\ \frac{-1-i}{2} & \frac{1-i}{2} \end{bmatrix}$,

$\quad Q^* = \begin{bmatrix} \frac{1-i}{2} & \frac{-1+i}{2} \\ \frac{1+i}{2} & \frac{1+i}{2} \end{bmatrix}$, $\quad QQ^* = I_2, \quad Q^*Q = I_2$;

(b) $\quad Q = \begin{bmatrix} \frac{1+i}{\sqrt{6}} & \frac{1+i}{\sqrt{6}} & \frac{1+i}{\sqrt{6}} \\ \frac{1-i}{2} & 0 & \frac{-1+i}{2} \\ \frac{1}{\sqrt{6}} & -\frac{2}{\sqrt{6}} & \frac{1}{\sqrt{6}} \end{bmatrix}$,

$\quad Q^* = \begin{bmatrix} \frac{1-i}{\sqrt{6}} & \frac{1+i}{2} & \frac{1}{\sqrt{6}} \\ \frac{1-i}{\sqrt{6}} & 0 & -\frac{2}{\sqrt{6}} \\ \frac{1-i}{\sqrt{6}} & \frac{-1-i}{2} & \frac{1}{\sqrt{6}} \end{bmatrix}$, $\quad QQ^* = I, \quad Q^*Q = I_3$.

Definition 4.3.13 (Semiunitary matrices). Let Q be an $n \times n$ unitary matrix such that $Q^*Q = I$, $QQ^* = I$. Consider m columns of Q, $m < n$, for example the first m columns $U_1, \ldots, U_m$. Let S be the matrix formed by these m vectors, $S = (U_1, \ldots, U_m)$. Then $S^*S = I_m$ and S is called a semiunitary matrix or an element in the Stiefel manifold.

Note that $SS^* \neq I_n$ and SS^* is an $n \times n$ matrix whereas S^*S is an $m \times m$ matrix. We could have also taken m row vectors $V_1, \ldots, V_m$ of Q and create the matrix $S_1 = \begin{pmatrix} V_1 \\ \vdots \\ V_m \end{pmatrix}$. Then $S_1 S_1^* = I_m$ whereas $S_1^* S_1 \neq I_n$. Thus S_1 is also a semiunitary matrix. Semiunitary matrix is a matrix formed with a subset of a full orthonormal system of vectors. In our illustrative examples above,

(1) $S = \begin{bmatrix} \frac{1+i}{\sqrt{6}} & \frac{1+i}{\sqrt{6}} \\ \frac{1-i}{2} & 0 \\ \frac{1}{\sqrt{6}} & -\frac{2}{\sqrt{6}} \end{bmatrix}$, $S^*S = I_2$, S is semiunitary;

(2) $S = \begin{bmatrix} \frac{1+i}{\sqrt{6}} & \frac{1+i}{\sqrt{6}} \\ \frac{1-i}{2} & \frac{-1+i}{2} \\ \frac{1}{\sqrt{6}} & \frac{1}{\sqrt{6}} \end{bmatrix}$, $S^*S = I_2$, S is semiunitary;

(3) $S_1 = \begin{bmatrix} \frac{1+i}{\sqrt{6}} & \frac{1+i}{\sqrt{6}} & \frac{1+i}{\sqrt{6}} \\ \frac{1-i}{2} & 0 & \frac{-1+i}{2} \end{bmatrix}$, $S_1 S_1^* = I_2$, S_1 is semiunitary;

(4) $S_1 = \begin{bmatrix} \frac{1-i}{2} & 0 & \frac{-1+i}{2} \\ \frac{1}{\sqrt{6}} & -\frac{2}{\sqrt{6}} & \frac{1}{\sqrt{6}} \end{bmatrix}$, $S_1 S_1^* = I_2$, S_1 is semiunitary.

Exercises 4.3

4.3.1. Mark the following complex numbers as points in an (x,y)-coordinate system. $2 + 4i, 2 - 4i, -2 + 4i, -2 - 4i, 1 + i, i, -2, -i$.

4.3.2. Compute the following: (a) $(z_1 + z_2) + z_3$, (b) $z_1 + (z_2 + z_3)$ where $z_1 = 1 + 3i$, $z_2 = 2 - i, z_3 = 2 + 4i$.

4.3.3. For z_1, z_2, z_3 in Exercise 4.3.2 compute the following:

(1) $z_1 z_2$,   (2) $z_1 z_2^c$,   (3) $z_1^c z_2$,   (4) $z_1^c z_2^c$,   (5) $z_1 z_2 z_3$,
(6) $z_1^c z_2^c z_3^c$,   (7) $[z_1 z_2 z_3]^c$,   (8) $\frac{z_1}{(z_2 z_3)^2}$,   (9) $\left(\frac{z_1 z_3}{z_2}\right)^{\frac{1}{6}}$,

all roots.

4.3.4. Compute the following for $z_1 = 2i, z_2 = 1 - i, z_3 = 4$:

(1) $(z_1 z_2 z_3)^{\frac{1}{5}}$,   all 5 roots,   (2) $(z_1^c z_2^2 z_3)^{\frac{2}{3}}$,   all roots,
(3) $\left(\frac{z_1 z_3}{z_2}\right)^4$,   (4) $\left(\frac{z_1 z_3}{z_2}\right)^{\frac{1}{6}}$,   all roots.

4.3.5. For the following vectors compute (1) the conjugate, (2) the conjugate transpose

(a) $[1 + 2i, 3i, 4]$,   (b) $\begin{bmatrix} -i \\ 2 + 3i \\ 4i \end{bmatrix}$,

(c) $[1, -1, 2i, 0, 4 + 2i]$.

4.3.6. Compute the lengths of the vectors in Exercise 4.3.5.

4.3.7. Mark the 5-th roots (all roots) of $1, -1, i, -i$ on the same graph.

4.3.8. Take the 6-th roots of 1. Let the roots be denoted by $1, w_1, w_2, \ldots, w_5$ where $w_1 = e^{i\pi/3}$. Then show the following:

(1) $w_j = w_1^j, \ j = 2, 3, 4, 5$,

(2) $1 + w_1 + \cdots + w_5 = 0$,

(3) plot all the 6 roots in the same graph.

4.3.9. With $w_1 = e^{i2\pi/n}$ show that the properties in Exercise 4.3.8 (1) and (2) hold for any n and plot the points $1, w_1, \ldots, w_{n-1}$ and examine their relations to a circle of unit radius in the complex plane.

4.3.10. Show that the Fourier matrix

$$F = \begin{bmatrix} 1 & 1 & 1 & 1 \\ 1 & i & i^2 & i^3 \\ 1 & i^2 & i^4 & i^6 \\ 1 & i^3 & i^6 & i^9 \end{bmatrix} \quad \text{has the inverse}$$

$$F^{-1} = \frac{1}{4} \begin{bmatrix} 1 & 1 & 1 & 1 \\ 1 & (-i) & (-i)^2 & (-i)^3 \\ 1 & (-i)^2 & (-i)^4 & (-i)^6 \\ 1 & (-i)^3 & (-i)^6 & (-i)^9 \end{bmatrix}.$$

4.3.11. Show the following: If

$$F_2 = \begin{bmatrix} 1 & 1 \\ 1 & w_2 \end{bmatrix} \quad \text{then } F_2^{-1} = \frac{1}{2} \begin{bmatrix} 1 & 1 \\ 1 & w_2^{-1} \end{bmatrix}, \quad w_n = e^{i2\pi/n},$$

$$F_3 = \begin{bmatrix} 1 & 1 & 1 \\ 1 & w_3 & w_3^2 \\ 1 & w_3^2 & w_3^4 \end{bmatrix} \quad \text{then } F_3^{-1} = \frac{1}{3} \begin{bmatrix} 1 & 1 & 1 \\ 1 & w_3^{-1} & w_3^{-2} \\ 1 & w_3^{-2} & w_3^{-4} \end{bmatrix},$$

$$F_n = \begin{bmatrix} 1 & 1 & 1 & \cdots & 1 \\ 1 & w_n & w_n^2 & \cdots & w_n^{n-1} \\ 1 & w_n^2 & w_n^4 & \cdots & w_n^{2(n-1)} \\ \vdots & \vdots & \vdots & \cdots & \vdots \\ 1 & w_n^{n-1} & w_n^{2(n-1)} & \cdots & w_n^{(n-1)^2} \end{bmatrix}$$

then

$$F_n^{-1} = \frac{1}{n} \begin{bmatrix} 1 & 1 & 1 & \cdots & 1 \\ 1 & w_n^{-1} & w_n^{-2} & \cdots & w_n^{-(n-1)} \\ 1 & w_n^{-2} & w_n^{-4} & \cdots & w_n^{-2(n-1)} \\ \vdots & \vdots & \vdots & \cdots & \vdots \\ 1 & w_n^{-(n-1)} & w_n^{-2(n-1)} & \cdots & w_n^{-(n-1)^2} \end{bmatrix}.$$

4.3.12. Check whether the following system of vectors are linearly independent:

(a) $U_1 = (1 - i, 1 + i, 2)$, $U_2 = (1 + i, 1 - i, 5)$,
$U_3 = (2 + i, 3, 1 - i)$;

(b) $U_1 = (1 - i, 1 + i, 2)$, $U_2 = (2 + 3i, 2 + i, 1 - i)$,
$U_3 = (4 + i, 4 + 2i, 5 - i)$.

4.3.13. Construct four 4×1 vectors in C^4 such that they form an orthonormal (unitary) system.

4.3.14. Construct 3 orthonormal 4×1 vectors U_1, U_2, U_3 in C^4 such that

$$(U_1, U_2, U_3)^* (U_1, U_2, U_3) = I_3.$$

4.3.15. Construct 3 examples each of (1) 2×2, (3) 3×3, (3) 4×4 Hermitian matrices.

4.3.16. Construct 3 examples each of (1) 2×2, (2) 3×3, (3) 4×4 skew Hermitian matrices.

4.3.17. Let A be an $n \times n$ matrix whose elements are in the complex field. Show that one can always construct a Hermitian matrix B as a function of A.

4.3.18. Let A be an $n \times n$ matrix whose elements are in the complex field. Show that one can always construct a skew Hermitian matrix C as a function of A.

4.3.19. If A is an $n \times n$ matrix and A^* its conjugate transpose then show that (1) $B = (A + A^*)/2$ is Hermitian and (2) $C = (A - A^*)/2$ is skew Hermitian.

4.3.20. Compute the eigenvalues and the corresponding eigenvectors of the matrix

$$A = \begin{bmatrix} 1 & 1-i \\ -1 & 2+i \end{bmatrix}.$$

4.3.21. Compute the eigenvalues of the matrices in one example each in (1) and (2) of Exercise 4.3.15. Is there any interesting property that you see for these eigenvalues?

4.3.22. Compute the eigenvalues of the matrices in one example each in (1) and (2) of Exercise 4.3.16. Is there any interesting property for these eigenvalues? What can you say about the singularity of the matrices in your examples?

4.3.23. If A is Hermitian show that A^k is Hermitian for k a positive integer.

4.3.24. If A^2 is Hermitian is A Hermitian? Justify your answers.

4.3.25. Find the rank of the following matrix:

$$A = \begin{bmatrix} 2+3i & 2 & 5-i \\ 2-i & 1+i & 3 \\ 2i & -4 & 1-i \end{bmatrix}$$

and if it is nonsingular then evaluate the regular inverse of A.

4.3.26. Find nonsingular matrices R and S such that $B = RAS$ where

$$A = \begin{bmatrix} 1+i & 2 \\ 3-i & i \end{bmatrix}, \quad B = \begin{bmatrix} 2i & -3 \\ 3+i & 1-i \end{bmatrix}.$$

4.3.27. Solve the following system of linear equations:

$$(2+i)x_1 - x_2 + (5+i)x_3 = 1-i$$
$$3ix_1 + (1+i)x_2 - x_3 = 2+i$$
$$x_1 - x_2 + (1+i)x_3 = 7.$$

4.3.28. Let A and B be $n \times n$ matrices. Show that AB and BA have the same characteristic polynomials and the same eigenvalues.

4.3.29. For square matrices A and B if $AB = BA$ then show that A and B have at least one common eigenvector.

4.3.30. If λ_0 is an eigenvalue of an $n \times n$ matrix A and if $P(\lambda)$ is any polynomial in λ then show that $P(\lambda_0)$ is an eigenvalue of $P(A)$. If

$$P(\lambda) = b_0 + b_1\lambda + \cdots + b_k\lambda^k \quad \text{then } P(A) = b_0 I + b_1 A + \cdots + b_k A^k.$$

4.4 More properties of matrices in the complex field

Since we have introduced vectors and matrices whose elements are in the complex field we are in a better position to derive more properties of matrices. One of the properties that we would like to investigate is concerned with the eigenvalues of symmetric and Hermitian matrices. What about the eigenvalues of a real symmetric matrix? From our numerical examples in Sections 4.1 and 4.2 we have seen that even if the elements of a matrix are real their eigenvalues as well as eigenvectors could be in the complex space.

4.4.1 Eigenvalues of symmetric and Hermitian matrices

If we confine our discussion to symmetric matrices with real elements (real symmetric matrices) do we have any interesting results? Let $A = A'$ and real. Let λ_1 be an eigenvalue of A and X_1 a corresponding eigenvector. Then

$$AX_1 = \lambda_1 X_1 \implies X_1^* A^* = \lambda_1^c X_1^* \tag{a}$$

where a $*$ indicates the conjugate transpose and c indicates the complex conjugate. But since A is real and symmetric $A^* = A$ itself. Let us postmultiply (a) by X_1. Then

$$X_1^* A X_1 = \lambda_1^c X_1^* X_1. \tag{b}$$

But $X_1^* X_1 = \|X_1\|^2 =$ a real nonzero quantity. Now premultiply $AX_1 = \lambda_1 X_1$ by X_1^* to obtain

$$X_1^* AX_1 = \lambda_1 X_1^* X_1. \tag{c}$$

From (b) and (c)

$$(\lambda_1 - \lambda_1^c) X_1^* X_1 = 0 \implies \lambda_1 = \lambda_1^c$$

since $X_1^* X_1 = \|X_1\|^2 \neq 0$. Then $\lambda_1 = \lambda_1^c$ means that λ_1 is real. Thus we have the following important property.

(i) The eigenvalues of a real symmetric matrix are all real.

Since Hermitian matrices in the complex field and the real symmetric matrices have many parallel properties let us look at the eigenvalues of a Hermitian matrix A. If A is Hermitian then $A = A^*$. Let

$$A = A_1 + iA_2 \quad \text{then } A = A^* \implies A_1 = A_1', \quad A_2' = -A_2.$$

Let λ_1 be an eigenvalue of A and X_1 a corresponding eigenvector, λ_1^c the complex conjugate of λ_1 and X_1^* the conjugate transpose of X_1. Then

$$AX_1 = \lambda_1 X_1 \implies X_1^* A^* = \lambda_1^c X_1^*,$$
$$\implies X_1^* A = \lambda_1^c X_1^* \quad \text{since } A = A^*$$
$$X_1^* AX_1 = \lambda_1^c X_1^* X_1. \tag{d}$$

From $AX_1 = \lambda_1 X_1$ we have

$$X_1^* AX_1 = \lambda_1 X_1^* X_1. \tag{e}$$

From (d) and (e), $\lambda_1 = \lambda_1^c$ or λ_1 is real. Thus we have another important result. Note that the step in (d) above holds if the matrix A is real symmetric or Hermitian symmetric. If A is symmetric but with some of the elements complex then (d) and (e) need not hold. Then when you take the conjugate transpose it need not be equal to the original matrix. Hence symmetry is not sufficient. Either A should be real and symmetric or Hermitian. Then only we can guarantee that the eigenvalues are real.

(ii) The eigenvalues of a Hermitian matrix are all real.

Example 4.4.1. Compute the eigenvalues and the eigenvectors of

$$A = \begin{bmatrix} 2 & 1+i \\ 1-i & 3 \end{bmatrix}.$$

Solution 4.4.1. Note that A is Hermitian since $A = A^*$, and hence we can expect real eigenvalues. Consider

$$|A - \lambda I| = 0 \implies \begin{vmatrix} 2 - \lambda & 1 + i \\ 1 - i & 3 - \lambda \end{vmatrix} = 0$$

$$\implies (2 - \lambda)(3 - \lambda) - (1 + i)(1 - i) = 0$$

$$\implies \lambda^2 - 5\lambda + 4 = 0 \quad \text{or} \quad \lambda_1 = 1, \quad \lambda_2 = 4$$

are the eigenvalues. For $\lambda_1 = 1$,

$$(A - \lambda_1 I)X = 0 \implies \begin{bmatrix} 2 - 1 & 1 + i \\ 1 - i & 3 - 1 \end{bmatrix} \begin{bmatrix} x_1 \\ x_2 \end{bmatrix} = \begin{bmatrix} 0 \\ 0 \end{bmatrix}$$

$$\implies \begin{bmatrix} 1 & 1 + i \\ 1 - i & 2 \end{bmatrix} \begin{bmatrix} x_1 \\ x_2 \end{bmatrix} = \begin{bmatrix} 0 \\ 0 \end{bmatrix}.$$

Are the two rows of $A - \lambda_1 I$ linearly dependent? They must be, otherwise we made some computational errors somewhere. Multiply the first row by $(1 - i)$, that is,

$$[1 - i, (1 - i)(1 + i)] = [1 - i, 2]$$

which is the second row. Taking the first equation

$$x_1 + (1 + i)x_2 = 0 \implies X_1 = \begin{pmatrix} -1 - i \\ 1 \end{pmatrix}$$

is one solution. Now, consider $\lambda_2 = 4$.

$$(A - \lambda_2 I)X = 0 \implies \begin{bmatrix} 2 - 4 & 1 + i \\ 1 - i & 3 - 4 \end{bmatrix} \begin{bmatrix} x_1 \\ x_2 \end{bmatrix} = \begin{bmatrix} 0 \\ 0 \end{bmatrix}$$

$$\implies (1 - i)x_1 - x_2 = 0$$

$$\implies X_2 = \begin{pmatrix} 1 \\ 1 - i \end{pmatrix}$$

is a vector corresponding to λ_2. A matrix of eigenvectors in this case is given by

$$Q = \begin{bmatrix} -1 - i & 1 \\ 1 & 1 - i \end{bmatrix} \quad \text{with its inverse } Q^{-1} = \begin{bmatrix} \frac{-1+i}{3} & \frac{1}{3} \\ \frac{1}{3} & \frac{1+i}{3} \end{bmatrix}.$$

Hence in this case we can have a representation of the form $A = Q \Lambda Q^{-1}$, that is,

$$\begin{bmatrix} 2 & 1 + i \\ 1 - i & 3 \end{bmatrix} = \begin{bmatrix} -1 - i & 1 \\ 1 & 1 - i \end{bmatrix} \begin{bmatrix} 1 & 0 \\ 0 & 4 \end{bmatrix} \begin{bmatrix} \frac{-1+i}{3} & \frac{1}{3} \\ \frac{1}{3} & \frac{1+i}{3} \end{bmatrix}.$$

Example 4.4.2. Compute the eigenvalues and eigenvectors of

$$A = \begin{bmatrix} 2i & 1+i \\ -1+i & 3i \end{bmatrix}.$$

Solution 4.4.2. Note that our matrix A here is skew Hermitian. Consider

$$|A - \lambda I| = 0 \Rightarrow \begin{vmatrix} 2i - \lambda & 1+i \\ -1+i & 3i - \lambda \end{vmatrix} = 0$$

$$\Rightarrow \lambda^2 - 5i\lambda - 4 = 0$$

$$\Rightarrow \lambda = \frac{5i \pm \sqrt{(5i)^2 - (4)(-4)}}{2} = \frac{5i \pm 3i}{2}$$

$$\Rightarrow \lambda_1 = i, \quad \lambda_2 = 4i$$

are the eigenvalues. An eigenvector corresponding to $\lambda_1 = i$ is available from

$$(A - \lambda_1 I)X = 0 \Rightarrow \begin{bmatrix} 2i - i & 1+i \\ -1+i & 3i - i \end{bmatrix} \begin{bmatrix} x_1 \\ x_2 \end{bmatrix} = \begin{bmatrix} 0 \\ 0 \end{bmatrix}$$

$$\Rightarrow ix_1 + (1+i)x_2 = 0.$$

For $x_2 = 1$, $x_1 = -\frac{1+i}{i} = -\frac{(1+i)(i)}{i^2} = -1 + i$. Hence one vector is

$$X_1 = \begin{pmatrix} -1+i \\ 1 \end{pmatrix}.$$

Now, consider

$$(A - \lambda_2 I)X = 0 \Rightarrow (-1+i)x_1 - ix_2 = 0.$$

For $x_2 = 1$, $x_1 = \frac{i}{-1+i} = \frac{i(-1-i)}{2} = \frac{1-i}{2}$. One vector corresponding to $\lambda_2 = 4i$ is then

$$X_2 = \begin{pmatrix} \frac{1-i}{2} \\ 1 \end{pmatrix}.$$

A matrix of eigenvectors is therefore

$$Q = \begin{bmatrix} -1+i & \frac{1-i}{2} \\ 1 & 1 \end{bmatrix} \quad \text{with } Q^{-1} = \frac{2}{3} \begin{bmatrix} -\frac{1}{1-i} & \frac{1}{2} \\ \frac{1}{1-i} & 1 \end{bmatrix}.$$

Note that A can be written as $A = Q \Lambda Q^{-1}$ in this case also. (Verification is left to the student.)

One observation can be made from this example. We get the eigenvalues of this skew Hermitian matrix as purely imaginary. Is this a general property? Let us investigate this further. Let A be a general skew Hermitian matrix. Then

$$A^* = -A, \quad A = A_1 + iA_2 \Rightarrow A_1' = -A_1, \quad A_2' = A_2.$$

Let X_1 be an eigenvector of A corresponding to an eigenvalue λ_1. Then

$$AX_1 = \lambda_1 X_1 \Rightarrow X_1^* AX_1 = \lambda_1 X_1^* X_1. \tag{a}$$

$$(AX_1)^* = X_1^* A^* = -X_1^* A \Rightarrow -X_1^* AX_1 = \lambda_1^c X_1^* X_1. \tag{b}$$

From (a) and (b) we have, since $X_1^* X_1 = \|X_1\|^2 \neq 0$,

$$\lambda_1 + \lambda_1^c = 0.$$

This means the real part of λ_1 is zero or $\lambda_1 = 0$. Thus the eigenvalues of a skew Hermitian matrix have to be purely imaginary or zero.

(iii) The eigenvalues of a skew Hermitian matrix are purely imaginary or zero.
(iv) The eigenvalues of a real skew symmetric matrix are purely imaginary or zero.

Since the complex roots appear in pairs we have an interesting result. There cannot be an odd number of purely imaginary roots. Then if the order n is odd and if the matrix is skew symmetric or skew Hermitian then it must be singular because at least one root must be zero.

(v) If an $n \times n$ matrix A is real skew symmetric or skew Hermitian then A is singular if n is odd and the determinant of A, being product of eigenvalues, is the square of real number if n is even. $I + A$ and $I - A$ are nonsingular.

Let us examine the eigenvalues of a unitary matrix. Let Q be a unitary matrix. Then

$$QQ^* = I, \quad Q^*Q = I.$$

Let λ_1 be an eigenvalue of Q and X_1 an eigenvector corresponding to λ_1. Then

$$QX_1 = \lambda_1 X_1 \Rightarrow X_1^* Q^* = \lambda_1^c X_1^* \tag{α}$$

$$\Rightarrow X_1^* Q^* X_1 = \lambda_1^c X_1^* X_1.$$

From (α), premultiplying the first part by Q^*, we have

$$X_1 = \lambda_1 Q^* X_1 \Rightarrow X_1^* X_1 = \lambda_1 X_1^* Q^* X_1 = \lambda_1 \lambda_1^c X_1^* X_1$$

$$\Rightarrow \lambda_1 \lambda_1^c = 1 \Rightarrow |\lambda_1|^2 = 1.$$

(vi) The eigenvalues of a unitary matrix are such that the absolute value of the roots are 1, that is, if λ is a root then $\lambda \lambda^c = 1$.
(vii) Every real square matrix can be written as a sum of a symmetric and a skew symmetric matrix. [Let $B = \frac{A+A'}{2}$ and $C = \frac{A-A'}{2}$, $B' = B$, $C' = -C$.]

(viii) Every square matrix can be written as the sum of a Hermitian and a skew Hermitian matrix.
(ix) Every square matrix can be written as a sum of two nonsingular matrices.

We had seen from Chapter 2 that every square matrix can be written as $A = PDQ$ through elementary operations, where P and Q are nonsingular, D is diagonal and A is $n \times n$. Let $D = \text{diag}(d_1, \dots, d_r, 0, \dots, 0)$ where $d_j \neq 0, j = 1, \dots, r$. Write, for example,

$$D = D_1 + D_2, \quad D_1 = \text{diag}\left(\frac{d_1}{2}, \dots, \frac{d_r}{2}, d_{r+1}, \dots, d_n\right),$$

$$D_2 = \text{diag}\left(\frac{d_1}{2}, \dots, \frac{d_r}{2}, -d_{r+1}, \dots, -d_n\right)$$

where $d_j \neq 0, j = r + 1, \dots, n$. Then both PD_1Q and PD_2Q are nonsingular and the sum is A.

(x) If P and Q are unitary $n \times n$ matrices then PQ as well as QP are unitary matrices.
(xi) If the determinant of A, that is, $|A| = a + ib$ then $|A^*| = a - ib$ which means that $|A|$ is real if $A = A^*$ (Hermitian).
(xii) Complex eigenvalues appear in pairs. If $a + ib$ is an eigenvalue of a given matrix A then $a - ib$ is also an eigenvalue of A.

Example 4.4.3. Express A as a sum of a symmetric and a skew symmetric matrix and B as the sum of a Hermitian and a skew Hermitian matrix, where

$$A = \begin{bmatrix} 1 & 2 & -2 \\ 3 & 1 & 5 \\ 2 & 4 & -2 \end{bmatrix}, \quad B = \begin{bmatrix} 2+3i & 1-i & 2+i \\ 1+2i & 3+i & 1+2i \\ 2-i & 2+5i & 7i \end{bmatrix}.$$

Solution 4.4.3. Let

$$A_1 = \frac{1}{2}[A + A'] = \frac{1}{2}\begin{bmatrix} 1 & 2 & -2 \\ 3 & 1 & 5 \\ 2 & 4 & -2 \end{bmatrix} + \frac{1}{2}\begin{bmatrix} 1 & 3 & 2 \\ 2 & 1 & 4 \\ -2 & 5 & -2 \end{bmatrix}$$

$$= \begin{bmatrix} 1 & \frac{5}{2} & 0 \\ \frac{5}{2} & 1 & \frac{9}{2} \\ 0 & \frac{9}{2} & -2 \end{bmatrix} = A_1'.$$

Let

$$A_2 = \frac{1}{2}[A - A'] = \frac{1}{2}\begin{bmatrix} 1 & 2 & -2 \\ 3 & 1 & 5 \\ 2 & 4 & -2 \end{bmatrix} - \frac{1}{2}\begin{bmatrix} 1 & 3 & 2 \\ 2 & 1 & 4 \\ -2 & 5 & -2 \end{bmatrix}$$

$$= \begin{bmatrix} 0 & -\frac{1}{2} & -2 \\ \frac{1}{2} & 0 & \frac{1}{2} \\ 2 & -\frac{1}{2} & 0 \end{bmatrix} = -A_2', \quad A = A_1 + A_2.$$

Let

$$B_1 = \frac{1}{2}[B + B^*]$$

$$= \frac{1}{2}\begin{bmatrix} 2+3i & 1-i & 2+i \\ 1+2i & 3+i & 1+2i \\ 2-i & 2+5i & 7i \end{bmatrix} + \frac{1}{2}\begin{bmatrix} 2-3i & 1-2i & 2+i \\ 1+i & 3-i & 2-5i \\ 2-i & 1-2i & -7i \end{bmatrix}$$

$$= \begin{bmatrix} 2 & 1-\frac{3}{2}i & 2+i \\ 1+\frac{3}{2}i & 3 & \frac{3}{2}-\frac{3}{2}i \\ 2-i & \frac{3}{2}+\frac{3}{2}i & 0 \end{bmatrix} = B_1^*.$$

Let

$$B_2 = \frac{1}{2}[B - B^*] = \begin{bmatrix} 3i & \frac{1}{2}i & 0 \\ \frac{1}{2}i & i & -\frac{1}{2}+\frac{7}{2}i \\ 0 & \frac{1}{2}+\frac{7}{2}i & 7i \end{bmatrix} = -B_2^*, \quad B = B_1 + B_2.$$

Example 4.4.4. Write the following matrices as the sum of two nonsingular matrices:

$$A = \begin{bmatrix} 1 & 0 & 1 \\ 2 & 1 & -1 \\ 0 & 1 & 2 \end{bmatrix}, \quad B = \begin{bmatrix} 1 & 0 & 1 \\ 2 & 1 & -1 \\ 3 & 1 & 0 \end{bmatrix}.$$

Solution 4.4.4. By inspection $|A| \neq 0$, $|B| = 0$. That is, A is nonsingular and B is singular. Then A can always be written as

$$A = A_1 + A_2, \quad A_1 = \alpha A, \quad A_2 = (1-\alpha)A, \quad \alpha \neq 0,$$

where both A_1 and A_2 are nonsingular. Hence we look at B. One way of doing it is to look for a representation $B = QDQ^{-1}$ where Q is a matrix of eigenvectors. For this procedure we need the eigenvectors and we cannot tell in advance whether Q will be nonsingular. Hence consider pre and post multiplications by elementary matrices. This process will always produce two nonsingular matrices. Let

$$F_1 = \begin{bmatrix} 1 & 0 & 0 \\ -2 & 1 & 0 \\ 0 & 0 & 1 \end{bmatrix}, \quad F_1^{-1} = \begin{bmatrix} 1 & 0 & 0 \\ 2 & 1 & 0 \\ 0 & 0 & 1 \end{bmatrix},$$

$$F_1 B = \begin{bmatrix} 1 & 0 & 1 \\ 0 & 1 & -3 \\ 3 & 1 & 0 \end{bmatrix}, \quad F_2 = \begin{bmatrix} 1 & 0 & 0 \\ 0 & 1 & 0 \\ -3 & 0 & 1 \end{bmatrix},$$

$$F_2^{-1} = \begin{bmatrix} 1 & 0 & 0 \\ 0 & 1 & 0 \\ 3 & 0 & 1 \end{bmatrix}, \quad F_2 F_1 B = \begin{bmatrix} 1 & 0 & 1 \\ 0 & 1 & -3 \\ 0 & 1 & -3 \end{bmatrix}$$

$$F_3 = \begin{bmatrix} 1 & 0 & 0 \\ 0 & 1 & 0 \\ 0 & -1 & 1 \end{bmatrix}, \quad F_3^{-1} = \begin{bmatrix} 1 & 0 & 0 \\ 0 & 1 & 0 \\ 0 & 1 & 1 \end{bmatrix},$$

$$F_3F_2F_1B = \begin{bmatrix} 1 & 0 & 1 \\ 0 & 1 & -3 \\ 0 & 0 & 0 \end{bmatrix},$$

$$F_4 = \begin{bmatrix} 1 & 0 & -1 \\ 0 & 1 & 0 \\ 0 & 0 & 1 \end{bmatrix}, \quad F_4^{-1} = \begin{bmatrix} 1 & 0 & 1 \\ 0 & 1 & 0 \\ 0 & 0 & 1 \end{bmatrix},$$

$$F_3F_2F_1BF_4 = \begin{bmatrix} 1 & 0 & 0 \\ 0 & 1 & -3 \\ 0 & 0 & 0 \end{bmatrix}, \quad F_5 = \begin{bmatrix} 1 & 0 & 0 \\ 0 & 1 & 3 \\ 0 & 0 & 1 \end{bmatrix},$$

$$F_5^{-1} = \begin{bmatrix} 1 & 0 & 0 \\ 0 & 1 & -3 \\ 0 & 0 & 1e \end{bmatrix},$$

$$F_3F_2F_1BF_4F_5 = \begin{bmatrix} 1 & 0 & 0 \\ 0 & 1 & 0 \\ 0 & 0 & 0 \end{bmatrix} = D.$$

Then

$$B = F_1^{-1}F_2^{-1}F_3^{-1}DF_5^{-1}F_4^{-1}.$$

Let

$$P = F_1^{-1}F_2^{-1}F_3^{-1} = \begin{bmatrix} 1 & 0 & 0 \\ 2 & 1 & 0 \\ 0 & 0 & 1 \end{bmatrix}\begin{bmatrix} 1 & 0 & 0 \\ 0 & 1 & 0 \\ 3 & 0 & 1 \end{bmatrix}\begin{bmatrix} 1 & 0 & 0 \\ 0 & 1 & 0 \\ 0 & 1 & 1 \end{bmatrix}$$

$$= \begin{bmatrix} 1 & 0 & 0 \\ 2 & 1 & 0 \\ 3 & 1 & 1 \end{bmatrix},$$

$$Q = F_5^{-1}F_4^{-1} = \begin{bmatrix} 1 & 0 & 0 \\ 0 & 1 & -3 \\ 0 & 0 & 1 \end{bmatrix}\begin{bmatrix} 1 & 0 & 1 \\ 0 & 1 & 0 \\ 0 & 0 & 1 \end{bmatrix} = \begin{bmatrix} 1 & 0 & 1 \\ 0 & 1 & -3 \\ 0 & 0 & 1 \end{bmatrix}.$$

Here P and Q are nonsingular since they are products of the basic elementary matrices. But D can be written in many ways as the sum to two nonsingular matrices. For example,

$$D = D_1 + D_2, \quad D_1 = \begin{bmatrix} 3 & 0 & 0 \\ 0 & 2 & 0 \\ 0 & 0 & 4 \end{bmatrix}, \quad D_2 = \begin{bmatrix} -2 & 0 & 0 \\ 0 & -1 & 0 \\ 0 & 0 & -4 \end{bmatrix}.$$

Then

$$B = B_1 + B_2, \quad B_1 = PD_1Q, \quad B_2 = PD_2Q.$$

Example 4.4.5. Show that one eigenvalue of a singly stochastic matrix (Markov matrix) is 1 and illustrate it by computing the eigenvalues of

$$A = \begin{bmatrix} 0.2 & 0.5 & 0.3 \\ 0.5 & 0 & 0.5 \\ 0.4 & 0.2 & 0.4 \end{bmatrix}.$$

Solution 4.4.5. Consider the equation $|A - \lambda I| = 0$. Add all the columns of $A - \lambda I$ to the first column, the determinant remains the same. The first column becomes $1 - \lambda$ repeated. Take out $1 - \lambda$. [If the columns have this property then add all rows to the first row. Then take out $1 - \lambda$ from the first row.] Thus, in general, when the matrix is a Markov matrix one eigenvalue is 1. For the example above

$$|A - \lambda I| = \begin{vmatrix} 0.2 - \lambda & 0.5 & 0.3 \\ 0.5 & -\lambda & 0.5 \\ 0.4 & 0.2 & 0.4 - \lambda \end{vmatrix} = \begin{vmatrix} 1 - \lambda & 0.5 & 0.3 \\ 1 - \lambda & -\lambda & 0.5 \\ 1 - \lambda & 0.2 & 0.4 - \lambda \end{vmatrix}$$

$$= (1 - \lambda) \begin{vmatrix} 1 & 0.5 & 0.3 \\ 1 & -\lambda & 0.5 \\ 1 & 0.2 & 0.4 - \lambda \end{vmatrix}.$$

Add -0.5 times the first column to the second column and -0.3 times the first column to the third column. Then

$$|A - \lambda I| = (1 - \lambda) \begin{vmatrix} 1 & 0 & 0 \\ 1 & -\lambda - 0.5 & 0.2 \\ 1 & -0.3 & 0.1 - \lambda \end{vmatrix}$$

$$= (1 - \lambda) \begin{vmatrix} -\lambda - 0.5 & 0.2 \\ -0.3 & 0.1 - \lambda \end{vmatrix}$$

$$= (1 - \lambda)[(-\lambda - 0.5)(0.1 - \lambda) - (-0.3)(0.2)]$$

$$= (1 - \lambda)(\lambda^2 + 0.4\lambda + 0.01).$$

The roots are $\lambda_1 = 1$, $\lambda_2 = -0.2 + \sqrt{0.03}$, $\lambda_3 = -0.2 - \sqrt{0.03}$. Note that $|\lambda_2| < 1$, $|\lambda_3| < 1$.

(xiii) One eigenvalue of a singly stochastic matrix (Markov matrix) is 1.

Note that the sum of the eigenvalues equals the trace. But the diagonal elements of a Markov matrix are non-negative and less than or equal to 1. The maximum value attainable for the trace of an $n \times n$ matrix is n, out of which one eigenvalue is 1. Hence the maximum value for the sum of the remaining eigenvalues is only $n - 1$. We can show by using the property that powers of Markov matrices are also Markov matrices that the remaining roots satisfy the condition $|\lambda_j| \le 1$ for all j.

4.4.2 Definiteness of matrices

So far we have given three definitions for the definiteness of a real symmetric matrix, one definition in terms of quadratic forms, another in terms of determinants of the leading submatrices and a third in terms of the eigenvalues. Are these three definitions equivalent? For example an $n \times n$ matrix A with real elements and symmetric is positive definite if

(1) the quadratic form $X'AX$ remains positive for all possible non-null X (definition 1);
(2) the leading or principal minors of A are all positive (definition 2);
(3) the eigenvalues are all positive (definition 3).

Are definitions (1) and (3) equivalent? When A is real symmetric it is trivial to show that there exists a full set of orthonormal eigenvectors (part of it is established in property (x) of Section 4.2 and for the remaining part see Exercise 4.4.6 (7)) so that

$$A = PDP', \quad PP' = I, \quad P'P = I$$

and D is a diagonal matrix with the diagonal elements being the eigenvalues of A. Then

$$X'AX = X'PDP'X = Y'DY$$
$$= \lambda_1 y_1^2 + \cdots + \lambda_n y_n^2, \quad Y = P'X,$$
$$D = \mathrm{diag}(\lambda_1, \ldots, \lambda_n).$$

For arbitrary real X which means for arbitrary real Y if $X'AX > 0$ then

$$\lambda_1 y_1^2 + \cdots + \lambda_n y_n^2 > 0$$

for all $Y' = (y_1, \ldots, y_n)$. Put $y_1 = 0, \ldots, y_{j-1} = 0, y_{j+1} = 0, \ldots, y_n = 0$. Then $\lambda_j y_j^2 > 0$. Since y_j is real, this means that $\lambda_j > 0$, $j = 1, \ldots, n$. By retracing the steps the converse is true. Note that the matrix appearing in the quadratic form can be taken as symmetric without any loss of generality.

Now let us look at the definitions (1) and (2). Assume that $X'AX > 0$ for all possible non-null X, $A = (a_{ij}) = A'$. Let $X' = (x_1, \ldots, x_n)$. Put $x_2 = 0 = \cdots = x_n$. Then $a_{11} x_1^2 > 0$. Then $a_{11} > 0$ since $x_1^2 > 0$ for real nonzero x_1. Now put $x_3 = 0 = \cdots = x_n$. Then

$$[x_1, x_2] \begin{bmatrix} a_{11} & a_{12} \\ a_{12} & a_{22} \end{bmatrix} \begin{bmatrix} x_1 \\ x_2 \end{bmatrix} > 0$$

for all x_1 and x_2. This means

$$A_2 = \begin{bmatrix} a_{11} & a_{12} \\ a_{12} & a_{22} \end{bmatrix}$$

is positive definite. Then by definition (3) the eigenvalues of A_2 must be positive and hence its determinant is positive, $|A_2| > 0$. Continuing like this the determinants of all the leading sub-matrices are positive.

When looking at negative definiteness note that when all the eigenvalues of a matrix are negative product of an odd number of them will be negative and the product of an even number of them will be positive. We will be looking at the eigenvalues of the leading sub-matrices $A_1, A_2, \ldots, A_n$ when following through the above arguments to show the equivalence of all the definitions for negative definiteness.

(xiv) If an $n \times n$ matrix $A = A' = (a_{ij})$ is real and positive definite then $a_{ii} > 0, i = 1, \ldots, n$.
(xv) When the $n \times n$ matrix A is real symmetric then there exist a full set of n orthonormal eigenvectors or there exists a matrix P such that $P'AP = \Lambda$, $PP' = I$, $P'P = I$ where Λ is diagonal with the diagonal elements being the eigenvalues of A.
(xvi) When the $n \times n$ matrix A is Hermitian there exists a full set of n orthonormal eigenvectors or there exists a matrix Q such that $Q^*AQ = \Lambda$, $QQ^* = I$, $Q^*Q = I$.
(xvii) Eigenvectors corresponding to real eigenvalues of a real matrix are real.

Observe that when the elements in a matrix are real one can have the eigenvalues real, rational, irrational or complex. Then the eigenvectors corresponding to irrational eigenvalues will be irrational and eigenvectors corresponding to complex eigenvalues will be complex. When a matrix has complex elements then also the eigenvalues can be real, rational, irrational or complex.

Definition 4.4.1 (Hermitian definiteness). A Hermitian matrix is said to be (1) positive definite, (2) positive semi-definite, (3) negative definite, (4) negative semi-definite, (5) indefinite if the eigenvalues are (1) all positive, (2) non-negative, (3) negative, (4) negative or zero, (5) some are positive and some are negative and at least one in each set. [Remember that the eigenvalues of a Hermitian matrix are real.]

Example 4.4.6. Reduce the following real quadratic form $u = x_1^2 - 2x_1x_2 + 2x_2^2$ to its canonical form (linear combination of squares) through the eigenvalues as well as through a different method.

Solution 4.4.6. Writing the quadratic form with a symmetric matrix, we have

$$u = X'AX = [x_1, x_2] \begin{bmatrix} 1 & -1 \\ -1 & 2 \end{bmatrix} \begin{bmatrix} x_1 \\ x_2 \end{bmatrix},$$

$$A = \begin{bmatrix} 1 & -1 \\ -1 & 2 \end{bmatrix}.$$

Let us evaluate the eigenvalues and eigenvectors of A. Consider the equation

$$|A - \lambda I| = 0 \implies (1 - \lambda)(2 - \lambda) - 1 = 0$$
$$\implies \lambda^2 - 3\lambda + 1 = 0$$
$$\implies \lambda_1 = \frac{3}{2} + \frac{\sqrt{5}}{2}, \quad \lambda_2 = \frac{3}{2} - \frac{\sqrt{5}}{2}$$

are the eigenvalues. An eigenvector corresponding to λ_1 is given by

$$(A - \lambda_1 I)X = O \implies \begin{bmatrix} 1 - (\frac{3}{2} + \frac{\sqrt{5}}{2}) & -1 \\ -1 & 2 - (\frac{3}{2} + \frac{\sqrt{5}}{2}) \end{bmatrix} \begin{bmatrix} x_1 \\ x_2 \end{bmatrix} = \begin{bmatrix} 0 \\ 0 \end{bmatrix}$$
$$\implies X_1 = \begin{bmatrix} 1 \\ -\frac{1}{2}(1 + \sqrt{5}) \end{bmatrix}$$

with $\|X_1\| = \alpha$, say, is an eigenvector. Then the normalized X_1 is $Y_1 = \frac{1}{\alpha}X_1$. An eigenvector corresponding to λ_2 is given by

$$(A - \lambda_2 I)X = O \implies X_2 = \begin{bmatrix} 1 \\ -\frac{1}{2}(1 - \sqrt{5}) \end{bmatrix}$$

is an eigenvector. The normalized vector is $Y_2 = \frac{1}{\beta}X_2$ where $\beta = \|X_2\|$. Let

$$Q = (Y_1, Y_2).$$

Since $Y_1' Y_2 = 0$, $Y_1' Y_1 = 1$, $Y_2' Y_2 = 1$ the matrix Q is orthonormal and its inverse is its transpose. Also

$$A = Q \Lambda Q' \implies Q'AQ = \Lambda = \text{diag}(\lambda_1, \lambda_2).$$

Writing

$$u = X'AX = X'Q\Lambda Q'X = Z' \Lambda Z, \quad Z = Q'X, \quad Z' = (z_1, z_2).$$
$$u = \lambda_1 z_1^2 + \lambda_2 z_2^2$$
$$= \left(\frac{3 + \sqrt{5}}{2} \right) z_1^2 + \left(\frac{3 - \sqrt{5}}{2} \right) z_2^2.$$

This is one representation through the eigenvalues. Let us consider another procedure. Since A is symmetric try to reduce A to a diagonal form by elementary operations:

$$\begin{bmatrix} 1 & 0 \\ 1 & 1 \end{bmatrix} \begin{bmatrix} 1 & -1 \\ -1 & 2 \end{bmatrix} \begin{bmatrix} 1 & 1 \\ 0 & 1 \end{bmatrix} = \begin{bmatrix} 1 & 0 \\ 0 & 1 \end{bmatrix} \implies$$
$$\begin{bmatrix} 1 & -1 \\ -1 & 2 \end{bmatrix} = \begin{bmatrix} 1 & 0 \\ -1 & 1 \end{bmatrix} \begin{bmatrix} 1 & 0 \\ 0 & 1 \end{bmatrix} \begin{bmatrix} 1 & -1 \\ 0 & 1 \end{bmatrix}.$$

Let

$$Y = \begin{bmatrix} 1 & -1 \\ 0 & 1 \end{bmatrix} X$$

$$= \begin{bmatrix} 1 & -1 \\ 0 & 1 \end{bmatrix} \begin{bmatrix} x_1 \\ x_2 \end{bmatrix} = \begin{bmatrix} x_1 - x_2 \\ x_2 \end{bmatrix}.$$

Then

$$u = X'AX = Y'DY, \quad D = \begin{pmatrix} 1 & 0 \\ 0 & 1 \end{pmatrix} \Rightarrow$$

$$u = y_1^2 + y_2^2, \quad y_1 = x_1 - x_2, \quad y_2 = x_2.$$

Note that the two representations are not the same.

4.4.3 Commutative matrices

One property, repeatedly used in many branches of statistics, econometrics, engineering and other areas is the simultaneous reduction of two matrices to diagonal forms. Let A and B be two $n \times n$ matrices. If there exists a Q such that

$$QAQ^{-1} = D_1 \quad \text{and} \quad QBQ^{-1} = D_2,$$

where D_1 and D_2 are diagonal, that is, the same matrix Q diagonalizes both A and B, then what should be the conditions on A and B? We will investigate this aspect a little bit further. If there exists a Q then

$$(QAQ^{-1})(QBQ^{-1}) = D_1D_2 = D_2D_1$$
$$\Rightarrow QABQ^{-1} = D_1D_2$$
$$\Rightarrow AB = Q^{-1}D_2D_1Q = (Q^{-1}D_2Q)(Q^{-1}D_1Q) = BA.$$

That means A and B commute. Now, suppose A and B commute, that is, $AB = BA$. Does there exist a Q such that $QAQ^{-1} = D_1$ and $QBQ^{-1} = D_2$? For simplicity let us assume that the eigenvalues of A are distinct so that there exists a full set of linearly independent eigenvectors for A which means there exists a Q with its regular inverse Q^{-1}. [The result can also be proved without this restriction, the steps will be longer.] Then

$$QAQ^{-1} = D_1.$$

Let λ_1 be an eigenvalue of A with X_1 a corresponding eigenvector. Then

$$AX_1 = \lambda_1 X_1 \Rightarrow BAX_1 = \lambda_1 BX_1.$$

But $BA = AB$ which means

$$ABX_1 = \lambda_1 BX_1 \implies A(BX_1) = \lambda_1(BX_1).$$

This shows that X_1 and BX_1 are eigenvectors of A for the same eigenvalue λ_1. By assumption the eigenvalues are distinct. Then the null space of $(A - \lambda_1 I)X = O$ has rank 1 which means X_1 and BX_1 are scalar multiples of each other or $BX_1 = \mu_1 X_1$. This shows that A and B have the same eigenvectors. Thus the same Q will diagonalize B also. Hence we have a very important result.

(xviii) There exists a Q such that QAQ^{-1} is diagonal and QBQ^{-1} is diagonal simultaneously if and only if A and B commute, that is, iff $AB = BA$. If A and B are symmetric then $Q^{-1} = Q'$ the transpose of Q. If A and B are Hermitian then $Q^{-1} = Q^*$ the conjugate transpose of Q.
(xix) If the eigenvalues of a matrix A are all distinct then the null space of $(A - \lambda_j I)X = O$ has rank 1 where λ_j is any eigenvalue of A.

Chisquaredness and independence of quadratic forms in real Gaussian random variables are the two fundamental results in statistical inference problems connected with regression analysis, analysis of variance, analysis of covariance, model building and many related areas. Without the statistical terminology we will illustrate the results here.

Example 4.4.7. Let $A = A'$ be a real $n \times n$ matrix. Consider the real quadratic form $X'AX$. Then show that $X'AX$ can be written as

$$X'AX = y_1^2 + \cdots + y_r^2, \quad r \leq n,$$

that is as a sum of squares, if and only if A is idempotent of rank r. [This corresponds to the result on chisquaredness in statistics.]

Solution 4.4.7. When $A = A'$ there exists an orthonormal matrix P such that

$$P'AP = D = \operatorname{diag}(\lambda_1, \dots, \lambda_n)$$

where $\lambda_1, \dots, \lambda_n$ are the eigenvalues of A. Let $P'X = Y$, $Y' = (y_1, \dots, y_n)$. Then

$$X'AX = X'PDP'X = Y'DY = \lambda_1 y_1^2 + \cdots + \lambda_n y_n^2. \tag{a}$$

If $A = A^2$ (idempotent) of rank r then r of the λ_j's are 1 each and the remaining ones are zeros. Then

$$X'AX = y_1^2 + \cdots + y_r^2.$$

Now, assume that $X'AX = y_1^2 + \cdots + y_r^2$ holds. Then from (a) it follows that r of the eigenvalues are 1 each and the remaining are zeros. Since A is symmetric eigenvalues being 1's and zeros imply that A is idempotent. [If A is not symmetric then eigenvalues being 1's and zeros need not imply that A is idempotent.]

Example 4.4.8. Let A and B be real symmetric $n \times n$ matrices. Let $X'AX$ and $X'BX$ be two real quadratic forms. Then show that

$$X'AX = \lambda_1 y_1^2 + \cdots + \lambda_r y_r^2 \quad \text{and}$$
$$X'BX = \lambda_{r+1} y_{r+1}^2 + \cdots + \lambda_k y_k^2, \quad r < n, \; k \leq n,$$

(where the y_j's appearing in $X'AX$ do not appear in $X'BX$) iff $AB = O$. [This result corresponds to the result on statistical independence of quadratic forms.]

Solution 4.4.8. Assume that $AB = O$. Then $(AB)' = B'A' = BA = O$. That is, $AB = BA$. Since both matrices are real symmetric and since they commute there exists an orthonormal matrix P such that

$$P'AP = D_1, \quad P'BP = D_2$$

where D_1 and D_2 are diagonal. But

$$AB = O \Rightarrow P'ABP = O \Rightarrow P'APP'BP = O \Rightarrow D_1 D_2 = O.$$

That is, if there is a nonzero diagonal element in D_1 the corresponding diagonal element in D_2 is zero and vice versa. Let $Y = P'X$. Then

$$X'AX = Y'D_1 Y$$

a linear function of y_j^2's for r of the y_j's if r is the rank of A. Writing them as $y_1^2, \ldots, y_r^2$ we have

$$X'AX = \lambda_1 y_1^2 + \cdots + \lambda_r y_r^2. \tag{a}$$

If s is the rank of B then s of the y_j's, which are not in the $X'AX$ representation, will be present in $X'BX$. That is, $X'BX$ will be of the form

$$X'BX = \lambda_{r+1} y_{r+1}^2 + \cdots + \lambda_{r+s} y_{r+s}^2. \tag{b}$$

For convenience, all the nonzero diagonal elements are denoted as $\lambda_1, \ldots, \lambda_{r+s}$. This establishes one part. Now if we assume that the physical separation of the variables as in (a) and (b) above then by retracing the steps we can show that AB must be null when $A = A'$ and $B = B'$ and real. [The statistical property of independence need not imply physical separation of the variables as in (a) and (b) above. Hence the converse, that is, to show that $AB = O$ from the property of independence involves a few more steps from the implications of statistical independence. Therefore we will not discuss the proof here. For a proof see [7].]

(xx) Let $A_1, \ldots, A_k$ be $n \times n$ matrices so that $A_i A_j = A_j A_i$ for all i and j. Then there exists the same matrix Q such that $Q^{-1} A_j Q = D_j, j = 1, \ldots, k$ where D_j is diagonal.

Definition 4.4.2 (A Hermitian form). Let X be an $n \times 1$ vector and A an $n \times n$ matrix whose elements are in the complex field. Further, let $A = A^*$ (Hermitian). Then $u = X^* A X$ is called a Hermitian form, analogous to a quadratic form in the real case.

A Hermitian form is always real since u is 1×1 and further

$$u^* = (X^* A X)^* = X^* A^* (X^*)^* = X^* A X = u.$$

If $u = a + ib$ then $u^* = a - ib$. If $u = u^*$ then $b = 0$ or u is real.

Example 4.4.9. Reduce the following Hermitian form to its canonical form.

$$u = 2x_1^c x_1 + (1 - i)x_3^c x_1 + 2x_2^c x_2 + (1 - i)x_3^c x_2$$
$$+ (1 + i)x_1^c x_3 + (1 + i)x_2^c x_3 + x_3^c x_3.$$

Solution 4.4.9. Writing in the standard form $X^* A X$ we have

$$X^* A X = [x_1^c, x_2^c, x_3^c] \begin{bmatrix} 2 & 0 & 1+i \\ 0 & 2 & 1+i \\ 1-i & 1-i & 1 \end{bmatrix} \begin{bmatrix} x_1 \\ x_2 \\ x_3 \end{bmatrix}.$$

Let us find the eigenvalues of A. Consider

$$|A - \lambda I| = 0 \Rightarrow \begin{vmatrix} 2 - \lambda & 0 & 1+i \\ 0 & 2 - \lambda & 1+i \\ 1-i & 1-i & 1-\lambda \end{vmatrix} = 0$$

$$\Rightarrow \lambda_1 = 2, \quad \lambda_2 = \frac{3}{2} + \frac{\sqrt{17}}{2}, \quad \lambda_3 = \frac{3}{2} - \frac{\sqrt{17}}{2}.$$

Let us compute the eigenvectors.

$$(A - \lambda_1 I)X = O \Rightarrow \begin{bmatrix} 0 & 0 & 1+i \\ 0 & 0 & 1+i \\ 1-i & 1-i & -1 \end{bmatrix} \begin{bmatrix} x_1 \\ x_2 \\ x_3 \end{bmatrix} = \begin{bmatrix} 0 \\ 0 \\ 0 \end{bmatrix}$$

$$\Rightarrow X_1 = \begin{bmatrix} 1 \\ -1 \\ 0 \end{bmatrix}$$

is one eigenvector. Consider

$$(A - \lambda_2 I)X = O \Rightarrow \begin{bmatrix} \frac{1}{2} - \frac{\sqrt{17}}{2} & 0 & 1+i \\ 0 & \frac{1}{2} - \frac{\sqrt{17}}{2} & 1+i \\ 1-i & 1-i & -\frac{1}{2} - \frac{\sqrt{17}}{2} \end{bmatrix} \begin{bmatrix} x_1 \\ x_2 \\ x_3 \end{bmatrix} = \begin{bmatrix} 0 \\ 0 \\ 0 \end{bmatrix}$$

$$\Rightarrow X_2 = \begin{bmatrix} (1+i)\frac{(1+\sqrt{17})}{8} \\ (1+i)\frac{(1+\sqrt{17})}{8} \\ 1 \end{bmatrix}$$

is one such vector. From $(A - \lambda_3 I)X = O$ an X_3 is given by

$$X_3 = \begin{bmatrix} (1+i)\frac{(1-\sqrt{17})}{8} \\ (1+i)\frac{(1-\sqrt{17})}{8} \\ 1 \end{bmatrix}.$$

Note that $X_1^* X_2 = 0$, $X_1^* X_3 = 0$, $X_2^* X_3 = 0$. Let $\alpha = \|X_1\|$, $\beta = \|X_2\|$ and $\gamma = \|X_3\|$ be the lengths. Let the normalized vectors be

$$Y_1 = \frac{1}{\alpha}X_1, \quad Y_2 = \frac{1}{\beta}X_2, \quad Y_3 = \frac{1}{\gamma}X_3 \quad \text{and} \quad Q = (Y_1, Y_2, Y_3).$$

Then

$$A = Q\Lambda Q^*, \quad Q^*Q = I, \quad QQ^* = I.$$

Let

$$Y = Q^*X = \begin{bmatrix} \frac{1}{\alpha}(1,-1,0) \\ \frac{1}{\beta}((1-i)\frac{(1+\sqrt{17})}{8}, (1-i)\frac{(1+\sqrt{17})}{8}, 1) \\ \frac{1}{\gamma}((1-i)\frac{(1-\sqrt{17})}{8}, (1-i)\frac{(1-\sqrt{17})}{8}, 1) \end{bmatrix} \begin{bmatrix} x_1 \\ x_2 \\ x_3 \end{bmatrix} \Rightarrow$$

$$y_1 = \frac{1}{\alpha}(x_1 - x_2),$$

$$y_2 = \frac{1}{\beta}\left[(1-i)\frac{(1+\sqrt{17})}{8}(x_1 + x_2) + x_3\right],$$

$$y_3 = \frac{1}{\gamma}\left[(1-i)\frac{(1-\sqrt{17})}{8}(x_1 + x_2) + x_3\right].$$

Then

$$X^* AX = 2|y_1|^2 + \frac{1}{2}(3 + \sqrt{17})|y_2|^2 + \frac{1}{2}(3 - \sqrt{17})|y_3|^2.$$

Verifications of the various steps are left to the student.

Definition 4.4.3 (The canonical form). The canonical form of a Hermitian form is given by

$$X^* AX = \lambda_1 |y_1|^2 + \cdots + \lambda_n |y_n|^2$$

where $|y_j|$ denotes the absolute value of y_j, $j = 1, \ldots, n$, $Y' = (y_1, \ldots, y_n)$ and $\lambda_1, \ldots, \lambda_n$ are scalars.

Example 4.4.10. Reduce the Hermitian form in Example 4.4.9 to its canonical form by a sweep-out process (by elementary operations).

Solution 4.4.10. Consider the following elementary matrices. F_1, F_1^*, F_2, F_2^* where

$$F_1 = \begin{bmatrix} 1 & 0 & 0 \\ 0 & 1 & 0 \\ -\frac{1}{2}(1-i) & 0 & 1 \end{bmatrix} \Rightarrow F_1^{-1} = \begin{bmatrix} 1 & 0 & 0 \\ 0 & 1 & 0 \\ \frac{1}{2}(1-i) & 0 & 1 \end{bmatrix},$$

$$F_2 = \begin{bmatrix} 1 & 0 & 0 \\ 0 & 1 & 0 \\ 0 & -\frac{1}{2}(1-i) & 1 \end{bmatrix} \Rightarrow F_2^{-1} = \begin{bmatrix} 1 & 0 & 0 \\ 0 & 1 & 0 \\ 0 & \frac{1}{2}(1-i) & 1 \end{bmatrix}.$$

Since A in this case is Hermitian we operate on the left of A by F_1 and on the right by F_1^*, the conjugate transpose of F_1. (In the real symmetric case we operate on the right with the transpose of the elementary matrix. In the complex case we operate on the right with the conjugate transpose.) Thus

$$F_1 A F_1^* = \begin{bmatrix} 1 & 0 & 0 \\ 0 & 1 & 0 \\ -\frac{1}{2}(1-i) & 0 & 1 \end{bmatrix} \begin{bmatrix} 2 & 0 & 1+i \\ 0 & 2 & 1+i \\ 1-i & 1-i & 1 \end{bmatrix} \begin{bmatrix} 1 & 0 & \frac{1}{2}(1+i) \\ 0 & 1 & 0 \\ 0 & 0 & 1 \end{bmatrix}$$

$$= \begin{bmatrix} 2 & 0 & 0 \\ 0 & 2 & 1+i \\ 0 & 1-i & 0 \end{bmatrix}.$$

Now let $B = F_2 F_1 A F_1^* F_2^*$ then

$$B = \begin{bmatrix} 1 & 0 & 0 \\ 0 & 1 & 0 \\ 0 & -\frac{1}{2}(1-i) & 1 \end{bmatrix} \begin{bmatrix} 2 & 0 & 0 \\ 0 & 2 & 1+i \\ 0 & 1-i & 0 \end{bmatrix} \begin{bmatrix} 1 & 0 & 0 \\ 0 & 1 & -\frac{1}{2}(1+i) \\ 0 & 0 & 1 \end{bmatrix}$$

$$= \begin{bmatrix} 2 & 0 & 0 \\ 0 & 2 & 0 \\ 0 & 0 & -1 \end{bmatrix} = D, \quad \text{say.}$$

Then

$$A = F_1^{-1} F_2^{-1} D (F_1^{-1} F_2^{-1})^* = QDQ^*$$

$$= \begin{bmatrix} 1 & 0 & 0 \\ 0 & 1 & 0 \\ \frac{1}{2}(1-i) & \frac{1}{2}(1-i) & 1 \end{bmatrix} \begin{bmatrix} 2 & 0 & 0 \\ 0 & 2 & 0 \\ 0 & 0 & -1 \end{bmatrix} \begin{bmatrix} 1 & 0 & \frac{1}{2}(1+i) \\ 0 & 1 & \frac{1}{2}(1+i) \\ 0 & 0 & 1 \end{bmatrix}.$$

By straight multiplication of the matrices on the right we can verify the result. Consider the transformation $Y = Q^*X$. That is,

$$Y = \begin{bmatrix} y_1 \\ y_2 \\ y_3 \end{bmatrix} = \begin{bmatrix} 1 & 0 & \frac{1}{2}(1+i) \\ 0 & 1 & \frac{1}{2}(1+i) \\ 0 & 0 & 1 \end{bmatrix} \begin{bmatrix} x_1 \\ x_2 \\ x_3 \end{bmatrix} \Rightarrow$$

$$y_1 = x_1 + \frac{1}{2}(1+i)x_3, \quad y_2 = x_2 + \frac{1}{2}(1+i)x_3, \quad y_3 = x_3.$$

Then

$$X^*AX = Y^*DY = 2|y_1|^2 + 2|y_2|^2 - |y_3|^2.$$

For example,

$$|y_1|^2 = y_1 y_1^c = \left[x_1 + \frac{1}{2}(1+i)x_3 \right]\left[x_1^c + \frac{1}{2}(1-i)x_3^c \right]$$

$$= x_1 x_1^c + \frac{1}{2}(1+i)x_3 x_1^c + \frac{1}{2}(1-i)x_1 x_3^c + \frac{1}{4}(1+i)(1-i)x_3 x_3^c$$

$$= |x_1|^2 + \frac{1}{2}|x_3|^2 + \frac{1}{2}(1+i)x_3 x_1^c + \frac{1}{2}(1-i)x_1 x_3^c.$$

Similarly computing $|y_2|^2, |y_3|^2$ and substituting in $2|y_1|^2 + 2|y_2|^2 - |y_3|^2$ we can easily verify that we get back the Hermitian form given in Example 4.4.9.

Note that if a real quadratic form in a real symmetric matrix A or if a Hermitian form in a Hermitian matrix A is to be reduced to a linear function of squares (canonical form), with the coefficients of the squares not necessarily the eigenvalues of A, then the easier method would be to use elementary operations on the left and on the right of A (pre and post multiplications by elementary matrices) rather than going through the eigenvalues of A. In the Hermitian case if A is premultiplied by G_1, a product of the basic elementary matrices, then A is to be postmultiplied by G_1^*, the conjugate transpose of G_1. (In the real symmetric case we postmultiply by G_1' only.) Such successive multiplications will reduce a Hermitian form into the following form:

$$X^*AX = X^*QDQ^*X = Y^*DY,$$

$$Y = Q^*X, \quad D = \text{diag}(d_1, \dots, d_n),$$

$$X^*AX = d_1|y_1|^2 + \cdots + d_n|y_n|^2$$

where $|y_j|$ is the absolute value of y_j, Q is the product of all elementary matrices used on the left. By definition, Q will be nonsingular. But in this case QQ' or QQ^* need not be an identity matrix. In other words the linear transformation $Y = Q^*X$ need not be an orthogonal or unitary transformation. The procedure through eigenvalues will give an orthogonal transformation when A is real symmetric and a unitary transformation when A is Hermitian.

(xxi) If $A = A^*$ then X^*AX is real for all X. [Take the conjugate transpose. A 1×1 matrix with its conjugate transpose equal to itself is real.]

(xxii) If $A = A'$ and real and if $A = P\Lambda P'$, $\Lambda = \text{diag}(\lambda_1, \ldots, \lambda_n)$ then

$$A = \lambda_1 P_1 P_1' + \cdots + \lambda_n P_n P_n'$$

where $P_1, \ldots, P_n$ are the columns of P.

(xxiii) If $A = A^*$ and if $A = Q\Lambda Q^*$ then

$$A = \lambda_1 Q_1 Q_1^* + \cdots + \lambda_n Q_n Q_n^*$$

where $Q_1, \ldots, Q_n$ are the columns of Q.

(xxiv) If A is Hermitian then iA is skew Hermitian.

Definition 4.4.4 (Definiteness of Hermitian forms). A Hermitian form X^*AX, $A = A^*$ is positive definite if for all possible non-null X, $X^*AX > 0$, positive semi-definite if $X^*AX \geq 0$, negative definite if $X^*AX < 0$, negative semi-definite if $X^*AX \leq 0$ and indefinite if for some X, $X^*AX > 0$ and for some other X it is negative.

Note that when $A = A^*$ we have X^*AX a real quantity. Also we know that when $A = A^*$ all the eigenvalues of A are real. It is easy to show that all definitions of definiteness of a Hermitian form are equivalent. The proofs are parallel to those in the real symmetric case.

Example 4.4.11. Show that the following Hermitian form can be written as a quadratic form in real variables.

$$h = [x_1^c, x_2^c] \begin{bmatrix} 3 & 2+i \\ 2-i & 5 \end{bmatrix} \begin{bmatrix} x_1 \\ x_2 \end{bmatrix}$$
$$= 3x_1^c x_1 + 5x_2^c x_2 + (2+i)x_1^c x_2 + (2-i)x_2^c x_1.$$

Solution 4.4.11. Let $x_1 = u + iv$, $x_2 = x + iy$ where u, v, x, y real and $i = \sqrt{-1}$. Then

$$x_1^c x_1 = (u - iv)(u + iv) = u^2 + v^2, \quad x_2^c x_2 = x^2 + y^2,$$
$$x_1^c x_2 = (u - iv)(x + iy) = ux + vy + i(uy - vx),$$
$$(2+i)x_1^c x_2 + (2-i)x_2^c x_1 = (2+i)[ux + vy + i(uy - vx)]$$
$$+ (2-i)[ux + vy + i(xv - yu)]$$
$$= 4(ux + vy) + 2(-uy + vx).$$

Hence the Hermitian form

$$h = 3(u^2 + v^2) + 5(x^2 + y^2) + 4ux + 4vy - 2uy + 2vx$$

$$= [u, v, x, y] \begin{bmatrix} 3 & 0 & 2 & -1 \\ 0 & 3 & 1 & 2 \\ 2 & 1 & 5 & 0 \\ -1 & 2 & 0 & 5 \end{bmatrix} \begin{bmatrix} u \\ v \\ x \\ y \end{bmatrix}$$

$$= [u, v] \begin{bmatrix} 3 & 0 \\ 0 & 3 \end{bmatrix} \begin{bmatrix} u \\ v \end{bmatrix} + [x, y] \begin{bmatrix} 5 & 0 \\ 0 & 5 \end{bmatrix} \begin{bmatrix} x \\ y \end{bmatrix}$$

$$+ 2[u, v] \begin{bmatrix} 2 & -1 \\ 1 & 2 \end{bmatrix} \begin{bmatrix} x \\ y \end{bmatrix}.$$

This, in fact, is a general result.

> **(xxv)** A Hermitian form in n complex variables is equivalent to a quadratic form in $2n$ real variables or equivalent to two quadratic forms in n real variables each plus two bilinear forms in n real variables each.

This result is frequently used when extending the theory of real Gaussian multivariate statistical distribution to the corresponding multivariate Gaussian distribution in complex random variables.

Another result which is frequently used in various applications involving a real quadratic form, or a Hermitian form in complex variables, is a representation of the matrix of the quadratic or Hermitian form in terms of a square root when the matrix is positive definite or positive semi-definite. Consider a real quadratic form u and a Hermitian form v where

$$u = X'AX, \quad A = A' \quad \text{and} \quad v = Z^* BZ, \quad B = B^*$$

where X is an $n \times 1$ real vector and Z is an $n \times 1$ vector in the complex space. We have already seen that there exist a full set of orthonormal vectors, and orthogonal and unitary matrices P and Q such that

$$P'AP = \Lambda \quad \text{and} \quad Q^*BQ = \mu$$

where $\Lambda = \text{diag}(\lambda_1, \ldots, \lambda_n)$, $\mu = \text{diag}(\mu_1, \ldots, \mu_n)$ with the λ_j's the eigenvalues of A and the μ_j's the eigenvalues of B. If A and B are positive definite then $\lambda_j > 0$, $\mu_j > 0$, $j = 1, \ldots, n$. We can write $\lambda_j = \sqrt{\lambda_j}\sqrt{\lambda_j}$ and $\mu_j = \sqrt{\mu_j}\sqrt{\mu_j}$. Define

$$\Lambda^{\frac{1}{2}} = \text{diag}(\sqrt{\lambda_1}, \ldots, \sqrt{\lambda_j}) \quad \text{and} \quad \mu^{\frac{1}{2}} = \text{diag}(\sqrt{\mu_1}, \ldots, \sqrt{\mu_n}).$$

Then

$$A = P\Lambda P' = P\Lambda^{\frac{1}{2}}\Lambda^{\frac{1}{2}}P' = P\Lambda^{\frac{1}{2}}P' P\Lambda^{\frac{1}{2}}P'$$

$$= A_1^2, \quad A_1 = P\Lambda^{\frac{1}{2}}P'$$

and similarly

$$B = B_1^2, \quad B_1 = Q^*\mu^{\frac{1}{2}}Q$$

where A_1 is called the positive definite square root of the positive definite matrix A and B_1 is called the Hermitian positive definite square root of the Hermitian positive definite matrix B. Note that if A and B are positive definite or positive semi-definite one can express both A and B as $A = A_2 A_2'$ and $B = B_2 B_2^*$ for some matrices A_2 and B_2 by following through the above procedure. What about the converses? If a real matrix A can be written as $A = CC'$ where C may be $n \times m$, n need not be equal to m, is A going to be at least positive semi-definite? Let us consider an arbitrary quadratic form

$$X'AX = X'CC'X = Y'Y, \quad Y = C'X$$
$$= y_1^2 + \cdots + y_m^2 \geq 0, \quad Y' = (y_1, \ldots, y_m)$$

for all real non-null X and A. This means that the matrix A is either positive definite or positive semi-definite. If C is of full rank (rank is m if $m \leq n$ or n if $n \leq m$) then A is positive definite. The Hermitian case is parallel. If any matrix B can be written in the form $B = GG^*$ then the Hermitian form

$$Z^*AZ = Z^*GG^*Z = Y^*Y, \quad Y = G^*Z$$
$$= |y_1|^2 + \cdots + |y_m|^2 \geq 0.$$

That is, B is at least positive semi-definite.

(xxvi) Any positive definite or positive semi-definite (or Hermitian positive definite or Hermitian positive semi-definite) matrix A can be written as $A = CC'$ (or $A = CC^*$) and conversely, any matrix A which can be written as $A = CC'$ (or $A = CC^*$), where C may be rectangular, is at least positive semi-definite.

Definition 4.4.5 (Square roots). A symmetric positive definite (or Hermitian positive definite) square root of a symmetric positive definite (or Hermitian positive definite) matrix A is $C = P\Lambda^{\frac{1}{2}}P'$ (or $C = P\Lambda^{\frac{1}{2}}P^*$) where P is the matrix of normalized eigenvectors of A, Λ is the diagonal matrix of the eigenvalues of A and $\Lambda^{\frac{1}{2}}$ denotes the diagonal matrix with the diagonal elements being the positive square roots of the diagonal elements in Λ.

Exercises 4.4

4.4.1. Compute the eigenvalues and eigenvectors of

$$A = \begin{bmatrix} 1 & 1+i & -2i \\ 1-i & 2+i & 2-i \\ 2i & 2+i & 3 \end{bmatrix}, \quad B = \begin{bmatrix} 1 & 1+i & -2i \\ -1+i & 3 & 1-i \\ -2i & -1-i & 4 \end{bmatrix}.$$

4.4.2. Write the following matrices as the sum of symmetric and skew symmetric matrices:

$$A = \begin{bmatrix} 1 & 5 \\ 3 & 7 \end{bmatrix}, \quad B = \begin{bmatrix} 2 & 1 & -1 \\ 1 & 0 & 2 \\ 3 & 1 & 1 \end{bmatrix}, \quad C = \begin{bmatrix} 3 & 2 & -2 & 1 \\ 4 & 1 & -1 & 2 \\ 1 & -1 & 0 & 4 \\ 5 & 0 & -1 & 6 \end{bmatrix}.$$

4.4.3. Write the matrices in Exercise 4.4.2 as the sum of two nonsingular matrices.

4.4.4. Write the following matrices as the sum of Hermitian and skew Hermitian matrices:

$$A = \begin{bmatrix} 2-i & 1+2i & 3i \\ 2i & 3-i & 2+5i \\ 1+i & 4i & 2-i \end{bmatrix},$$

$$B = \begin{bmatrix} 1+2i & 2-i & 3+2i & 1+i \\ 1-i & 3+2i & 4i & -5i \\ 2+i & 2+5i & 1-i & 1+i \\ 3+5i & 2-i & 1+i & 5i \end{bmatrix}.$$

4.4.5. Compute the eigenvalues and eigenvectors of the following stochastic matrices:

$$A = \begin{bmatrix} 0.1 & 0.7 & 0.2 \\ 0.3 & 0.5 & 0.2 \\ 0.3 & 0.2 & 0.5 \end{bmatrix}, \quad B = \begin{bmatrix} 0.2 & 0.4 & 0.2 & 0.2 \\ 0.3 & 0.1 & 0.4 & 0.2 \\ 0.2 & 0.4 & 0.1 & 0.2 \\ 0.3 & 0.1 & 0.3 & 0.4 \end{bmatrix}.$$

4.4.6. Are the following statements true, give a counter example if false and prove the results if true.

(1) If the eigenvalues are real then the matrix is symmetric or Hermitian.

(2) If the eigenvalues are all purely imaginary then the matrix is skew symmetric or skew Hermitian.

(3) If the eigenvalues are 1's and 0's then the matrix is idempotent.

(4) If the eigenvalues are ±1 then the matrix is orthonormal.

(5) If the eigenvalues are such that $\lambda\lambda^c = 1$ then the matrix is unitary.

(6) Since the eigenvalues of A are the eigenvalues of A' then the eigenvalues of A and $(A + A')/2$ are the same.

(7) If A is $n \times n$ and real symmetric then there is a full set of n mutually orthogonal eigenvectors whether some eigenvalues are zero or repeated.

(8) If λ is an eigenvalue and X a normalized eigenvector corresponding to λ of a matrix A then $\lambda = X'AX$ if X is real, $\lambda = X^*AX$ is X is in the complex space.

(9) If the eigenvalues of a matrix are real then the eigenvectors are also real.

4.4.7. Let

$$A = \begin{bmatrix} 1 & 1 & 1 & 1 \\ 1 & 1 & 1 & 1 \\ 1 & 1 & 1 & 1 \\ 1 & 1 & 1 & 1 \end{bmatrix}, \quad B = \begin{bmatrix} -1 & -1 & 1 & 1 \\ -1 & 0 & 0 & 1 \\ 1 & 0 & 0 & -1 \\ 1 & 1 & -1 & -1 \end{bmatrix}.$$

If possible, reduce A and B simultaneously to diagonal forms. Compute that orthonormal matrix which will achieve this.

4.4.8. Repeat Exercise 4.4.7 if A and B are $n \times n$ matrices where

$$A = I - C, \quad C = \frac{1}{n} \begin{bmatrix} 1 & 1 & \cdots & 1 \\ \vdots & \vdots & \cdots & \vdots \\ 1 & 1 & \cdots & 1 \end{bmatrix}.$$

4.4.9. Let A_1, A_2 be $n \times n$ real symmetric idempotent matrices such that $A_1 + A_2 = I_n$. Show that (1) both A_1 and A_2 can be simultaneously reduced to diagonal forms, (2) rank of A_1 plus rank of A_2 is n, (3) A_1 and A_2 commute.

4.4.10. Let A_1 and A_2 be $n \times n$ real symmetric matrices so that $A_1 + A_2 = I$ with $A_1 A_2 = O$. Show that A_1 and A_2 commute.

4.4.11. Let A_1 and A_2 be $n \times n$ real symmetric matrices so that $A_1 + A_2 = I$ and the rank of A_1 plus the rank of A_2 is n. Show that (1) $A_1 A_2 = O$, (2) A_1 and A_2 are both idempotent.

4.4.12. Generalize Exercises 4.4.9, 4.4.10 and 4.4.11 and establish the corresponding results if k such $n \times n$ matrices are involved, $k \geq 2$ such that $A_1 + \cdots + A_k = I$. Orthogonality condition to be interpreted as $A_i A_j = O$ for all $i \neq j$.

4.4.13. Reduce the following Hermitian forms to their canonical forms by using elementary operations:

(a) $2x_1 x_1^c + 3x_2 x_2^c + 2x_3 x_3^c + (1+i)x_1^c x_3 + (1+i)x_3^c x_1 + (2-i)x_2^c x_3$
$$+ (1-i)x_3^c x_1 + (2+i)x_3^c x_2 + (1-i)x_1^c x_2,$$

(b) $3x_1 x_1^c + 4x_2 x_2^c + 2x_3 x_3^c + ix_1^c x_2 + 2ix_1^c x_3 - ix_2^c x_1$
$$+ (1-i)x_2^c x_3 - 2ix_3^c x_1 + (1+i)x_3^c x_2.$$

4.4.14. Repeat Exercise 4.4.13 by using the procedure through eigenvalues.

4.4.15. Check the definiteness of the Hermitian forms in Exercise 4.4.13.

4.4.16. Heisenberg's uncertainty principle in quantum mechanics. Consider the position matrix P, which is symmetric, and momentum matrix Q which is skew symmetric. These P and Q satisfy the equation $QP - PQ = I$. By using Cauchy–Schwartz inequality, or otherwise, show that

$$1 \le 2\frac{\|QX\|}{\|X\|}\frac{\|PX\|}{\|X\|}.$$

Hint: $\|X\|^2 = X'X = X'IX$.

4.4.17. If A is a real skew symmetric or skew Hermitian matrix then show that the determinant of A is either zero or positive.

4.4.18. Compute the symmetric positive definite square root of A and the Hermitian positive definite square root of B where

$$A = \begin{bmatrix} 2 & 1 \\ 1 & 3 \end{bmatrix}, \quad B = \begin{bmatrix} 2 & 1+i \\ 1-i & 3 \end{bmatrix}.$$

4.4.19. For the Kronecker product defined in Exercise 2.6.8 where $A = (a_{ij})$ and $p \times p$, $B = (b_{ij})$ and $q \times q$, and the Kronecker product denoted by $A \otimes B$, show that

$$(1) \qquad |A \otimes B| = \left(\prod_{i=1}^{p} \lambda_i \right)^q \left(\prod_{j=1}^{q} v_j \right)^p$$

where the λ_i and the v_j's are the eigenvalues of A and B respectively;
  (2) The eigenvalues of $A \otimes B$ are $\lambda_i v_j$ for all i and j;
  (3) What are the eigenvectors of $A \otimes B$ in terms of the eigenvectors of λ_i and v_j?

4.4.20. Let A be skew symmetric. Construct a matrix B in terms of $I + A$ and $I - A$ so that B is orthonormal.

4.4.21. Show that the characteristic polynomial of the matrix

$$A = \begin{bmatrix} A_1 & A_2 \\ O & A_3 \end{bmatrix}$$

is the product of the characteristic polynomials of A_1 and A_3.

4.4.22. Show that the eigenvalues of A and A' coincide whereas the eigenvalues of A and A^* are complex conjugates of each other.

4.4.23. Similar matrices. Square matrices A and B are said to be similar (notation: $A \sim B$) if there exists a nonsingular matrix Q such that $A = QBQ^{-1}$. If $A \sim B$ then show that
(i) $A' \sim B'$
(ii) $A^k \sim B^k$
(iii) $A - \lambda I \sim B - \lambda I$
(iv) $|A| = |B|$

(v) $\mathrm{rank}(A) = \mathrm{rank}(B)$

(vi) $P(A) \sim P(B)$

where $P(\lambda)$ is a polynomial in λ.

4.4.24. Show that the following two matrices are similar. That is,

$$\begin{bmatrix} B_1 & O \\ O & B_2 \end{bmatrix} \sim \begin{bmatrix} B_2 & O \\ O & B_1 \end{bmatrix}.$$

4.4.25. If $A \sim B$ then show that $\mathrm{tr}(A) = \mathrm{tr}(B)$.

4.4.26. If $A \sim B$ then show that (i) if A is idempotent then B is idempotent, (ii) if A is nilpotent of degree r then B is also nilpotent of degree r.

4.4.27. Kernel of the matrix A. Notation: Ker(A). Consider the homogeneous system of linear equations $AX = O$. Then the set of all solutions $\{X\}$ is the *null space* or the *right null space* of A. This right null space is also called the kernel of A.

Image of the matrix A. Notation: Im(A). Let A be an $m \times n$ matrix. Consider the set of all $m \times 1$ vectors Y such that $Y = AX$ for some $n \times 1$ vector X. This set is called the *image* or *range* of A. That is, $\mathrm{Im}(A) = \{Y : Y = AX$ for some $X\}$. If the matrix product AB is defined then show that

$$\mathrm{Ker}(AB) \supset \mathrm{Ker}(B)$$

with equality if A^{-1} exists, and

$$\mathrm{Im}(AB) \subset \mathrm{Im}(A)$$

with equality if B^{-1} exists.

The following are some problems on ranks posed by Dr R. B. Bapat of the Indian Statistical Institute, New Delhi, India.

4.4.28. Let A, B be $n \times n$ matrices. Show that

$$\mathrm{rank} \begin{bmatrix} A+B & A \\ A & A \end{bmatrix} = \mathrm{rank}(A) + \mathrm{rank}(B).$$

4.4.29. Let A be $m \times n$ and suppose $A = \begin{bmatrix} B & C \\ O & D \end{bmatrix}$. Show that $\mathrm{rank}\, A \geq \mathrm{rank}(B) + \mathrm{rank}(D)$ and that the inequality may be strict.

4.4.30. Let A be $n \times n$ nonsingular matrix of rank $n-1$ and suppose that each row sum and each column sum of A is zero. Show that every submatrix of A of order $(n-1) \times (n-1)$ is nonsingular.

4.4.31. Let A and B be $n \times n$ nonsingular matrices. Show that rank $(A-B) = \mathrm{rank}(A^{-1} - B^{-1})$.

4.4.32. Let A be $n \times n$ of rank 1. Show that $|I + A| = 1 + \mathrm{tr}(A)$.

4.4.33. Let A be $m \times n$, where the elements could be complex also. Show that rank $(A) = \mathrm{rank}(AA^*) = \mathrm{rank}(A^*A)$. Is it true that rank $(A) = \mathrm{rank}(AA')$?

4.4.34. Let A and B be $n \times n$ where B is positive semidefinite. Show that rank $(AB) = \mathrm{rank}(AB^{\frac{1}{2}})$.

4.4.35. Let A and B be $n \times n$. Show that rank $\left[\begin{smallmatrix} A & I \\ I & B \end{smallmatrix}\right] = n$ if and only if $B = A^{-1}$.

4.4.36. Show that rank $\left[\begin{smallmatrix} O & A \\ B & I_n \end{smallmatrix}\right] = n + \mathrm{rank}(AB)$.

4.4.37. Let A be $n \times n$. If rank $(A) = \mathrm{rank}(A^2)$ then show that rank $(A) = \mathrm{rank}(A^m)$ for all $m \geq 1$.

4.4.38. Let A be $n \times n$ and let rank $(A) = 1$. Show that rank $(A^2) = 1$ if and only if $\mathrm{tr}(A) \neq 0$.

4.4.39. Let A be $n \times n$. If rank $(A^k) = \mathrm{rank}(A^{k+1})$, then show that rank $(A^k) = \mathrm{rank}(A^m)$ for all $m \geq k$.

4.4.40. Let A be $n \times n$. Show that $A = A^2$ if and only if rank $(A) = \mathrm{rank}(I - A) = n$.

5 Some applications of matrices and determinants

5.0 Introduction

A few applications in solving difference and differential equations, applications in evaluating Jacobians of matrix transformations, optimization problems, probability measures and Markov processes and some topics in statistics will be discussed in this chapter.

5.1 Difference and differential equations

In order to introduce the idea of how eigenvalues can be used to solve difference and differential equations a few illustrative examples will be done here.

5.1.1 Fibonacci sequence and difference equations

The famous Fibonacci sequence is the following:

$$0, 1, 1, 2, 3, 5, 8, 13, 21, \ldots$$

where the sum of two consecutive numbers is the next number. Surprisingly, this sequence appears in very many places in nature. Consider a living micro organism such as a cell which is reproducing in the following fashion: To start with there is one mother. The mother cell needs only one unit of time to reproduce. Each mother produces only one daughter cell. The daughter cell needs one unit of time to grow and then one unit of time to reproduce. Let us examine the population size at each stage:

stage1	number $= 1$	one mother at the first unit of time	
stage 2	number $= 1$	1 mother only	
stage 3	number $= 2$	1 mother +1 young daughter	
stage 4	number $= 3$	1 mother, 1 mature and 1 young daughters	
stage 5	number $= 5$	2 mothers, 1 mature and 2 young daughters	

and so on. The population size follows the sequence $1, 1, 2, 3, 5, 8, \ldots$ the famous Fibonacci sequence.

If you look at the capitulum of a sunflower the florets, or the seeds when the florets become seeds, seem to be arranged along spirals starting from the periphery and going inward. You will see one set of such radial spirals going in one direction and another set of radial spirals going in the opposite direction. These numbers are always two successive numbers from the Fibonacci sequence. In a small sunflower it may be $(3, 5)$, in a slightly larger flower it may be $(5, 8)$ and so on. Arrangement of florets on a pineapple, thorns on certain cactus head, leaves on certain palm trees, petals in dhalias and in very many such divergent places one meets Fibonacci sequence. A theory of growth and forms, explanation for the emergence of Fibonacci sequence and a mathematically reconstructed sunflower head and many other details can be seen from the paper [5]. Incidently the journal, *Mathematical Biosciences*, has adapted the above mathematically reconstructed sunflower model as its cover design from 1976 onward.

If the Fibonacci number at the k-th stage is denoted by F_k then the number at the $(k+2)$-th stage is

$$F_{k+2} = F_{k+1} + F_k. \tag{5.1.1}$$

This is a difference equation of order 2. $F_{k+1} - F_k$ is a first order difference and $F_{k+2} - F_{k+1}$ is again a first order difference. Then going from F_k to F_{k+2} is a second order difference. That is, (5.1.1) is a second order difference equation. One way of computing F_k for any k, k may be 10 385, is to go through properties of matrices. In order to write a matrix equation let us introduce dummy equations such as $F_k = F_k$, $F_{k+1} = F_{k+1}$ and so on. Consider the equations

$$F_{k+2} = F_{k+1} + F_k$$
$$F_{k+1} = F_{k+1} \tag{5.1.2}$$

and

$$V_k = \begin{bmatrix} F_{k+1} \\ F_k \end{bmatrix} \implies V_{k+1} = \begin{bmatrix} F_{k+2} \\ F_{k+1} \end{bmatrix}.$$

Then the two equations in (5.1.2) can be written as

$$V_{k+1} = AV_k, \quad A = \begin{bmatrix} 1 & 1 \\ 1 & 0 \end{bmatrix}. \tag{5.1.3}$$

Let us assume $F_0 = 0$ and $F_1 = 1$ which means $V_0 = \begin{bmatrix} 1 \\ 0 \end{bmatrix}$. Then from (5.1.3) we have

$$V_1 = AV_0, \quad V_2 = AV_1 = A(AV_0) = A^2 V_0, \quad \ldots, \quad V_k = A^k V_0.$$

In order to compute V_k we need to compute only A^k since V_0 is known. Straight multiplication of A with A for a total of 10 385 times is not an easy process. We will use the property that the eigenvalues of A^k are the k-th powers of the eigenvalues of A, sharing the same eigenvectors of A. Let us compute the eigenvalues and the eigenvectors of A.

$$|A - \lambda I| = 0 \implies \begin{vmatrix} 1 - \lambda & 1 \\ 1 & -\lambda \end{vmatrix} = 0$$

$$\implies \lambda^2 - \lambda - 1 = 0$$

$$\implies \lambda_1 = \frac{1 + \sqrt{5}}{2}, \quad \lambda_2 = \frac{1 - \sqrt{5}}{2}.$$

An eigenvector corresponding to $\lambda_1 = \frac{1+\sqrt{5}}{2}$ is given by

$$\begin{bmatrix} 1 - \frac{(1+\sqrt{5})}{2} & 1 \\ 1 & -\frac{(1+\sqrt{5})}{2} \end{bmatrix} \begin{bmatrix} x_1 \\ x_2 \end{bmatrix} = \begin{bmatrix} 0 \\ 0 \end{bmatrix} \Rightarrow$$

$$x_1 = \frac{1+\sqrt{5}}{2}, \quad x_2 = 1, \text{ or}$$

$$X_1 = \begin{bmatrix} \frac{1+\sqrt{5}}{2} \\ 1 \end{bmatrix} = \begin{bmatrix} \lambda_1 \\ 1 \end{bmatrix}$$

is an eigenvector. Similarly an eigenvector corresponding to $\lambda_2 = \frac{1-\sqrt{5}}{2}$ is given by

$$X_2 = \begin{bmatrix} \frac{1-\sqrt{5}}{2} \\ 1 \end{bmatrix} = \begin{bmatrix} \lambda_2 \\ 1 \end{bmatrix}.$$

Let

$$Q = (X_1, X_2) = \begin{bmatrix} \lambda_1 & \lambda_2 \\ 1 & 1 \end{bmatrix} \Rightarrow$$

$$Q^{-1} = \frac{1}{(\lambda_1 - \lambda_2)} \begin{bmatrix} 1 & -\lambda_2 \\ -1 & \lambda_1 \end{bmatrix}.$$

Therefore

$$A = \begin{bmatrix} \lambda_1 & \lambda_2 \\ 1 & 1 \end{bmatrix} \begin{bmatrix} \lambda_1 & 0 \\ 0 & \lambda_2 \end{bmatrix} \left\{ \frac{1}{(\lambda_1 - \lambda_2)} \begin{bmatrix} 1 & -\lambda_2 \\ -1 & \lambda_1 \end{bmatrix} \right\}.$$

Since A and A^k share the same eigenvectors we have

$$A^k = \frac{1}{(\lambda_1 - \lambda_2)} \begin{bmatrix} \lambda_1 & \lambda_2 \\ 1 & 1 \end{bmatrix} \begin{bmatrix} \lambda_1^k & 0 \\ 0 & \lambda_2^k \end{bmatrix} \begin{bmatrix} 1 & -\lambda_2 \\ -1 & \lambda_1 \end{bmatrix}$$

$$= \frac{1}{(\lambda_1 - \lambda_2)} \begin{bmatrix} \lambda_1^{k+1} - \lambda_2^{k+1} & -\lambda_2 \lambda_1^{k+1} + \lambda_1 \lambda_2^{k+1} \\ \lambda_1^k - \lambda_2^k & -\lambda_2 \lambda_1^k + \lambda_1 \lambda_2^k \end{bmatrix}.$$

Hence

$$V_k = A^k V_0 = \frac{1}{(\lambda_1 - \lambda_2)} \begin{bmatrix} \lambda_1^{k+1} - \lambda_2^{k+1} \\ \lambda_1^k - \lambda_2^k \end{bmatrix}$$

$$= \frac{1}{\sqrt{5}} \begin{bmatrix} \lambda_1^{k+1} - \lambda_2^{k+1} \\ \lambda_1^k - \lambda_2^k \end{bmatrix}, \quad V_k = \begin{bmatrix} F_{k+1} \\ F_k \end{bmatrix}.$$

Therefore

$$F_k = \frac{1}{\sqrt{5}} (\lambda_1^k - \lambda_2^k) = \frac{1}{\sqrt{5}} \left\{ \left(\frac{1+\sqrt{5}}{2} \right)^k - \left(\frac{1-\sqrt{5}}{2} \right)^k \right\}.$$

Since $|\frac{1-\sqrt{5}}{2}| < 1$, λ_2^k will approach zero when k is large.

$$\lim_{k\to\infty} \frac{F_{k+1}}{F_k} = \lambda_1 = \frac{1+\sqrt{5}}{2} \approx 1.618. \tag{5.1.4}$$

Evidently, one has to take only powers of λ_1 when k is large. That is, $F_k \approx \frac{1}{\sqrt{5}}\lambda_1^k$ for large k. This number $\frac{1+\sqrt{5}}{2}$ is known as the "*golden ratio*" which appears in nature at many places.

One general observation that can be made is that we have an equation of the type

$$V_k = A^k V_0 = Q\Lambda^k Q^{-1} V_0$$

for an $n \times n$ matrix A where $\Lambda = \text{diag}(\lambda_1, \ldots, \lambda_n)$ and Q is the matrix of eigenvectors of A, assuming $|Q| \neq 0$. Then setting $Q^{-1}V_0 = C$, $C' = (c_1, \ldots, c_n)$ we have

$$\Lambda^k C = (\lambda_1^k c_1, \ldots, \lambda_n^k c_n)'.$$

If $X_1, \ldots, X_n$ are the eigenvectors of A, constituting $Q = (X_1, \ldots, X_n)$, then

$$V_k = Q(\Lambda^k Q^{-1} V_0) = (X_1, \ldots, X_n) \begin{bmatrix} \lambda_1^k c_1 \\ \vdots \\ \lambda_n^k c_n \end{bmatrix}$$

$$= c_1 \lambda_1^k X_1 + \cdots + c_n \lambda_n^k X_n$$

which is a linear function of $\lambda_i^k X_i$, $i = 1, \ldots, n$ or a linear combination of the so called pure solutions $\lambda_i^k X_i$.

Example 5.1.1. Suppose that a system is growing in the following fashion. The first stage size plus 3 times the second stage size plus the third stage size is the fourth stage size. Let $F_0 = 0$, $F_1 = 1$, $F_2 = 1$ be the initial conditions. Then

$$F_3 = 0 + 3(1) + 1 = 4, \quad F_4 = 1 + 3(1) + 4 = 7,$$
$$F_5 = 1 + 3(4) + 7 = 19,$$

and so on. Then for any k we have

$$F_k + 3F_{k+1} + F_{k+2} = F_{k+3}.$$

Compute F_k for $k = 100$.

Solution 5.1.1. Consider the following set of equations:

$$F_k + 3F_{k+1} + F_{k+2} = F_{k+3}$$
$$F_{k+2} = F_{k+2}$$

$$F_{k+1} = F_{k+1} \Rightarrow$$

$$\begin{bmatrix} 1 & 3 & 1 \\ 1 & 0 & 0 \\ 0 & 1 & 0 \end{bmatrix} \begin{bmatrix} F_{k+2} \\ F_{k+1} \\ F_k \end{bmatrix} = \begin{bmatrix} F_{k+3} \\ F_{k+2} \\ F_{k+1} \end{bmatrix}.$$

Let $U_k = \begin{bmatrix} F_{k+2} \\ F_{k+1} \\ F_k \end{bmatrix}$ and $A = \begin{bmatrix} 1 & 3 & 1 \\ 1 & 0 & 0 \\ 0 & 1 & 0 \end{bmatrix}$. Then we have

$$AU_k = U_{k+1} \quad \text{and} \quad U_0 = \begin{bmatrix} F_2 \\ F_1 \\ F_0 \end{bmatrix} = \begin{bmatrix} 1 \\ 1 \\ 0 \end{bmatrix};$$

$$AU_0 = U_1, \quad U_2 = A^2 U_0, \quad \dots.$$

Then

$$U_k = A^k U_0.$$

Let us compute the eigenvalues of A. Consider the equation

$$|A - \lambda I| = 0 \Rightarrow \begin{vmatrix} 1-\lambda & 3 & 1 \\ 1 & -\lambda & 0 \\ 0 & 1 & -\lambda \end{vmatrix} = 0$$

$$\Rightarrow -\lambda^3 + \lambda^2 + 3\lambda + 1 = 0.$$

Obviously $\lambda = -1$ is one root. Dividing $-\lambda^3 + \lambda^2 + 3\lambda + 1$ by $\lambda + 1$ we have $-\lambda^2 + 2\lambda + 1$. The other two roots are $[1 \pm \sqrt{2}]$. Then $\lambda_1 = -1$, $\lambda_2 = 1 + \sqrt{2}$, $\lambda_3 = 1 - \sqrt{2}$ are the roots. Let us compute some eigenvectors corresponding to these roots. For $\lambda = -1$

$$(A - \lambda_1 I)X = O \Rightarrow$$

$$\begin{bmatrix} 2 & 3 & 1 \\ 1 & 1 & 0 \\ 0 & 1 & 1 \end{bmatrix} \begin{bmatrix} x_1 \\ x_2 \\ x_3 \end{bmatrix} = \begin{bmatrix} 0 \\ 0 \\ 0 \end{bmatrix} \Rightarrow X_1 = \begin{bmatrix} 1 \\ -1 \\ 1 \end{bmatrix}$$

is one vector. For $\lambda_2 = 1 + \sqrt{2}$

$$(A - \lambda_2 I)X = O \Rightarrow$$

$$\begin{bmatrix} -\sqrt{2} & 3 & 1 \\ 1 & -1-\sqrt{2} & 0 \\ 0 & 1 & -1-\sqrt{2} \end{bmatrix} \begin{bmatrix} x_1 \\ x_2 \\ x_3 \end{bmatrix} = \begin{bmatrix} 0 \\ 0 \\ 0 \end{bmatrix} \Rightarrow$$

$$X_2 = \begin{bmatrix} 3+2\sqrt{2} \\ 1+\sqrt{2} \\ 1 \end{bmatrix}$$

is one vector. For $\lambda_3 = 1 - \sqrt{2}$

$$(A - \lambda_3 I)X = O \Rightarrow$$

$$\begin{bmatrix} \sqrt{2} & 3 & 1 \\ 1 & -1+\sqrt{2} & 0 \\ 0 & 1 & -1+\sqrt{2} \end{bmatrix} \begin{bmatrix} x_1 \\ x_2 \\ x_3 \end{bmatrix} = \begin{bmatrix} 0 \\ 0 \\ 0 \end{bmatrix} \Rightarrow$$

$$X_3 = \begin{bmatrix} 3-2\sqrt{2} \\ 1-\sqrt{2} \\ 1 \end{bmatrix}.$$

is one vector. Let

$$Q = (X_1, X_2, X_3) = \begin{bmatrix} 1 & 3+2\sqrt{2} & 3-2\sqrt{2} \\ -1 & 1+\sqrt{2} & 1-\sqrt{2} \\ 1 & 1 & 1 \end{bmatrix},$$

$$Q^{-1} = \frac{1}{4} \begin{bmatrix} 2 & -4 & -2 \\ \sqrt{2}-1 & 2-\sqrt{2} & 3-2\sqrt{2} \\ -\sqrt{2}-1 & 2+\sqrt{2} & 3+2\sqrt{2} \end{bmatrix}.$$

Therefore

$$U_k = A^k U_0$$

$$= Q \begin{bmatrix} (-1)^k & 0 & 0 \\ 0 & (1+\sqrt{2})^k & 0 \\ 0 & 0 & (1-\sqrt{2})^k \end{bmatrix} Q^{-1} \begin{bmatrix} 1 \\ 1 \\ 0 \end{bmatrix}.$$

But

$$Q^{-1} \begin{bmatrix} 1 \\ 1 \\ 0 \end{bmatrix} = \frac{1}{4} \begin{bmatrix} -2 \\ 1 \\ 1 \end{bmatrix}$$

and

$$U_k = \frac{1}{4} \begin{bmatrix} 2(-1)^{k+1} + (3+2\sqrt{2})(1+\sqrt{2})^k + (3-2\sqrt{2})(1-\sqrt{2})^k \\ 2(-1)^{k+2} + (1+\sqrt{2})^{k+1} + (1-\sqrt{2})^{k+1} \\ 2(-1)^{k+1} + (1+\sqrt{2})^k + (1-\sqrt{2})^k \end{bmatrix}$$

$$= \begin{bmatrix} F_{k+2} \\ F_{k+1} \\ F_k \end{bmatrix}.$$

When $k \to \infty$ we have $(1-\sqrt{2})^k \to 0$. Thus for k large a good approximation to U_k is the following:

$$U_k \approx \frac{1}{4} \begin{bmatrix} 2(-1)^{k+1} + (3+2\sqrt{2})(1+\sqrt{2})^k \\ 2(-1)^{k+2} + (1+\sqrt{2})^{k+1} \\ 2(-1)^{k+1} + (1+\sqrt{2})^k \end{bmatrix} = \begin{bmatrix} F_{k+2} \\ F_{k+1} \\ F_k \end{bmatrix}.$$

Hence for $k = 100$

$$F_{100} \approx \frac{1}{4}[-2 + (1 + \sqrt{2})^{100}], \quad F_{101} \approx \frac{1}{4}[2 + (1 + \sqrt{2})^{101}],$$
$$F_{102} \approx \frac{1}{4}[-2 + (3 + 2\sqrt{2})(1 + \sqrt{2})^{100}].$$

5.1.2 Population growth

Consider, for example competing populations of foxes and rabbits in a given region. If there is no rabbit available to eat the foxes die out. If rabbits are available then for every kill the population of foxes has a chance of increasing. Suppose that the observations are made at the end of every six months, call them stages $0, 1, 2, \ldots$ where stage 0 means the starting number. Let F_i and R_i denote the fox and rabbit populations at stage i. Suppose that the growth of fox population is governed by the difference equation

$$F_{i+1} = 0.6F_i + 0.2R_i.$$

Left alone the rabbits multiply. Thus the rabbit population is influenced by the natural growth minus the ones killed by the foxes. Suppose that the rabbit population is given by the equation

$$R_{i+1} = 1.5R_i - pF_i$$

where p is some number. We will look at the problem for various values of p. Suppose that the initial populations of foxes and rabbits are 10 and 100 respectively. Let us denote by

$$X_0 = \begin{pmatrix} F_0 \\ R_0 \end{pmatrix} = \begin{pmatrix} 10 \\ 100 \end{pmatrix}, \quad X_i = \begin{pmatrix} F_i \\ R_i \end{pmatrix}.$$

Then the above difference equations can be written as

$$\begin{pmatrix} F_{i+1} \\ R_{i+1} \end{pmatrix} = \begin{pmatrix} 0.6 & 0.2 \\ -p & 1.5 \end{pmatrix} \begin{pmatrix} F_i \\ R_i \end{pmatrix} \quad \text{or}$$

$$X_{i+1} = AX_i, \quad A = \begin{pmatrix} 0.6 & 0.2 \\ -p & 1.5 \end{pmatrix}.$$

Thus

$$X_1 = AX_0, \quad X_2 = AX_1 = A^2X_0, \quad \ldots, \quad X_k = A^kX_0.$$

For example, at the first observation period the population sizes are given by

$$X_1 = AX_0 = \begin{pmatrix} 0.6 & 0.2 \\ -p & 1.5 \end{pmatrix} \begin{pmatrix} 10 \\ 100 \end{pmatrix}.$$

For example, for $p = 1$, the numbers are

$$F_1 = (0.6)(10) + (0.2)(100) = 26$$

and

$$R_1 = -1(10) + 1.5(100) = 140.$$

Let us see what happens in the second stage with the same p, that is for $k = 2$, $p = 1$. This can be computed either from the first stage values and by using A or from the initial values and by using A^2. That is,

$$\begin{pmatrix} F_2 \\ R_2 \end{pmatrix} = A^2 \begin{pmatrix} F_0 \\ R_0 \end{pmatrix} = A \begin{pmatrix} F_1 \\ R_1 \end{pmatrix}$$

$$= \begin{pmatrix} 0.6 & 0.2 \\ -1 & 1.5 \end{pmatrix} \begin{pmatrix} 26 \\ 140 \end{pmatrix} = \begin{pmatrix} 43.6 \\ 184 \end{pmatrix}, \quad F_2 \approx 44, \quad R_2 = 184.$$

Note that for $p = 1$ the fox population and the rabbit population will explode eventually. Let us see what happens if $p = 5$. Then

$$X_1 = AX_0 = \begin{pmatrix} 0.6 & 0.2 \\ -5 & 1.5 \end{pmatrix} \begin{pmatrix} 10 \\ 100 \end{pmatrix} = \begin{pmatrix} 26 \\ 100 \end{pmatrix}, \quad F_1 = 26, \quad R_1 = 100.$$

$$X_2 = A^2 X_0 = AX_1 = \begin{pmatrix} 35.6 \\ 20 \end{pmatrix}, \quad F_2 = 35.6, \quad R_2 = 20.$$

Note that at the next stage the rabbits will disappear and from then on the fox population will start decreasing at each stage.

Growths of interdependent species of animals, insects, plants and so on are governed by difference equations of the above types. If there are three competing populations involved then the coefficient matrix A will be 3×3 and if there are n such populations then A will be $n \times n$, $n \geq 1$. The long-term behavior of the populations can be studied by looking at the eigenvalues of A because when A is representable as

$$A = QDQ^{-1}, \quad D = \text{diag}(\lambda_1, \dots, \lambda_n) \implies A^k = QD^k Q^{-1}$$

where $\lambda_1, \dots, \lambda_n$ are the eigenvalues of the $n \times n$ matrix A. Then $\lambda^k \to 0$ as $k \to \infty$ when $|\lambda| < 1$ and $\lambda^k \to \infty$ for $\lambda > 1$ as $k \to \infty$. Thus the eventual extinction or explosion or stability of the populations is decided by the eigenvalues of A.

5.1.3 Differential equations and their solutions

Consider a system of total differential equations of the linear homogeneous type with constant coefficients. Suppose that a supermarket has barrels of almonds and pecans (two competing types of nuts as far as demand is concerned). Let u denote the amount of stock, in kilograms (kg) of almonds and v that of pecans. The store fills up the barrels according to the sales. The store finds that the rate of change of u over time is a linear function of u and v, so also the rate of change of v over time t. Suppose that the following are the equations.

$$\frac{du}{dt} = 2u + v$$

$$\frac{dv}{dt} = u + 2v$$

which means

$$\frac{dW}{dt} = AW, \quad W = \begin{bmatrix} u \\ v \end{bmatrix}, \quad A = \begin{bmatrix} 2 & 1 \\ 1 & 2 \end{bmatrix}. \tag{5.1.5}$$

At the start of the observations, $t = 0$, suppose that the stock is $u = 500$ kg and $v = 200$ kg. If W is the vector $W = \begin{bmatrix} u \\ v \end{bmatrix}$ then we say that the initial value of W, denoted by W_0, is $W_0 = \begin{bmatrix} 500 \\ 200 \end{bmatrix}$. We want to solve (5.1.5) with this initial value. The differential equations in (5.1.5) are linear and homogeneous in u and v with u and v having constant (free of t) coefficients. The method that we will describe here will work for n equations in n variables $u_1, \ldots, u_n$, where each is a function of another independent variable such as t, $u_i = u_i(t)$, $i = 1, \ldots, n$, and when the right sides are linear homogeneous with constant coefficients. For simplicity we consider only a two variables case.

If there was only one equation in one variable of the type in (5.1.5) then the equation would be of the form

$$\frac{du}{dt} = au$$

where a is a known number. Then the solution is

$$u = e^{at} u_0 \quad \text{if } u = u_0 \text{ at } t = 0 \quad \text{(initial value).} \tag{5.1.6}$$

Then in the case of two equations as in (5.1.5) we can search for solutions of the type in (5.1.6). Let us assume that

$$u = e^{\lambda t} x_1 \quad \text{and} \quad v = e^{\lambda t} x_2,$$

$$W = \begin{pmatrix} u \\ v \end{pmatrix} = \begin{pmatrix} e^{\lambda t} x_1 \\ e^{\lambda t} x_2 \end{pmatrix} = e^{\lambda t} X,$$

$$X = \begin{pmatrix} x_1 \\ x_2 \end{pmatrix} \tag{5.1.7}$$

for some unknown λ, the same λ for both u and v, x_1 and x_2 are some parameters free of t. Substituting these in (5.1.5) we obtain

$$\lambda e^{\lambda t} x_1 = 2 e^{\lambda t} x_1 + e^{\lambda t} x_2,$$

$$\lambda e^{\lambda t} x_2 = e^{\lambda t} x_1 + 2 e^{\lambda t} x_2. \tag{5.1.8}$$

Canceling $e^{\lambda t}$ and writing the equations in matrix form we have

$$AX = \lambda X, \quad A = \begin{bmatrix} 2 & 1 \\ 1 & 2 \end{bmatrix}, \quad X = \begin{bmatrix} x_1 \\ x_2 \end{bmatrix}. \tag{5.1.9}$$

The problem reduces to that of finding the eigenvalues and eigenvectors of A. The eigenvalues are given by

$$|A - \lambda I| = 0 \Rightarrow \begin{vmatrix} 2 - \lambda & 1 \\ 1 & 2 - \lambda \end{vmatrix} = 0$$

$$\Rightarrow \lambda^2 - 4\lambda + 3 = 0$$

$$\Rightarrow \lambda_1 = 1, \quad \lambda_2 = 3.$$

An eigenvector corresponding to $\lambda = 1$ is given by

$$\begin{bmatrix} 2 - 1 & 1 \\ 1 & 2 - 1 \end{bmatrix} \begin{bmatrix} x_1 \\ x_2 \end{bmatrix} = \begin{bmatrix} 0 \\ 0 \end{bmatrix} \Rightarrow x_1 = -x_2 \Rightarrow X_1 = \begin{bmatrix} 1 \\ -1 \end{bmatrix}$$

is one vector. Corresponding to $\lambda_2 = 3$,

$$\begin{bmatrix} 2 - 3 & 1 \\ 1 & 2 - 3 \end{bmatrix} \begin{bmatrix} x_1 \\ x_2 \end{bmatrix} = \begin{bmatrix} 0 \\ 0 \end{bmatrix} \Rightarrow X_2 = \begin{bmatrix} 1 \\ 1 \end{bmatrix}$$

is one vector. For $\lambda = \lambda_1 = 1$ a solution for W is

$$W_1 = e^{\lambda t} X = e^t X_1 = e^t \begin{bmatrix} 1 \\ -1 \end{bmatrix}. \tag{5.1.10}$$

For $\lambda = \lambda_2 = 3$ a solution for W is

$$W_2 = e^{\lambda t} X = e^{3t} X_2 = e^{3t} \begin{bmatrix} 1 \\ 1 \end{bmatrix}. \tag{5.1.11}$$

Any linear function of W_1 and W_2 is again a solution for W. Hence a general solution for W is

$$W = c_1 W_1 + c_2 W_2 = c_1 e^t \begin{bmatrix} 1 \\ -1 \end{bmatrix} + c_2 e^{3t} \begin{bmatrix} 1 \\ 1 \end{bmatrix} \tag{5.1.12}$$

where c_1 and c_2 are arbitrary constants. Let us try to choose c_1 and c_2 to satisfy the initial condition, $W_0 = \begin{bmatrix} 500 \\ 200 \end{bmatrix}$ for $t = 0$. Letting $t = 0$ in (5.1.12) we have

$$c_1 \begin{bmatrix} 1 \\ -1 \end{bmatrix} + c_2 \begin{bmatrix} 1 \\ 1 \end{bmatrix} = W_0 = \begin{bmatrix} 500 \\ 200 \end{bmatrix}$$

$$\Rightarrow c_1 + c_2 = 500, \quad -c_1 + c_2 = 200$$

$$\Rightarrow c_2 = 350, \quad c_1 = 150.$$

Then the solution to the equation in (5.1.5) is

$$W = \begin{bmatrix} u \\ v \end{bmatrix} = 150e^t \begin{bmatrix} 1 \\ -1 \end{bmatrix} + 350e^{3t} \begin{bmatrix} 1 \\ 1 \end{bmatrix} \Rightarrow$$

$$u = 150e^t + 350e^{3t}, \quad v = -150e^t + 350e^{3t}.$$

Since the exponents are positive, $e^{bt} \to \infty$ as $t \to \infty$ when $b > 0$, u and v both increase with time. In fact, the eigenvalues $\lambda_1 = 1$ and $\lambda_2 = 3$, appearing in the exponents, measure the rate of growth. This can be noticed from the pure solutions in (5.1.10) and (5.1.11). A mixture of these pure solutions is what is given in (5.1.12). If an eigenvalue λ is positive, as in (5.1.10) and (5.1.11), then $e^{\lambda t} \to \infty$ as $t \to \infty$. In this case we say that the equations are *unstable*. If $\lambda = 0$ the equations are said to be *neutrally stable*. When $\lambda < 0, e^{\lambda t} \to 0$ as $t \to \infty$. In this case we say that the equations are *stable*. In our example above, the pure solutions for both $\lambda_1 = 1$ and $\lambda_2 = 3$, as seen from (5.1.10) and (5.1.11), are unstable.

A slightly more general situation arises if there are some constant coefficients for $\frac{du}{dt}$ and $\frac{dv}{dt}$ in (5.1.5).

Example 5.1.2. Solve the following system of differential equations if u and v are functions of t and when $t = 0, u = 100 = u_0$ and $v = 200 = v_0$:

$$2\frac{du}{dt} = 2u + v,$$

$$3\frac{dv}{dt} = u + 2v. \tag{5.1.13}$$

Solution 5.1.2. Divide the first equation by 2 and the second equation by 3. Then the problem reduces to that in (5.1.5). But if we want to avoid fractions at the beginning stage itself of solving the system, or to solve the system as they are in (5.1.13), then we look for a solution of the type

$$u = e^{\lambda t} x_1, \quad v = e^{\lambda t} x_2$$

for some λ and for some constants x_1 and x_2. [Observe that if the original system of equations has some fractional coefficients then multiply the system by appropriate numbers to make the coefficients non-fractional. Then the following procedure can be applied.] Then the equations in (5.1.13) reduce to the following form:

$$2\lambda e^{\lambda t} x_1 = 2e^{\lambda t} x_1 + e^{\lambda t} x_2,$$
$$3\lambda e^{\lambda t} x_2 = e^{\lambda t} x_1 + 2e^{\lambda t} x_2.$$

Canceling $e^{\lambda t}$ and writing

$$X = \begin{pmatrix} x_1 \\ x_2 \end{pmatrix} \Rightarrow W = \begin{pmatrix} u \\ v \end{pmatrix} = e^{\lambda t} X,$$

we have,

$$\begin{bmatrix} 2\lambda & 0 \\ 0 & 3\lambda \end{bmatrix} X = \begin{bmatrix} 2 & 1 \\ 1 & 2 \end{bmatrix} X \implies \begin{bmatrix} 2-2\lambda & 1 \\ 1 & 2-3\lambda \end{bmatrix} X = O.$$

If this equation has a non-null solution then

$$\begin{vmatrix} 2-2\lambda & 1 \\ 1 & 2-3\lambda \end{vmatrix} = 0 \implies 6\lambda^2 - 10\lambda + 3 = 0$$

$$\implies \lambda_1 = \frac{5+\sqrt{7}}{6}, \quad \lambda_2 = \frac{5-\sqrt{7}}{6}.$$

Let us compute X corresponding to λ_1 and λ_2. For $\lambda_1 = \frac{5+\sqrt{7}}{6}$

$$\begin{bmatrix} 2-2(\frac{5+\sqrt{7}}{6}) & 1 \\ 1 & 2-3(\frac{5+\sqrt{7}}{6}) \end{bmatrix} \begin{bmatrix} x_1 \\ x_2 \end{bmatrix} = \begin{bmatrix} 0 \\ 0 \end{bmatrix}.$$

One solution for X is

$$X_1 = \begin{bmatrix} 1 \\ -\frac{1}{3} + \frac{\sqrt{7}}{3} \end{bmatrix}.$$

Similarly for $\lambda_2 = \frac{5-\sqrt{7}}{6}$ one solution is

$$X_2 = \begin{bmatrix} 1 \\ -\frac{1}{3} - \frac{\sqrt{7}}{3} \end{bmatrix}.$$

For $\lambda = \lambda_1$ one solution for W is

$$W_1 = e^{\lambda_1 t} X_1 = e^{(\frac{5+\sqrt{7}}{6})t} \begin{bmatrix} 1 \\ -\frac{1}{3} + \frac{\sqrt{7}}{3} \end{bmatrix}$$

and for $\lambda = \lambda_2$ the solution for W is

$$W_2 = e^{\lambda_2 t} X_2 = e^{(\frac{5-\sqrt{7}}{6})t} \begin{bmatrix} 1 \\ -\frac{1}{3} - \frac{\sqrt{7}}{3} \end{bmatrix}.$$

Thus a general solution for W is $W = c_1 W_1 + c_2 W_2$ where c_1 and c_2 are arbitrary constants. That is,

$$W = c_1 e^{(\frac{5+\sqrt{7}}{6})t} \begin{bmatrix} 1 \\ -\frac{1}{3} + \frac{\sqrt{7}}{3} \end{bmatrix} + c_2 e^{(\frac{5-\sqrt{7}}{6})t} \begin{bmatrix} 1 \\ -\frac{1}{3} - \frac{\sqrt{7}}{3} \end{bmatrix} \implies$$

$$u = c_1 e^{(\frac{5+\sqrt{7}}{6})t} + c_2 e^{(\frac{5-\sqrt{7}}{6})t}$$

and

$$v = c_1\left(-\frac{1}{3} + \frac{\sqrt{7}}{3}\right)e^{\left(\frac{5+\sqrt{7}}{6}\right)t} + c_2\left(-\frac{1}{3} - \frac{\sqrt{7}}{3}\right)e^{\left(\frac{5-\sqrt{7}}{6}\right)t}.$$

But for $t = 0$, $u = u_0 = 100$ and for $t = 0$, $v = v_0 = 200$. That is,

$$100 = c_1 + c_2$$
$$200 = c_1\left(-\frac{1}{3} + \frac{\sqrt{7}}{3}\right) + c_2\left(-\frac{1}{3} - \frac{\sqrt{7}}{3}\right).$$

Solving for c_1 and c_2 we have

$$c_1 = 50(1 + \sqrt{7}) \quad \text{and} \quad c_2 = 50(1 - \sqrt{7}).$$

Hence the general solution is,

$$u = 50(1 + \sqrt{7})e^{\left(\frac{5+\sqrt{7}}{6}\right)t} + 50(1 - \sqrt{7})e^{\left(\frac{5-\sqrt{7}}{6}\right)t}$$
$$v = 50(1 + \sqrt{7})\left(-\frac{1}{3} + \frac{\sqrt{7}}{3}\right)e^{\left(\frac{5+\sqrt{7}}{6}\right)t} - 50(1 - \sqrt{7})\left(\frac{1 + \sqrt{7}}{3}\right)e^{\left(\frac{5-\sqrt{7}}{6}\right)t}$$
$$= 100e^{\left(\frac{5+\sqrt{7}}{6}\right)t} + 100e^{\left(\frac{5-\sqrt{7}}{6}\right)t}.$$

Note that the same procedure works if we have m-th order equations of the type

$$b_1\frac{d^m u_1}{dt^m} = a_{11}u_1 + \cdots + a_{1k}u_k$$
$$\vdots \quad = \quad \vdots$$
$$b_k\frac{d^m u_k}{dt^m} = a_{k1}u_1 + \cdots + a_{kk}u_k \tag{5.1.14}$$

where $b_1, \ldots, b_k$ and a_{ij}'s are all constants and u_j, $j = 1, \ldots, k$ are functions of t. In this case look for a solution of the type $u_j = e^{\mu t}x_j$, $j = 1, \ldots, k$ with the same μ and x_j's are some quantities free of t. Then the left sides of (5.1.14) will contain μ^m. Put $\lambda = \mu^m$. Then the problem reduces to the one in Example 5.1.2.

Higher order differential equations can also be solved by using the same technique as above. In order to illustrate the procedure we will do a simple example here.

Example 5.1.3. Let y be a function of t and let y', y'', y''' denote the first order, second order and third order derivatives respectively. Solve the following differential equation by using eigenvalue method:

$$y''' - 4y'' + 3y' = 0.$$

Solution 5.1.3. The classical way of doing the problem is to search for an exponential solution of the type $y = e^{\lambda t}$. Then we get the characteristic equation

$$\lambda^3 - 4\lambda^2 + 3\lambda = 0 \implies \lambda_1 = 0, \quad \lambda_2 = 1, \quad \lambda_3 = 3$$

are the solutions of this characteristic equation. Hence the three pure exponential solutions are $e^{0t} = 1$, e^t, e^{3t}. Now let us do the same problem by using eigenvalues. Let

$$u = y', \quad v = y'' = u', \quad v' = 4v - 3u \quad \text{and}$$

$$W = \begin{bmatrix} y \\ u \\ v \end{bmatrix} \implies W' = \begin{bmatrix} y' \\ u' \\ v' \end{bmatrix} = \begin{bmatrix} y' \\ y'' \\ y''' \end{bmatrix}.$$

Writing the above three equations in terms of the vector W and its first derivative we have

$$\frac{dW}{dt} = W' = \begin{bmatrix} 0 & 1 & 0 \\ 0 & 0 & 1 \\ 0 & -3 & 4 \end{bmatrix} \begin{bmatrix} y \\ u \\ v \end{bmatrix} = AW,$$

$$A = \begin{bmatrix} 0 & 1 & 0 \\ 0 & 0 & 1 \\ 0 & -3 & 4 \end{bmatrix}. \tag{5.1.15}$$

Now, compare with (5.1.5). We have a first order system in W. Let $y = e^{\lambda t} x_1$, $u = e^{\lambda t} x_2$, $v = e^{\lambda t} x_3$ for some x_1, x_2, x_3 free of t. Then substituting in (5.1.15) and canceling $e^{\lambda t}$ the equation $W' = AW$ reduces to the form

$$AX = \lambda X, \quad X = \begin{bmatrix} x_1 \\ x_2 \\ x_3 \end{bmatrix} \tag{5.1.16}$$

or the problem reduces to an eigenvalue problem. The eigenvalues of A are $\lambda_1 = 0$, $\lambda_2 = 1$, $\lambda_3 = 3$. Some eigenvectors corresponding to these eigenvalues are the following:

$$X_1 = \begin{bmatrix} 1 \\ 0 \\ 0 \end{bmatrix}, \quad X_2 = \begin{bmatrix} 1 \\ 1 \\ 1 \end{bmatrix}, \quad X_3 = \begin{bmatrix} 1 \\ 3 \\ 9 \end{bmatrix}$$

which gives

$$W_1 = e^{\lambda_1 t} X_1 = \begin{bmatrix} 1 \\ 0 \\ 0 \end{bmatrix}, \quad W_2 = e^{\lambda_2 t} X_2 = \begin{bmatrix} e^t \\ e^t \\ e^t \end{bmatrix},$$

$$W_3 = e^{\lambda_3 t} X_3 = \begin{bmatrix} e^{3t} \\ 3e^{3t} \\ 9e^{3t} \end{bmatrix}.$$

Thus the pure solutions for y are $1, e^t$ and e^{3t}. A general solution for y is then

$$y = c_1 + c_2 e^t + c_3 e^{3t} \tag{5.1.17}$$

where c_1, c_2, c_3 are arbitrary constants.

Exercises 5.1

5.1.1. Suppose that a population of bacteria colony increases by the following laws. If the population at the k-th stage is a_k then

$$a_k + 2a_{k+1} = a_{k+2}.$$

Compute the population size at the 100-th stage ($k = 100$) if the initial numbers are $a_0 = 1$, $a_1 = 1$.

5.1.2. Prove that every third Fibonacci number is even, starting with $1, 1$.

5.1.3. Show that the Fibonacci sequence $F_k + F_{k+1} = F_{k+2}$ with $F_0 = 0, F_1 = 1$ is such that

$$\lim_{k \to \infty} \frac{F_{k+1}}{F_k} = \frac{1 + \sqrt{5}}{2} = \text{golden ratio.}$$

5.1.4. Consider a sequence $F_k + F_{k+1} = F_{k+2}$ with $F_0 = 0, F_1 = a > 0$. Show that

$$\lim_{k \to \infty} \frac{F_{k+1}}{F_k} = \frac{1 + \sqrt{5}}{2} = \text{golden ratio.}$$

5.1.5. Find the limiting value, $\lim_{k \to \infty} W_k = \lim_{k \to \infty} \binom{u_k}{v_k}$ for the following systems:

$$
\begin{aligned}
\text{(a)} \quad & u_{k+1} = \frac{1}{2} u_k + \frac{1}{2} v_k, \quad u_0 = 2, \\
& v_{k+1} = \frac{1}{2} u_k + \frac{1}{2} v_k, \quad v_0 = 1; \\
\text{(b)} \quad & u_{k+1} = 0.4 u_k + 0.3 v_k, \quad u_0 = \frac{1}{2}, \\
& v_{k+1} = 0.6 u_k + 0.7 v_k, \quad v_0 = \frac{2}{3}.
\end{aligned}
$$

5.1.6. For the sequence $F_{k+2} = \frac{1}{2}[F_k + F_{k+1}]$ compute $\lim_{k \to \infty} F_k$.

5.1.7. Solve the following systems of differential equations by using matrix methods:

$$
\begin{aligned}
\text{(a)} \quad & \frac{du}{dt} = 2u + v, \quad \frac{dv}{dt} = u + 2v - w, \quad \frac{dw}{dt} = -v + 2w. \\
\text{(v)} \quad & \frac{du}{dt} = u + 2v - w, \quad \frac{dv}{dt} = 2u + 3v + 2w, \quad \frac{dw}{dt} = 3u + 2v + w.
\end{aligned}
$$

5.1.8. In physics, an oscillating system with unequal masses m_1 and m_2 is governed by the following system of differential equations:

$$m_1 \frac{d^2u}{dt^2} = -2u + v, \quad m_2 \frac{d^2v}{dt^2} = u - 2v$$

where u and v are functions of t. Solve the system for $m_1 = 1$, $m_2 = 2$ and with the initial conditions, at $t = 0$, $u = u_0 = -1$ and $v = v_0 = 1$.

5.1.9. A system of differential equations governing the diffusion of a chemical in two different concentrations is the following:

$$\frac{du}{dt} = (v - u) + (0 - u)$$

$$\frac{dv}{dt} = (0 - v) + (u - v)$$

where $u = u(t)$ and $v = v(t)$ denote the concentrations. Solve the system when u_0 and v_0 are the concentrations at $t = 0$ respectively.

5.1.10. Let

$$W = \begin{pmatrix} u \\ v \end{pmatrix}, \quad \frac{dW}{dt} = \begin{pmatrix} \frac{du}{dt} \\ \frac{dv}{dt} \end{pmatrix}.$$

Solve the following systems of differential equations:

(a) $\quad \dfrac{dW}{dt} = \begin{bmatrix} -1 & 1 \\ 1 & -1 \end{bmatrix} W, \quad W_0 = \begin{bmatrix} 2 \\ 1 \end{bmatrix}$

(b) $\quad \dfrac{dW}{dt} = \begin{bmatrix} 1 & -1 \\ -1 & 1 \end{bmatrix} W, \quad W_0 = \begin{bmatrix} 2 \\ 1 \end{bmatrix}.$

5.1.11. Solve the following system of differential equations by using eigenvalue method, if possible, where y is a function of t and primes denote the derivatives.
(a) y'''-5y''+4y'=0;
(b) y''+y=0;
(c) y''=0;
(d) y''+ay'+by=0,

where a and b are constants.

5.1.12. A biologist has found that the owl population u and the mice population v in a particular area are governed by the following system of differential equations, where t denotes time:

$$\frac{du}{dt} = 2u + 2v$$

$$\frac{dv}{dt} = u + 2v.$$

(a) Solve the system if the initial values are $u_0 = 2$, $v_0 = 100$. (b) Is the system, stable, neutrally stable, or unstable? (c) What will be the proportions of mice and owl in the long run, that is, when $t \to \infty$?

5.1.13. In Exercise 5.1.12 suppose that the inventory of owl population u and the mice population v are taken at the beginning of every year. At the i-th year let these be u_i and v_i respectively. At the beginning of the observation period let the population sizes be $u_0 = 5, V_0 = 60$, that is, for $i = 0$. Suppose that it is found that the population growth is governed by the following difference equations

$$u_{i+1} = 4u_i - 2v_i \quad \text{and} \quad v_{i+1} = -5u_i + 2v_i.$$

Compute u_i and v_i for $i = 1, 2, 3, \infty$.

5.1.14. Suppose that two falcons are introduced into the same region of the owl and mice habitat of Exercise 5.1.13. Thus the initial population sizes of falcon, owl and mice are $f_0 = 2, u_0 = 5, v_0 = 60$. The falcons kill both owl and mice and the owls kill mice and not falcons. Suppose that the difference equations governing the population growth are the following:

$$f_{i+1} = f_i + 2u_i - 2v_i,$$
$$u_{1+1} = 2f_i + 3u_i - 3v_i,$$
$$v_{i+1} = -2f_i - 3u_i + 3v_i.$$

Compute the population sizes of falcon, owl and mice at $i = 1, 2, 3, 4, \infty$.

5.1.15. In Exercise 5.1.14 if the observation period t is continuous and if the rate of change of the falcon, owl and mice populations, with respect to t, are governed by the differential equations

$$\frac{df}{dt} = f + u + v, \quad \frac{du}{dt} = -f + u + 2v, \quad \frac{dv}{dt} = f - u - 2v$$

with the initial populations, at $t = 0$, respectively $f_0 = 2, u_0 = 5, v_0 = 60$, then compute the eventual, $t \to \infty$, population sizes of falcon, owl and mice in that region.

5.2 Jacobians of matrix transformations and functions of matrix argument

This field is very vast and hence we will select a few topics on Jacobians and introduce some functions of matrix argument for the purpose of illustrating the possible applications of matrices and determinants.

5.2.1 Jacobians of matrix transformations

Jacobians in some linear and non-linear transformations are discussed in Section 3.3.3 of Chapter 3. One linear transformation appearing as Exercise 3.3.7 will be restated here because it is concerned with a real symmetric matrix. Let $X = X'$ be a real symmetric $p \times p$ matrix of functionally independent (distinct) $p(p + 1)/2$ real scalar variables and let A be a $p \times p$ nonsingular matrix of constants. Then, ignoring the sign, we have the following result:

(i)
$$Y = AXA' \;\Rightarrow\; \mathrm{d}Y = \begin{cases} |A|^{p+1}\mathrm{d}X, & \text{for } X = X', \quad |A| \neq 0 \\ |A|^{2p}\mathrm{d}X, & \text{for a general } X, \quad |A| \neq 0. \end{cases}$$

The second part for a general X of p^2 distinct elements follows as a particular case of the multi-linear transformation of Section 3.3.3.

Example 5.2.1. Let X be a $p \times 1$ vector of real scalar variables, V a $p \times p$ real symmetric positive definite matrix of constants. Consider the function

$$f(X) = c\, e^{-\frac{1}{2}(X-\mu)'V^{-1}(X-\mu)}$$

where μ is a $p \times 1$ vector of constants and c is a constant. If $f(X)$ is a statistical density then evaluate c.

Solution 5.2.1. For $f(X)$ to be a density, two conditions are to be satisfied: (1) $f(X) \geq 0$ for all X; (2) $\int_X f(X)\mathrm{d}X = 1$ where $\int_X$ denotes the integral over X and $\mathrm{d}X = \wedge_{j=1}^{p} \mathrm{d}x_{ij}$. Let us check the conditions. Since the exponential function cannot be negative, condition (1) is satisfied if $c > 0$. Let us look for the total integral. Since $V = V' > O$ (positive definite) we can write V^{-1} in the form $V^{-1} = BB'$ for some nonsingular matrix B. Then

$$(X - \mu)' V^{-1}(X - \mu) = (X - \mu)' BB'(X - \mu).$$

Let

$$Y = B'(X - \mu) \;\Rightarrow\; \mathrm{d}Y = |B'|\mathrm{d}(X - \mu) = |B'|\mathrm{d}X$$

since μ is a constant vector, $\mathrm{d}(X - \mu) = \mathrm{d}X$, and $|B'| = |B|$ from property (i) of Section 3.3. But

$$|V^{-1}| = |BB'| = |B||B'| = |B|^2 = |V^{-1}|^{\frac{1}{2}} = |V|^{-\frac{1}{2}}.$$

Now, consider the integral

$$\int_X f(X)\mathrm{d}X = c \int_X e^{-\frac{1}{2}(X-\mu)'V^{-1}(X-\mu)}\mathrm{d}X$$

$$= c|V|^{\frac{1}{2}} \int_Y e^{-\frac{1}{2}Y'Y}\mathrm{d}Y$$

where $Y = B'(X - \mu)$, $V^{-1} = BB'$. But $Y'Y = y_1^2 + \cdots + y_p^2$ where y_j is the j-th element in Y. Thus, the integral reduces to the form

$$\int_Y e^{-\frac{1}{2}Y'Y} dy = \prod_{j=1}^p \int_{-\infty}^{\infty} e^{-\frac{1}{2}y_j^2} dy_j$$

$$= \prod_{j=1}^p \sqrt{2\pi} = (2\pi)^{p/2}.$$

This integral is evaluated by using a gamma integral

$$\int_{-\infty}^{\infty} e^{-u^2} du = 2\int_0^{\infty} e^{-u^2} du$$

due to the property of an even function, provided $\int_0^{\infty} e^{-u^2} du$ is convergent. $u^2 = z \Rightarrow u = z^{\frac{1}{2}}$ since $u > 0$ and $du = \frac{1}{2}z^{-\frac{1}{2}} dz$. From gamma integral, one has

$$\int_0^{\infty} x^{\alpha-1} e^{-x} dx = \Gamma(\alpha), \quad \mathbb{R}(\alpha) > 0$$

and $\Gamma(\frac{1}{2}) = \sqrt{\pi}$. Hence $\int_{-\infty}^{\infty} e^{-\frac{1}{2}y_j^2} dy_j = \sqrt{2\pi}$. That is,

$$\prod_{j=1}^p \int_{-\infty}^{\infty} e^{-\frac{1}{2}y_j^2} dy_j = (2\pi)^{p/2}.$$

For $f(X)$ to be a statistical density, the total integral has to be unity which means that for

$$c = \frac{1}{(2\pi)^{p/2}|V|^{1/2}}$$

this $f(X)$ is a density. Then

$$f(X) = \frac{e^{-\frac{1}{2}(X-\mu)V^{-1}(X-\mu)}}{(2\pi)^{p/2}|V|^{1/2}}, \quad V > O$$

is a density and it called the p-variate nonsingular normal or Gaussian density, for $-\infty < x_j < \infty, -\infty < \mu_j < \infty, X' = (x_1, \ldots, x_p), \mu' = (\mu_1, \ldots, \mu_p), V > O$. It is usually written as $X \sim N_p(\mu, V)$. [X is distributed as a p-variate real Gaussian or normal density with the parameters μ (the mean value vector) and V (the covariance matrix).] This is the fundamental density in multivariate statistical analysis.

Example 5.2.2. Show that if $X \sim N_p(\mu, V), V > O$ then the mean value of X or the expected value of X or $E(X)$ is μ and V is the covariance matrix of X.

Solution 5.2.2. The real nonsingular p-variate Gaussian density is given by

$$f(X) = \frac{e^{-\frac{1}{2}(X-\mu)'V^{-1}(X-\mu)}}{(2\pi)^{p/2}|V|^{1/2}}.$$

Then the expected value of X, denoted by $E(X)$, is given by the integral

$$E(X) = \int_X Xf(X)\mathrm{d}X.$$

For convenience, let us write $X = (X-\mu) + \mu$ and $Y = B'(X-\mu)$, $V^{-1} = BB'$. Then $(X-\mu) = (B')^{-1}Y$, $\mathrm{d}Y = |B|\mathrm{d}X = |V|^{-\frac{1}{2}}\mathrm{d}X$. That is

$$E(X) = \mu \int_X f(X)\mathrm{d}X + \frac{(B')^{-1}}{|V|^{\frac{1}{2}}} \int_Y Ye^{-\frac{1}{2}Y'Y}\mathrm{d}Y.$$

But from Example 5.2.1, $\int_X f(X)\mathrm{d}X = 1$ and each integral in the second term is zero because each element in $Ye^{-\frac{1}{2}Y'Y}$ is an odd function of the type $y_je^{-\frac{1}{2}Y'Y}$ where $Y = \begin{bmatrix} y_1 \\ \vdots \\ y_p \end{bmatrix}$. Then the integral over y_j will be of the following form, for example, for $j = 1$:

$$\int_Y y_1 e^{-\frac{1}{2}Y'Y}\mathrm{d}Y = \left\{ \int_{-\infty}^{\infty} y_1 e^{-\frac{1}{2}y_1^2}\mathrm{d}y_1 \right\} \left\{ \prod_{j=1}^{p} e^{-\frac{1}{2}(y_2^2 + \cdots + y_p^2)}\mathrm{d}y_2 \wedge \cdots \wedge \mathrm{d}y_p \right\}. \tag{a}$$

But the integral over y_1 is zero due to $y_1 e^{-\frac{1}{2}y_1^2}$ being odd and the integral existing. The second factor in (a) above is only a finite constant, namely $\sqrt{2\pi}^{p-1}$. Thus, the integral over each element in Y will be zero. Hence $E(X) = \mu$. Now, consider the covariance of X. By definition

$$\mathrm{Cov}(X) = E\{[X - E(X)][X - E(X)]'\} = E[(X-\mu)(X-\mu)']$$
$$= E\{(B')^{-1}YY'B^{-1}\}$$

for $Y = B'(X_\mu) \Rightarrow \mathrm{d}Y = |V|^{-\frac{1}{2}}\mathrm{d}X$.

$$\mathrm{Cov}(X) = (B')^{-1}\left\{ \int_Y \frac{(YY')}{(2\pi)^{p/2}} e^{-\frac{1}{2}Y'Y} \right\}B^{-1}, \quad |V|^{\frac{1}{2}} \text{ is canceled.}$$

$$YY' = \begin{bmatrix} y_1 \\ \vdots \\ y_p \end{bmatrix} [y_1, \dots, y_p]$$

$$= \begin{bmatrix} y_1^2 & \cdots & y_1 y_p \\ \vdots & \vdots & \vdots \\ y_p y_1 & \cdots & y_p^2 \end{bmatrix}.$$

Integrals for the diagonal elements, say the first diagonal element, is of the form

$$\int_{-\infty}^{\infty} y_1^2 e^{-\frac{1}{2}y_1^2}\mathrm{d}y_1 \times \int_{y_2} \cdots \int_{y_p} e^{-\frac{1}{2}(y_2^2 + \cdots + y_p^2)}\mathrm{d}y_2 \wedge \cdots \wedge \mathrm{d}y_p. \tag{i}$$

Consider,

$$\int_{-\infty}^{\infty} y_1^2 e^{-\frac{1}{2}y_1^2} dy_1 = 2 \int_0^{\infty} y_1^2 e^{-\frac{1}{2}y_1^2} dy_1 \tag{ii}$$

since $y_1^2 e^{-\frac{1}{2}y_1^2}$ is and even function. Put $u = \frac{1}{2} y_1^2 \Rightarrow \sqrt{2}\frac{1}{2} u^{-\frac{1}{2}} du = dy_1$ because y_1 in the integral on the right in (ii) is positive. Now, the right side of (ii) becomes

$$\sqrt{2} \int_0^{\infty} u^{\frac{1}{2}-1} e^{-u} du = \sqrt{2}\Gamma\left(\frac{1}{2}\right) = \sqrt{2}\sqrt{\pi} = \sqrt{2\pi}.$$

This means the first diagonal element in the integral in (i), namely,

$$\int_Y y_1^2 e^{-\frac{1}{2}Y'Y} dY = \sqrt{2\pi} \times \prod_{j=2}^{p}\left[\int_{-\infty}^{\infty} e^{-\frac{1}{2}y_j^2} dy_j\right] = (2\pi)^{\frac{p}{2}}$$

since $\int_{-\infty}^{\infty} e^{-\frac{1}{2}y_j^2} dy_j = \sqrt{2\pi}$ for each $j = 2,\dots,p$. Thus, all diagonal elements in $\int_Y (YY')e^{-\frac{1}{2}Y'Y} dY$ are of the form $(2\pi)^{\frac{p}{2}}$. Now, consider one non-diagonal element, say the first row second column element. This will be of the form

$$\int_Y (y_1 y_2) e^{-\frac{1}{2}Y'Y} dY = A \int_{-\infty}^{\infty}\int_{-\infty}^{\infty} y_1 y_2 e^{-\frac{1}{2}(y_1^2 + y_2^2)} dy_1 \wedge dy_2 \tag{iii}$$

where

$$A = \prod_{j=3}^{p} \int_{-\infty}^{\infty} e^{-\frac{1}{2}y_j^2} dy_j. \tag{iv}$$

In (iii) the integrand for each of y_1 and y_2 is an odd function and since $\int_0^{\infty} y_j e^{-\frac{1}{2}y_j^2} dy_j < \infty$ (finite) the integrals over y_1 and y_2 in (iii) are zeros due to odd function property. Note that (iv) only produces a finite quantity, namely, $(\sqrt{2\pi})^{p-2}$. Thus, each non-diagonal element in the integral $\int_Y YY' e^{-\frac{1}{2}(Y'Y)} dY$ is zero, or one can write

$$\frac{1}{(2\pi)^{\frac{p}{2}}} \int_Y (YY') e^{-\frac{1}{2}(Y'Y)} dY = I$$

where I is the identity matrix of order p. Then the covariance matrix becomes

$$\text{Cov}(X) = (B')^{-1} I B^{-1} = (B')^{-1} B^{-1} = (BB')^{-1} = V$$

or the parameter V in the density is the covariance matrix there.

Example 5.2.3. If the following function $f(X)$, where X is $m \times n$ matrix of mn distinct real scalar variables, A is $m \times m$ and B is $n \times n$ real positive definite constant matrices:

$$f(X) = c\, e^{-\text{tr}(AXBX')}$$

for $X = (x_{ij}), -\infty < x_{ij} < \infty, i = 1,\dots,m, j = 1,\dots,n$ where $\text{tr}(\cdot)$ denotes the trace of the matrix $(\cdot)$, is a statistical density then evaluate c.

Solution 5.2.3. Since $A = A' > O, B = B' > O$ we can write $A = CC'$, $B = GG'$ where C and G are nonsingular matrices. Then

$$\text{tr}(AXBX') = \text{tr}(CC'XGG'X') = \text{tr}(C'XGG'X'C)$$
$$= \text{tr}(YY'), \quad Y = C'XG.$$

In writing this we have used the property that for two matrices P and Q, whenever PQ and QP are defined, $\text{tr}(PQ) = \text{tr}(QP)$ where PQ need not be equal to QP. From Section 3.3 we have

$$Y = C'XG \implies \text{d}Y = |C'|^n |G|^m \text{d}X.$$

Note that $|A| = |CC'| = |C| |C'| = |C|^2$ or $|C| = |A|^{\frac{1}{2}}$. Similarly, $|G| = |B|^{\frac{1}{2}}$. Therefore

$$\text{d}Y = |A|^{\frac{n}{2}} |B|^{\frac{m}{2}} \text{d}X. \tag{i}$$

But for any matrix Z, $\text{tr}(ZZ') = $ sum of squares of all elements in Z. Hence

$$\text{tr}(YY') = \sum_{i=1}^{m} \sum_{j=1}^{n} y_{ij}^2$$

or

$$\int_Y e^{-\frac{1}{2} \text{tr}(YY')} \text{d}Y = \prod_{i=1}^{m} \prod_{j=1}^{n} \int_{-\infty}^{\infty} e^{-\frac{1}{2} y_{ij}^2} \text{d}y_{ij} = (\sqrt{2\pi})^{mn}.$$

The total integral in $f(X)$ being one implies that

$$c = \frac{|A|^{\frac{n}{2}} |B|^{\frac{m}{2}}}{(2\pi)^{\frac{mn}{2}}}$$

or

$$f(X) = \frac{|A|^{\frac{n}{2}} |B|^{\frac{m}{2}}}{(2\pi)^{\frac{mn}{2}}} e^{-\frac{1}{2} \text{tr}(AXBX')}$$

for $X = (x_{ij})$, $-\infty < x_{ij} < \infty$ for all i and j, $A > O, B > O$, is known as the real matrix-variate Gaussian or normal density. In the above case $E(X) = O$ (null). If we replace X by $X - M$, where M is an $m \times n$ constant matrix then $E(X) = M$ and there will be three parameter matrices A, B, M.

Jacobians in one non-linear transformation in the form of a triangular reduction was considered in Section 3.3. Now let us consider a few more non-linear transformations. A very important non-linear transformation is the nonsingular matrix going to its inverse. Let X be a nonsingular $p \times p$ matrix and let $Y = X^{-1}$ be its regular inverse. Then what is the relationship between $\text{d}Y$ and $\text{d}X$? This can be achieved by the following procedure. Let $X = (x_{ij})$. We have $XX^{-1} = I$, the $p \times p$ identity matrix. If we differentiate both sides with respect to some θ we get

$$O = \left(\frac{\partial}{\partial\theta}X\right) + X\left(\frac{\partial}{\partial\theta}X^{-1}\right)$$
$$\Rightarrow \left(\frac{\partial}{\partial\theta}X^{-1}\right) = -X^{-1}\left(\frac{\partial}{\partial\theta}X\right)X^{-1}.$$

Hence if we consider the differentials dx_{ij} and the matrix of differentials, denoted by $(d\mathbf{X}) = (dx_{ij})$ then we have the relationship

$$(d\mathbf{X}^{-1}) = -X^{-1}(d\mathbf{X})X^{-1}.$$

Let $V = (v_{ij}) = (d\mathbf{X}^{-1})$ and $U = (u_{ij}) = (d\mathbf{X})$ then we have

$$V = -X^{-1}UX^{-1}. \tag{a}$$

Noe that in (a) only U and V contains differentials and X^{-1} does not contain any differential or X^{-1} acts as a constant matrix. Now, by taking the wedge product of these differentials and using property (iv) of Section 3.3 we have

$$dV = \begin{cases} |X|^{-2p}dU & \text{for a general real } X \\ |X|^{-(p+1)}dU & \text{for } X = X' \text{ and real.} \end{cases}$$

Thus we have the following result:

(ii)
$$Y = X^{-1} \Rightarrow dY = \begin{cases} |X|^{-2p}dX & \text{for a general real } X \\ |X|^{-(p+1)}dX & \text{for } X = X' \text{ and real.} \end{cases}$$

Example 5.2.4. A real $p \times p$ matrix-variate gamma density is given by

$$f(X) = \begin{cases} \dfrac{|X|^{\alpha-\frac{p+1}{2}}e^{-\text{tr}(X)}}{\Gamma_p(\alpha)}, & X = X' > O, \quad \mathbb{R}(\alpha) > \frac{p-1}{2} \\ 0, & \text{elsewhere} \end{cases} \tag{5.2.1}$$

where the real matrix-variate gamma function is given by

$$\Gamma_p(\alpha) = \pi^{\frac{p(p-1)}{4}}\Gamma(\alpha)\Gamma\left(\alpha-\frac{1}{2}\right)\cdots\Gamma\left(\alpha-\frac{p-1}{2}\right), \quad \mathbb{R}(\alpha) > \frac{p-1}{2}. \tag{5.2.2}$$

Evaluate the density of $Y = X^{-1}$.

Solution 5.2.4. Here $X = X' > O$ and hence $dY = |X|^{-(p+1)}dX$ or $dX = |X|^{p+1}dY = |Y|^{-(p+1)}dY$. Since the transformation X to X^{-1} is one-to-one if the density of Y is denoted by $g(Y)$ then $g(Y)\mathrm{d}Y = f(X)\mathrm{d}X$. That is,

$$g(Y) = f(Y^{-1})|Y|^{-(p+1)}$$

$$= \frac{|Y|^{-\alpha-\frac{p+1}{2}}\mathrm{e}^{-\operatorname{tr}(Y^{-1})}}{\Gamma_p(\alpha)}, \quad \Re(\alpha) > \frac{p-1}{2}$$

$$= \begin{cases} \frac{|Y|^{-\alpha-\frac{p+1}{2}}\mathrm{e}^{-\operatorname{tr}(Y^{-1})}}{\Gamma_p(\alpha)}, & Y = Y' > O, \ \Re(\alpha) > \frac{p-1}{2} \\ 0, & \text{elsewhere.} \end{cases}$$

5.2.2 Functions of matrix argument

A real matrix-variate gamma function was introduced in Section 3.3.4 of Chapter 3. A corresponding gamma function in the complex space is defined in terms of a Hermitian positive definite matrix. [A Hermitian positive definite matrix means a matrix $\tilde{X}$ which is Hermitian, $\tilde{X} = \tilde{X}^*$ where $\tilde{X}^*$ denotes the conjugate transpose of $\tilde{X}$ and all the eigenvalues of $\tilde{X}$ are positive. Note that Hermitian means that all eigenvalues will be real.] Let $\tilde{X}$ be Hermitian positive definite and $p \times p$, denoted by $\tilde{X} > O$.

Definition 5.2.1 (A complex matrix-variate gamma). Notation $\tilde{\Gamma}_p(\alpha)$: It is defined as

$$\tilde{\Gamma}_p(\alpha) = \pi^{\frac{p(p-1)}{2}} \Gamma(\alpha)\Gamma(\alpha-1) \cdots \Gamma(\alpha-p+1), \Re(\alpha) > p-1 \tag{5.2.3}$$

and it has the integral representation

$$\tilde{\Gamma}_p(\alpha) = \int_{\tilde{X}>O} |\tilde{X}|^{\alpha-p}\mathrm{e}^{-\operatorname{tr}(\tilde{X})}\mathrm{d}\tilde{X}. \tag{5.2.4}$$

A complex matrix-variate gamma density is associated with a complex matrix-variate gamma and a Hermitian positive definite matrix random variable.

Definition 5.2.2 (A complex matrix-variate gamma density).

$$f(\tilde{X}) = \begin{cases} \frac{|B|^\alpha \mathrm{e}^{-\operatorname{tr}(B\tilde{X})}}{\tilde{\Gamma}_p(\alpha)}, & \tilde{X} > O, \ \Re(\alpha) > p-1 \\ 0, & \text{elsewhere} \end{cases}$$

where $B = B^* > O$ is a constant matrix, free of $\tilde{X}$, and all matrices are $p \times p$.

Another basic matrix-variate function is the beta function. They are associated with type-1, and type-2 (also known as inverted) beta integrals.

Definition 5.2.3 (A $p \times p$ matrix-variate beta in the real case). Notation $B_p(\alpha, \beta)$: It is defined as

$$B_p(\alpha, \beta) = \frac{\Gamma_p(\alpha)\Gamma_p(\beta)}{\Gamma_p(\alpha + \beta)}, \quad \Re(\alpha) > \frac{p-1}{2}, \quad \Re(\beta) > \frac{p-1}{2}. \tag{5.2.5}$$

It has the following two types of integral representations:

$$B_p(\alpha, \beta) = \int_{O<X<I} |X|^{\alpha - \frac{p+1}{2}} |I - X|^{\beta - \frac{p+1}{2}} \, dX \quad \text{(type-1 beta integral)},$$

$$B_p(\alpha, \beta) = \int_{Y>O} |Y|^{\alpha - \frac{p+1}{2}} |I + Y|^{-(\alpha + \beta)} \quad \text{(type-2 beta integral)}.$$

Here $\int_{O<X<I}$ means the integral over all $p \times p$ matrices X such that $X > O, I - X > O$ or the symmetric matrix X is positive definite and $I - X$ is also positive definite, where I is the $p \times p$ identity matrix. This also means that all the eigenvalues of X are in the open interval $0 < \lambda < 1$ where λ is an eigenvalue. When X is symmetric or Hermitian we can show that all eigenvalues are real.

Definition 5.2.4 (A $p \times p$ matrix-variate beta density, real case).

$$f_1(X) = \begin{cases} \frac{|X|^{\alpha - \frac{p+1}{2}} |I-X|^{\beta - \frac{p+1}{2}}}{B_p(\alpha, \beta)}, & \Re(\alpha) > \frac{p-1}{2}, \Re(\beta) > \frac{p-1}{2} \, (\text{type-1}) \\ 0, & \text{elsewhere;} \end{cases} \tag{5.2.6}$$

$$f_2(Y) = \begin{cases} \frac{|Y|^{\alpha - \frac{p+1}{2}} |I+Y|^{-(\alpha + \beta)}}{B_p(\alpha, \beta)}, & Y > O, \Re(\alpha) > \frac{p-1}{2}, \Re(\beta) > \frac{p-1}{2} \, (\text{type-2}) \\ 0, & \text{elsewhere.} \end{cases} \tag{5.2.7}$$

Corresponding densities in the complex case can also be defined analogous to the real case. More on Jacobians of matrix transformations and functions of matrix argument can be read from the books [3, 2].

Example 5.2.5. Show that the two types of integrals defining the $p \times p$ matrix-variate beta functions in the real case give rise to the same quantity.

Solution 5.2.5. Consider the type-1 beta integral. Call it I_1. Then

$$I_1 = \int_{O<X<I} |X|^{\alpha - \frac{p+1}{2}} |I - X|^{\beta - \frac{p+1}{2}} \, dX.$$

Make the transformation

$$X = (I + Y)^{-\frac{1}{2}} Y (I + Y)^{-\frac{1}{2}} = (I + Y^{-1})^{-\frac{1}{2}} (I + Y^{-1})^{-\frac{1}{2}} = (I + Y^{-1})^{-1}$$

where $Y^{\frac{1}{2}}$ and $(I + Y)^{\frac{1}{2}}$ denote the symmetric positive definite square roots of the symmetric positive definite matrices Y and $I + Y$ respectively. Note that for defining definiteness the matrix has to be symmetric when real and Hermitian when in the complex domain. The word "symmetric" is repeated in order to stress the point. [There are no non-symmetric positive definite matrices.] Applying the Jacobian in (ii) above twice we have

$$X = (I + Y^{-1})^{-1} \Rightarrow dX = |I + Y^{-1}|^{-(p+1)}|Y|^{-(p+1)}dY$$
$$= |I + Y|^{-(p+1)}dY.$$

Note that

$$|X|^{\alpha - \frac{p+1}{2}} = |I + Y^{-1}|^{-\alpha + \frac{p+1}{2}} = |Y|^{\alpha - \frac{p+1}{2}}|I + Y|^{-\alpha + \frac{p+1}{2}}$$
$$|I - X|^{\beta - \frac{p+1}{2}} = |I - (I + Y^{-1})^{-1}|^{\beta - \frac{p+1}{2}}$$
$$= |(I + Y^{-1})^{-1}[I - (I + Y^{-1})]|^{\beta - \frac{p+1}{2}}$$
$$= |I + Y|^{-\beta + \frac{p+1}{2}}.$$

Therefore

$$|X|^{\alpha - \frac{p+1}{2}}|I - X|^{\beta - \frac{p+1}{2}}dX = |Y|^{\alpha - \frac{p+1}{2}}|I + Y|^{-(\alpha + \beta)}dY.$$

Hence

$$I_1 = \int_{Y>O} |Y|^{\alpha - \frac{p+1}{2}}|I + Y|^{-(\alpha + \beta)}dY = I_2$$

and hence the result.

We can also see two results by making the transformations $U = I - X$ and $V = (I + Y)^{-1}$ in I_2. These are the following:

$$\int_{O<X<I} |X|^{\alpha - \frac{p+1}{2}}|I - X|^{\beta - \frac{p+1}{2}}dX$$
$$= \int_{O<U<I} |U|^{\beta - \frac{p+1}{2}}|I - U|^{\alpha - \frac{p+1}{2}}dU$$
$$= B_p(\alpha, \beta), \quad \mathbb{R}(\alpha) > \frac{p-1}{2}, \quad \mathbb{R}(\beta) > \frac{p-1}{2}$$

or the parameters α and β are interchanged. Similarly

$$\int_{Y>O} |Y|^{\alpha - \frac{p+1}{2}}|I + Y|^{-(\alpha + \beta)}dY$$
$$= \int_{V>O} |V|^{\beta - \frac{p+1}{2}}|I + V|^{-(\alpha + \beta)}V$$
$$= B_p(\alpha, \beta), \quad \mathbb{R}(\alpha) > \frac{p-1}{2}, \quad \mathbb{R}(\beta) > \frac{p-1}{2}.$$

Another very important Jacobian is obtained when a general real $p \times n$ matrix X, $n \geq p$, is uniquely represented in terms of a lower triangular matrix with positive diagonal elements $T = (t_{ij}), t_{ij} = 0, i < j, t_{jj} > 0, j = 1, \dots, p$ and a semi-orthonormal matrix U where U is $p \times n$, $UU' = I_p$. That is,

$$X = TU, \quad n \geq p.$$

It can be shown, going through the steps in establishing property (ii) above, that:

(iii) $$X = TU \implies dX = \left\{ \prod_{j=1}^{p} t_{jj}^{n-j} \right\} dT \, dG, \quad n \geq p$$

where

$$dG = \bigwedge_{i=1}^{p} \bigwedge_{i<j=1}^{n} u_i'(du_j)$$

where (du_j) is the j-th column vector of the differentials of U and u_i is the i-th column vector of U.

Here U is an element of the Stiefel manifold $V_{p,n}, n \geq p$, of all semi-orthonormal $p \times n$ matrices U, such that $UU' = I_p$.

Example 5.2.6. Evaluate $\int_{V_{p,n}} dG$.

Solution 5.2.6. Let X be a $p \times n$ matrix of np distinct real scalar variables, $n \geq p$. Consider the integral $\int_X e^{-\text{tr}(XX')} dX$. Note that $\text{tr}(XX')$ is the sum of squares of all the np elements in X. Then integrating directly we have

$$\int_X e^{-\text{tr}(XX')} dX = \prod_{i=1}^{p} \prod_{j=1}^{n} \int_{-\infty}^{\infty} e^{-x_{ij}^2} dx_{ij} = \pi^{\frac{np}{2}}. \tag{a}$$

Let us apply the transformation in (iii). Note that $XX' = TUU'T' = TT'$ and

$$X = TU \implies dX = \left\{ \prod_{j=1}^{p} t_{jj}^{n-j} \right\} dT \, dG.$$

Then

$$\int_X e^{-\text{tr}(XX')} dX = \int_T \left\{ \prod_{j=1}^{p} t_{jj}^{n-j} \right\} e^{-\sum_{i \geq j=1}^{p} t_{ij}^2} dT \int_{V_{n,p}} dG. \tag{b}$$

But

$$\int_0^{\infty} t_{jj}^{n-j} e^{-t_{jj}^2} dt_{jj} = \frac{1}{2} \int_0^{\infty} y_j^{\frac{n-j+1}{2}-1} e^{-y_j} dy_j, \quad y_j = t_j^2$$

$$= \frac{1}{2}\Gamma\left(\frac{n-j+1}{2}\right), \quad \Re\left(\frac{n-j+1}{2}\right) > 0;$$

$$\int_{-\infty}^{\infty} e^{-t_{ij}^2} dt_{ij} = \sqrt{\pi} \implies \prod_{i>j} \int_{-\infty}^{\infty} e^{-t_{ij}^2} dt_{ij} = \pi^{\frac{p(p-1)}{4}}.$$

Then

$$\int_T \left\{ \prod_{j=1}^{p} t_{jj}^{n-j} \right\} e^{-\operatorname{tr}(TT')} dT = \frac{1}{2^p} \pi^{\frac{p(p-1)}{4}} \prod_{j=1}^{p} \Gamma\left(\frac{n-j+1}{2}\right)$$

$$= \frac{1}{2^p} \Gamma_p\left(\frac{n}{2}\right)$$

see the notation from (5.2.2). Comparing (a) and (b) above we have the following result:

(iv) $$\int_{V_{p,n}} dG = \frac{2^p \pi^{\frac{np}{2}}}{\Gamma_p\left(\frac{n}{2}\right)}, n \geq p. \tag{5.2.8}$$

Thus, the integral of dG, the volume elements, over the Stiefel manifold gives the result in (iv). This is a very important result in the theory of functions of matrix argument where the integral is over the full Stiefel manifold (not in any subset there). When $n = p$ we have the full orthogonal group denoted by $O(p)$. Then putting $n = p$ in (iv) we have another very important result:

(v) $$\int_{O(p)} dG = \frac{2^p \pi^{\frac{p^2}{2}}}{\Gamma_p\left(\frac{p}{2}\right)}. \tag{5.2.9}$$

Exercises 5.2

5.2.1. Show that a real symmetric positive definite matrix A can be written as $A = BB'$ where B is nonsingular.

5.2.2. Show that a real symmetric positive semi-definite matrix A can be written as $A = CC'$, where C is a rectangular matrix, and that if C is of full rank then A is positive definite.

5.2.3. Show that a Hermitian positive definite matrix A can be written as $A = BB^*$ where B is nonsingular.

5.2.4. Construct the positive definite square root of (a) a positive definite matrix A, (b) a Hermitian positive definite matrix A.

5.2.5. For real positive definite or Hermitian positive definite matrices A and B show that, denoting the positive definite square roots by $A^{\frac{1}{2}}$ and $B^{\frac{1}{2}}$,

$$|I - AB| = |I - BA| = |I - A^{\frac{1}{2}} B A^{\frac{1}{2}}| = |I - B^{\frac{1}{2}} A B^{\frac{1}{2}}|.$$

5.2.6. Let X_1 and X_2 be $p \times p$ matrix-variate real gamma variables having densities in (5.2.1) with parameters α_1 and α_2 respectively. Let X_1 and X_2 be statistically independently distributed (the joint density of X_1 and X_2 is the product of the individual densities of X_1 and X_2). Then show that

$$Y = X_1 + X_2 \tag{a}$$

is a matrix-variate gamma with parameter $\alpha_1 + \alpha_2$;

$$Y_1 = Y^{-\frac{1}{2}} X_1 Y^{-\frac{1}{2}} \tag{b}$$

has a matrix-variate type-1 beta density;

$$Y_2 = Y^{-\frac{1}{2}} X_2 Y^{-\frac{1}{2}} \tag{c}$$

has a matrix-variate type-1 beta density;

$$Y_3 = X_2^{-\frac{1}{2}} X_1 X_2^{-\frac{1}{2}} \quad \text{and} \quad Y_4 = X_1^{-\frac{1}{2}} X_2 X_1^{-\frac{1}{2}} \tag{d}$$

have matrix-variate type-2 beta densities;

$$Y \text{ and } Y_1 \text{ as well as } Y \text{ and } Y_2 \tag{e}$$

are independently distributed.

5.2.7. Let X be a $p \times 1$ real vector random variable and T a $p \times 1$ vector of parameters (free of X). Then

$$X'T = T'X = t_1 x_1 + \cdots + t_p x_p$$

where $X' = (x_1, \ldots, x_p)$ and $T' = (t_1, \ldots, t_p)$. Then the expected value of $e^{T'X}$, denoted by $M_X(T)$, that is,

$$M_X(T) = E[e^{T'X}] = \int_X e^{T'X} f(X) dX$$

where $f(X)$ is the density of X, is called the moment generating function of X when X is continuous. Evaluate the moment generating function of X when $X \sim N_p(\mu, V), V > O$ of Example 5.2.1.

5.2.8. Consider a $p \times p$ real positive definite matrix random variable X with the density $f(X)$. Let $T = (\hat{t}_{ij}) = T' > O, \hat{t}_{ii} = t_{ii}, \hat{t}_{ij} = \frac{1}{2} t_{ij}, i \neq j, t_{ij} = t_{ji}$ for all i and j, be $p \times p$ parameter matrix. T is free of X. Then the moment generating function of X, denoted by $M_X(T)$, is given by

$$M_X(T) = E[e^{T'X}] = \int_{X>O} e^{\mathrm{tr}(T'X)} f(X) dX.$$

Evaluate $M_X(T)$ for (a) the real matrix-variate gamma variable of (5.2.1); (b) the real type-2 matrix-variate beta variable of (5.2.7). Does this exist?

5.2.9. Consider a $p \times n$ matrix $X, n \geq p$, with the columns of X independently distributed as $N_p(O, I)$ of Example 5.2.1. Independently distributed means the density of X, denoted by $f(X)$, is available by taking the product of the densities of $X_j \sim N_p(O, I), j = 1, \ldots, n$. Let $Y = XX'$. With the help of (iii), or otherwise, show that the density of Y is a particular case of a matrix-variate gamma density given in (5.2.1).

5.2.10. Repeat Exercise 5.2.9 if the columns of X are independently distributed as $N_p(O, V), V > O$.

5.2.11. By using moment generating function show that if the $p \times p$ matrix X has a matrix-variate gamma density of (5.2.1) then every leading sub-matrix of X also has a density in the same family of gamma densities.

5.2.12. By partitioning matrices and integrating out the remaining variables (not using the moment generating function) establish the same result in Exercise 5.2.11.

5.2.13. Let $X = X'$ and $T = T'$ be two $p \times p$ matrices. Show that

$$\text{tr}(XT) = \text{tr}(TX) \neq \sum_{i=1}^{p} \sum_{j=1}^{p} t_{ij} x_{ij}.$$

5.2.14. For $X = X'$ and $p \times p$ construct a $p \times p$ matrix T such that $T = T'$ and $\text{tr}(XT) = \sum_{i=1}^{p} \sum_{j=1}^{p} t_{ij} x_{ij}$ so that $E[e^{-\text{tr}(TX)}]$ can act as the Laplace transform of a real-valued scalar function $f(X)$ of $X > O$ where $T > O$.

5.3 Some topics from statistics

In almost all branches of theoretical and applied statistics involving more than one random variable (real or complex) vectors, matrices, determinants and the associated properties play vital roles. Here we will list a few of those topics for the sake of illustration.

5.3.1 Principal components analysis

In a practical experimental study a scientist may make measurements on hundreds of characteristics of a given specimen. For example, if the aim is to identify the skeletal remains of 10 individuals and classify them as coming from the some groups (ethnic, racial or other) then all characteristics which may have some relevance to the study are measured on each skeleton. Initially the experimenter does not know which characteristics are relevant. If the experimenter has made measurements on 100 characteristics such as the length of thigh bone, dimension of the skull (several measurements), nasal cavity and so on then with 100 such characteristics the analysis of the data becomes too involved. Then the idea is to cut down the number of variables (characteristics on which measurements are already made). If the aim is classification of a skeleton into one of k racial groups then variables which do not have much dispersion (variation) are not that important. One of such variables may be sufficient. Hence the variables which have more scatter in them (squared scatter is measured in terms of variances) are very important. Thus one way of reducing the number of variables involved is to look for variables that have larger variances. These are very important variables as far as classification is concerned. Since linear functions also contain individual variables it is more convenient to look at linear functions (all possible linear functions) and select that linear function with maximum variance, second largest variance and so on. Such an analysis of variable reduction process is called the *principal components analysis*.

Let $x_1, \ldots, x_p$ be p real scalar random variables with mean value zero. Since variance is invariant under relocation of the variables the assumption that the mean value is zero can be made without any loss of generality. Consider an arbitrary linear function:

$$u = a_1 x_1 + \cdots + a_p x_p = a'X = X'a,$$

$$a = \begin{pmatrix} a_1 \\ \vdots \\ a_p \end{pmatrix}, \quad X = \begin{pmatrix} x_1 \\ \vdots \\ x_p \end{pmatrix} \tag{5.3.1}$$

where $a_1, \ldots, a_p$ are constants (free of $x_1, \ldots, x_p$). Then the variance of u, denoted by $\text{Var}(u)$, is given by

$$
\begin{aligned}
\text{Var}(u) = \text{Var}(a'X) &= E(a'X)^2 = E(a'X)(a'X)' \\
&= E(a'XX'a) = a'[E(XX')]a = a'Va
\end{aligned}
\tag{5.3.2}
$$

where E denotes the expected value, V is the covariance matrix of the $p \times 1$ vector random variable X. [Note that since $a'X$ is a scalar quantity its square $(a'X)^2$ can also be written as it times its transpose or $(a'X)^2 = (a'X)(a'X)'$.] Here $V = (v_{ij})$, $v_{ii} = \text{Var}(x_i)$, $v_{ij} = \text{Cov}(x_i, x_j) =$ the covariance between x_i and x_j. Note that (5.3.2) with unrestricted a can go to $+\infty$ since $a'Va \geq 0$ and then maximization does not make sense. Let us restrict a to the boundary of a p-sphere of unit radius, that is, $a'a = 1$. Going through Lagrangian multiplier let

$$\phi = a'Va - \lambda(a'a - 1) \tag{5.3.3}$$

where λ is a Lagrangian multiplier. Differentiating ϕ partially with respect to a (see vector derivatives from Chapter 1) we have

$$\frac{\partial \phi}{\partial a} = O \implies 2Va - 2\lambda a = O \implies Va = \lambda a. \tag{5.3.4}$$

Thus a vector a maximizing $a'Va$ must satisfy (5.3.4). A non-null a satisfying (5.3.4) must have its coefficient matrix singular or its determinant zero or $|V - \lambda I| = 0$. That is, a is an eigenvector of V corresponding to the eigenvalue λ. The equation

$$|V - \lambda I| = 0$$

has p roots, $\lambda_1 \le \lambda_2 \le \cdots \le \lambda_p$. When the variables $x_1, \ldots, x_p$ are not linearly dependent (no experimenter will include a variable which is linearly dependent on other variables under study because no additional information is conveyed by such a variable) V is real symmetric positive definite and we have taken $a'a = 1$. Then from (5.3.4)

$$\lambda_1 = \alpha_1' V\alpha_1, \quad V\alpha_1 = \lambda\alpha_1, \quad \alpha_1'\alpha_1 = 1, \quad \alpha_1 = \begin{pmatrix} \alpha_{11} \\ \vdots \\ \alpha_{p1} \end{pmatrix}$$

where α_1 is the eigenvector corresponding to the largest eigenvalue of V. Thus the first principal component is

$$u_1 = \alpha_1'X = \alpha_{11}x_1 + \alpha_{21}x_2 + \cdots + \alpha_{p1}x_p$$

with variance, $\text{Var}(u_1) = \alpha_1' V\alpha_1 = \lambda_1$. Now take the second largest eigenvalue λ_2 and a corresponding eigenvector α_2 such that $\alpha_2'\alpha_2 = 1$. Then

$$u_2 = \alpha_2'X = \alpha_{12}x_1 + \alpha_{22}x_2 + \cdots + \alpha_{p2}x_p$$

is the second principal component with the variance λ_2 and so on. Since $V = V'$, real symmetric, the eigenvectors for different eigenvalues will be orthogonal. Thus the principal components constructed at the r-th stage will be mutually orthogonal to all others (the coefficient vectors are mutually orthogonal).

Example 5.3.1. Show that the following V can represent a covariance matrix and compute the principal components where

$$V = \begin{bmatrix} 2 & 0 & 1 \\ 0 & 2 & 1 \\ 1 & 1 & 2 \end{bmatrix}.$$

Solution 5.3.1. In order for V to be the covariance matrix of some 3×1 real vector random variable, V must be symmetric and at least positive semi-definite. Let us check the leading minors.

$$2 > 0, \quad \begin{vmatrix} 2 & 0 \\ 0 & 2 \end{vmatrix} = 4 > 0, \quad \begin{vmatrix} 2 & 0 & 1 \\ 0 & 2 & 1 \\ 1 & 1 & 2 \end{vmatrix} = 4 > 0$$

and $V = V'$. Hence $V = V' > 0$ (symmetric positive definite). In order to compute the principal components one needs to compute the eigenvalues of V. Consider

$$|V - \lambda I| = 0 \Rightarrow \begin{vmatrix} 2-\lambda & 0 & 1 \\ 0 & 2-\lambda & 1 \\ 1 & 1 & 2-\lambda \end{vmatrix} = 0$$

$$\Rightarrow (2-\lambda)(\lambda^2 - 4\lambda + 2) = 0$$

$$\Rightarrow \lambda_1 = 2 + \sqrt{2}, \quad \lambda_2 = 2, \quad \lambda_3 = 2 - \sqrt{2}$$

are the three roots. Now, consider the largest one, namely $\lambda_1 = 2 + \sqrt{2}$. An eigenvector corresponding to λ_1 is given by

$$(A - \lambda_1 I)X = O \Rightarrow \begin{bmatrix} -\sqrt{2} & 0 & 1 \\ 0 & -\sqrt{2} & 1 \\ 1 & 1 & -\sqrt{2} \end{bmatrix} \begin{bmatrix} x_1 \\ x_2 \\ x_3 \end{bmatrix} = \begin{bmatrix} 0 \\ 0 \\ 0 \end{bmatrix}$$

$$\Rightarrow X_1 = \begin{bmatrix} 1 \\ 1 \\ \sqrt{2} \end{bmatrix}.$$

Normalizing it we have

$$\alpha_1 = \begin{pmatrix} \frac{1}{2} \\ \frac{1}{2} \\ \frac{\sqrt{2}}{2} \end{pmatrix}, \quad \alpha_1' \alpha_1 = 1$$

is the normalized X_1. Hence the first principal component is

$$u_1 = \alpha_1' X = \frac{1}{2}x_1 + \frac{1}{2}x_2 + \frac{\sqrt{2}}{2}x_3.$$

Now take $\lambda_2 = 2$. $(A - \lambda_2 I)X = O$ gives an

$$X_2 = \begin{pmatrix} 1 \\ -1 \\ 0 \end{pmatrix} \quad \text{and then} \quad \alpha_2 = \begin{pmatrix} \frac{1}{\sqrt{2}} \\ -\frac{1}{\sqrt{2}} \\ 0 \end{pmatrix}$$

and therefore the second principal component is

$$u_2 = \alpha_2' X = \frac{1}{\sqrt{2}}x_1 - \frac{1}{\sqrt{2}}x_2.$$

Now take the third eigenvalue λ_3 and consider $(A - \lambda_3 I)X = O$. This gives

$$X_3 = \begin{pmatrix} 1 \\ 1 \\ -\sqrt{2} \end{pmatrix} \quad \text{and then} \quad \alpha_3 = \begin{pmatrix} \frac{1}{2} \\ \frac{1}{2} \\ -\frac{\sqrt{2}}{2} \end{pmatrix}.$$

Therefore the third principal component is

$$u_3 = \alpha_3' X = \frac{1}{2}x_1 + \frac{1}{2}x_2 - \frac{\sqrt{2}}{2}x_3.$$

Note that $\alpha_1' \alpha_2 = 0$, $\alpha_1' \alpha_3 = 0$, $\alpha_2' \alpha_3 = 0$, $\alpha_i' \alpha_i = 1$, $i = 1, 2, 3$. $\text{Var}(u_1) = \alpha_1' V \alpha_1 = \lambda_1 = 2 + \sqrt{2}$, $\text{Var}(u_2) = 2$, $\text{Var}(u_3) = 2 - \sqrt{2}$. $\text{Var}(u_1) > \text{Var}(u_2) > \text{Var}(u_3)$.

Observe that if one or more eigenvalues are zeros then the corresponding principal components need not be calculated since when $\lambda = 0$, variance $= 0$ and then the variable is degenerate (a constant) having no scatter. No λ can be negative since the covariance matrix is at least positive semi-definite. Computations need be carried out only as long as there are positive eigenvalues above a preassigned threshold number. The variance falling below which (eigenvalues below the threshold number) may not be of any interest to the experimenter. In practice what is done is to have a cutoff point for $\lambda_j = \text{Var}(u_j)$, $j = 1, \dots, p$ and include all principal components with λ_j bigger than or equal to the cutoff point in the study.

When theoretical knowledge about the variables $x_1, \dots, x_p$ is not available then instead of V we consider the sample covariance matrix S which is an estimate of V and work with S. We get estimates of the principal components and their variances. One drawback of the procedure of principal components analysis is that our initial aim was to reduce the number of variables p when p is large. In order to apply the above procedure we need the eigenvalues and eigenvectors of a $p \times p$ matrix with p large. Hence it is questionable whether computation-wise anything tangible is achieved unless the eigenvalues are so far apart that the number of principal components is only a handful when p, in fact, is really large. Since the problem is relevant only when p is large an illustrative example here is not feasible. What is done in Example 5.3.1 is to illustrate the steps.

5.3.2 Regression analysis and model building

One of the frequent activities in applied statistics, econometrics and other areas is to predict a variable by either observing other variables or by prefixing other variables. Let us call the variable to be predicted as the dependent variable y and the variables which are used, say $x_1, \dots, x_k$, to predict y as free (not a proper term in this respect) variables. As examples would be (1) $y =$ market price of the stock of a particular product, $x_1 =$ market demand for that product, $x_2 =$ price of a competing product, $x_3 =$ amount demanded of the company through law suits against the company, and so on or $x_1, x_2, \dots$ are factors which have some relevance on y, (2) $y =$ rate of inflation, $x_1 =$ unit price of gasoline, $x_2 =$ unit price of staple foods, $x_3 =$ rent, and so on, (3) $y =$ weight of a beef cow, $x_1 =$ the age, $x_2 =$ amount of food item 1 consumed, $x_2 =$ amount of green fodder consumed, and so on.

We can prove that the best predictor, best in the minimum mean square sense, is the conditional expectation $E(y|x_1, \dots, x_k)$ where y is the variable to be predicted and $x_1, \dots, x_k$ are the free variables or the variables to be preassigned. For obtaining the best predictor, one needs the conditional distribution of y at given $x_1, \dots, x_k$ and the conditional expectation $E(y|x_1, \dots, x_k)$ existing. If we do not have the conditional distribution then what one can do is only to guess the nature of $E(y|x_1, \dots, x_k)$ and assume a functional form. Then try to estimate that function. If we do not have the conditional distribution and if we suspect that the conditional expectation is a linear function of the conditioned variables $x_1, \ldots, x_k$ then we may take a model $E(y|x_1, \ldots, x_k) = a_0 + a_1 x_1 + \cdots + a_k x_k$ a general linear function.

Suppose we assume that the expected value of y at preassigned values of $x_1, \ldots, x_k$ is a linear function of the type,

$$E(y|x_1, \ldots, x_k) = a_0 + a_1 x_1 + \cdots + a_k x_k \tag{5.3.5}$$

where $E(y|(\cdot))$ denotes the conditional expectation of y given $(\cdot)$, $x_1, \ldots, x_k$ are preassigned, and hence known, and the unknowns are the coefficients $a_0, \ldots, a_k$. Hence if (5.3.5) is treated as a model that we are setting up then linearity or nonlinearity are decided by the linearity of the unknowns, $a_0, a_1, \ldots, a_k$ in the model. If it is treated as a predictor function of $x_1, \ldots, x_k$ then linearity or nonlinearity is decided as a function of $x_1, \ldots, x_k$. This is the essential difference between a predictor function and a model set up to estimate the predictor function. Since the equation (5.3.5) is linear in the unknowns we say that we have a linear model for y in (5.3.5). If we had a regression function (conditional expectation of y given $x_1, \ldots, x_k$, which is the best predictor of y, best in the minimum mean square sense) of the form

$$E(y|x_1, \ldots, x_k) = a_0 a_1^{x_1} e^{-(a_2 x_2 + \cdots + a_k x_k)} \tag{5.3.6}$$

then we have a nonlinear predictor of y since the right side in (5.3.6) is nonlinear in $x_1, \ldots, x_k$ and if (5.3.6) is treated as a model set up to estimate a regression function then the model is nonlinear because it is nonlinear in the unknowns $a_0, \ldots, a_k$.

Consider a linear model of the regression type such as the one in (5.3.5). Our first aim is to estimate the unknowns $a_0, \ldots, a_k$. One distribution-free method (a method that does not depend on the statistical distributions of the variables involved) that is frequently used is the *method of least squares*. This needs some data points, observations or preassigned values on $(x_1, \ldots, x_k)$ and the corresponding observations on y. Let the j-th preassigned value on $(x_1, \ldots, x_k)$ be $(x_{1j}, \ldots, x_{kj})$ and the corresponding observation on y be y_j. Since (5.3.5) is not a mathematical equation we cannot expect every data point $(x_{1j}, \ldots, x_{kj})$ substituted on the right to give exactly y_j. (The model is simply assumed. There may or may not be such a linear relationship.) Write

$$y_j = a_0 + a_1 x_{1j} + \cdots + a_k x_{kj} + e_j$$

where e_j is the error in using $a_0 + a_1 x_{1j} + \cdots + a_k x_{kj}$ to estimate y_j. Then

$$e_j = y_j - a_0 - a_1 x_{1j} - \cdots - a_k x_{kj}, \quad j = 1, \ldots, n$$

if there are n data points. Since $k + 1$ parameters are to be estimated we will take n to be at least $k + 1$, $n \geq k + 1$. In matrix notation we have

$$e = Y - X\beta,$$

$$e = \begin{bmatrix} e_1 \\ \vdots \\ e_n \end{bmatrix}, \quad Y = \begin{bmatrix} y_1 \\ \vdots \\ y_n \end{bmatrix}, \quad \beta = \begin{bmatrix} a_0 \\ \vdots \\ a_k \end{bmatrix},$$

$$X = \begin{bmatrix} 1 & x_{11} & \cdots & x_{k1} \\ \vdots & \vdots & \cdots & \vdots \\ 1 & x_{1n} & \cdots & x_{kn} \end{bmatrix}$$

where e and Y are $n \times 1$, X is $n \times (k+1)$ and β is $(k+1) \times 1$. The error sum of squares is then

$$e_1^2 + \cdots + e_n^2 = e'e = (Y - X\beta)'(Y - X\beta). \tag{5.3.7}$$

If the parameters in β are estimated by minimizing the error sum of squares then the method is called the method of least squares. Differentiating (5.3.7) with respect to β and equating to a null vector we have (see the vector derivatives from Chapter 1)

$$\frac{\partial}{\partial \beta} e'e = O \implies -2X'(Y - X\beta) = O$$

$$\implies X'X\hat{\beta} = X'Y \tag{5.3.8}$$

where $\hat{\beta}$ denotes the estimated β. In the theory of least square analysis the minimizing equation in (5.3.8) is called the *normal equation* (nothing to do with Gaussian or normal distribution). If $X'X$ is nonsingular which happens when X is of full rank, that is, the rank of X is $k + 1 \le n$, then

$$\hat{\beta} = (X'X)^{-1}X'Y. \tag{5.3.9}$$

In a regression type model the final model is going to be used at preassigned points $(x_1, \ldots, x_k)$. Naturally, one would not be taking linearly dependent rows for X. Even if $x_1, \ldots, x_k$ are not linear functions of each other when actual observations are made on $(x_1, \ldots, x_k)$ there is a possibility of near singularity for $X'X$. In a regression type model, more or less one can assume $X'X$ to be nonsingular. There are other models such as design models where by the nature of the design itself $X'X$ is singular. Since $X'Y$ in (5.3.8) is a linear function of the columns of X' and since $X'X$ is also of the same type the linear system in (5.3.8) is always consistent when $n \ge k + 1$. The least square minimum from (5.3.9), usually denoted by s^2, is given by

$$s^2 = (Y - X\hat{\beta})'(Y - X\hat{\beta}) = Y'(Y - X\hat{\beta})$$

since $-X'(Y - X\hat{\beta}) = O$, normal equations;

$$s^2 = Y'Y - Y'X(X'X)^{-1}X'Y = Y'[I - X(X'X)^{-1}X']Y. \tag{5.3.10}$$

Note that the matrices

$$A = I - X(X'X)^{-1}X', \quad B = X(X'X)^{-1}X'$$

are idempotent, $A = A^2$ and $B = B^2$, and further, $AB = O$, that is, they are orthogonal to each other. If $\beta = O$ then the least square minimum is

$$s_0^2 = Y'Y.$$

Thus

$$s_0^2 - s^2 = Y'Y - Y'[I - X(X'X)^{-1}X']Y = Y'X(X'X)^{-1}X'Y$$

is the sum of squares due to the presence of the parameter vector β. Comparing the relative significance of $s_0^2 - s^2$ with s^2 is the basis of testing statistical hypotheses on β.

If the parameter a_0 is to be separated then we can consider the vector

$$\begin{bmatrix} y_1 - \bar{y} \\ \vdots \\ y_n - \bar{y} \end{bmatrix} = \begin{bmatrix} x_{11} - \bar{x}_1 & \cdots & x_{k1} - \bar{x}_k \\ \vdots & \cdots & \vdots \\ x_{1n} - \bar{x}_1 & \cdots & x_{kn} - \bar{x}_k \end{bmatrix} \begin{bmatrix} a_1 \\ \vdots \\ a_k \end{bmatrix} + \begin{bmatrix} e_1 - \bar{e} \\ \vdots \\ e_n - \bar{e} \end{bmatrix} \quad (5.3.11)$$

where $\bar{y} = \frac{1}{n}(y_1 + \cdots + y_n)$, $\bar{x}_i = \frac{1}{n}\sum_{j=1}^{n} x_{ij}$, $\bar{e} = \frac{1}{n}(e_1 + \cdots + e_n)$ where $\bar{e}$ can be taken to be zero without much loss of generality. The least square estimate of β in this case, where $\beta' = (a_1, \ldots, a_k)$, will have the same structure as in (5.3.9) but the Y and X are to be replaced by $Y - \bar{Y}$ and $X - \bar{X}$ respectively, where

$$\bar{Y} = \begin{bmatrix} \bar{y} \\ \vdots \\ \bar{y} \end{bmatrix}, \quad \bar{X} = \begin{bmatrix} \bar{x}_1 & \bar{x}_2 & \cdots & \bar{x}_k \\ \vdots & \vdots & \cdots & \vdots \\ \bar{x}_1 & \bar{x}_2 & \cdots & \bar{x}_k \end{bmatrix}.$$

Example 5.3.2. If the expected value of y at preassigned values of x_1 and x_2 is suspected to be a function of the form $a_0 + a_1 x_1^2 + a_2 x_1 x_2$ estimate the prediction function by the method of least squares and estimate y at $(x_1, x_2) = (2, 1)$ and at $(5, 7)$ respectively by using the following data points:

x_1	0	1	1	−1	2	0	1
x_2	0	1	−1	1	0	2	2
y	2	4	1	1	5	2	5

Solution 5.3.2. Writing in matrix notation we have

$$Y = \begin{bmatrix} 2 \\ 4 \\ 1 \\ 1 \\ 5 \\ 2 \\ 5 \end{bmatrix}, \quad \beta = \begin{bmatrix} a_0 \\ a_1 \\ a_2 \end{bmatrix}, \quad X = \begin{bmatrix} 1 & 0 & 0 \\ 1 & 1 & 1 \\ 1 & 1 & -1 \\ 1 & 1 & -1 \\ 1 & 4 & 0 \\ 1 & 0 & 0 \\ 1 & 1 & 2 \end{bmatrix}$$

where the columns of X correspond to $1, x_1^2, x_1 x_2$;

$$X'X = \begin{bmatrix} 7 & 8 & 1 \\ 8 & 20 & 1 \\ 1 & 1 & 7 \end{bmatrix},$$

$$(X'X)^{-1} = \frac{1}{521} \begin{bmatrix} 139 & -55 & -12 \\ -55 & 48 & 1 \\ -12 & 1 & 76 \end{bmatrix},$$

$$X'Y = \begin{bmatrix} 20 \\ 31 \\ 12 \end{bmatrix}, \quad (X'X)^{-1}X'Y = \frac{1}{521} \begin{bmatrix} 931 \\ 400 \\ 703 \end{bmatrix}.$$

Therefore

$$\hat{a}_0 = \frac{931}{521}, \quad \hat{a}_1 = \frac{400}{521}, \quad \hat{a}_2 = \frac{703}{521}$$

and the estimated model is

$$y = \frac{931}{521} + \frac{400}{521}x_1^2 + \frac{703}{521}x_1 x_2.$$

The predicted y at $(2, 1)$, denoted by $\hat{y}$, is given by

$$\hat{y} = \frac{931}{521} + \frac{400}{521}(2)^2 + \frac{703}{521}(2)(1) \approx 7.5566.$$

Since the point $(5, 7)$ is too far out of the range of observations, based on which the model is constructed, it is not reasonable to predict at $(5, 7)$ by using this model unless it is certain that the behavior of the function is the same for all points on the (x, y)-plane.

When the model is linear in the parameters (unknowns) the above procedure can be adopted but when the model is nonlinear in the parameters one has to go through a nonlinear least squares procedure. Many such algorithms are available in the literature. One such algorithm may be seen from [6].

5.3.3 Design type models

Various types of models appear in the area of statistical design of experiments. A special case of a two-way analysis with fixed effect model, applicable in randomized block designs is the following: Suppose that a controlled experiment is conducted to study the effects of r fertilizers on s varieties of corn, for example 2 fertilizers on 3 varieties of corn. Suppose that rs experimental plots of land of the same size which are homogeneous with respect to all known factors of variation are selected and each fertilizer is applied to randomly assigned s plots where a particular variety of corn is planted. Suppose we have one observation y_{ij} (yield of corn) corresponding to each (fertilizer, variety) combination. Then y_{ij} can be contributed by some general effect, deviation from the general effect due to the i-th fertilizer and the j-th variety, and a random part. The simplest model of this type is

$$y_{ij} = \mu + \alpha_i + \beta_j + e_{ij}, \quad i = 1, \dots, r, \quad j = 1, \dots, s \tag{5.3.12}$$

where μ is the general effect, α_i is the deviation from the general effect due to the i-th fertilizer, β_j is the deviation from the general effect due to the j-th variety and e_{ij} is the random part which is the sum total contributions from all unknown factors which could not be controlled through the design. The model in (5.3.12), when μ, α_i's and β_j's are fixed, but unknown, and e_{ij}'s are random, is known as a linear, additive, fixed effect two-way classification model without interaction (one of the simplest models one can consider under this situation). Writing (5.3.12) in matrix notation we have

$$Y = X\beta + e,$$

where

$$
Y = \begin{bmatrix} y_{11} \\ \vdots \\ y_{1s} \\ y_{21} \\ \vdots \\ y_{2s} \\ \vdots \\ y_{r1} \\ \vdots \\ y_{rs} \end{bmatrix}, \quad
\beta = \begin{bmatrix} \mu \\ \alpha_1 \\ \vdots \\ \alpha_r \\ \beta_1 \\ \vdots \\ \beta_s \end{bmatrix}, \quad
e = \begin{bmatrix} e_{11} \\ \vdots \\ e_{1s} \\ e_{21} \\ \vdots \\ e_{2s} \\ \vdots \\ e_{r1} \\ \vdots \\ e_{rs} \end{bmatrix},
$$

$$
X = \begin{bmatrix}
1 & 1 & 0 & \cdots & 0 & 1 & 0 & \cdots & 0 \\
1 & 1 & 0 & \cdots & 0 & 0 & 1 & \cdots & 0 \\
\vdots & \vdots & \vdots & \vdots & \vdots & \vdots & \vdots & \vdots & \vdots \\
1 & 1 & 0 & \cdots & 0 & 0 & 0 & \cdots & 1 \\
1 & 0 & 1 & \cdots & 0 & 1 & 0 & \cdots & 0 \\
1 & 0 & 1 & \cdots & 0 & 0 & 1 & \cdots & 0 \\
\vdots & \vdots & \vdots & \vdots & \vdots & \vdots & \vdots & \vdots & \vdots \\
1 & 0 & 1 & \cdots & 0 & 0 & 0 & \cdots & 1 \\
\vdots & \vdots & \vdots & \vdots & \vdots & \vdots & \vdots & \vdots & \vdots \\
1 & 0 & 0 & \cdots & 0 & 1 & 0 & \cdots & 1
\end{bmatrix}.
$$

Since the sum of the last $r + s$ columns is equal to the first column the matrix X is not of full rank. This is due to the design itself. The matrix X here is called the design matrix. Thus $X'X$ is singular. Hence if we apply matrix method we could only come to an equation corresponding to (5.3.8) and then go for other methods of solving that equation or compute a g-inverse (generalized inverse) $(X'X)^-$ of $X'X$ so that one solution of the equation corresponding to (5.3.8) is

$$\hat{\beta} = (X'X)^- X'Y.$$

The theory of g-inverses is already available for dealing with such problems. But in the fields of design of experiments, analysis of variance and analysis of covariance, matrix methods will be less efficient compared to computing the error sum of squares as a sum, computing the derivatives directly and then solving the resulting system of normal equations one by one. Hence we will not elaborate on this topic any further.

5.3.4 Canonical correlation analysis

In Section 5.3.2 we dealt with the problem of predicting one real scalar variable by using one set of variables. We generalize this problem here. We would like to predict real scalar variables in one set by using another set of real scalar variables. If the set to be predicted contains only one variable then it is the case of Section 5.3.2. Instead of treating them as two sets of variables we consider all possible linear functions in each set. Then use a principle of maximizing a measure of scale-free joint dispersion known as correlation (nothing to do with relationships, does not measure relationships) for constructing such predictors. Let all the variables in the two sets be denoted by

$$X = \begin{bmatrix} X_{(1)} \\ X_{(2)} \end{bmatrix}, \quad X_{(1)} = \begin{bmatrix} x_1 \\ \vdots \\ x_{p_1} \end{bmatrix}, \quad X_{(2)} = \begin{bmatrix} x_{p_1+1} \\ \vdots \\ x_{p_1+p_2} \end{bmatrix}, \quad p = p_1 + p_2.$$

For convenience let us assume that $p_1 \le p_2$. The covariance matrix of X, denoted by V, is

$$V = E\{[X - E(X)][X - E(X)]'\}.$$

Let us partition V conformally with X. That is,

$$V = \begin{bmatrix} V_{11} & V_{12} \\ V_{21} & V_{22} \end{bmatrix}$$

where V_{11} is $p_1 \times p_1$, $V_{21} = V_{12}'$ since V is symmetric. Let us consider arbitrary linear functions of $X_{(1)}$ and $X_{(2)}$. Let $u = \alpha' X_{(1)}$, $w = \beta' X_{(2)}$ where $\alpha' = (\alpha_1, \ldots, \alpha_{p_1})$ and $\beta' = (\beta_{p_1+1}, \ldots, \beta_{p_1+p_2})$ are arbitrary constants. Since correlation is invariant under scaling and relocation of the variables involved we can assume, without any loss of generality, that $\text{Var}(u) = 1$ and $\text{Var}(w) = 1$. That is,

$$1 = \text{Var}(u) = \alpha' V_{11} \alpha, \quad 1 = \text{Var}(w) = \beta' V_{22} \beta.$$

The correlation between u and w is

$$\frac{\text{Cov}(u,w)}{[\text{Var}(u)\,\text{Var}(w)]^{\frac{1}{2}}} = \text{Cov}(u,w) = \alpha' V_{12}\beta$$

since $\text{Var}(u) = 1$ and $\text{Var}(w) = 1$. Then the principle of maximizing correlation reduces to maximizing $\alpha' V_{12}\beta$ subject to the conditions $\alpha' V_{11}\alpha = 1$ and $\beta' V_{22}\beta = 1$. Using Lagrangian multipliers $-\frac{1}{2}\lambda$ and $-\frac{1}{2}\mu$ the function to be maximized is given by

$$\phi = \alpha' V_{12}\beta - \frac{1}{2}\lambda(\alpha' V_{11}\alpha - 1) - \frac{1}{2}\mu(\beta' V_{22}\beta - 1).$$

The partial derivatives equated to null give,

$$\frac{\partial \phi}{\partial \alpha} = 0, \quad \frac{\partial \phi}{\partial \beta} = 0 \Rightarrow$$

$$V_{12}\beta - \lambda V_{11}\alpha = 0, \tag{a}$$

$$-\mu V_{22}\beta + V_{21}\alpha = 0. \tag{b}$$

Since $\alpha' V_{11}\alpha = 1 = \beta' V_{22}\beta$ we have $\lambda = \mu = \alpha' V_{12}\beta$. Then (a) and (b) reduce to

$$\begin{bmatrix} -\lambda V_{11} & V_{12} \\ V_{21} & -\lambda V_{22} \end{bmatrix} \begin{bmatrix} \alpha \\ \beta \end{bmatrix} = 0.$$

For a non-null solution we must have the coefficient matrix singular or the determinant zero. That is

$$\begin{vmatrix} -\lambda V_{11} & V_{12} \\ V_{21} & -\lambda V_{22} \end{vmatrix} = 0. \tag{c}$$

The determinant on the left is a polynomial of degree $p = p_1 + p_2$. Let $\lambda_1 \geq \lambda_2 \geq \cdots \geq \lambda_{p_1+p_1}$ be the roots of the determinantal equation (c). Since $\lambda = \alpha' V_{12}\beta =$ correlation between u and w, the maximum correlation is available for the largest root λ_1. With this λ_1 solve (a) and (b) for α and β. Let the solution be $\alpha_{(1)}$, $\beta_{(1)}$. Normalize by using $\alpha'_{(1)} V_{11}\alpha_{(1)} = 1$ and $\beta'_{(1)} V_{22}\beta_{(1)} = 1$ to obtain the corresponding normalized vectors, denoted by $\alpha^{(1)}$ and $\beta^{(1)}$. Then the first pair of canonical variables is

$$(u_1, w_1) = (\alpha^{(1)'} X_{(1)}, \beta^{(1)'} X_{(2)}).$$

Note that $\alpha^{(1)}$ and $\beta^{(1)}$ are not only the solutions of (a) and (b) but also satisfy the conditions $\alpha^{(1)'} V_{11}\alpha^{(1)} = 1$ and $\beta^{(1)'} V_{22}\beta^{(1)} = 1$. Thus $\beta^{(1)'} X_{(2)}$ is the best predictor, in the sense of maximum correlation, for predicting the linear function $\alpha^{(1)'} X_{(1)}$ and vice versa. When looking for the second pair of canonical variables we can impose the additional conditions that the second pair (u, w) should be such that u is non-correlated with u_1 and w_1 and w is non-correlated with u_1 and w_1 and at the same time u and

w have the maximum correlation between them, subject to the normalization conditions $u'V_{11}u = 1$ and $w'V_{22}w = 1$. We can continue requiring the new pair to be non-correlated with each of the earlier pairs as well as with maximum correlation and normalized. Using more Lagrangian multipliers the function to be maximized at the $(r+1)$-st stage is,

$$\phi_{r+1} = \alpha'V_{12}\beta - \frac{1}{2}\lambda(\alpha'V_{11}\alpha - 1) - \frac{1}{2}\mu(\beta'V_{22}\beta - 1)$$
$$+ \sum_{i=1}^{r}\nu_i\alpha'V_{11}\alpha^{(i)} + \sum_{i=1}^{r}\theta_i\beta'V_{22}\beta^{(i)}.$$

Then

$$\frac{\partial\phi_{r+1}}{\partial\alpha} = O, \quad \frac{\partial\phi_{r+1}}{\partial\beta} = 0 \Rightarrow$$

$$V_{12}\beta - \lambda V_{11}\alpha + \sum_{i=1}^{r}\nu_i V_{11}\alpha^{(i)} = O, \tag{d}$$

$$V_{21}\alpha - \mu V_{22}\beta + \sum_{i=1}^{r}\theta_i V_{22}\beta^{(i)} = O. \tag{e}$$

Premultiplying (d) by $\alpha^{(j)'}$ and (e) by $\beta^{(j)'}$ we have

$$0 = \nu_j\alpha^{(j)'}V_{11}\alpha^{(j)} = \nu_j$$

and

$$0 = \theta_j\beta^{(j)'}V_{22}\beta^{(j)} = \theta_j.$$

But $\alpha^{(j)'}V_{11}\alpha^{(j)} - 1 = 0$ and $\beta^{(j)'}V_{22}\beta^{(j)} - 1 = 0$. Thus the equations go back to the original equations (a) and (b). Therefore take the second largest root of (c), take solutions $\alpha^{(2)}$ and $\beta^{(2)}$ from (a) and (b) which will be such that $\alpha^{(2)'}V_{11}\alpha^{(2)} = 1$, $\beta^{(2)'}V_{22}\beta^{(2)} = 1$, $\alpha^{(2)'}V_{11}\alpha^{(1)} = 0$, $\alpha^{(2)'}V_{12}\beta^{(1)} = 0$, $\beta^{(2)'}V_{22}\beta^{(1)} = 0$, $\beta^{(2)'}V_{21}\alpha^{(1)} = 0$. Then

$$(u_2, w_2) = (\alpha^{(2)'}X_{(1)}, \beta^{(2)'}X_{(2)})$$

is the second pair of canonical variables and so on. For more properties on canonical variables and canonical correlations see books on multivariate statistical analysis.

When computing the roots of the determinantal equation (c) the following observations will be helpful. From (a) and (b) for $\lambda = \mu$, we can eliminate one of the vectors α or β and write separate equations, one in α alone and one in β alone. For example, multiply (b), with $\mu = \lambda$, by λ and (a) by $V_{11}^{-1}V_{21}$ when $|V_{11}| \neq 0$ and add the two to obtain

$$(V_{21}V_{11}^{-1}V_{12} - \lambda^2 V_{22})\beta = O \Rightarrow$$
$$|V_{21}V_{11}^{-1}V_{12} - \lambda^2 V_{22}| = 0 \Rightarrow$$
$$|V_{22}^{-1}V_{21}V_{11}^{-1}V_{12} - \lambda^2 I| = 0 \tag{f}$$

and similarly

$$(V_{12}V_{22}^{-1}V_{21} - \lambda^2 V_{11})\alpha = 0 \implies$$
$$|V_{11}^{-1}V_{12}V_{22}^{-1}V_{21} - \lambda^2 I| = 0. \tag{g}$$

From (f) and (g) note that λ^2 can be taken as the eigenvalues of $V_{22}^{-1}V_{21}V_{11}^{-1}V_{12}$ or of $V_{11}^{-1}V_{12}V_{22}^{-1}V_{21}$, or as the roots of the determinantal equations (f) and (g), and α and β the corresponding eigenvectors. Thus the problem again reduces to an eigenvalue problem.

Example 5.3.3. (1) Show that the following matrix V can be the covariance matrix of the real random vector $X' = (x_1, x_2, x_3, x_4)$. (2) Let $X'_{(1)} = (x_1, x_2)$ and $X'_{(2)} = (x_3, x_4)$. Suppose we wish to predict linear functions of x_1 and x_2 by using linear functions of x_3 and x_4. Construct the best predictors, best in the sense of maximizing correlations.

$$V = \begin{bmatrix} 1 & 1 & 1 & 1 \\ 1 & 2 & -1 & 0 \\ 1 & -1 & 6 & 4 \\ 1 & 0 & 4 & 4 \end{bmatrix}.$$

Solution 5.3.3. Let us consider the leading minors of V. Note that

$$1 > 0, \quad \begin{vmatrix} 1 & 1 \\ 1 & 2 \end{vmatrix} = 1 > 0,$$

$$\begin{vmatrix} 1 & 1 & 1 \\ 1 & 2 & -1 \\ 1 & -1 & 6 \end{vmatrix} = 1, \quad |V| > 0.$$

V is symmetric and positive definite and hence can represent a covariance matrix. Let

$$V = \begin{bmatrix} V_{11} & V_{12} \\ V_{21} & V_{22} \end{bmatrix}, \quad V_{11} = \begin{bmatrix} 1 & 1 \\ 1 & 2 \end{bmatrix},$$

$$V_{12} = V'_{21} = \begin{bmatrix} 1 & 1 \\ -1 & 0 \end{bmatrix}, \quad V_{22} = \begin{bmatrix} 6 & 4 \\ 4 & 4 \end{bmatrix},$$

$$X_{(1)} = \begin{pmatrix} x_1 \\ x_2 \end{pmatrix}, \quad X_{(2)} = \begin{pmatrix} x_3 \\ x_4 \end{pmatrix},$$

$$V_{11}^{-1} = \begin{bmatrix} 2 & -1 \\ -1 & 1 \end{bmatrix}, \quad V_{22}^{-1} = \begin{bmatrix} \frac{1}{2} & -\frac{1}{2} \\ -\frac{1}{2} & \frac{3}{4} \end{bmatrix}.$$

We are looking for the pairs of canonical variables. Consider the equation

$$|V_{21}V_{11}^{-1}V_{12} - \nu V_{22}| = 0, \quad \nu = \lambda^2, \implies$$

$$\left| \begin{bmatrix} 1 & -1 \\ 1 & 0 \end{bmatrix} \begin{bmatrix} 2 & -1 \\ -1 & 1 \end{bmatrix} \begin{bmatrix} 1 & 1 \\ -1 & 0 \end{bmatrix} - \nu \begin{bmatrix} 6 & 4 \\ 4 & 4 \end{bmatrix} \right| = 0 \implies$$

$$\left\| \begin{bmatrix} 5 & 3 \\ 3 & 2 \end{bmatrix} - v \begin{bmatrix} 6 & 4 \\ 4 & 4 \end{bmatrix} \right\| = 0 \Rightarrow$$

$$8v^2 - 8v + 1 = 0 \Rightarrow v = \frac{1}{2} \pm \frac{1}{4}\sqrt{2}.$$

Let $v_1 = \frac{1}{2} + \frac{\sqrt{2}}{4}$ and $v_2 = \frac{1}{2} - \frac{\sqrt{2}}{4}$. For v_1 consider the equation

$$[V_{21}V_{11}^{-1}V_{12} - v_1 V_{22}]\beta = 0 \Rightarrow$$

$$\left[\begin{pmatrix} 5 & 3 \\ 3 & 2 \end{pmatrix} - \left(\frac{1}{2} + \frac{\sqrt{2}}{4} \right) \begin{pmatrix} 6 & 4 \\ 4 & 4 \end{pmatrix} \right] \begin{bmatrix} \beta_1 \\ \beta_2 \end{bmatrix} = \begin{bmatrix} 0 \\ 0 \end{bmatrix} \Rightarrow$$

$$\begin{bmatrix} 2 - \frac{3}{2}\sqrt{2} & 1 - \sqrt{2} \\ 1 - \sqrt{2} & -\sqrt{2} \end{bmatrix} \begin{bmatrix} \beta_1 \\ \beta_2 \end{bmatrix} = \begin{bmatrix} 0 \\ 0 \end{bmatrix} \Rightarrow$$

$$b_1 = \begin{pmatrix} -2 - \sqrt{2} \\ 1 \end{pmatrix}$$

is one solution. Let us normalize through the relation $1 = b_1' V_{22} b_1$. That is,

$$(-2 - \sqrt{2}, 1) \begin{bmatrix} 6 & 4 \\ 4 & 4 \end{bmatrix} \begin{bmatrix} -2 - \sqrt{2} \\ 1 \end{bmatrix} = 4(2 + \sqrt{2})^2 = d.$$

Then

$$\beta^{(1)} = \frac{b_1}{\sqrt{d}} = \frac{1}{2(2 + \sqrt{2})} \begin{bmatrix} -(2 + \sqrt{2}) \\ 1 \end{bmatrix} = \begin{bmatrix} -\frac{1}{2} \\ \frac{1}{2(2+\sqrt{2})} \end{bmatrix}.$$

Now consider

$$(V_{12}V_{22}^{-1}V_{21} - v_1 V_{11})\alpha = 0 \Rightarrow$$

$$\left[\begin{bmatrix} 1 & 1 \\ -1 & 0 \end{bmatrix} \begin{bmatrix} \frac{1}{2} & -\frac{1}{2} \\ -\frac{1}{2} & \frac{3}{4} \end{bmatrix} \begin{bmatrix} 1 & -1 \\ 1 & 0 \end{bmatrix} - \left(\frac{1}{2} + \frac{\sqrt{2}}{4} \right) \begin{bmatrix} 1 & 1 \\ 1 & 2 \end{bmatrix} \right] \begin{bmatrix} \alpha_1 \\ \alpha_2 \end{bmatrix} = \begin{bmatrix} 0 \\ 0 \end{bmatrix} \Rightarrow$$

$$\begin{bmatrix} 1 + \sqrt{2} & 2 + \sqrt{2} \\ 2 + \sqrt{2} & 2(1 + \sqrt{2}) \end{bmatrix} \begin{bmatrix} \alpha_1 \\ \alpha_2 \end{bmatrix} = \begin{bmatrix} 0 \\ 0 \end{bmatrix} \Rightarrow$$

$$a_1 = \begin{pmatrix} -\sqrt{2} \\ 1 \end{pmatrix}$$

is one solution. Consider

$$a_1' V_{11} a_1 = [-\sqrt{2}, 1] \begin{bmatrix} 1 & 1 \\ 1 & 2 \end{bmatrix} \begin{bmatrix} -\sqrt{2} \\ 1 \end{bmatrix} = 4 - 2\sqrt{2}.$$

Then

$$\alpha^{(1)} = \frac{1}{\sqrt{(4 - 2\sqrt{2})}} \begin{pmatrix} -\sqrt{2} \\ 1 \end{pmatrix}.$$

The first pair of canonical variables is then

$$(u_1, w_1), \quad u_1 = \alpha^{(1)\prime} X_{(1)} = \frac{1}{\sqrt{(4 - 2\sqrt{2})}} (-\sqrt{2}x_1 + x_2),$$

$$w_1 = \beta^{(1)\prime} X_{(2)} = -\frac{1}{2}x_3 + \frac{1}{2(2 + \sqrt{2})} x_4.$$

w_1 is the best predictor of u_1 and vice versa, best in the sense of maximum correlation. Now let us look for the second pair of canonical variables. This is available from the second root $v_2 = \frac{1}{2} - \frac{\sqrt{2}}{4}$. Consider the equation

$$(V_{21} V_{11}^{-1} V_{12} - v_2 V_{22})\beta = 0 \Rightarrow$$

$$\left[\begin{bmatrix} 5 & 3 \\ 3 & 2 \end{bmatrix} - \left(\frac{1}{2} - \frac{\sqrt{2}}{4} \right) \begin{bmatrix} 6 & 4 \\ 4 & 4 \end{bmatrix} \right] \begin{bmatrix} \beta_1 \\ \beta_2 \end{bmatrix} = \begin{bmatrix} 0 \\ 0 \end{bmatrix} \Rightarrow$$

$$b_2 = \begin{pmatrix} \sqrt{2} - 2 \\ 1 \end{pmatrix}$$

is one solution. Let us normalize. Consider

$$b_2' V_{22} b_2 = [\sqrt{2} - 2, 1] \begin{bmatrix} 6 & 4 \\ 4 & 4 \end{bmatrix} \begin{bmatrix} \sqrt{2} - 2 \\ 1 \end{bmatrix}$$

$$= [2(2 - \sqrt{2})]^2 \Rightarrow$$

$$\beta^{(2)} = \frac{1}{2(2 - \sqrt{2})} \begin{pmatrix} \sqrt{2} - 2 \\ 1 \end{pmatrix} = \begin{pmatrix} -\frac{1}{2} \\ \frac{1}{2(2 - \sqrt{2})} \end{pmatrix}.$$

Let us look for $\alpha^{(2)}$. Consider the equation

$$(V_{12} V_{22}^{-1} V_{21} - v_2 V_{11})\alpha = 0 \Rightarrow$$

$$\left[\begin{bmatrix} 1 & 1 \\ -1 & 0 \end{bmatrix} \begin{bmatrix} \frac{1}{2} & -\frac{1}{2} \\ -\frac{1}{2} & \frac{3}{4} \end{bmatrix} \begin{bmatrix} 1 & -1 \\ 1 & 0 \end{bmatrix} - \left(\frac{1}{2} - \frac{\sqrt{2}}{4} \right) \begin{bmatrix} 1 & 1 \\ 1 & 2 \end{bmatrix} \right] \begin{bmatrix} \alpha_1 \\ \alpha_2 \end{bmatrix} = \begin{bmatrix} 0 \\ 0 \end{bmatrix} \Rightarrow$$

$$\begin{bmatrix} -1 + \sqrt{2} & -2 + \sqrt{2} \\ -2 + \sqrt{2} & -2 + 2\sqrt{2} \end{bmatrix} \begin{bmatrix} \alpha_1 \\ \alpha_2 \end{bmatrix} = \begin{bmatrix} 0 \\ 0 \end{bmatrix} \Rightarrow$$

$$a_2 = \begin{pmatrix} \sqrt{2} \\ 1 \end{pmatrix}$$

is one such vector. Let us normalize it. Consider

$$a_2' V_{11} a_2 = (\sqrt{2}, 1) \begin{pmatrix} 1 & 1 \\ 1 & 2 \end{pmatrix} \begin{pmatrix} \sqrt{2} \\ 1 \end{pmatrix} = 4 + 2\sqrt{2}.$$

Then

$$\alpha^{(2)} = \frac{1}{\sqrt{(4 + 2\sqrt{2})}} \begin{pmatrix} \sqrt{2} \\ 1 \end{pmatrix}.$$

The second pair of canonical variables is then

$$(u_2, w_2) = (\alpha^{(2)'} X_{(1)}, \beta^{(2)'} X_{(2)}),$$

$$u_2 = \frac{1}{\sqrt{(4 + 2\sqrt{2})}}(\sqrt{2}x_1 + x_2), \quad w_2 = -\frac{1}{2}x_3 + \frac{1}{2(2 - \sqrt{2})}x_4.$$

It is easy to verify that $Cov(u_1, u_2) = 0$, $Cov(u_1, w_2) = 0$, $Cov(w_1, w_2) = 0$, $Cov(w_1, u_2) = 0$. Here w_1 is the best predictor of u_1 and w_2 is the best predictor of u_2, (w_1, w_2) is the best predictor of (u_1, u_2) and vice versa.

It is worth noting that the maximum number of such pairs possible is p_1 if $p_1 \le p_2$ or p_2 if $p_2 \le p_1$.

Factor analysis is another topic which is widely used in psychology, educational testing problems and applied statistics. This topic boils down to discussing some structural properties of matrices. In a large variety of testing problems of statistical hypotheses on the parameters of one or more multivariate Gaussian distributions the likelihood ratio test statistics reduce to ratios of determinants which often reduce to functions of eigenvalues of certain matrices. Thus the testing problem reduces to the determination of the null and non-null distributions of ratios of determinants or functions of eigenvalues. Generalized analysis of variance problems and generalized linear model problems essentially reduce to the study of certain determinants.

Exercises 5.3

5.3.1. Evaluate the principal components in $X' = (x_1, x_2, x_3)$ with the covariance matrix

$$V = \begin{bmatrix} 2 & 0 & 1 \\ 0 & 2 & 2 \\ 1 & 2 & 4 \end{bmatrix}.$$

5.3.2. Check whether the following V can represent a covariance matrix. If so evaluate the principal components in the corresponding vector, $X' = (x_1, x_2, x_3)$.

$$V = \begin{bmatrix} 1 & -1 & 1 \\ -1 & 2 & 0 \\ 1 & 0 & 3 \end{bmatrix}.$$

5.3.3. If the regression of y on x_1 and x_2 is of the form

$$E(y|x_1, x_2) = a_0 + a_1 x_1^2 + a_2 x_1 x_2 + a_3 x_2^2$$

estimate the regression function based on the following data:

y	1	5	1	2	4	4	5
x_1	0	1	-1	1	2	0	0
x_2	0	1	1	-1	0	2	-2

5.3.4. If the regression of y on x is

$$E(y|x) = b_0 + b_1 x + b_2 x^2 + b_3 x^3$$

estimate the regression function based on the following data:

x	0	1	−1	2	−2	3
y	1	5	1	16	−5	40

and estimate y at $x = 1.5$ and at $x = 2.5$.

5.3.5. Three groups of students are subjected to 3 different methods of teaching. The experiment is conducted according to a completely randomized design so that the grades obtained by the j-th student under the i-th method, x_{ij}, can be written as

$$x_{ij} = \mu + \alpha_i + e_{ij}, \quad i = 1, 2, 3, \quad j = 1, \ldots, n_i$$

where n_i is the size of group i, μ is a general effect, α_i is the deviation from the general effect due to the i-th method and e_{ij} is the random part. Evaluate the least square estimates of $\alpha_1, \alpha_2, \alpha_3$ and the sum of squares due to the α_j's based on the following data where $\mu, \alpha_1, \alpha_2, \alpha_3$ are constants. Grades obtained by the students are the following:
Method 1: (grades $80, 85, 90, 70, 75, 60, 70$);
Method 2: (grades $90, 90, 85, 80, 85, 70, 75, 60, 65, 70$);
Method 3: (grades $40, 50, 70, 60, 65, 50, 60, 65$).

5.3.6. Compute the canonical correlation between $\{x_1, x_3\}$ and $\{x_2, x_4\}$ if the covariance matrix V of $X' = (x_1, x_2, x_3, x_4)$ is given by the following:

$$V = \begin{bmatrix} 1 & -1 & 1 & 1 \\ -1 & 4 & 0 & -1 \\ 1 & 0 & 2 & 1 \\ 1 & -1 & 1 & 3 \end{bmatrix}.$$

5.3.7. Check whether the following matrix can be a covariance matrix of the vector random variable $X' = (x_1, x_2, x_3, x_4)$. If so compute the canonical correlations between $\{x_1, x_4\}$ and $\{x_2, x_3\}$.

$$V = \begin{bmatrix} 3 & 1 & -1 & 1 \\ 1 & 2 & 0 & -1 \\ -1 & 0 & 2 & 1 \\ 1 & -1 & 1 & 3 \end{bmatrix}.$$

5.4 Probability measures and Markov processes

In many areas of measure theory, geometrical probability and stochastic processes, matrices, determinants and eigenvalues play important roles. Two such typical examples will be presented here, one from invariance properties of probability measures, applicable in geometrical probability problems, and another one from discrete time Markov processes.

5.4.1 Invariance of probability measures

A random plane in an Euclidean k-space R^k can be given in Cartesian coordinates $x_1, \ldots, x_k$ as follows:

$$u_1 x_1 + \cdots + u_k x_k + 1 = 0. \tag{5.4.1}$$

This plane does not pass through the origin. When the coefficients $u_1, \ldots, u_k$ are real random variables we call (5.4.1) a random plane. We can write the plane in vector notation as

$$U'X + 1 = 0, \quad U' = (u_1, \ldots, u_k), \quad X' = (x_1, \ldots, x_k). \tag{5.4.2}$$

Let us consider a rotation of the axes of coordinates. A rotation of the axes can, in general, be represented by an orthonormal matrix Q, $QQ' = I, Q'Q = I$ where I is the identity matrix. Let $A' = (a_1, \ldots, a_k)$ be a translation of U. Let the new point X be denoted by X_1. Then

$$X_1 = Q(X - Q'A) \implies Q'X_1 + Q'A = X, \quad Q^{-1} = Q'$$

and then

$$U'X + 1 = 0 \implies U'[Q'X_1 + Q'A] + 1 = 0$$
$$\implies \frac{U'Q'X_1}{U'Q'A + 1} + 1 = 0 \tag{5.4.3}$$

where $U'Q'A + 1 \neq 0$ almost surely (a.s.). If the plane in (5.4.3) is denoted by $U_1'X_1 + 1 = 0$ then

$$U_1' = \frac{U'Q'}{U'Q'A + 1}. \tag{5.4.4}$$

An event B on this plane is a function of the parameter vector U. Let a measure on B, denoted by $m(B)$, be given by the integral

$$m(B) = \int_B f(U) dU, \quad dU = du_1 \wedge \cdots \wedge du_k.$$

Under a translation and rotation let the resulting $f(U)$ and B be denoted by $f_1(U_1)$ and B_1 respectively. Let the corresponding measure be $m_1(B_1)$. Then

$$m_1(B_1) = \int_{B_1} f_1(U_1) dU_1.$$

Invariance property of the measure under Euclidean motion implies that $m(B)$ and $m_1(B_1)$ are the same. What should be the condition so that $m(B) = m_1(B_1)$? Let us examine this a little further.

$$U_1' = \frac{U'Q'}{U'Q'A+1} = \frac{V'}{V'A+1},$$
$$V' = U'Q' = (v_1, \ldots, v_k).$$

Then the first element in this row vector is $\frac{v_1}{V'A+1}$ and its partial derivatives with respect to $v_1, \ldots, v_k$ yield

$$(1 + V'A - v_1a_1, -v_1a_2, \ldots, -v_1a_k)(1 + V'A)^{-2}.$$

Then the Jacobian is the following determinant:

$$(1 + V'A)^{-(2k)}$$

$$\times \begin{vmatrix} (1+V'A) - v_1a_1 & -v_1a_2 & \cdots & -v_1a_k \\ -v_2a_1 & (1+V'A) - v_2a_2 & \cdots & -v_2a_k \\ \vdots & \vdots & \cdots & \vdots \\ -v_ka_1 & -v_ka_2 & \cdots & (1+V'A) - v_ka_k \end{vmatrix}$$

$$= (1 + V'A)^{-(k+1)}.$$

Then

$$f(U) = f_1(U_1)\left|\frac{\partial U_1}{\partial U}\right| = f_1(U_1)(1 + V'A)^{-(k+1)}; \tag{5.4.5}$$

$$U_1'U_1 = \frac{U'U}{[1+V'A]^2}, \quad (1 + V'A)^{-(k+1)} = \left[\frac{U_1'U_1}{U'U}\right]^{\frac{k+1}{2}}.$$

Therefore

$$\frac{f(U)}{f_1(U_1)} = \left[\frac{U_1'U_1}{U'U}\right]^{\frac{k+1}{2}}.$$

But

$$(U'U)^{-\frac{(k+1)}{2}} = (u_1^2 + \cdots + u_k^2)^{-\frac{(k+1)}{2}}.$$

Thus $f(U)$ is proportional to $(u_1^2 + \cdots + u_k^2)^{-\frac{(k+1)}{2}}$. Therefore the invariant measure is

$$m(B) = c_k \int_B \frac{1}{(u_1^2 + \cdots + u_k^2)^{\frac{k+1}{2}}} dU \tag{5.4.6}$$

where c_k is a constant.

Example 5.4.1. Compute the invariant measure in Cartesian coordinates, invariant under Euclidean motion, the invariant density and the element of the invariant measure for a plane in 3-space R^3.

Solution 5.4.1. From (5.4.6) for $k = 3$ we have the invariant measure for a set B given by

$$m(B) = c_3 \int_B \frac{1}{(u_1^2 + u_2^2 + u_3^2)^2} du_1 \wedge du_2 \wedge du_3$$

where c_3 is a constant. The invariant density is then

$$f(u_1, u_2, u_3) = c_3(u_1^2 + u_2^2 + u_3^2)^{-2}$$

where c_3 is a normalizing constant so that the total volume under f is unity. Therefore the element of the invariant measure, denoted by dm, is given by

$$dm = c_3(u_1^2 + u_2^2 + u_3^2)^{-2} du_1 \wedge du_2 \wedge du_3$$

where $(u_1, u_2, u_3) \neq (0, 0, 0)$ a.s.

For more on measures, invariance and other topics such as random areas and volumes in higher dimensional Euclidean space where matrices and determinants play dominant roles see the book [4].

5.4.2 Discrete time Markov processes and transition probabilities

In Example 2.2.6 of Section 2.2 in Chapter 2 we have considered a transition probability matrix which is singly stochastic in the sense that all elements are non-negative and further, either the sum of the elements in each row is 1 or the sum of the elements in each column is 1. If this property holds for both rows and columns then the matrix is called a doubly stochastic matrix. For example

$$A = \begin{bmatrix} \frac{1}{2} & \frac{1}{2} \\ \frac{1}{2} & \frac{1}{2} \end{bmatrix}$$

is doubly stochastic whereas

$$B = \begin{bmatrix} 0.8 & 0.2 \\ 0.5 & 0.5 \end{bmatrix} \quad \text{and} \quad C = \begin{bmatrix} 0.3 & 0.4 \\ 0.7 & 0.6 \end{bmatrix}$$

are singly stochastic.

Consider a problem of the following type: Consider a fishing spot in a river such as a pool area in the river. Some fish move into the pool area from outside and some fish from the pool move out every day. Suppose that the following is the process every day. 70% of fish who are outside stay outside and 30% move into the pool. 50% of the fish in the pool stay there and 50% move out of the pool. If stage 1 is "outside" and stage 2 is "inside" then we have the following transition proportion matrix, if the columns represent the transitions:

$$
\begin{array}{ccc}
 & \text{outside} = 1 & \text{inside} = 2 \\
\text{outside} = 1 & 0.7 & 0.5 \\
\text{inside} = 2 & 0.3 & 0.5
\end{array}
\quad , \quad
A = (a_{ij}) = \begin{pmatrix} 0.7 & 0.5 \\ 0.3 & 0.5 \end{pmatrix},
$$

a_{ij} = transition proportion from the j-th stage to the i-th stage. [If a_{ij} represents the transition proportion from the i-th stage to the j-th stage then we have the transpose of the above matrix A.] For convenience of other interpretations later on we take a_{ij} to denote the transition from the j-th stage to the i-th stage so that the sum of the elements in each column is 1 in the above singly stochastic matrix. If this process is repeated every day then at the end of the first day A is the situation, by the end of the second day A^2 is the situation (see the details of the argument in Example 2.2.6) and at the end of the k-th day the situation is A^k. What is A^k in this case? In order to compute A^k let us compute the eigenvalues and the matrix of eigenvectors. Consider $|A - \lambda I| = 0$. That is,

$$
\begin{vmatrix} 0.7 - \lambda & 0.5 \\ r0.3 & 0.5 - \lambda \end{vmatrix} = 0 \Rightarrow \lambda_1 = 1, \quad \lambda_2 = 0.2.
$$

If we add all the rows to the first row then the first row becomes $1 - \lambda, 1 - \lambda$. Then $1 - \lambda$ can be taken out. Then $\lambda_1 = 1$ is an eigenvalue for any singly stochastic matrix.

(i) One eigenvalue of any singly stochastic matrix is 1.

Computing the eigenvectors for our matrix A we have for $\lambda_1 = 1$,

$$
(A - \lambda_1 I)X = O \Rightarrow \begin{bmatrix} 0.7 - 1 & 0.5 \\ 0.3 & 0.5 - 1 \end{bmatrix} \begin{bmatrix} x_1 \\ x_2 \end{bmatrix} = \begin{bmatrix} 0 \\ 0 \end{bmatrix}
$$

$$
\Rightarrow X_1 = \begin{bmatrix} 1 \\ 0.6 \end{bmatrix}
$$

is one eigenvector. For $\lambda_2 = 0.2$,

$$
(A - \lambda_2 I)X = O \Rightarrow X_2 = \begin{bmatrix} 1 \\ -1 \end{bmatrix}
$$

is an eigenvector. Let

$$Q = (X_1, X_2) = \begin{bmatrix} 1 & 1 \\ 0.6 & -1 \end{bmatrix} \Rightarrow$$

$$Q^{-1} = \frac{1}{1.6} \begin{bmatrix} 1 & 1 \\ 0.6 & -1 \end{bmatrix}.$$

Then

$$A = Q\Lambda Q^{-1}, \quad \Lambda = \begin{bmatrix} 1 & 0 \\ 0 & 0.2 \end{bmatrix}.$$

Therefore

$$A^k = (Q\Lambda Q^{-1}) \cdots (Q\Lambda Q^{-1}) = Q\Lambda^k Q^{-1}$$

$$= \begin{bmatrix} 1 & 1 \\ 0.6 & -1 \end{bmatrix} \begin{bmatrix} 1^k & 0 \\ 0 & (0.2)^k \end{bmatrix} \frac{1}{1.6} \begin{bmatrix} 1 & 1 \\ 0.6 & -1 \end{bmatrix}.$$

If $k \to \infty$ then $(0.2)^k \to 0$. Then

$$A^k \to \begin{bmatrix} 1 & 1 \\ 0.6 & -1 \end{bmatrix} \begin{bmatrix} 1 & 0 \\ 0 & 0 \end{bmatrix} \frac{1}{1.6} \begin{bmatrix} 1 & 1 \\ 0.6 & -1 \end{bmatrix}$$

$$= \frac{1}{1.6} \begin{bmatrix} 1 & 1 \\ 0.6 & 0.6 \end{bmatrix}.$$

$A^\infty = \lim_{k \to \infty} A^k$ represents the *steady state*. Suppose there were 10 000 fish outside the pool and 500 inside the pool to start with. Then at the end of the first day the numbers will be the following:

$$\begin{bmatrix} 0.7 & 0.5 \\ 0.3 & 0.5 \end{bmatrix} \begin{bmatrix} 10\,000 \\ 500 \end{bmatrix} = \begin{bmatrix} 7\,250 \\ 3\,250 \end{bmatrix}.$$

That is, 7 250 fish outside the pool and 3 250 inside the pool. At the end of the second day the numbers are

$$\begin{bmatrix} 0.7 & 0.5 \\ 0.3 & 0.5 \end{bmatrix}^2 \begin{bmatrix} 10\,000 \\ 500 \end{bmatrix} = \begin{bmatrix} 0.64 & 0.60 \\ 0.36 & 0.40 \end{bmatrix} \begin{bmatrix} 10\,000 \\ 500 \end{bmatrix}$$

$$= \begin{bmatrix} 6\,700 \\ 3\,800 \end{bmatrix}.$$

That is, 6 700 fish outside the pool and 3 800 inside the pool. Evidently, in the long run the numbers will be

$$A^\infty \begin{bmatrix} 10\,000 \\ 500 \end{bmatrix} = \frac{1}{1.6} \begin{bmatrix} 1 & 1 \\ 0.6 & 0.6 \end{bmatrix} \begin{bmatrix} 10\,000 \\ 500 \end{bmatrix} = \begin{bmatrix} 6\,562.5 \\ 3\,937.5 \end{bmatrix}.$$

Even though 0.5 fish does not make sense this vector is the eventual limiting vector with 6 562.5 outside and 3 937.5 inside.

Let us look at the general situation. Let $P = (p_{ij})$ be the transition probability matrix.

$$P = \begin{bmatrix} p_{11} & p_{12} & \cdots & p_{1n} \\ \vdots & \vdots & \cdots & \vdots \\ p_{n1} & p_{n2} & \cdots & p_{nn} \end{bmatrix}$$

with the sum of the elements in each column 1, that is, $\sum_{i=1}^{n} p_{ij} = 1$ for each $j = 1, \ldots, n$. Then, as we have already seen, one eigenvalue of P is 1. What about the other eigenvalues of P? Note that the sum of the eigenvalues of P is the trace of P. That is,

$$\text{tr}(P) = p_{11} + \cdots + p_{nn} = \lambda_1 + \cdots + \lambda_n$$

where $\lambda_1, \ldots, \lambda_n$ are the eigenvalues of P with $\lambda_1 = 1$. But $P^2, P^3, \ldots$ being transition probability matrices, all have the same property that the sum of the elements in each column is 1. Hence $\text{tr}(P^k)$ cannot exceed n because $p_{11}^{(k)}, \ldots, p_{nn}^{(k)}$ are all probabilities, where $p_{11}^{(k)}, \ldots, p_{nn}^{(k)}$ are the diagonal elements in P^k. Note that $p_{ii}^{(k)} \neq p_{ii}^k$. But the eigenvalues of P^k are $\lambda_1^k = 1$, $\lambda_2^k, \ldots, \lambda_n^k$. Then

$$\lambda_2^k + \cdots + \lambda_n^k \leq n - 1 \tag{5.4.7}$$

where n is fixed and k could be arbitrarily large. But (5.4.7) can hold only if $|\lambda_j| \leq 1$ for all $j = 1, \ldots, n$.

(ii) The eigenvalues of a singly stochastic matrix are all less than or equal to 1 in absolute value.

Example 5.4.2. Suppose that a flu virus is going around in a big school system. The children there are only of 3 types, healthy (unaffected), infected, seriously ill. The probability that a healthy child remains healthy at the end of the day is 0.2, becomes infected is 0.5, becomes seriously ill is 0.3 and suppose the following is the transition probability matrix:

$$P = \begin{bmatrix} 0.2 & 0.1 & 0.1 \\ 0.5 & 0.5 & 0.6 \\ 0.3 & 0.4 & 0.3 \end{bmatrix}.$$

Suppose that this P remains the same from day to day. Compute the following: (1) The transition probability matrix after 10 days; (2) The transition probability matrix eventually; (3) If there are 1 000 children in the healthy category, 500 in the infected category and 100 in the seriously ill category at the start of the observation period (zeroth day) what will be these numbers eventually?

Solution 5.4.2. In order to answer all the above questions we need the eigenvalues and eigenvectors of P. Consider $|P - \lambda I| = 0$. That is,

$$0 = \begin{vmatrix} 0.2 - \lambda & 0.1 & 0.1 \\ 0.5 & 0.5 - \lambda & 0.6 \\ 0.3 & 0.4 & 0.3 - \lambda \end{vmatrix} = (1 - \lambda) \begin{vmatrix} 1 & 1 & 1 \\ 0.5 & 0.5 - \lambda & 0.6 \\ 0.3 & 0.4 & 0.3 - \lambda \end{vmatrix}$$

$$= (1 - \lambda)[\lambda^2 - (0.1)^2].$$

The roots are $\lambda_1 = 1$, $\lambda_2 = 0.1$, $\lambda_3 = -0.1$. For $\lambda_1 = 1$,

$$(P - \lambda_1 I)X = O \implies X_1 = \begin{bmatrix} 2.2/7 \\ 10.6/7 \\ 1 \end{bmatrix}.$$

For $\lambda_2 = 0.1$ and for $\lambda_3 = -0.1$ we have

$$X_2 = \begin{bmatrix} -2 \\ 1 \\ 1 \end{bmatrix}, \quad X_3 = \begin{bmatrix} 0 \\ 1 \\ -1 \end{bmatrix}.$$

Let

$$Q = \begin{bmatrix} 2.2/7 & -2 & 0 \\ 10.6/7 & 1 & 1 \\ 1 & 1 & -1 \end{bmatrix}.$$

Then

$$P = Q\Lambda Q^{-1}, \quad \Lambda = \begin{bmatrix} 1 & 0 & 0 \\ 0 & 0.1 & 0 \\ 0 & 0 & -0.1 \end{bmatrix},$$

$$Q^{-1} = \frac{1}{39.6} \begin{bmatrix} 14 & 14 & 14 \\ -17.6 & 2.2 & 2.2 \\ -3.6 & 16.2 & -23.4 \end{bmatrix}.$$

Now we can answer all the questions.

$$P^{10} = Q\Lambda^{10} Q^{-1}$$

$$= Q \begin{bmatrix} 1 & 0 & 0 \\ 0 & (0.1)^{10} & 0 \\ 0 & 0 & (-0.1)^{10} \end{bmatrix} Q^{-1}, \quad (\pm 0.1)^{10} \approx 0,$$

$$\approx \frac{14}{39.6} \begin{bmatrix} \frac{2.2}{7} & \frac{2.2}{7} & \frac{2.2}{7} \\ \frac{10.6}{7} & \frac{10.6}{7} & \frac{10.6}{7} \\ 1 & 1 & 1 \end{bmatrix} = P^\infty.$$

If the initial vector is $X_0' = (1\,000, 500, 100)$ then the eventual situation is $P^\infty X_0$. That is,

$$P^\infty X_0 = \frac{14}{39.6} \begin{bmatrix} \frac{2.2}{7} & \frac{2.2}{7} & \frac{2.2}{7} \\ \frac{10.6}{7} & \frac{10.6}{7} & \frac{10.6}{7} \\ 1 & 1 & 1 \end{bmatrix} \begin{bmatrix} 1000 \\ 500 \\ 100 \end{bmatrix}$$

$$= \begin{bmatrix} 1\,600/9 \\ 53(1\,600)/99 \\ 35(1\,600)/99 \end{bmatrix} \approx \begin{bmatrix} 177.78 \\ 856.57 \\ 565.65 \end{bmatrix}.$$

Thus, eventually one can expect 178 children in the healthy category, 856 in the infected category and 566 in the very ill category according to the transition probability matrix P.

In the two examples above, we have noted that the steady state or the eventual state of the initial vector X_0, that is $P^\infty X_0$, is nothing but a scalar multiple of the eigenvector corresponding to the eigenvalue $\lambda_1 = 1$, that is X_1 in our notation.

$$P^\infty X_0 = (\text{sum of the elements in } X_0)X_1.$$

This, in fact, is a general result if all the elements in P are strictly positive, that is, no element in P is zero. Then all other eigenvalues will be strictly less than 1 in absolute value also.

(iii) If all the elements in a singly stochastic matrix P are strictly positive then one eigenvalue of P is 1 and all other eigenvalues of P are less than 1 in absolute value. Then the steady state is the sum of the elements in the initial vector multiplied by X_1, the eigenvector corresponding to the eigenvalue $\lambda_1 = 1$.

Exercises 5.4

5.4.1. Show that a plane in 3-space R^3 can be given by the equation

$$x \sin\phi \cos\theta + y \sin\phi \sin\theta + z \cos\phi = \rho,$$

for $0 \le \theta \le 2\pi$, $0 \le \phi \le \pi$, $0 \le \rho < \infty$ where ρ is the perpendicular distance of the plane from the origin, and θ and ϕ are the polar angles.

5.4.2. Compute the Jacobian in the transformation

$$x = \rho \sin\phi \cos\theta$$
$$y = \rho \sin\phi \sin\theta$$
$$z = \rho \cos\phi.$$

5.4.3. Show that the element of the invariant density in R^3 in polar coordinates is given by

$$dm = \lambda |\sin \phi| d\theta \wedge d\phi \wedge d\rho, \quad \lambda \text{ a constant.}$$

5.4.4. Consider the following general polar coordinate transformation:

$$x_1 = \rho \sin \theta_1 \sin \theta_2 \cdots \sin \theta_{k-2} \sin \theta_{k-1}$$
$$x_2 = \rho \sin \theta_1 \sin \theta_2 \cdots \sin \theta_{k-2} \cos \theta_{k-1}$$
$$x_3 = \rho \sin \theta_1 \sin \theta_2 \cdots \cos \theta_{k-2}$$
$$\vdots$$
$$x_{k-1} = \rho \sin \theta_1 \cos \theta_2$$
$$x_k = \rho \cos \theta_1$$

for $0 < \theta_j \leq \pi$, $j = 1, \ldots, k-2$, $0 < \theta_{k-1} \leq 2\pi$, $0 \leq \rho < \infty$, (1) Compute the Jacobian; (2) Show that the invariant density, invariant under Euclidean motion, is given by

$$f(\rho, \theta_1, \ldots, \theta_{k-1}) = \lambda_k \prod_{j=1}^{k-1} |\sin \theta_j|^{k-j-1}, \quad \lambda_k \text{ a constant.}$$

5.4.5. Mr. Good's job requires frequent travels abroad on behalf of his company in California. If he is in California the chance that he will stay in California on the same day is only 10% and the chance that he will be outside California is 90%. If he is outside California the chance that he will stay outside on the same day is 80% and that he will come to California is 20%. This is the daily routine. (1) Compute the transition probability matrix, P, for any given day; (2) Compute the transition probability matrix for the 10-th day of observation, P^{10}; (3) Compute the eventual behavior of the transition probability matrix, P^∞.

5.4.6. For a terminally ill patient suppose that there are only two possible stages of transition, dead or terminally ill for any given day. Suppose that the chance that the patient is still ill the next day is 0.5. Compute (1) the transition probability matrix P; (2) Compute P^5; (3) Compute P^∞.

5.4.7. For the following transition probability matrices compute the steady state if the initial state vector is X_0:

$$P_1 = \begin{bmatrix} 0.4 & 0.7 \\ 0.6 & 0.3 \end{bmatrix}, \quad P_2 = \begin{bmatrix} 0.2 & 0.4 & 0.3 \\ 0.3 & 0.5 & 0.2 \\ 0.5 & 0.1 & 0.5 \end{bmatrix}, \quad P_3 = \begin{bmatrix} 1 & 0.5 & 0.2 \\ 0 & 0 & 0.4 \\ 0 & 0.5 & 0.4 \end{bmatrix}.$$

5.4.8. Let P be a transition probability matrix with the sum of the elements in each column 1. Let $PX = \lambda X$. Show that if $\lambda \neq 1$ then the sum of the elements in X is zero, that is, if X_j is an eigenvector corresponding to $\lambda_j \neq 1$ then $J'X_j = 0$, $J' = (1, \ldots, 1)$.

5.4.9. Compute the steady states for the matrices P_1, P_2, P_3 in Exercise 5.4.7 if the initial states are the following:

(i) $X_0 = \begin{pmatrix} 1 \\ 0 \end{pmatrix}$, (ii) $X_0 = \begin{pmatrix} 0 \\ 1 \end{pmatrix}$ for P_1;

(i) $X_0 = \begin{bmatrix} 1 \\ 0 \\ 0 \end{bmatrix}$, (ii) $X_0 = \begin{bmatrix} 0 \\ 1 \\ 0 \end{bmatrix}$, (iii) $X_0 = \begin{bmatrix} 0 \\ 0 \\ 1 \end{bmatrix}$ for P_2;

(i) $X_0 = \begin{bmatrix} 20 \\ 30 \\ 10 \end{bmatrix}$, (ii) $X_0 = \begin{bmatrix} 10 \\ 10 \\ 10 \end{bmatrix}$ for P_3.

5.4.10. In a particular township there are 3 grocery stores. The number of households in that township is fixed, only 200. Initially the three stores have $100, 50, 50$ households respectively as customers. The stores started weekly sales. Depending upon the good sales the customers started moving from store to store. Suppose that the chances of customer of store 1 to remain or move to store 2 or 3 are respectively $0.7, 0.2, 0.1$. Suppose the weekly transition probability matrix is

$$A = \begin{bmatrix} 0.7 & 0.5 & 0.6 \\ 0.2 & 0.4 & 0.2 \\ 0.1 & 0.1 & 0.2 \end{bmatrix}.$$

What will be the numbers in column one of the transition matrix after 3 weeks? When can we expect store 3 to close if less than 20 customers is not a viable operation?

5.4.11. Suppose there are four popular brands of detergents in the market. These brands have initially $40\%, 30\%, 20\%, 10\%$ of the customers respectively. It is found that the customers move from brand to brand at the end of every month. Answer the following questions if (1) A and (2) B is the transition matrix for every month:

$$A = \begin{bmatrix} 0.4 & 0.5 & 0.4 & 0.4 \\ 0.2 & 0.4 & 0.4 & 0.5 \\ 0.2 & 0.1 & 0.1 & 0.0 \\ 0.2 & 0.0 & 0.1 & 0.1 \end{bmatrix}, \quad B = \begin{bmatrix} 0 & 0 & 0 & 1 \\ 0 & 0 & 1 & 0 \\ 0 & 1 & 0 & 0 \\ 1 & 0 & 0 & 0 \end{bmatrix}.$$

What will be the percentages of customers after (i) two months, (ii) three months, and (iii) eventually, for each A and B?

5.5 Maxima/minima problems

The basic maxima/minima problems were already discussed in earlier chapters. Here we will consider a unified way of treating the problem and then look into some more applications.

5.5.1 Taylor series

From elementary calculus the student may be familiar with the Taylor series expansion for one real variable x. The expansion of a function $f(x)$ at $x = a$ is given by

$$f(x) = f(a) + \frac{(x-a)}{1!}f^{(1)}(a) + \frac{(x-a)^2}{2!}f^{(2)}(a) + \cdots$$

where $f^{(r)}(a)$ denotes the r-th derivative of $f(x)$ with respect to x and then evaluated at $x = a$. Let us denote D_0^r as the r-th derivative operator evaluated at a given point so that $D_0^r f$ will indicate $f^{(r)}(a)$. Then consider the exponential series

$$e^{(x-a)D_0} = (x-a)^0 D_0^0 + \frac{(x-a)}{1!}D_0 + \frac{(x-a)^2}{2!}D_0^2 + \cdots$$

so that $e^{(x-a)D_0}$ operating on $f(x)$ gives

$$f(a) + \frac{(x-a)}{1!}f^{(1)}(a) + \frac{(x-a)^2}{2!}f^{(2)}(a) + \cdots$$

where $(x-a)^0 D_0^0 f = D_0^0 f = f(a)$. This is the expansion for $f(x)$ at $x = a$. For a function of two real variables, $f(x_1, x_2)$, if the Taylor series expansion is needed at the point $(x_1, x_2) = (a_1, a_2)$ we can achieve it by considering the operators $D_{i(0)}$ = partial derivative operator with respect to x_i evaluated at the point (a_1, a_2) for $i = 1, 2$, and the linear form

$$\delta = (x_1 - a_1)D_{1(0)} + (x_2 - a_2)D_{2(0)}.$$

Consider the exponential series

$$e^\delta = \left[\delta^0 + \frac{\delta^1}{1!} + \frac{\delta^2}{2!} + \cdots \right].$$

Then this operator, operating on $f(x_1, x_2)$ is given by

$$f(a_1, a_2) + \frac{1}{1!}\left[(x_1 - a_1)\frac{\partial}{\partial x_1}f(a_1, a_2) + (x_2 - a_2)\frac{\partial}{\partial x_2}f(a_1, a_2) \right]$$
$$+ \frac{1}{2!}\left[(x_1 - a_1)^2 \frac{\partial^2}{\partial x_1^2}f(a_1, a_2) + (x_2 - a_2)^2 \frac{\partial^2}{\partial x_2^2}f(a_1, a_2) \right.$$
$$+ 2(x_1 - a_1)(x_2 - a_2)\frac{\partial^2}{\partial x_1 \partial x_2}f(a_1, a_2) \Big]$$
$$+ \frac{1}{3!}[\cdots] + \cdots = f(x_1, x_2),$$

where, for example, $\frac{\partial^r}{\partial x_j^r}f(a_1, a_2)$ indicates the r-th partial derivative of f with respect to x_j, evaluated at the point (a_1, a_2). If we have a function of k real variables, $f(x_1, \ldots, x_k)$, and if a Taylor series expansion at $(a_1, \ldots, a_k)$ is needed then it is available from the operator

$$e^\delta, \quad \delta = (x_1 - a_1)D_{1(0)} + \cdots + (x_k - a_k)D_{k(0)}$$

operating on f. Denoting $a = (a_1, \ldots, a_k)$ we have

$$f(x_1, \ldots, x_k) = f(a)$$
$$+ \frac{1}{1!}\left[(x_1 - a_1)\frac{\partial}{\partial x_1}f(a) + \cdots + (x_k - a_k)\frac{\partial}{\partial x_k}f(a)\right]$$
$$+ \frac{1}{2!}\left[(x_1 - a_1)^2\frac{\partial^2}{\partial x_1^2}f(a) + \cdots + (x_k - a_k)^2\frac{\partial^2}{\partial x_k^2}f(a)\right.$$
$$+ 2(x_1 - a_1)(x_2 - a_2)\frac{\partial^2}{\partial x_1 \partial x_2}f(a) + \cdots$$
$$\left. + 2(x_{k-1} - a_{k-1})(x_k - a_k)\frac{\partial^2}{\partial x_{k-1}\partial x_k}f(a)\right]$$
$$+ \frac{1}{3!}[\cdots] + \cdots$$

If $a = (a_1, \ldots, a_k)$ is a critical point for $f(x_1, \ldots, x_k)$ then $\frac{\partial}{\partial x_j}f(a) = 0$, $j = 1, \ldots, k$. Then in the neighborhood of the point a, that is, $x_i - a_i = \Delta x_i$, $i = 1, \ldots, k$, where Δx_i's are infinitesimally small, we have

$$f(x_1, \ldots, x_k) - f(a_1, \ldots, a_k) = \frac{1}{2!}\alpha + \frac{1}{3!}[\cdots]$$
$$\approx \frac{1}{2!}\alpha$$

where

$$\alpha = (\Delta x_1)^2\frac{\partial^2}{\partial x_1^2}f(a) + \cdots + 2(\Delta x_{k-1})(\Delta x_k)\frac{\partial^2}{\partial x_{k-1}\partial x_k}f(a).$$

But α is the following quadratic form:

$$\alpha = [\Delta x_1, \ldots, \Delta x_k]\begin{bmatrix} \frac{\partial^2}{\partial x_1^2}f(a) & \cdots & \frac{\partial^2}{\partial x_1 \partial x_k}f(a) \\ \vdots & \ddots & \vdots \\ \frac{\partial^2}{\partial x_k \partial x_1}f(a) & \cdots & \frac{\partial^2}{\partial x_k^2}f(a) \end{bmatrix}\begin{bmatrix} \Delta x_1 \\ \vdots \\ \Delta x_k \end{bmatrix}.$$

This term decides the sign of $f(x_1, \ldots, x_k) - f(a_1, \ldots, a_k)$ in the neighborhood of $a = (a_1, \ldots, a_k)$. This term will remain positive for all $\Delta x_1, \ldots, \Delta x_k$ if the matrix of this quadratic form, namely,

$$A_0 = \frac{\partial}{\partial X}\frac{\partial f(a)}{\partial X'} = \begin{bmatrix} \frac{\partial^2 f(a)}{\partial x_1^2} & \cdots & \frac{\partial^2 f(a)}{\partial x_1 \partial x_k} \\ \vdots & \cdots & \vdots \\ \frac{\partial^2 f(a)}{\partial x_k \partial x_1} & \cdots & \frac{\partial^2 f(a)}{\partial x_k^2} \end{bmatrix}$$

is positive definite. That is, in the neighborhood of $a = (a_1, \ldots, a_k)$ the function is increasing or a corresponds to a local minimum. Similarly if A_0 is negative definite then a corresponds to a local maximum. If A_0 is indefinite or semi-definite then we say that a is a saddle point. We have already given three definitions for the definiteness of a real symmetric or Hermitian matrix. The one in terms of the leading determinants (leading minors) of A_0 is the most convenient one to apply in this case. Thus at the point a we have a

local minimum if all the leading minors of A_0 are positive,
local maximum if the leading minors of A_0 are alternately negative, positive, negative,
...

Example 5.5.1. Check for maxima/minima of $f(x, y) = 2x^2 + y^2 + 2xy - 3x - 4y + 5$.

Solution 5.5.1. Consider the equations $\frac{\partial f}{\partial x} = 0$ and $\frac{\partial f}{\partial y} = 0$.

$$\frac{\partial f}{\partial x} = 0 \implies 4x + 2y - 3 = 0$$

$$\frac{\partial f}{\partial y} = 0 \implies 2x + 2y - 4 = 0$$

$$\implies x = -\frac{1}{2}, \quad y = \frac{5}{2}.$$

There is only one critical point $(x, y) = (-\frac{1}{2}, \frac{5}{2})$. Now, consider the matrix of second order derivatives, evaluated at this point.

$$A_0 = \begin{bmatrix} \frac{\partial^2 f}{\partial x^2} & \frac{\partial^2 f}{\partial x \partial y} \\ \frac{\partial^2 f}{\partial y \partial x} & \frac{\partial^2 f}{\partial^2 y} \end{bmatrix} = \begin{bmatrix} 4 & 2 \\ 2 & 2 \end{bmatrix}.$$

In this case since the second order derivatives are free of the variables there is no need to evaluate at the critical point. The leading minors of this matrix A_0 are the following:

$$4 > 0, \quad \begin{vmatrix} 4 & 2 \\ 2 & 2 \end{vmatrix} = 8 - 4 > 0.$$

Hence the point $(-\frac{1}{2}, \frac{5}{2})$ corresponds to a minimum.

In this example $f(x, y)$ is of the form of a quadratic form $2x^2 + y^2 + 2xy$ plus a linear form $-3x - 4y$ plus a constant 5. When the matrix of the quadratic form is positive definite then we can devise a general procedure without using calculus. Consider a quadratic expression of the type

$$u = X'AX - 2b'X + c \tag{a}$$

where $A = A' \geq 0$, b, c are known and $X' = (x_1, \ldots, x_n)$, $b' = (b_1, \ldots, b_n)$. Let us minimize u without using calculus. In order to illustrate the method let us open up a quadratic form of the following type where β is an $n \times 1$ vector:

$$(X - \beta)' A (X - \beta) = X'AX - 2\beta'AX + \beta'A\beta. \tag{b}$$

Comparing (b) with (a) we note that for $\beta' = b'A^{-1}$ we can write

$$u = (X - A^{-1}b)' A (X - A^{-1}b) - b'A^{-1}b + c. \tag{c}$$

But $b'A^{-1}b$ and c are free of X. Hence maxima/minima of u depends upon the maxima/minima of the quadratic form $(X - A^{-1}b)A(X - A^{-1}b)$ which is positive definite. Hence the maximum is at $+\infty$ and a minimum is achieved when

$$X - A^{-1}b = O \;\Rightarrow\; X = A^{-1}b.$$

In Example 5.5.1 this point $A^{-1}b$ is what we got since in that example

$$A = \begin{bmatrix} 2 & 1 \\ 1 & 1 \end{bmatrix}, \quad b = \begin{bmatrix} 3/2 \\ 2 \end{bmatrix}, \quad c = 5.$$

Then

$$A^{-1}b = \begin{bmatrix} 1 & -1 \\ -1 & 2 \end{bmatrix} \begin{bmatrix} 3/2 \\ 2 \end{bmatrix} = \begin{bmatrix} -1/2 \\ 5/2 \end{bmatrix}.$$

This type of a technique is usually used in a calculus-free course on model building and other statistical procedures.

Example 5.5.2. n measurements $x_1, \ldots, x_n$ are made on an unknown quantity a. Estimate a by minimizing the sum of squares of the measurement errors, and without using calculus.

Solution 5.5.2. The measurement errors are $x_1 - a, \ldots, x_n - a$. The sum of squares of the measurement errors is then $\sum_{i=1}^{n}(x_i - a)^2$. What should be a so that $\sum_{i=1}^{n}(x_i - a)^2$ is the minimum? If we are using calculus then we differentiate with respect to a, equate to zero and solve. Without using calculus we may proceed as follows:

$$\sum_{i=1}^{n}(x_i - a)^2 = \sum_{i=1}^{n}(x_i - \bar{x} + \bar{x} - a)^2, \quad \bar{x} = \frac{x_1 + \cdots + x_n}{n}$$
$$= \sum_{i=1}^{n}(x_i - \bar{x})^2 + n(\bar{x} - a)^2$$

since for any set of numbers

$$\sum_{i=1}^{n}(x_i - \bar{x}) = \sum_{i=1}^{n}x_i - \sum_{i=1}^{n}x_i = 0.$$

But for real numbers

$$\sum_{i=1}^{n}(x_i - \bar{x})^2 \geq 0 \quad \text{and} \quad n(\bar{x} - a)^2 \geq 0.$$

Only the term $n(\bar{x} - a)^2$ contains a. Hence the minimum is achieved when $n(\bar{x} - a)^2 = 0 \Rightarrow a = \bar{x}$. This, in fact, is a special case of a general result.

(i) The mean squared deviations is least when the deviations are taken from the mean value.

This same procedure of completing a quadratic form can be used for general model building also. Consider the general linear model of the regression type considered in Section 2.7.5 of Chapter 2. There the error sum of squares, denoted by $e'e$, can be written in the form

$$e'e = (Y - X\beta)'(Y - X\beta)$$

where Y is an $n \times 1$ vector of known observations, X is $n \times m$, $n \geq m$, and known, and β is $m \times 1$ and unknown. This parameter vector β is to be estimated by minimizing the error sum of squares. A calculus-free procedure, using only matrix methods, is the following when X is of full rank, that is, the rank of X is m or when $X'X$ is nonsingular: Note that

$$e'e = Y'Y - 2\beta'X'Y + \beta'X'X\beta.$$

Among terms on the right, $Y'Y$ does not contain β. Hence write

$$\beta'X'X\beta - 2\beta'X'Y = [\beta - (X'X)^{-1}X'Y]'(X'X)[\beta - (X'X)^{-1}X'Y]$$
$$- Y'X(X'X)^{-1}X'Y.$$

The only term on the right containing β is the quadratic form where the matrix of the quadratic form, namely $X'X$, is positive definite. Hence the minimum is achieved when this part is zero or when

$$\beta - (X'X)^{-1}X'Y = O \Rightarrow \beta = (X'X)^{-1}X'Y.$$

This is the least square solution obtained through calculus in Section 2.7.5 and the least square minimum, denoted by s^2, is then

$$s^2 = Y'Y - Y'X(X'X)^{-1}X'Y = Y'[I - X(X'X)^{-1}X']Y.$$

5.5.2 Optimization of quadratic forms

Consider a real quadratic form $q = X'AX$, $A = A'$ where X is $n \times 1$ and A is $n \times n$. We have seen that there exists an orthonormal matrix P such that $A = P\Lambda P'$ where $\Lambda = \text{diag}(\lambda_1, \dots, \lambda_n)$ with $\lambda_1, \dots, \lambda_n$ being the eigenvalues of A. Then

$$q = X'AX = X'P\Lambda P'X = Y'\Lambda Y, \quad Y = P'X$$
$$= \lambda_1 y_1^2 + \cdots + \lambda_n y_n^2, \quad Y' = (y_1, \dots, y_n). \tag{5.5.1}$$

If $\lambda_j \geq 0$, $j = 1, \dots, n$ then $q = \lambda_1 y_1^2 + \cdots + \lambda_n y_n^2$, $q > 0$, represents a hyperellipsoid in r-space where r is the number of nonzero λ_j's. If $\lambda_j = 1$, $j = 1, \dots, r$, $\lambda_{r+1} = 0 = \cdots = \lambda_n$ then q is a hypersphere in r-space with radius $\sqrt{q}$.

Definition 5.5.1 (Hyperellipsoid). The equation $X'AX \leq c$ when $c > 0$, $A = A' > 0$ (positive definite) represents all points inside and on a hyperellipsoid.

Its standard form is available by rotating the axes of coordinates or by an orthogonal transformation $Y = P'X$, $PP' = I$, $P'P = I$ where

$$A = P\Lambda P' = \text{diag}(\lambda_1, \dots, \lambda_n)$$

where $\lambda_1, \dots, \lambda_n$ are the eigenvalues of A. When $A > 0$ all λ_j's are positive. The standard form of a hyperellipsoid is

$$c = X'AX = \lambda_1 y_1^2 + \cdots + \lambda_n y_n^2$$

where $\sqrt{c/\lambda_1}, \dots, \sqrt{c/\lambda_n}$ are the semi-axes.

If there are no conditions or restrictions on X and A in $q = X'AX$ then from (5.5.1) we can see that q can go to $+\infty$ or $-\infty$ or to an indeterminate form depending upon the eigenvalues $\lambda_1, \dots, \lambda_n$. Hence the unrestricted maxima/minima problem is not that meaningful. Let us restrict X to a hypersphere of radius unity, that is $X'X = 1$, or,

$$x_1^2 + x_2^2 + \cdots + x_n^2 = 1, \quad X' = (x_1, \dots, x_n).$$

Let us try to optimize $q = X'AX$ subject to the condition $X'X = 1$. Let λ be a Lagrangian multiplier and consider the function

$$q_1 = X'AX - \lambda(X'X - 1).$$

Since $X'X - 1 = 0$ we have not made any change in q, $q_1 = q$, where λ is an arbitrary scalar. Using the vector derivative operator $\frac{\partial}{\partial X}$ (see Chapter 1)

$$\frac{\partial}{\partial X} q_1 = O \implies \frac{\partial}{\partial X}[X'AX - \lambda(X'X - 1)] = O$$
$$\implies 2AX - 2\lambda X = O$$
$$\implies AX = \lambda X \tag{5.5.2}$$

(see the results on the operator $\frac{\partial}{\partial X}$ operating on linear and quadratic forms from Chapter 1). From (5.5.2) it is clear that λ is an eigenvalue of A and X the corresponding eigenvector. Premultiply (5.5.2) by X' to obtain

$$X'AX = \lambda X'X = \lambda \quad \text{when } X'X = 1.$$

Therefore the maximum value of $X'AX$ is the largest eigenvalue of A and the minimum value of $X'AX$ is the smallest eigenvalue of A. We can state these as follows:

(ii)
$$\max_{X'X=1}[X'AX] = \lambda_n = \text{largest eigenvalue of } A.$$

(iii)
$$\min_{X'X=1}[X'AX] = \lambda_1 = \text{smallest eigenvalue of } A.$$

If we did not restrict A to a hypersphere of radius 1 let us see what happens if we simply say that $X'X < \infty$ (the length is finite). Let $X'X = c$ for some $c < \infty$. Then proceeding as before we end up with (5.5.2). Then premultiplying both sides by X' we have

$$\frac{X'AX}{X'X} = \lambda.$$

Therefore we have the following results:

(iv)
$$\max\left[\frac{X'AX}{X'X}\right] = \lambda_n = \text{largest eigenvalue of } A \text{ if } X'X < \infty.$$

(v)
$$\min\left[\frac{X'AX}{X'X}\right] = \lambda_1 = \text{smallest eigenvalue of } A \text{ if } X'X < \infty.$$

(vi)
$$\lambda_1 \leq \frac{X'AX}{X'X} = \frac{\lambda_1 y_1^2 + \cdots + \lambda_n y_n^2}{y_1^2 + \cdots + y_n^2} \leq \lambda_n \text{ for } X'X < \infty$$

where $Y = P'X$, $PP' = I$, $P'P = I$, $Y' = (y_1, \ldots, y_n)$.

For convenience, λ_1 is taken as the smallest and λ_n the largest eigenvalue of A. The results in (iii) and (iv) are known as *Rayleigh's principle* and $\frac{X'AX}{X'X}$ the *Rayleigh's quotient* in physics.

Example 5.5.3. Evaluate the Rayleigh's quotient for the quadratic form $X'AX$, $A = (a_{ij})$ if (a) $X' = (1, 0, \ldots, 0)$, (b) $X' = (0, 0, \ldots, 1, 0, \ldots, 0)$, j-th element is 1.

Solution 5.5.3. (a) When $X' = (1, 0, \ldots, 0)$, $X'AX = a_{11}x_1^2$ and $X'X = x_1^2$. Then

$$\frac{X'AX}{X'X} = a_{11}.$$

Similarly for (b)

$$\frac{X'AX}{X'X} = a_{jj}.$$

5.5.3 Optimization of a quadratic form with quadratic form constraints

Let us generalize the above problem a little further. Suppose we wish to optimize $X'AX$ subject to the condition $X'BX = c < \infty$. Consider

$$\phi = X'AX - \lambda(X'BX - c)$$

where λ is a Lagrangian multiplier.

$$\frac{\partial \phi}{\partial X} = O \Rightarrow AX = \lambda BX. \tag{5.5.3}$$

This means that for a non-null solution X we must have

$$|A - \lambda B| = 0 \Rightarrow |B^{-1}A - \lambda I| = 0 \quad \text{if } |B| \neq 0 \text{ or}$$
$$|A^{-1}B - (1/\lambda)I| = 0 \quad \text{if } |A| \neq 0.$$

Thus the maximum value of $X'AX$ occurs at the largest root, λ_n, of the determinantal equation $|A - \lambda B| = 0$ which is also the largest eigenvalue of $B^{-1}A$ if $|B| \neq 0$. From (5.5.3),

$$X'AX = \lambda X'BX = \lambda c$$

and thus we have the following results:

(vii) $$\max_{X'BX=c}[X'AX] = \lambda_n c, \quad \lambda_n = \text{largest root of } |A - \lambda B| = 0.$$

(viii) $$\min_{X'BX=c}[X'AX] = \lambda_1 c, \quad \lambda_1 = \text{smallest root of } |A - \lambda B| = 0.$$

Observe that if A and B are at least positive semi-definite then we are looking at the slicing of the hyperellipsoid $X'BX = c$ with an arbitrary hyperellipsoid $X'AX = q$ for all q. Note also that the above procedure and the results (vii) and (viii) correspond to the canonical correlation analysis when $B = B' > 0$ and $A = A' > 0$.

Example 5.5.4. Optimize $2x_1^2 + 3x_2^2 + 2x_1x_2$ subject to the condition $x_1^2 + 2x_2^2 + 2x_1x_2 = 3$.

Solution 5.5.4. Writing the quadratic forms in matrix notation we have

$$2x_1^2 + 3x_2^2 + 2x_1x_2 = [x_1, x_1]\begin{bmatrix} 2 & 1 \\ 1 & 3 \end{bmatrix}\begin{bmatrix} x_1 \\ x_2 \end{bmatrix}$$

$$= X'AX, \quad A = A' = \begin{bmatrix} 2 & 1 \\ 1 & 3 \end{bmatrix}, \quad X = \begin{bmatrix} x_1 \\ x_2 \end{bmatrix};$$

$$x_1^2 + 2x_2^2 + 2x_1x_2 = [x_1, x_2]\begin{bmatrix} 1 & 1 \\ 1 & 2 \end{bmatrix}\begin{bmatrix} x_1 \\ x_2 \end{bmatrix}$$

$$= X'BX = 3, \quad B = B' = \begin{bmatrix} 1 & 1 \\ 1 & 2 \end{bmatrix}.$$

Consider the determinantal equation

$$|A - \lambda B| = 0 \implies \left| \begin{bmatrix} 2 & 1 \\ 1 & 3 \end{bmatrix} - \lambda \begin{bmatrix} 1 & 1 \\ 1 & 2 \end{bmatrix} \right| = 0$$

$$\implies (2 - \lambda)(3 - 2\lambda) - (1 - \lambda)^2 = 0$$

$$\implies \lambda^2 - 5\lambda + 5 = 0$$

$$\implies \lambda_1 = \frac{5 + \sqrt{5}}{2}, \quad \lambda_2 = \frac{5 - \sqrt{5}}{2}.$$

Therefore the maximum value of $X'AX$ is

$$\lambda_1 c = (3)\left(\frac{5 + \sqrt{5}}{2} \right) = \frac{3}{2}(5 + \sqrt{5})$$

and the minimum value of $X'AX$ is

$$\lambda_2 c = \frac{3}{2}(5 - \sqrt{5}).$$

5.5.4 Optimization of a quadratic form with linear constraints

Consider the optimization of $X'AX$ subject to the condition $X'b = c$ where b is an $n \times 1$ known vector and c is a known scalar. If $A = A' > 0$ then $q = X'AX$ with $q > 0$ a constant, is a hyperellipsoid and this ellipsoid is cut by the plane $X'b = c$. We get another ellipsoid in a lower dimension. For example if the ellipsoid is

$$q = 3x_1^2 + x_2^2 + x_3^2 - 2x_1x_2$$

$$= [x_1, x_2, x_3] \begin{bmatrix} 3 & -1 & 0 \\ -1 & 1 & 0 \\ 0 & 0 & 1 \end{bmatrix} \begin{bmatrix} x_1 \\ x_2 \\ x_3 \end{bmatrix} = X'AX,$$

$$X = \begin{bmatrix} x_1 \\ x_2 \\ x_3 \end{bmatrix}, \quad A = \begin{bmatrix} 3 & -1 & 0 \\ -1 & 1 & 0 \\ 0 & 0 & 1 \end{bmatrix}$$

and if the condition is

$$x_1 + x_2 + x_3 = 1 \implies X'b = 1,$$

$$b = \begin{bmatrix} 1 \\ 1 \\ 1 \end{bmatrix}, \quad c = 1,$$

then from the condition we have $x_3 = 1 - x_1 - x_2$. Substituting in $X'AX$ we have

$$q = 3x_1^2 + x_2^2 + (1 - x_1 - x_2)^2 - 2x_1x_2$$
$$= 4x_1^2 + 2x_2^2 - 2x_1 - 2x_2 + 1$$
$$= 4\left(x_1 - \frac{1}{4}\right)^2 + 2\left(x_2 - \frac{1}{2}\right)^2 + \frac{1}{4}.$$

It is an ellipse with the center at $(\frac{1}{4}, \frac{1}{2})$ and the semi-axes proportional to $\frac{1}{2}$ and $\frac{1}{\sqrt{2}}$. For example if $q - \frac{1}{4} = 1$ then the semi-axes are $\frac{1}{2}$ and $\frac{1}{\sqrt{2}}$. Then the critical point is $(\frac{1}{4}, \frac{1}{2})$ and the maximum value is $\frac{1}{2} + \frac{1}{\sqrt{2}}$ and the minimum value is $\frac{1}{2} - \frac{1}{\sqrt{2}}$.

But our original aim was to optimize $X'AX$ and then q is a function of x_1, x_2, say $q(x_1, x_2)$. Then

$$q(x_1, x_2) = 4\left(x_1 - \frac{1}{4}\right)^2 + 2\left(x_2 - \frac{1}{2}\right)^2 + \frac{1}{4}.$$

Then going through the usual maximization process we see that there is a minimum at $(\frac{1}{4}, \frac{1}{2})$ for finite X and the minimum value is $\frac{1}{4}$.

Substitution may not be always convenient and hence we need a general procedure to handle such problems. Consider the method of Lagrangian multipliers and consider

$$\phi = X'AX - 2\lambda(X'b - c)$$

where 2λ is a Lagrangian multiplier. Then

$$\frac{\partial \phi}{\partial X} = O \Rightarrow 2AX - 2\lambda b = O \Rightarrow AX = \lambda b.$$

If $|A| \neq 0$ then

$$X = \lambda A^{-1}b \Rightarrow X'AX = \lambda X'b = \lambda c,$$
$$\Rightarrow \lambda = \frac{c}{b'A^{-1}b}$$
$$\Rightarrow X'AX = \frac{c^2}{b'A^{-1}b}.$$

When A is positive definite the maximum is at ∞ and then for finite X we have only a minimum for $X'AX$ and the minimum value is $\frac{c^2}{b'A^{-1}b}$. We can verify this result for our example above. In our illustrative example

$$A^{-1} = \frac{1}{2}\begin{bmatrix} 1 & 1 & 0 \\ 1 & 3 & 0 \\ 0 & 0 & 2 \end{bmatrix}$$

and then $b'A^{-1}b = 4$. Since $c = 1$ we get the minimum value as $\frac{1}{4}$ which is what is seen earlier.

(ix) $$\min_{X'b=c}[X'AX] = \frac{c^2}{b'A^{-1}b} \quad \text{when } A = A' > 0.$$

5.5.5 Optimization of bilinear forms with quadratic constraints

Consider a bilinear form $X'AY$ where X is $p \times 1$, Y is $q \times 1$ and A is $p \times q$. For convenience let $p \leq q$. Consider the positive definite quadratic form constraints in X and Y. Without loss of generality the constraints can be written as $X'BX = 1$, $Y'CY = 1$, $B = B' > 0$, $C = C' > 0$. Our aim is to optimize $X'AY$ subject to the conditions $X'BX = 1$ and $Y'CY = 1$. Let

$$\phi = X'AY - \frac{1}{2}\lambda_1(X'BX - 1) - \frac{1}{2}\lambda_2(Y'CY - 1)$$

where $\frac{1}{2}\lambda_1$ and $\frac{1}{2}\lambda_2$ are Lagrangian multipliers. Then

$$\frac{\partial \phi}{\partial X} = O \;\Rightarrow\; AY - \lambda_1 BX = O$$
$$\Rightarrow AY = \lambda_1 BX$$
$$\Rightarrow X'AY = \lambda_1 X'BX = \lambda_1.$$

For taking the partial derivative with respect to Y write $X'AY = Y'A'X$.

$$\frac{\partial \phi}{\partial Y} = O \;\Rightarrow\; A'X - \lambda_2 CY = O$$
$$\Rightarrow A'X = \lambda_2 CY$$
$$\Rightarrow Y'A'X = \lambda_2 Y'CY = \lambda_2.$$

That is, $\lambda_1 = \lambda_2 (= \lambda$, say). The maximum or minimum is given by λ. Writing the equations once again we have

$$-\lambda BX + AY = O \quad \text{and} \quad A'X - \lambda CY = O$$

$$\Rightarrow \begin{bmatrix} -\lambda B & A \\ A' & -\lambda C \end{bmatrix} \begin{bmatrix} X \\ Y \end{bmatrix} = O.$$

In order for this to have a non-null solution for $\binom{X}{Y}$ the coefficient matrix must be singular. That is,

$$\begin{vmatrix} -\lambda B & A \\ A' & -\lambda C \end{vmatrix} = 0.$$

Evaluating the left side with the help of partitioned matrices (see Section 2.7 in Chapter 2) we have

$$| - \lambda B| |-\lambda C - A'(-\lambda B)^{-1}A| = 0 \Rightarrow$$

$$\left| -\lambda C + \frac{1}{\lambda}A'B^{-1}A \right| = 0 \Rightarrow$$

$$|A'B^{-1}A - \lambda^2 C| = 0. \qquad (\alpha)$$

There $v = \lambda^2$ is a root of the determinantal equation (α) above. Thus we have the following results:

(x)
$$\max_{\substack{X'BX=1,\ Y'CY=1 \\ B=B'>0,\ C=C'>0}} [|X'AY|] = |\lambda_p|,$$

where λ_p^2 is the largest root of the equation (α).

(xi)
$$\min_{\substack{X'BX=1,\ Y'CY=1 \\ B=B'>0,\ C=C'>0}} [|X'AY|] = |\lambda_1|,$$

where λ_1^2 is the smallest root of the equation (α).

Note that λ^2 can also be written as an eigenvalue problem. Equations can be written as

$$|C^{-1}A'B^{-1}A - \lambda^2 I| = 0 \quad \text{or}$$

$$|C^{-\frac{1}{2}}A'B^{-1}AC^{-\frac{1}{2}} - \lambda^2 I| = 0$$

where $C^{\frac{1}{2}}$ is the symmetric positive definite square root of $C = C' > 0$. Either λ^2 can be looked upon as an eigenvalue of $C^{-1}A'B^{-1}A$ or of $C^{-\frac{1}{2}}A'B^{-1}AC^{-\frac{1}{2}}$. From symmetry it follows that λ^2 is also an eigenvalue of $B^{-1}AC^{-1}A'$ or of $B^{-\frac{1}{2}}AC^{-1}A'B^{-\frac{1}{2}}$.

(xii)
$$\max_{\substack{X'BX=1,\ Y'CY=1 \\ B=B'>0,\ C=C'>0}} [|X'AY|] = |\lambda_p|$$

where λ_p^2 is the largest eigenvalue of $C^{-1}A'B^{-1}A$ or that of $B^{-1}AC^{-1}A'$ or of $C^{-\frac{1}{2}}A'B^{-1}AC^{-\frac{1}{2}}$ or of $B^{-\frac{1}{2}}AC^{-1}A'B^{-\frac{1}{2}}$ or the root of the determinantal equations

$$|A'B^{-1}A - \lambda^2 C| = 0$$

or of

$$|AC^{-1}A' - \lambda^2 B| = 0.$$

(xiii)
$$\min_{\substack{X'BX=1,\ Y'CY=1 \\ B=B'>0,\ C=C'>0}} [|X'AY|] = |\lambda_1|$$

where λ_1^2 is the smallest eigenvalue of the matrices or the smallest root of the determinantal equations as in (xii) above.

An important application of the results in (xii) and (xiii) is already discussed in connection with the canonical correlation analysis and a particular case of which is the multiple correlation analysis. The structure in partial correlation analysis in statistics is also more or less the same.

Example 5.5.5. Look for maxima/minima of

$$x_1y_1 + 2x_1y_2 - x_1y_3 + x_2y_1 + x_2y_2 + x_2y_3$$

subject to the conditions

$$x_1^2 + 2x_2^2 + 2x_1x_2 = 1 \quad \text{and} \quad 2y_1^2 + y_2^2 + y_3^2 + 2y_1y_3 = 1.$$

Solution 5.5.5. Writing these in matrix notations, the quantity to be optimized is

$$X'AY, \quad A = \begin{bmatrix} 1 & 2 & -1 \\ 1 & 1 & 1 \end{bmatrix}, \quad X = \begin{bmatrix} x_1 \\ x_2 \end{bmatrix}, \quad Y = \begin{bmatrix} y_1 \\ y_2 \\ y_3 \end{bmatrix}$$

and the conditions are

$$X'BX = 1, \quad B = \begin{bmatrix} 1 & 1 \\ 1 & 2 \end{bmatrix} \quad \text{and}$$

$$Y'CY = 1, \quad C = \begin{bmatrix} 2 & 0 & 1 \\ 0 & 1 & 0 \\ 1 & 0 & 1 \end{bmatrix}, \quad C^{-1} = \begin{bmatrix} 1 & 0 & -1 \\ 0 & 1 & 0 \\ -1 & 0 & 2 \end{bmatrix}.$$

Since the order of B is 2 or B is 2×2 whereas C is 3×3 we consider the determinantal equation

$$|AC^{-1}A' - \lambda^2 B| = 0,$$

$$AC^{-1}A' = \begin{bmatrix} 1 & 2 & -1 \\ 1 & 1 & 1 \end{bmatrix} \begin{bmatrix} 1 & 0 & -1 \\ 0 & 1 & 0 \\ -1 & 0 & 2 \end{bmatrix} \begin{bmatrix} 1 & 1 \\ 2 & 1 \\ -1 & 1 \end{bmatrix} = \begin{bmatrix} 9 & 1 \\ 1 & 2 \end{bmatrix};$$

$$|AC^{-1}A' - \lambda^2 B = 0 \implies$$

$$\left| \begin{bmatrix} 9 & 1 \\ 1 & 2 \end{bmatrix} - v \begin{bmatrix} 1 & 1 \\ 1 & 2 \end{bmatrix} \right| = 0, \quad v = \lambda^2$$

$$\implies \begin{vmatrix} 9 - v & 1 - v \\ 1 - v & 2 - 2v \end{vmatrix} = 0$$

$$\implies (1 - v)(17 - v) = 0.$$

Therefore $v_1 = 17$ and $v_2 = 1$ are the roots. The largest λ is $\sqrt{17}$ and the smallest positive value of $\lambda = 1$. Hence the maximum value of $X'AY$ is $\sqrt{17}$ and minimum absolute value of $X'AY$ is 1.

In other problems in statistics especially in conditional distributions and when searching for minimum variance estimators in a conditional space the problem is something like to maximizing a bilinear form of the type $X'AY$, where one of the vectors, either X or Y, is fixed (given by the conditionality assumption), subject to a quadratic form condition involving the other vector, usually something like the variance of a linear function in that vector is 1 which reduces to the form $Y'CY = 1$, $C = C' > 0$ if Y is the variable vector. For achieving such a maximization we can use a different approach based on Cauchy–Schwartz inequality also. Suppose we wish to maximize $X'AY$ subject to the condition $Y'CY = 1$, $C = C' > 0$ and X fixed. Then

$$X'AY = (X'AC^{-\frac{1}{2}})(C^{\frac{1}{2}}Y) \le \sqrt{X'AC^{-1}A'X}\sqrt{Y'CY}$$
$$= \sqrt{X'AC^{-1}A'X}$$

since $Y'CY = 1$, by Cauchy–Schwartz inequality, where $C^{\frac{1}{2}}$ is the symmetric positive definite square root of C. Thus we have the following result:

(xiv)
$$\max_{Y'CY=1, C=C'>0, X \text{ fixed}} [X'AY] = \sqrt{X'AC^{-1}A'X}.$$

(xv)
$$\max_{X'BX=1, B=B'>0, Y \text{ fixed}} [X'AY] = \sqrt{Y'A'B^{-1}AY}.$$

Example 5.5.6 (Best linear predictors). A farmer suspects that the yield of corn, y, is a linear function of the amount of a certain fertilizer used, x_1, and the amount of water supplied x_2. What is the best linear function of x_1 and x_2, $u = a_1x_1 + a_2x_2$, to predict y, best in the sense that the variance of u is 1 and the covariance of u with y is maximum. Evaluate this best linear predictor as well as the maximum covariance if the following items are available from past experience. $\mathrm{Var}(x_1) = 2$, $\mathrm{Var}(x_2) = 1$, $\mathrm{Cov}(x_1, x_2) = 1$, $\mathrm{Cov}(y, x_1) = 1$, $\mathrm{Cov}(y, x_2) = -1$, $\mathrm{Var}(y) = 3$ where $\mathrm{Var}(\cdot)$ and $\mathrm{Cov}(\cdot)$ denote the variance of $(\cdot)$ and covariance of $(\cdot)$ respectively.

Solution 5.5.6. Let

$$a = \begin{bmatrix} a_1 \\ a_2 \end{bmatrix}, \quad X = \begin{bmatrix} x_1 \\ x_2 \end{bmatrix}$$

where a is an arbitrary coefficient vector for an arbitrary linear function $u = a'X = a_1x_1 + a_2x_2$. Since we are dealing with variances and covariances, assume without loss of generality, that the mean values are zeros. Then

$$\mathrm{Var}(u) = [a_1, a_2] \begin{bmatrix} \mathrm{Var}(x_1) & \mathrm{Cov}(x_1, x_2) \\ \mathrm{Cov}(x_2, x_1) & \mathrm{Var}(x_2) \end{bmatrix} \begin{bmatrix} a_1 \\ a_2 \end{bmatrix}$$

$$= a'Ba, \quad B = \begin{bmatrix} 2 & 1 \\ 1 & 1 \end{bmatrix},$$

$$\mathrm{Cov}(y, u) = \mathrm{Cov}(y, a'X) = a'\,\mathrm{Cov}(y, X)$$

$$= a'\begin{bmatrix} \mathrm{Cov}(y, x_1) \\ \mathrm{Cov}(y, x_2) \end{bmatrix} = a'\begin{bmatrix} 1 \\ -1 \end{bmatrix} = a_1 - a_2.$$

Our problem is to maximize $\mathrm{Cov}(y, u)$ subject to the condition $\mathrm{Var}(u) = 1$. For convenience let

$$V_{21} = \begin{bmatrix} 1 \\ -1 \end{bmatrix}, \quad V_{22} = \begin{bmatrix} 2 & 1 \\ 1 & 1 \end{bmatrix},$$

then

$$\mathrm{Var}(u) = a'\, V_{22}\, a = 1$$

and

$$\mathrm{Cov}(y, u) = a'\, V_{21} = (a'\, V_{22}^{\frac{1}{2}})(V_{22}^{-\frac{1}{2}}\, V_{21})$$

$$\leq \sqrt{a'\, V_{22}\, a}\,\sqrt{V_{21}'\, V_{22}^{-1}\, V_{21}}$$

$$= \sqrt{V_{21}'\, V_{22}^{-1}\, V_{21}}$$

since $a'\, V_{22}\, a = 1$. Hence the maximum covariance is

$$\sqrt{V_{21}'\, V_{22}^{-1}\, V_{21}} = \left\{ [1, -1]\begin{bmatrix} 2 & 1 \\ 1 & 1 \end{bmatrix}^{-1}\begin{bmatrix} 1 \\ -1 \end{bmatrix} \right\}^{\frac{1}{2}} = \sqrt{5}.$$

From Cauchy–Schwartz inequality the maximum is attained when $(a'\, V_{22}^{\frac{1}{2}})$ and $(V_{22}^{\frac{1}{2}}\, V_{21})$ are linear functions of each other. That is,

$$a'\, V_{22}^{\frac{1}{2}} = k_1 V_{22}^{-\frac{1}{2}}\, V_{21} + k_2$$

where k_1 is a scalar and k_2 is a vector and the best predictor is

$$a'X = V_{22}^{-\frac{1}{2}}[k_1 V_{22}^{-\frac{1}{2}}\, V_{21} + k_2]X.$$

With the help of the conditions that $E(X) = O$, $\mathrm{Var}(a'X) = a'\, V_{22}\, a = 1$ and the maximum covariance is $\mathrm{Cov}(y, u) = \sqrt{V_{21}'\, V_{22}^{-1}\, V_{21}}$ we have $k_1 = 1$ and $k_2 = O$. Thus the best predictor is given by

$$u = a'X = V_{22}^{-1}\, V_{21}X$$

and the maximum covariance is given by $\sqrt{V_{21}'\, V_{22}^{-1}\, V_{21}}$.

Exercises 5.5

5.5.1. Look for maxima/minima of the following functions:

$$f(x, y) = 3x^2 + 2y^2 + 2xy - 4x - 5y + 10,$$
$$g(x, y) = 2x^2 + y^2 - 2xy - x - y + 5.$$

5.5.2. Repeat Exercise 5.5.1 without using calculus.

5.5.3. Look for maxima/minima of the following functions:

(a) $3x_1^2 + x_2^2 + 2x_3^2 - 2x_1x_3$

subject to $x_1^2 + x_2^2 + x_3^2 = 1$,

(b) $3x_1^2 + x_2^2 + 2x_3^2 - 2x_1x_3$

subject to $2x_1^2 + x_2^2 + 2x_3^2 = 1$.

5.5.4. Let $x_1, \ldots, x_n$ be n real numbers and α an unknown quantity. What is α so that $\sum_{i=1}^{n} |x_i - \alpha|$ is a minimum? [This principle can be stated as "the mean absolute deviations is least when the deviations are taken from (?)".]

5.5.5. Let $U_1, \ldots, U_k$ and V be $n \times 1$ vectors of real numbers. Let there exist an X such that $U_j' X \geq 0$ for $j = 1, \ldots, k$. Then show that the necessary and sufficient condition that $V'X \geq 0$ for all X such that $U_j' X \geq 0$ is that V can be written as $V = \sum_{j=1}^{k} a_j U_j$, $a_j \geq 0$. [This is known as Farka's lemma.]

5.5.6. Gramian matrix. Any matrix A which can be written as $A = B'B$ is called a Gramian matrix. For a Gramian matrix A and $n \times 1$ real vectors X and Y show that

$$(X'AY)^2 \leq (X'AX)(Y'AY)$$

with equality when X is proportional to Y, and

$$(X'Y)^2 \leq (X'AX)(Y'A^{-1}Y)$$

if A^{-1} exists, with equality when X is proportional to $A^{-1}Y$.

5.5.7. Let $X' = (x_1, \ldots, x_n)$ and $Y' = (y_1, \ldots, y_n)$ then show that

$$\left(\sum_{j=1}^{n} x_j^2\right)\left(\sum_{j=1}^{n} y_j^2\right) - \left(\sum_{j=1}^{n} x_j y_j\right)^2 = \sum_{i<j}(x_i y_j - x_j y_i)^2.$$

[This is known as Lagrange identity.]

5.5.8. For a nonsingular $n \times n$ matrix $A = (a_{ij})$ show that

$$|A| \leq 1, \quad \text{if } \sum_{j=1}^{n} a_{ij}^2 = 1, \quad i = 1, \ldots, n$$

and that

$$|A| \leq \delta^n n^{n/2} \quad \text{if } |a_{ij}| \leq \delta \text{ for all } i \text{ and } j.$$

5.5.9. Let $A = (a_{ij})$ be an $n \times n$ real symmetric positive definite matrix. Show that

$$|A| \leq a_{11}(\text{cof}(a_{11}))$$

where $(\text{cof}(a_{11}))$ denotes the cofactor of a_{11}, and

$$|A| \leq a_{11}a_{22} \cdots a_{nn}.$$

5.5.10. Let A be an $n \times n$ real symmetric positive definite matrix and U and X be n-vectors. Show that

$$\max_X \frac{(U'X)^2}{X'AX} = U'A^{-1}U.$$

5.5.11. Let A be as in Exercise 5.5.10 and B any $n \times n$ matrix. Then show that

$$\max_X \frac{(U'BX)^2}{X'AX} = U'BA^{-1}B'U.$$

5.5.12. Let $X_1, \ldots, X_k$ be mutually orthogonal $n \times 1$ vectors and $A = A'$ a real symmetric matrix. Then show that

$$\max_{X_1,\ldots,X_k} \sum_{i=1}^{k} \frac{X_i'AX_i}{X_i'X_i} = \sum_{i=1}^{k} \lambda_i$$

where $\lambda_1 \geq \lambda_2 \geq \cdots \geq \lambda_n$ are the eigenvalues of A.

5.6 Linear programming and nonlinear least squares

In the previous sections we dealt with optimizations of nonlinear functions with linear or nonlinear constraints. Suppose we wish to optimize (maximize/minimize) a linear function with linear constraints. Obviously methods based on calculus fail here. Geometrically a linear function represents a line in a 2-dimensional space (plane) or a plane or hyperplane in n-dimensional Euclidean space, $n \geq 3$. A line or plane stretch from $-\infty$ to ∞ and hence there is no optimum value if the whole line or plane is taken into consideration. But if we confine to a convex region in a plane or space in our search for an optimum then there is always a maximum and a minimum for an arbitrary line or plane passing through that region. For example, let us try to look for maxima/minima of the linear function $x + y$ in the region bounded by the following constraints: $x \geq 0$, $y \geq 0$, $x - 3y \leq -3$, $3x + y \leq 6$.

The conditions $x \geq 0$, $y \geq 0$ imply non-negative values or we are in the first quadrant. $x - 3y \leq -3$ means below the line $x - 3y = -3$ and $3x + y \leq 6$ means below the line $3x + y = 6$. Thus we are looking for maxima/minima of $x + y$ in the shaded region in Figure 5.6.1. Consider the equation $x + y = c$ for various values of c or move the line $x + y = 0$ parallel to itself. A few positions are shown in Figure 5.6.1. It is obvious that

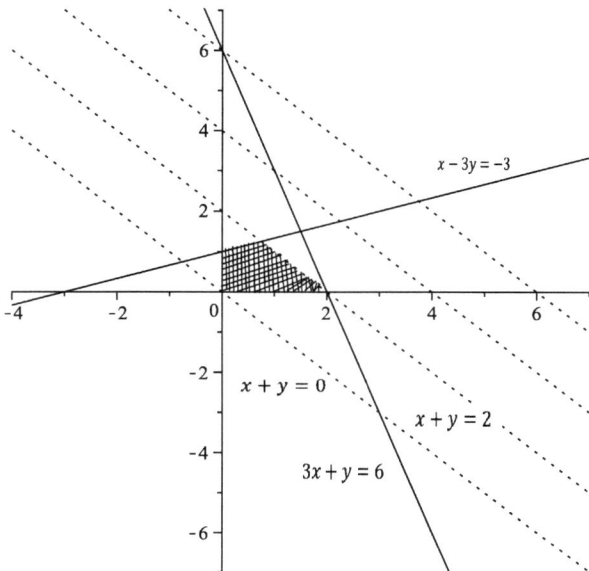

Figure 5.6.1: Optimization of a linear function.

the maximum or minimum value of c is obtained at one of the corner points of our convex region, the shaded region in Figure 5.6.1. This, in fact, is a general property when the region is convex. The corner points are $(0,0)$, $(0,1)$, $(1.5,1.5)$, $(2,0)$. The values of $x + y$ at these points are $0 + 0 = 0$, $0 + 1 = 1$, $1.5 + 1.5 = 3$, $2 + 0 = 2$ respectively. Hence the minimum value of $x + y$ is zero and the maximum value is 3.

In the above problem suppose we had one more condition that $3x + y \leq 2$. Note that our region remains the same or in other words this new condition is superfluous in our problem.

Example 5.6.1. A lunch counter in an office building plans to prepare two types of sandwiches for a particular day. Let x_1 and x_2 be the numbers. Then $x_1 \geq 0$ and $x_2 \geq 0$. The profit from the first type is $2 per sandwich and that from the second type is $3 per sandwich. It costs $1 and $2 each respectively to make these sandwiches and the operator does not want to allocate more than $100 for these two types of sandwiches for that day. That is, $x_1 + 2x_2 \leq 100$. It takes 2 minutes each to prepare these sandwiches and the operator does not want to spend more than 2 hours in preparing them. That is, $2x_1 + 2x_2 \leq 120$. What should be the numbers x_1 and x_2 so as to maximize the profit assuming that all the sandwiches will be sold.

Solution 5.6.1. The region where we want to maximize $2x_1 + 3x_2$ is the shaded region in Figure 5.6.2.

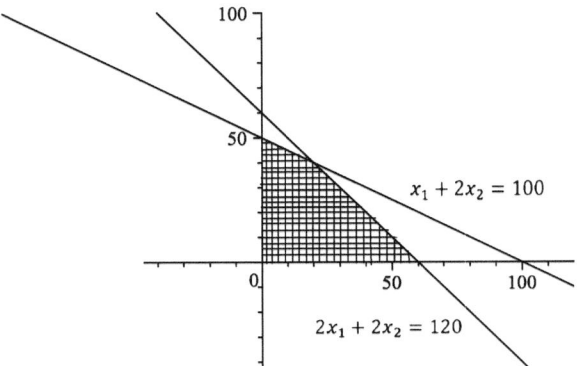

Figure 5.6.2: Linear function with linear constraints.

The values of $2x_1 + 3x_2$ at the corner points $(0,0)$, $(60,0)$, $(20,40)$, $(0,50)$ are respectively $0, 120, 160, 150$, and thus the maximum value is 160 occurring at $(20,40)$ and hence the operator should make 20 of type 1 and 40 of type 2 for that day.

From the above examples it is clear that if only two variables are involved then the optimization problem can be solved graphically. But if we are in an n-space, $n > 3$ or if more than 3 variables are involved then a graphical solution is not feasible. Even for $n = 3$ it is quite difficult to see graphically. Hence we need other methods. One such method, based on matrix considerations, is called a *simplex method*. This optimization problem involving linear functions with linear constraints is called a *linear programming problem*, nothing to do with any special computer programme or computer language. There are many ways of explaining the simplex method. First, observe that an inequality of the type $x - y \geq 2$ is the same as saying $-x + y \leq -2$. The inequality is reversed by multiplying with (-1) on both sides. Hence in some of the constraints if the inequality is the other way around it can be brought to the pattern of the remaining inequalities by the above procedure.

5.6.1 The simplex method

Let us consider a problem of the following type: Maximize

$$f = c_1 x_1 + \cdots + c_n x_n = C'X = X'C, \quad C' = (c_1, \ldots, c_n), \quad X' = (x_1, \ldots, x_n)$$

subject to the conditions

$$a_{11} x_1 + a_{12} x_2 + \cdots + a_{1n} x_n \leq b_1$$
$$a_{21} x_1 + a_{22} x_2 + \cdots + a_{2n} x_n \leq b_2 \tag{5.6.1}$$

$$\vdots$$
$$a_{m1}x_1 + a_{m2}x_2 + \cdots + a_{mn}x_n \leq b_m$$
$$x_1 \geq 0, \quad x_2 \geq 0, \ldots, x_n \geq 0$$

where c_j's, b_j's, a_{ij}'s are all known constants. For convenience, we may use the following standard notation. The inequalities in (5.6.1) will be written in matrix notation as

$$AX \leq b, \quad X \geq O, \quad b' = (b_1, \ldots, b_m), \quad A = (a_{ij}). \tag{5.6.2}$$

Note that earlier we used a notation of the type $G > 0$ or $G \geq 0$ to denote positive definiteness and positive semi-definiteness of a matrix G. In (5.6.2) we are not talking about definiteness but only a convenient way of writing all the inequalities in (5.6.1) together. Thus the problem can be stated as follows:

$$\text{Maximize } f = C'X \quad \text{subject to } AX \leq b, \quad X \geq O. \tag{5.6.3}$$

Note that an inequality can be made to an equality by adding or subtracting a certain amount to the inequality. For example,

$$a_{11}x_1 + \cdots + a_{1n}x_n \leq b_1 \implies a_{11}x_1 + \cdots + a_{1n}x_n + y_1 = b_1, \quad y_1 \geq 0.$$

Thus by adding a positive quantity y_1 to the inequality an equality is obtained. Note that y_1 is an unknown quantity. Thus by adding $y_1, y_2, \ldots, y_m$, the quantities may all be different, to the inequalities in (5.6.1) we get m linear equations:

$$a_{11}x_1 + \cdots + a_{1n}x_n + y_1 = b_1$$
$$a_{21}x_1 + a_{22}x_2 + \cdots + a_{2n}x_n + y_2 = b_2$$
$$\vdots$$
$$a_{m1}x_1 + a_{m2}x_2 + \cdots + a_{mn}x_n + y_m = b_m.$$

That is,

$$\begin{bmatrix} a_{11} & \cdots & a_{1n} & 1 & 0 & \cdots & 0 \\ a_{21} & \cdots & a_{2n} & 0 & 1 & \cdots & 0 \\ \vdots & \vdots & \vdots & \vdots & \vdots & \vdots & \vdots \\ a_{m1} & \cdots & a_{mn} & 0 & 0 & \cdots & 1 \end{bmatrix} \begin{bmatrix} x_1 \\ x_2 \\ \vdots \\ x_n \\ \vdots \\ y_1 \\ \vdots \\ y_m \end{bmatrix} = \begin{bmatrix} b_1 \\ \vdots \\ b_m \end{bmatrix}.$$

This can be written in partitioned form as follows, augmenting the equation $f = C'X$:

$$\begin{bmatrix} A & I_m \\ C' & O \end{bmatrix} \begin{bmatrix} X \\ Y \end{bmatrix} = \begin{bmatrix} b \\ f \end{bmatrix}, \quad Y = \begin{bmatrix} y_1 \\ \vdots \\ y_m \end{bmatrix}, \quad X \geq O, \quad Y \geq O. \tag{5.6.4}$$

Note that

$$f = C'X = c_1x_1 + \cdots + c_nx_n = c_1x_1 + \cdots + c_nx_n + 0y_1 + \cdots + 0y_m.$$

Now, it is a matter of solving the system of linear equations in (5.6.4) for X, Y and f. We have considered many methods of solving a system of linear equations. Let us try to solve the system in (5.6.4) by elementary operations on the left, that is, operating on the rows only. For this purpose we write only the coefficient matrix and the right side separated by dotted lines. We may also separate the sub-matrices in the coefficient matrix by dotted lines. Write

$$
\begin{array}{ccccc}
A & : & I_m & : & b \\
\cdots & \cdots & \cdots & \cdots & \cdots \\
C' & : & O & : & f
\end{array}
\tag{5.6.5}
$$

For the time being, let $m \le n$. In doing elementary operations on the rows our aim will be to reduce the above matrix format to the following form:

$$
\begin{array}{ccccc}
A & : & I_m & : & b \\
\cdots & \cdots & \cdots & \cdots & \cdots \\
C' & : & O & : & f
\end{array}
\Rightarrow
\begin{array}{ccccc}
B, I_m & : & G & : & b_1 \\
\cdots & \cdots & \cdots & \cdots & \cdots \\
-C_1', O & : & -C_2' & : & f - d
\end{array}
\tag{5.6.6}
$$

In the original position of A there is a matrix B which may or may not be null and an $m \times m$ identity matrix or of the form I_m, B. Remember that the original A was $m \times n$, $m \le n$. In the original position of I_m there is an $m \times m$ matrix G. In the original position of the row vector C' there is a row vector $-C_1'$ augmented with the null vector O or the other way around. In the original position of the null vector there is a vector $-C_2'$ where the elements in C_1 and C_2 are non-negative. Original f has gone to $f - d$ where d is a known number. Original vector b has changed to the known vector b_1 where the elements in b_1 are non-negative. We will show that the maximum value of $C'X$ is d and the corresponding solution of X is such that the first $n - m$ elements are zeros and the last m elements are the last m elements of b_1. If the final form is I_m, B instead of B, I_m then take the last $n - m$ elements in X zeros and the first m as that of b_1. Before interpreting the form in (5.6.6) let us do a numerical example.

Example 5.6.2. Redo the problem in Example 5.6.1 by using matrix method and by reducing to the format in (5.6.6).

Solution 5.6.2. Our problem is to maximize $2x_1 + 3x_2$ subject to the conditions

$$x_1 + 2x_2 \le 100 \quad \text{and} \quad x_1 + x_2 \le 60.$$

Writing in matrix notations, we want to maximize

$$f = C'X = 2x_1 + 3x_2, \quad C' = (2, 3), \quad X' = (x_1, x_2)$$

subject to the conditions

$$AX \le b, \quad A = \begin{bmatrix} 1 & 2 \\ 1 & 1 \end{bmatrix}, \quad b = \begin{bmatrix} 100 \\ 60 \end{bmatrix}.$$

Introduce the dummy variables y_1 and y_2 to write

$$x_1 + 2x_2 + y_1 = 100$$
$$x_1 + x_2 + y_2 = 60$$
$$2x_1 + 3x_2 = f.$$

Writing the coefficient matrix and the right side we have

$$
\begin{array}{ccccccc}
1 & 2 & \vdots & 1 & 0 & \vdots & 100 \\
1 & 1 & \vdots & 0 & 1 & \vdots & 60 \\
\cdots & \cdots & \cdots & \cdots & \cdots & \cdots & \cdots \\
2 & 3 & \vdots & 0 & 0 & \vdots & f
\end{array}
=
\begin{array}{ccccc}
A & \vdots & I_2 & \vdots & b \\
\cdots & \cdots & \cdots & \cdots & \cdots \\
C' & \vdots & O & \vdots & f
\end{array}
$$

We will denote the elementary row operations with the help of our usual notation. Add (-1) times the first row to the second row, (-2) times the first row to the third row. That is,

$$-(1) + (2); \quad -2(1) + (3) \Longrightarrow
\begin{array}{ccccccc}
1 & 2 & \vdots & 1 & 0 & \vdots & 100 \\
0 & -1 & \vdots & -1 & 1 & \vdots & -40 \\
\cdots & \cdots & \cdots & \cdots & \cdots & \cdots & \cdots \\
0 & -1 & \vdots & -2 & 0 & \vdots & f - 200
\end{array}
$$

Now, add (-1) times the second row to the third row, 2 times the second row to the first row and then multiply the second row by (-1). That is,

$$-(2) + (3); \quad 2(2) + (1); \quad -(2) \Longrightarrow
\begin{array}{ccccccc}
1 & 0 & \vdots & -1 & 2 & \vdots & 20 \\
0 & 1 & \vdots & 1 & -1 & \vdots & 40 \\
\cdots & \cdots & \cdots & \cdots & \cdots & \cdots & \cdots \\
0 & 0 & \vdots & -1 & -1 & \vdots & f - 160
\end{array}
$$

$$
=
\begin{array}{ccccc}
I_2 & \vdots & G & \vdots & b_1 \\
\cdots & \cdots & \cdots & \cdots & \cdots \\
O & \vdots & -C'_2 & \vdots & f - d
\end{array}
$$

Writing back in terms of the original variables we have

$$x_1 - y_1 + 2y_2 = 20$$
$$x_2 + y_1 - y_2 = 40$$
$$-y_1 - y_2 = f - 160.$$

A feasible solution is $x_1 = 20$, $x_2 = 40$, $y_1 = 0$, $y_2 = 0$, $f - 160 = 0 \Rightarrow f = 160$. Thus the maximum is attained at $x_1 = 20$ and $x_2 = 40$ and the maximum value of f is 160. This is exactly what we had seen in Example 5.6.1. We may also observe that $C_2'b = 160 = f$ also.

Note that the form in (5.6.5) is brought to the form in (5.6.6) by elementary operations on the left then d is the maximum value for $f = C'X$. In general, we can show that the maximum of $f = C'X$ subject to the conditions $AX \le b$, $X \ge O$ is also the same as the minimum of $g = b'Y$ subject to the condition $A'Y \ge C$, $Y \ge O$. These two are called *duals of each other*. That is, for $m \times n$ matrix A, $n \times 1$ vector X, $m \times 1$ vector b, $n \times 1$ vector C, $m \times 1$ vector Y, with non-negative components in X and Y, the following two problems are equivalent:

$$\text{Maximize } f = C'X \quad \text{subject to } AX \le b, \quad X \ge O$$

is equivalent to

$$\text{minimize } g = b'Y \quad \text{subject to } A'Y \ge C, \quad Y \ge O$$

and the

$$\text{maximum of } f = \text{minimum of } g.$$

The proof of this result as well as other properties of the simplex method or other linear programming methods will not be pursued here. We will do one more example on linear programming and then look at some non-linear least squares problems before concluding this section.

In the above two examples we could easily achieve the format in (5.6.6) by row operations without taking into account any other factor. The problem is not as simple as it appears from the above examples. One way of keeping all elements in b_1 of (5.6.6) non-negative, when possible, is to adopt the following procedure. Do not start operating with the first row and do not interchange rows to start with. Divide each element in b of (5.6.5) by the corresponding elements in the first column of A. Look at the smallest of these. If this occurs at the i-th row element of A then operate with the i-th row of A first and reduce all elements in the first column of (5.6.5) to zeros. Now look at the resulting b and the resulting elements in the second column of A. Repeat the above process to reduce the second column elements in the resulting (5.6.5) to zeros, except the one we are operating with. Repeat the process with the resulting third, fourth,…, last columns of A. This may produce the elements of b_1 non-negative. Still this process may not guarantee the elements in C_1' and C_2' of (5.6.6) to be non-negative. Let us do an example to verify the above steps.

Example 5.6.3. Minimize $g = 3y_1 + 2y_2$ subject to the conditions $y_1 \geq 0$, $y_2 \geq 0$ and

$$2y_1 + y_2 \geq 8$$
$$y_1 + y_2 \geq 5$$
$$y_1 + 2y_2 \geq 8.$$

Solution 5.6.3. In order to bring the problem to the format of (5.6.5) let us consider the dual problem: Maximize $f = 8x_1 + 5x_2 + 8x_3$ subject to the conditions $x_1 \geq 0$, $x_2 \geq 0$, $x_3 \geq 0$, and

$$2x_1 + x_2 + x_3 \leq 3$$
$$x_1 + x_2 + 2x_3 \leq 2.$$

Form the sub-matrices as in (5.6.5) to obtain the following:

$$
\begin{array}{ccccc}
A & \vdots & I_m & \vdots & b \\
\cdots & \cdots & \cdots & \cdots & \cdots \\
C' & \vdots & O & \vdots & f
\end{array}
=
\begin{array}{ccccccc}
2 & 1 & 1 & \vdots & 1 & 0 & \vdots & 3 \\
1 & 1 & 2 & \vdots & 0 & 1 & \vdots & 2 \\
\cdots & \cdots & \cdots & \cdots & \cdots & \cdots & \cdots \\
8 & 5 & 8 & \vdots & 0 & 0 & \vdots & f
\end{array}
$$

Divide each element of the last column by the corresponding element in the first column of A. That is, $\frac{3}{2}$ and $\frac{2}{1}$. The smaller one is $\frac{3}{2}$ which occurs at the first row. Hence we operate with the first row to start with. Do the following operations to reduce the first column elements to zeros:

$$
\frac{1}{2}(1); \quad -(1) + (2); \quad -8(1) + (3) \Rightarrow
\begin{array}{ccccccc}
1 & \frac{1}{2} & \frac{1}{2} & \vdots & \frac{1}{2} & 0 & \vdots & \frac{3}{2} \\
0 & \frac{1}{2} & \frac{3}{2} & \vdots & -\frac{1}{2} & 1 & \vdots & \frac{1}{2} \\
\cdots & \cdots & \cdots & \cdots & \cdots & \cdots & \cdots \\
0 & 1 & 4 & \vdots & -4 & 0 & \vdots & f - 12
\end{array}
$$

Divide each element of the last column by the corresponding element in the second column of the resulting A. That is, $\frac{3}{2}/\frac{1}{2} = 3$ and $\frac{1}{2}/\frac{1}{2} = 1$. The smaller one occurs at the second row and hence we operate with the second row.

$$
-(2) + (1); \quad -2(2) + (3); \quad 2(2) \Rightarrow
\begin{array}{ccccccc}
1 & 0 & -1 & \vdots & 1 & -1 & \vdots & 1 \\
0 & 1 & 3 & \vdots & -1 & 2 & \vdots & 1 \\
\cdots & \cdots & \cdots & \cdots & \cdots & \cdots & \cdots \\
0 & 0 & 1 & \vdots & -3 & -2 & \vdots & f - 13
\end{array}
$$

Now it appears that a solution is reached because the last column, b_1, has non-negative elements, $-C_2'$ is such that C_2' has non-negative elements but $-C_1' = 1$ and hence the conditions are not met yet. Now divide the elements in the last column with the third column elements of the resulting A. The smaller one occurs at the second row, and now operate with the second row.

$$\frac{1}{3}(2);\quad -(2)+(1);\quad (2)+(3)\ \Rightarrow\quad
\begin{array}{ccccccccccc}
1 & \frac{1}{3} & 0 & : & \frac{2}{3} & -\frac{1}{3} & : & \frac{4}{3}\\
0 & \frac{1}{3} & 1 & : & -\frac{1}{3} & \frac{2}{3} & : & \frac{1}{3}\\
\cdots & \cdots & \cdots & \cdots & \cdots & \cdots & \cdots & \cdots\\
0 & -\frac{1}{3} & 0 & : & -\frac{8}{3} & -\frac{8}{3} & : & f-\frac{40}{3}
\end{array}$$

Now all elements are in the proper order as in (5.6.6) and hence the maximum of $f = 8x_1 + 5x_2 + 8x_3$, our dual problem, occurs at $x_1 = \frac{4}{3}$, $x_2 = 0$ and $x_3 = \frac{1}{3}$ and the maximum is

$$8\left(\frac{4}{3}\right) + 5(0) + 8\left(\frac{1}{3}\right) = \frac{40}{3}.$$

The minimum for our starting problem occurs at $y_1 = \frac{8}{3} = y_2$ and the minimum value of $g = 3y_1 + 2y_2$ is given by the following:

$$\text{Minimum of } g = 3\left(\frac{8}{3}\right) + 2\left(\frac{8}{3}\right) = \frac{40}{3} = \text{maximum of } f.$$

The student may start with our problem in Example 5.6.3 and try to reduce to the format in (5.6.6) without going through the above procedure and see what happens and may also construct examples where even after going through the above procedure a solution is not reached. Thus, many more points are involved in solving a linear programming problem. What is given above is only an introductory exposure to matrix methods in linear programming problems.

5.6.2 Nonlinear least squares

The method of least squares is already discussed in connection with regression and other model building problems. The problem is to predict a variable y such as the rainfall in a particular region during a particular month. This amount y depends on various factors such as the wind, pressure, temperature and such atmospheric conditions which can vary, and fixed quantities such as the topographic parameters. Let $x_1, \ldots, x_m$ be the variables which may have some relevance in predicting y and let the value of y that we can expect, $E(y)$, for a preassigned set of values of $x_1, \ldots, x_m$ be denoted by

$$E(y) = f(x_1, \ldots, x_m, a_1, \ldots, a_k) \tag{5.6.7}$$

where $a_1, \ldots, a_k$ are fixed but unknown parameters and $x_1, \ldots, x_m$ are observable variables. If the function f is linear in the unknowns $a_1, \ldots, a_k$ then we have a linear model and if f is nonlinear in the parameters then it is called a nonlinear model. What will happen to the method of least squares when f is nonlinear in the unknowns $a_1, \ldots, a_k$? This is what we will investigate here. As examples of linear, nonlinear models and the corresponding least square problem when there are n data points, consider the following:

$$E(y) = a + bx_1 + cx_2 \implies \min_{a,b,c} \sum_{i=1}^{n} (y_i - a - b\,x_{1i} - c\,x_{2i})^2, \tag{a}$$

$$E(y) = ab^x \implies \min_{a,b} \sum_{i=1}^{n} (y_i - a\,b^{x_i})^2, \tag{b}$$

$$E(y) = ae^{-(b\,x_1 + c\,x_2)} \implies \min_{a,b,c} \sum_{i=1}^{n} (y_i - a\,e^{-(b\,x_{1i} + c\,x_{2i})})^2, \tag{c}$$

$$E(y) = a + bx_1 x_2 + cx_2^2 \implies \min_{a,b,c} \sum_{i=1}^{n} (y_i - a - bx_{1i}x_{2i} - cx_{2i}^2)^2. \tag{d}$$

In (a) and (d) we have linear models whereas in (b) and (c) the models are nonlinear since they are nonlinear functions of the parameters. How to carry out the minimization over the parameters is what we will investigate here. If there are n data points in (5.6.7) then we have the following:

$$Y = \begin{bmatrix} y_1 \\ \vdots \\ y_n \end{bmatrix}, \quad X = \begin{bmatrix} x_{11} & x_{21} & \cdots & x_{m1} \\ x_{12} & x_{22} & \cdots & x_{m2} \\ \vdots & \vdots & \ddots & \vdots \\ x_{1n} & x_{2n} & \cdots & x_{mn} \end{bmatrix}, \quad a = \begin{bmatrix} a_1 \\ \vdots \\ a_k \end{bmatrix}$$

where the quantity to be minimized is

$$\psi = \sum_{i=1}^{n} [y_i - f(x_{1i}, x_{2i}, \ldots, x_{mi}, a_1, \ldots, a_k)]^2. \tag{5.6.8}$$

The usual approach in minimizing ψ with respect to the parameters $a_1, \ldots, a_k$ and when f is nonlinear in $a_1, \ldots, a_k$ is to reduce the problem into a linear one. Let $\alpha' = (\alpha_1, \ldots, \alpha_k)$ be a value of a for which ψ is a minimum. Let $\alpha + \delta$, $\delta' = (\delta_1, \ldots, \delta_k)$ be a neighborhood of α. Expanding y_i around α by using a Taylor series we have

$$f(x_{1i}, \ldots, x_{mi}, \alpha_1 + \delta_1, \ldots, \alpha_k + \delta_k) \approx f(x_{1i}, \ldots, x_{mi}, \alpha_1, \ldots, \alpha_k)$$

$$+ \sum_{j=1}^{k} \frac{\partial f_i}{\partial \alpha_j} \delta_j = \langle y_i \rangle, \quad \text{say}$$

$$f_i = f(x_{1i}, \ldots, x_{mi}, \alpha_1, \ldots, \alpha_k).$$

Then

$$\langle y \rangle = f_0 + P\delta \tag{5.6.9}$$

where

$$\delta = \begin{bmatrix} \delta_1 \\ \vdots \\ \delta_k \end{bmatrix}, \quad f_0 = \begin{bmatrix} f_1 \\ \vdots \\ f_n \end{bmatrix}, \quad P = \begin{bmatrix} \frac{\partial f_1}{\partial \alpha_1}, & \frac{\partial f_1}{\partial \alpha_2}, & \cdots, & \frac{\partial f_1}{\partial \alpha_k} \\ \vdots & \vdots & \ddots & \vdots \\ \frac{\partial f_n}{\partial \alpha_1}, & \frac{\partial f_n}{\partial \alpha_2}, & \cdots, & \frac{\partial f_n}{\partial \alpha_k} \end{bmatrix}.$$

Hence

$$\sum_{i=1}^{n} |y_i - \langle y_i \rangle|^2 = (Y - f_0 - P\delta)'(Y - f_0 - P\delta) = \langle \phi \rangle, \quad \text{say.} \tag{5.6.10}$$

Now (5.6.10) can be looked upon as a linear least squares problem of minimizing $\langle \phi \rangle$ to estimate δ. Differentiating partially with respect to the vector δ we have

$$\frac{\partial}{\partial \delta} \langle \phi \rangle = O \implies P'(Y - f_0 - P\delta) = O.$$

That is,

$$A\delta = g \tag{5.6.11}$$

where

$$A = P'P, \quad g' = (g_1, \ldots, g_k), \quad g_j = \sum_{i=1}^{n} (y_i - f_i) \frac{\partial f_i}{\partial \alpha_j}.$$

The classical Gauss method is to use (5.6.11) for successive iterations. This method is also often known as the gradient method.

5.6.3 Marquardt's method

Marquardt's modification to the gradient method is to minimize $\langle \phi \rangle$ on the sphere $\|\delta\|^2 = $ a constant. Under this minimization the modified equations are the following:

$$(A + \lambda I)\delta = g \quad \text{for } \lambda > 0 \tag{5.6.12}$$

where λ is an arbitrary constant to be chosen by the algorithm. Various authors have suggested modifications to Marquardt's method. One modification is to replace the various derivatives by the corresponding finite differences in forming equation (5.6.12). Another modification is to consider weighted least squares. Another modification is to take the increment vector in the iterations as an appropriate linear combination of δ and g.

Example 5.6.4. Illustrate the steps in (5.6.12) if the problem is to fit the model $y = a_1(a_2)^x$ based on the following data:

$$\begin{array}{lcccc} x = & 0 & 1 & 2 & 3 \\ y = & 3 & 5 & 12 & 22 \end{array}$$

Solution 5.6.4. Let $\alpha = \left(\begin{smallmatrix} \alpha_1 \\ \alpha_2 \end{smallmatrix}\right)$ be the point for $a = \left(\begin{smallmatrix} a_1 \\ a_2 \end{smallmatrix}\right)$ where the least square minimum is obtained. Then according to the notations in (5.6.12), observing that $(x_1, y_1) = (0, 3)$, $(x_2, y_2) = (1, 5)$, ..., $(x_4, y_4) = (3, 22)$, we have the following:

$$f_1 = \alpha_1 \alpha_2^0 = \alpha_1 \Rightarrow \frac{\partial f_1}{\partial \alpha_1} = 1, \quad \frac{\partial f_1}{\partial \alpha_2} = 0,$$

$$f_2 = \alpha_1(\alpha_2)^1 = \alpha_1 \alpha_2 \Rightarrow \frac{\partial f_2}{\partial \alpha_1} = \alpha_2, \quad \frac{\partial f_2}{\partial \alpha_2} = \alpha_1,$$

$$f_3 = \alpha_1(\alpha_2)^2 = \alpha_1 \alpha_2^2 \Rightarrow \frac{\partial f_3}{\partial \alpha_1} = \alpha_2^2, \quad \frac{\partial f_3}{\partial \alpha_2} = 2\alpha_1 \alpha_2,$$

$$f_4 = \alpha_1(\alpha_2)^3 = \alpha_1 \alpha_2^3 \Rightarrow \frac{\partial f_4}{\partial \alpha_1} = \alpha_2^3, \quad \frac{\partial f_4}{\partial \alpha_2} = 3\alpha_1 \alpha_2^2.$$

Then

$$g_1 = (3 - \alpha_1)(1) + (5 - \alpha_1 \alpha_2)(\alpha_2)$$
$$+ (12 - \alpha_1 \alpha_2^2)\alpha_2^2 + (22 - \alpha_1 \alpha_2^3)\alpha_2^3,$$
$$g_2 = (3 - \alpha_1)(0) + (5 - \alpha_1 \alpha_2)(\alpha_1)$$
$$+ (12 - \alpha_1 \alpha_2^2)(2\alpha_1 \alpha_2) + (22 - \alpha_1 \alpha_2^3)(3\alpha_1 \alpha_2^2).$$

The matrices P and A are given by the following:

$$P = \begin{bmatrix} 1 & 0 \\ \alpha_2 & \alpha_1 \\ \alpha_2^2 & 2\alpha_1 \alpha_2 \\ \alpha_2^3 & 3\alpha_1 \alpha_2^2 \end{bmatrix},$$

$$A = P'P = \begin{bmatrix} 1 + \alpha_2^2 + \alpha_2^4 + \alpha_2^6, & \alpha_1(\alpha_2 + 2\alpha_2^3 + 3\alpha_2^5) \\ \alpha_1(\alpha_2 + 2\alpha_2^3 + 3\alpha_2^5), & \alpha_1^2(1 + 4\alpha_2^2 + 9\alpha_2^4) \end{bmatrix}.$$

From (5.6.12) we have

$$\delta = (A + \lambda I)^{-1} g.$$

For Gauss' procedure $\lambda = 0$ and take a trial vector α, say $\alpha_{(0)}$. Compute g, A and $\delta_{(0)} = A^{-1} g$ for this trial vector. Then $\alpha_{(1)} = \alpha_{(0)} + \delta_{(0)}$ is the new trial value of α. The success of the iterative procedure depends upon guessing the first trial value of $\alpha_{(0)}$ very close to the true α because our expansion is valid only in the neighborhood of the true α. For applying Marquardt's procedure start with a trial value of λ also. Various techniques are available for selecting a λ_0 as well as an $\alpha_{(0)}$ and making adjustments at each iteration so that a convergence to the true optimal point is reached. Both the gradient method and Marquardt's method fail in many of the standard test problems in this field.

5.6.4 Mathai–Katiyar procedure

Here the equation for iteration is the following:

$$(A + \lambda B)\delta = g$$

where $A = P'P$ and $B = g'gI - gg'$. Here A is the same matrix as in Gauss' and Marquardt's procedure and B is obtained by incorporating the condition that the angle between the increment vector δ and the gradient vector g is zero when a minimum is reached, and then proceeding with the minimization as before. For the details and a flow-chart for executing the algorithm the interested reader may see [6]. It is shown that the algorithm always produce convergence to the optimal points at least in all the standard test problems in the field.

Exercises 5.6

Solve Exercises 5.6.1–5.6.4 graphically as well as by using matrix methods. Form the dual problem and solve by simplex method when a minimization is involved.

5.6.1. Maximize $x_1 + 2x_2$ subject to the conditions $x_1 \geq 0$, $x_2 \geq 0$ and

$$x_1 - x_2 \geq 4$$
$$x_1 + x_2 \leq 5$$
$$x_1 - x_2 \leq 2.$$

5.6.2. Minimize $x_1 + 2x_2$ subject to the conditions $x_1 \geq 0$, $x_2 \geq 0$ and

$$2x_1 + x_1 \geq 8$$
$$x_1 + x_2 \geq 6$$
$$x_1 + 2x_2 \geq 9.$$

5.6.3. Maximize $8x_1 + 15x_2$ subject to the conditions $x_1 \geq 0$, $x_2 \geq 0$ and

$$2x_1 + 3x_2 \geq 2$$
$$2x_1 + 5x_2 \geq 3.$$

5.6.4. Use simplex method to maximize $2x_1 + x_2 + 6x_3 + x_4$ subject to the conditions $x_1 \geq 0$, $x_2 \geq 0$, $x_3 \geq 0$, $x_4 \geq 0$ and

$$x_1 + x_3 + 2x_4 \leq 5$$
$$x_2 + x_3 \leq 2$$
$$x_1 + 3x_2 + x_3 + x_4 \leq 4.$$

5.6.5. A carpentry firm manufactures desks, tables and shelves. Each table requires one hour of labor, 10 square feet of wood and 4 quarts of varnish. Each desk requires 3 hours of labor, 35 square feet of wood and 1 quart of varnish. Each shelf needs 45 minutes of labor, 15 square feet of wood and one quart of varnish. At the firm's disposal there are at most 25 hours of labor, at most 350 square feet of wood and at most 55 quarts of varnish. Each table produces a profit of $5, each desk $4 and each shelf $3. How many of each item be produced so that the firm's profit is a maximum from this operation and what is that maximum profit?

5.6.6. A dealer packages and sells three types of mixtures of nuts. The dealer has 8 kg (kilograms) of cashews, 24 kg of almonds and 36 kg of peanuts. Mixture type 1 consists of 20% of cashews, 20% almonds and 60% peanuts and brings a profit of $2 per kg. Mixture type 2 contains 20% cashews, 40% almonds and 40% peanuts and brings a profit of $4 per kg. Mixture type 3 is of 10% cashews, 30% almonds and 60% peanuts and brings a profit of $3 per kg. How many kg of each mixture the dealer should make to maximize the profit from this operation? What is the maximum profit?

5.6.7. A farmer requires at least 4 000 kg (kilograms) of nitrogen, 2 000 kg of phosphoric acid and 2 000 kg of potash. The farmer can buy 50 kg bags of three types of fertilizers, types 1, 2, 3. Each type 1 bag contains 20 kg nitrogen, 15 kg phosphoric acid and 5 kg potash and costs $20 per bag. Each type 2 bag contains 10 kg nitrogen, 20 kg phosphoric acid and 25 kg potash and costs $30 per bag. Each type 3 bag contains 15 kg phosphoric acid and 20 kg potash and no nitrogen and costs $20 per bag. How many bags of each type of fertilizer should the farmer buy so that the farmer's cost is minimized, what is the minimum cost?

5.7 A list of some more problems from physical, engineering and social sciences

In order to do a real problem in one of the engineering areas or physical sciences it requires the knowledge of the technical terms and a clear understanding of the problem itself. This needs a lot of discussion and explanations but the reader may not be interested to invest that much time into it. Hence we will only indicate some of the problems where matrices, determinants and eigenvalues play vital roles in simplifying matters.

5.7.1 Turbulent flow of a viscous fluid

Analysis of turbulent flow of a viscous fluid through a pipe requires what is known as a dimensional matrix of the following form:

	F	L	T	t
P	1	-2	0	0
D	0	1	0	0
V	0	1	0	-1
ρ	1	-4	0	2
μ	1	-2	0	1

where P = the resistance per unit area, D = diameter of the pipe, V = fluid velocity, ρ = fluid density, μ = fluid viscosity, F = force, L = length, T = temperature, t = time. The above matrix is formed from the table of dimensions of gas-dynamic quantities.

5.7.2 Compressible flow of viscous fluids

When the velocity of gas flow is greater than one-half the speed of sound, or when thermal effects are appreciable, compressibility must be accounted for. In this case the dimensional matrix for the solution of this problem is the following:

	F	L	T	t
ρ	1	-4	0	2
L	0	1	0	0
V	0	1	0	-1
μ	1	-2	0	1
g	0	1	0	-2
a	0	1	0	-1
P	1	-2	0	0

where g = gravitational constant, a = acoustic velocity and other items remain as in the case of turbulent flow of Section 5.7.1.

5.7.3 Heat loss in a steel rod

In studying heat loss of a steel rod the dimensional matrix becomes the following, again constructed from the table of dimensions of gas-dynamic quantities:

	H	L	T	t
Q	1	0	0	-1
$T_s - T_a$	0	0	1	0
l	0	1	0	0
d	0	1	0	0
k_s	1	-1	-1	-1
k_a	1	-1	-1	-1

where H = thermal energy, L = length, T = temperature, t = time, d = diameter of the rod, l = length of the rod, Q = amount of heat rejected per unit time, T_a = temperature of steel, T_a = temperature of air, k_s = thermal conductivity of steel, k_a = thermal conductivity of air.

5.7.4 Small oscillations

Oscillations of mechanical systems or current or voltage in an electrical system and many such problems fall into this category. Suppose that the displacement from an equilibrium position of the system can be described by the $n \times 1$ vector X, $X' = (x_1, \dots, x_n)$ so that the equilibrium position is $X = O$. When the system performs small oscillations its kinetic energy T is represented by a quadratic form of the type

$$T = \frac{1}{2}\tilde{X}'A\tilde{X}$$

where $A = (a_{ij})$ is a real symmetric positive definite matrix, that is, $T > 0$ for all non-null X,

$$\tilde{X}' = \left(\frac{d}{dt}x_1, \dots, \frac{d}{dt}x_n\right) = \frac{d}{dt}X'$$

the derivative of the components in X with respect to the time variable t. The potential energy V for small oscillations from the equilibrium position can be shown to be represented by the quadratic form

$$V = \frac{1}{2}X'BX$$

where $B = B'$ is positive definite or at least positive semi-definite, that is, $V \geq 0$ for all non-null X. Equations of motion for the system, with no external forces acting, can be shown to give rise to the differential equation

$$AX^* + CX = O \tag{5.7.1}$$

where

$$X^* = \frac{d^2}{dt^2}X.$$

If a solution of the type $X = e^{i\mu t}Y$, μ real, $i = \sqrt{-1}$, is assumed for (5.7.1) then from (5.7.1) we obtain the equation (see Section 5.5)

$$(-\lambda A + C)Y = O \tag{5.7.2}$$

where $\lambda = \mu^2$. Note that (5.7.2) is an eigenvalue problem where λ will represent the natural frequencies of vibration. The largest frequency is then the largest eigenvalue λ_n. Writing (5.7.2) in the form

$$(\lambda I - A^{-1}C)Y = O \tag{5.7.3}$$

the λ's are the eigenvalues of the matrix $A^{-1}C$ or that of $A^{-\frac{1}{2}}CA^{-\frac{1}{2}} = G$ where $A^{\frac{1}{2}}$ is the real symmetric positive definite square root of the real symmetric positive definite matrix A. Simultaneous diagonalizations of A and C lead to the determination of the *normal mode* of the system. The *Rayleigh quotient* in this connection is given by

$$R_1(X) = \frac{Z'CZ}{Z'AZ}, \quad Z = A^{-\frac{1}{2}}X$$

and then

$$\lambda_n = \max_{X \neq O} R_1(X),$$

see Section 5.5. If X_j is an eigenvector of G then $Z_j = A^{-\frac{1}{2}}X_j$ is a normal mode vector for the small vibration problem. If external forces are taken into account then (5.7.1) will be modified with a function of t sitting on the right side. We will not elaborate on this problem any further.

5.7.5 Input–output analysis

Another important set of problems fall in the category of input–output analysis. Input–output type situations arise in a wide variety of fields. Let X, $X' = (x_1, \ldots, x_n)$ be the input variables. Suppose that these go through a process, denoted by a matrix of operators M, and the resulting quantity is the output, say, Y, $Y' = (y_1, \ldots, y_m)$. Then the system can be denoted by

$$Y = MX.$$

If M is a matrix of partial differential operators then the output is a system of differential equations. For example let

$$M = \begin{bmatrix} 2\frac{\partial}{\partial x_1} & -\frac{\partial}{\partial x_2} & 5\frac{\partial}{\partial x_3} \\ \frac{\partial}{\partial x_1} & 4\frac{\partial}{\partial x_2} & -\frac{\partial}{\partial x_3} \end{bmatrix}$$

then

$$MX = \begin{bmatrix} 2\frac{\partial}{\partial x_1}(x_1) - \frac{\partial}{\partial x_2}(x_2) + 5\frac{\partial}{\partial x_3}(x_3) \\ \frac{\partial}{\partial x_1}(x_1) + 4\frac{\partial}{\partial x_2}(x_2) - \frac{\partial}{\partial x_3}(x_3) \end{bmatrix} = \begin{bmatrix} 6 \\ 4 \end{bmatrix} = Y.$$

If M is a linear operator, designated by a constant matrix, then the output is the result of a linear transformation. For example if $k = 4 = n$ and if

$$M = \begin{bmatrix} 1 & -2 & 3 & 1 \\ 0 & 1 & -1 & 2 \\ 2 & 1 & 1 & -1 \\ 2 & 0 & 2 & 3 \end{bmatrix}$$

then the output is $Y = MX$ where

$$y_1 = x_1 - 2x_2 + 3x_3 + x_4, \quad y_2 = x_2 - x_3 + 2x_4,$$
$$y_3 = 2x_1 + x_2 + x_3 - x_4, \quad y_4 = 2x_1 + 2x_3 + 3x_4.$$

If x_1, x_2, x_3 are quantities of 3 items shipped by a firm to two different shops and if M represents the per unit sales prices of these items in these shops then the output is the vector of revenues from these two shops on these three items. Suppose that the unit price matrix is the following:

$$M = \begin{bmatrix} \$3 & \$2 & \$1 \\ \$2 & \$2 & \$2 \end{bmatrix}.$$

Then the output or the revenue vector is given by

$$Y = MX = \begin{bmatrix} 3x_1 + 2x_2 + x_3 \\ 2x_1 + 2x_2 + 2x_3 \end{bmatrix}.$$

For example, if $x_1 = 10$ kilograms (kg), $x_2 = 5$ kg, $x_3 = 20$ kg then the revenue vector is

$$Y = \begin{bmatrix} \$60 \\ \$70 \end{bmatrix}.$$

If M is a transition probability matrix and if X_0 is the initial vector (see Chapters 2 and 5) then the eventual behavior of the system is given by the output vector

$$Y = M^\infty X_0.$$

Several such input–output situations arise in different fields. The input X can be in the form of a vector or matrix, the operator M can also be in the form of a vector or matrix so that MX is defined then the output will be scalar, vector or matrix as determined by MX. Further analysis of such an input–output model will require properties of matrices and the nature of the problem in hand.

6 Matrix series and additional properties of matrices

6.0 Introduction

The ideas of sequences, polynomials, series, convergence and so on in scalar variables will be generalized to matrix variables in this chapter. We start with some basic properties of polynomials and then see what happens if the scalar variable in the polynomial is replaced by a square matrix.

6.1 Matrix polynomials

Here a "matrix polynomial" does not mean a matrix where the elements are polynomials in a scalar variable such as

$$B = \begin{bmatrix} 1+x & 2x+x^2 \\ 2-x+x^2 & x^3 \end{bmatrix}.$$

Such a matrix will be called a matrix of polynomials. The term "matrix polynomial" will be reserved for the situation where to start with we have a polynomial in a scalar variable and we are replacing the scalar variable by a square matrix to obtain a polynomial in a square matrix. For example, consider a polynomial of degree m in the scalar variable x,

$$p(x) = a_0 + a_1 x + \cdots + a_m x^m, \quad a_m \neq 0 \tag{6.1.1}$$

where $a_0, \ldots, a_m$ are known constants. For example,

$$p_1(x) = 4 + 2x - 3x^2, \quad \text{a polynomial in } x \text{ of degree 2};$$
$$p_2(x) = 2 + 5x, \quad \text{a polynomial in } x \text{ of degree 1};$$
$$p_3(x) = 7, \quad \text{a polynomial in } x \text{ of degree 0}.$$

Let

$$A = \begin{bmatrix} 1 & 0 & 0 \\ 0 & 1 & 1 \\ 1 & 1 & 1 \end{bmatrix}.$$

Let us try to construct polynomials $p_1(A), p_2(A), p_3(A)$ in the matrix A, corresponding to the scalar polynomials $p_1(x), p_2(x), p_3(x)$ above. When x in (6.1.1) is replaced by the matrix A then the constant term a_0 will be replaced by $a_0 I$, I the identity matrix. That is,

$$p(A) = a_0 I + a_1 A + \cdots + a_m A^m, \quad a_m \neq 0. \tag{6.1.2}$$

Thus for our illustrative examples we have

$$p_1(A) = 4 \begin{bmatrix} 1 & 0 & 0 \\ 0 & 1 & 0 \\ 0 & 0 & 1 \end{bmatrix} + 2 \begin{bmatrix} 1 & 0 & 0 \\ 0 & 1 & 1 \\ 1 & 1 & 1 \end{bmatrix} - 3 \begin{bmatrix} 1 & 0 & 0 \\ 0 & 1 & 1 \\ 1 & 1 & 1 \end{bmatrix}^2$$

$$= \begin{bmatrix} 3 & 0 & 0 \\ -3 & 0 & -4 \\ -4 & -4 & 0 \end{bmatrix}$$

which is again a 3×3 matrix. The following results are obviously true for matrix polynomials.

(i) If $p_1(x)$, $p_2(x)$, $q_1(x) = p_1(x) + p_2(x)$, $q_2(x) = p_1(x)p_2(x)$ are polynomials in the scalar x then for any square matrix A,

$$p_1(A) + p_2(A) = q_1(A), \quad p_1(A)p_2(A) = q_2(A).$$

We can note that the factorization properties also go through.

(ii) If

$$p(x) = (x - a)(x - b)$$

where x, a, b are scalars, a and b free of x, then for any square matrix A,

$$p(A) = (A - aI)(A - bI)$$

where I is the identity matrix.

Consider the characteristic polynomial of an $n \times n$ matrix A. That is,

$$p(\lambda) = |A - \lambda I| = (\lambda_1 - \lambda)(\lambda_2 - \lambda) \cdots (\lambda_n - \lambda)$$

where $\lambda_1, \dots, \lambda_n$ are the eigenvalues of A and $p(\lambda) = 0$ is the characteristic equation. Then it is easy to see that $p(A) = O$. That is,

(iii) every $n \times n$ matrix A satisfies its own characteristic equation or

$$(\lambda_1 I - A)(\lambda_2 I - A) \cdots (\lambda_n I - A) = O.$$

6.1.1 Lagrange interpolating polynomial

Consider the following polynomial where $\lambda_1, \dots, \lambda_n$ are distinct quantities free of λ and $a_1, \dots, a_n$ are constants:

$$p(\lambda) = a_1 \frac{(\lambda - \lambda_2)(\lambda - \lambda_3) \cdots (\lambda - \lambda_n)}{(\lambda_1 - \lambda_2)(\lambda_1 - \lambda_3) \cdots (\lambda_1 - \lambda_n)} + a_2 \frac{(\lambda - \lambda_1)(\lambda - \lambda_3) \cdots (\lambda - \lambda_n)}{(\lambda_2 - \lambda_1)(\lambda_2 - \lambda_3) \cdots (\lambda_2 - \lambda_n)}$$

$$+ \cdots + a_n \frac{(\lambda - \lambda_1) \cdots (\lambda - \lambda_{n-1})}{(\lambda_n - \lambda_1) \cdots (\lambda_n - \lambda_{n-1})}$$

$$= \sum_{j=1}^{n} a_j \left\{ \prod_{i=1,i\neq j}^{n} \frac{(\lambda - \lambda_i)}{(\lambda_j - \lambda_i)} \right\} \tag{6.1.3}$$

which is a polynomial of degree $n-1$ in λ. Put $\lambda = \lambda_1$ in (6.1.3). Then we have $p(\lambda_1) = a_1$. Similarly $p(\lambda_j) = a_j, j = 1, \ldots, n$. Therefore

$$p(\lambda) = \sum_{j=1}^{n} p(\lambda_j) \left\{ \prod_{i=1,i\neq j}^{n} \frac{(\lambda - \lambda_i)}{(\lambda_j - \lambda_i)} \right\}. \tag{6.1.4}$$

The polynomial in (6.1.4) is called *Lagrange interpolating polynomial*. A more general polynomial in this category, allowing multiplicities for $\lambda_1, \ldots, \lambda_n$ is *Hermite interpolating polynomial* which we will not discuss here. From (6.1.4) we have, for any square matrix A, and $p(\lambda)$ satisfying (6.1.4),

$$p(A) = \sum_{j=1}^{n} p(\lambda_j) \left\{ \prod_{i=1,i\neq j}^{n} \frac{(A - \lambda_i I)}{(\lambda_j - \lambda_i)} \right\}. \tag{6.1.5}$$

An interesting application of (6.1.5) is that if $\lambda_1, \ldots, \lambda_n$ are the distinct eigenvalues of any $n \times n$ matrix A and $p(\lambda)$ is any polynomial of the type in (6.1.4) then the matrix $p(A)$ has the representation in (6.1.5). Let us do an example to highlight this point.

Example 6.1.1. Compute e^{5A} where

$$A = \begin{bmatrix} 1 & 0 \\ 1 & 4 \end{bmatrix}.$$

Solution 6.1.1. The eigenvalues of A are obviously $\lambda_1 = 1, \lambda_2 = 4$. Let $p(\lambda) = e^{5\lambda}$. Then from (6.1.4)

$$p(\lambda) = e^{5\lambda} = p(\lambda_1) \frac{\lambda - \lambda_2}{\lambda_1 - \lambda_2} + p(\lambda_2) \frac{\lambda - \lambda_1}{\lambda_2 - \lambda_1}$$

$$= -\frac{e^5}{3}(\lambda - 4) + \frac{e^{20}}{3}(\lambda - 1).$$

Therefore from (6.1.5)

$$p(A) = -\frac{e^5}{3}(A - 4I) + \frac{e^{20}}{3}(A - I)$$

$$= -\frac{e^5}{3} \begin{pmatrix} -3 & 0 \\ 1 & 0 \end{pmatrix} + \frac{e^{20}}{3} \begin{pmatrix} 0 & 0 \\ 1 & 3 \end{pmatrix}$$

$$= \begin{pmatrix} e^5 & 0 \\ \frac{e^{20} - e^5}{3} & e^{20} \end{pmatrix} = e^{5A}.$$

6.1.2 A spectral decomposition of a matrix

We will consider the spectral decomposition of a matrix A when the eigenvalues are distinct. The results hold when some of the eigenvalues are repeated also. In the repeated case we will need Hermite interpolating polynomials to establish the results. When the eigenvalues of A are distinct we have the representation in (6.1.5) where $p(\lambda)$ is a polynomial defined on the set of distinct eigenvalues of A (*spectrum of A*). Let (6.1.5) be written as

$$p(A) = A_1 + \cdots + A_n. \tag{6.1.6}$$

Let us consider the product $A_1 A_2$. Excluding the constant parts, A_1 and A_2 are given by

$$A_1 \rightarrow (A - \lambda_2 I)(A - \lambda_3 I) \cdots (A - \lambda_n I)$$

and

$$A_2 \rightarrow (A - \lambda_1 I)(A - \lambda_3 I) \cdots (A - \lambda_n I).$$

Then

$$A_1 A_2 \rightarrow (A - \lambda_1 I)(A - \lambda_2 I)(A - \lambda_3 I)^2 \cdots (A - \lambda_n I)^2.$$

But from property (iii),

$$(\lambda_1 I - A)(\lambda_2 I - A) \cdots (\lambda_n I - A) = O$$

and hence $A_1 A_2 = O$. Similarly $A_i A_j = O$ for all i and j, $i \neq j$. Thus $A_1, \ldots, A_n$ are mutually orthogonal matrices and hence linearly independent. Taking $p(\lambda) = 1$ in (6.1.6) we have the relation

$$I = B_1 + \cdots + B_n \tag{6.1.7}$$

where

$$B_j = \frac{(A - \lambda_1 I) \cdots (A - \lambda_{j-1} I)(A - \lambda_{j+1} I) \cdots (A - \lambda_n I)}{(\lambda_j - \lambda_1) \cdots (\lambda_j - \lambda_{j-1})(\lambda_j - \lambda_{j+1}) \cdots (\lambda_j - \lambda_n)}$$

and

$$B_i B_j = O \quad \text{for all } i \neq j.$$

Then multiply both sides of (6.1.7) by B_j we have $B_j = B_j^2$ for each j, $j = 1, \ldots, n$.

(iv) In the spectral decomposition of an identity matrix, as given in (6.1.7), the B_j's are mutually orthogonal and each B_j is an idempotent matrix.

Taking $p(\lambda) = \lambda$ or $p(A) = A$ in (6.1.6) we have the following spectral decomposition for A:

(v) For any $n \times n$ matrix A with distinct eigenvalues $\lambda_1, \ldots, \lambda_n$,

$$A = \lambda_1 B_1 + \cdots + \lambda_n B_n \qquad (6.1.8)$$

where the B_j's are defined in (6.1.7).

This can be observed from property (iii) and (6.1.7). Note that

$$AB_j = (A - \lambda_j I + \lambda_j I)B_j = (A - \lambda_j I)B_j + \lambda_j B_j = \lambda_j B_j$$

since $(A - \lambda_j I)B_j = O$ by property (iii). Hence

$$\lambda_1 B_1 + \cdots + \lambda_n B_n = A(B_1 + \cdots + B_n) = A$$

since $B_1 + \cdots + B_n = I$ by (6.1.7). We can also notice some more interesting properties from (6.1.8):

$$B_i B_j = O = B_j B_i, \quad i \neq j$$

as well as

$$B_j A = AB_j = \lambda_j B_j^2 = \lambda_j B_j.$$

Thus the matrices $A, B_1, \ldots, B_n$ commute and hence all can be reduced to diagonal forms by a nonsingular matrix Q such that

$$D = \lambda_1 D_1 + \cdots + \lambda_n D_n \qquad (6.1.9)$$

where $QAQ^{-1} = D$, $QB_j Q^{-1} = D_j$ for all j, $D_i D_j = O$ for all $i \neq j$. The matrices $B_1, \ldots, B_n$ in (6.1.8) are also called the *idempotents of A*, different from idempotent matrices.

Example 6.1.2. For the matrix $A = \begin{bmatrix} 1 & 3 \\ 2 & 2 \end{bmatrix}$ verify (6.1.7) and (6.1.8).

Solution 6.1.2. The eigenvalues are $\lambda_1 = 4$ and $\lambda_2 = -1$. Two eigenvectors corresponding to λ_1 and λ_2 are

$$X_1 = \begin{pmatrix} 1 \\ 1 \end{pmatrix}, \quad X_2 = \begin{pmatrix} 3 \\ -2 \end{pmatrix}.$$

Let

$$Q = (X_1, X_2) = \begin{bmatrix} 1 & 3 \\ 1 & -2 \end{bmatrix},$$

$$Q^{-1} = \frac{1}{5} \begin{bmatrix} 2 & 3 \\ 1 & -1 \end{bmatrix},$$

$$QDQ^{-1} = \frac{1}{5} \begin{bmatrix} 1 & 3 \\ 1 & -2 \end{bmatrix} \begin{bmatrix} 4 & 0 \\ 0 & -1 \end{bmatrix} \begin{bmatrix} 2 & 3 \\ 1 & -1 \end{bmatrix} = \begin{bmatrix} 1 & 3 \\ 2 & 2 \end{bmatrix} = A.$$

$$B_1 = \frac{A - \lambda_2 I}{\lambda_1 - \lambda_2} = \frac{1}{5} \begin{bmatrix} 2 & 3 \\ 2 & 3 \end{bmatrix},$$

$$B_2 = \frac{A - \lambda_1 I}{\lambda_2 - \lambda_1} = -\frac{1}{5} \begin{bmatrix} -3 & 3 \\ 2 & -2 \end{bmatrix};$$

$$B_1 + B_2 = \frac{1}{5} \begin{bmatrix} 5 & 0 \\ 0 & 5 \end{bmatrix} = I;$$

$$\lambda_1 B_1 + \lambda_2 B_2 = \frac{4}{5} \begin{bmatrix} 2 & 3 \\ 2 & 3 \end{bmatrix} + \frac{1}{5} \begin{bmatrix} -3 & 3 \\ 2 & -2 \end{bmatrix}$$

$$= \begin{bmatrix} 1 & 3 \\ 2 & 2 \end{bmatrix} = A.$$

Example 6.1.3. For the matrix A in Example 6.1.2 compute Q such that $Q^{-1}AQ = $ diagonal. Also establish (6.1.9).

Solution 6.1.3. By straight multiplication

$$Q^{-1}B_1 Q = \begin{bmatrix} 1 & 0 \\ 0 & 0 \end{bmatrix}$$

and

$$Q^{-1}B_2 Q = \begin{bmatrix} 0 & 0 \\ 0 & 1 \end{bmatrix}.$$

Taking the linear combination (6.1.9) is established.

(vi) In the spectral representation of any $n \times n$ matrix A with distinct eigenvalues, as in (6.1.8), the rank of B_j for each j cannot exceed 1.

6.1.3 An application in statistics

In the spectral decomposition of an $n \times n$ matrix A, as given in (6.1.8), each B_j is idempotent. If A is real symmetric then $B_j, j = 1, \ldots, n$ are also real symmetric since the eigenvalues of a real symmetric matrix are real. If the eigenvalues of A are all distinct then each B_j is of rank 1. Consider X an $n \times 1$ real Gaussian vector random variable having a standard Gaussian distribution. In our notation $X \sim N_n(O, I)$ where I is an identity matrix. Consider the quadratic form $X'AX$. Then

$$X'AX = \lambda_1 X'B_1 X + \cdots + \lambda_n X'B_n X.$$

Since $B_1 = B_j' = B_j^2$ and since $X \sim N_n(O, I)$ it follows that $X'B_j X \sim \chi_1^2$, that is, $X'B_j X$ is a real chisquare random variable with one degree of freedom. Since $B_i B_j = O, i \neq j$ these chisquare random variables are mutually independently distributed. Thus one has a representation

$$X'AX = \lambda_1 y_1 + \cdots + \lambda_n y_n$$

where the $y_1, \ldots, y_n$ are mutually independently distributed chisquare random variables with one degree of freedom each when the λ_j's are distinct. One interesting aspect is that in each B_j all the eigenvalues of A are present.

Exercises 6.1

6.1.1. If A is symmetrically partitioned to the form

$$A = \begin{bmatrix} A_{11} & A_{12} \\ O & I \end{bmatrix}$$

then show that for any positive integer n,

$$A^n = \begin{bmatrix} A_{11}^n & p(A_{11})A_{12} \\ O & I \end{bmatrix}$$

where

$$p(x) = \frac{(x^n - 1)}{x - 1}.$$

6.1.2. Compute e^{-2A} where

$$A = \begin{bmatrix} 1 & 2 \\ 3 & 5 \end{bmatrix}.$$

6.1.3. Compute $\sin A$ where

$$A = \frac{\pi}{4} \begin{bmatrix} 2 & 0 & 0 \\ 4 & 1 & 0 \\ -2 & 5 & -2 \end{bmatrix}.$$

6.1.4. Spectrum of a matrix A. The spectrum of a matrix is the set of all distinct eigenvalues of A. If $B = QAQ^{-1}$ and if $f(\lambda)$ is a polynomial defined on the spectrum of A then show that

$$f(B) = Qf(A)Q^{-1}.$$

Prove the result when the eigenvalues are distinct. The result is also true when some eigenvalues are repeated.

6.1.5. If A is a block diagonal matrix, $A = \text{diag}(A_1, A_2, \ldots, A_k)$, and if $f(\lambda)$ is a polynomial defined on the spectrum of A then show that

$$f(A) = \text{diag}(f(A_1), f(A_2), \ldots, f(A_k)).$$

6.1.6. If $\lambda_1, \ldots, \lambda_n$ are the eigenvalues of an $n \times n$ matrix A and if $f(\lambda)$ is a polynomial defined on the spectrum of A then show that the eigenvalues of $f(A)$ are $f(\lambda_1), f(\lambda_2), \ldots, f(\lambda_n)$.

6.1.7. For any square matrix A show that e^{kA}, where k is a nonzero scalar, is a nonsingular matrix.

6.1.8. If A is a real symmetric positive definite matrix then show that there exists a unique Hermitian matrix B such that $A = e^B$.

6.1.9. By using the ideas from Exercise 6.1.3, or otherwise, show that for any $n \times n$ matrix A

$$e^{iA} = \cos A + i \sin A, \quad i = \sqrt{-1}.$$

6.1.10. For the matrix $A = \left(\begin{smallmatrix} 3 & 0 \\ 7 & 2 \end{smallmatrix} \right)$ compute $\ln A$, if it exists.

6.2 Matrix sequences and matrix series

We will introduce matrix sequences and matrix series and concepts analogous to convergence of series in scalar variables. A few properties of matrix sequences will be considered first. Then we will look at convergence of matrix series and we will also introduce a concept called "norm of a matrix", analogous to the concept of "distance" in scalar variables, for measuring rate of convergence of a matrix series.

6.2.1 Matrix sequences

Let $A_1, A_2, \ldots$ be a sequence of $m \times n$ matrices so that the k-th member in this sequence of matrices is A_k. Let the (i, j)-th element in A_k be denoted by $a_{ij}^{(k)}$ so that $A_k = (a_{ij}^{(k)})$, $k = 1, 2, \ldots$. The elements $a_{ij}^{(k)}$ are real or complex numbers.

Definition 6.2.1 (Convergence of a sequence of matrices). For scalar sequences we say that the limit of $a_{ij}^{(k)}$, as $k \to \infty$, is a_{ij} if there exists a finite number a_{ij} such that $a_{ij}^{(k)} \to a_{ij}$ when $k \to \infty$. Convergence of a matrix sequence is defined through element-wise convergence. Thus if $a_{ij}^{(k)} \to a_{ij}$ for all i and j when $k \to \infty$ we say that A_k converges to $A = (a_{ij})$ as $k \to \infty$.

Example 6.2.1. Check for the convergence of the sequence $A_1, A_2, \ldots$ as well as that of the sequence $B_1, B_2, \ldots$ where

$$A_k = \begin{bmatrix} \frac{1}{2^k} & \frac{k}{1+k} \\ -2+\frac{1}{k} & e^{-k} \end{bmatrix}, \quad B_k = \begin{bmatrix} (-1)^k & 0 \\ e^k & \frac{2k}{1+k} \end{bmatrix}.$$

Solution 6.2.1. Let us check the sequence $A_1, A_2, \ldots$. Here

$$a_{11}^{(k)} = \frac{1}{2^k}, \quad a_{12}^{(k)} = \frac{k}{1+k}, \quad a_{21}^{(k)} = -2+\frac{1}{k}, \quad a_{22}^{(k)} = e^{-k}.$$

$$\lim_{k\to\infty} a_{11}^{(k)} = \lim_{k\to\infty} \frac{1}{2^k} = 0, \quad \lim_{k\to\infty} a_{12}^{(k)} = \lim_{k\to\infty} \frac{k}{1+k} = 1,$$

$$\lim_{k\to\infty} a_{21}^{(k)} = \lim_{k\to\infty} \left[-2+\frac{1}{k}\right] = -2, \quad \lim_{k\to\infty} a_{22}^{(k)} = \lim_{k\to\infty} e^{-k} = 0.$$

Hence

$$\lim_{k\to\infty} A_k = A = \begin{bmatrix} 0 & 1 \\ -2 & 0 \end{bmatrix}$$

and the sequence is a convergent sequence. Now, consider $B_1, B_2, \ldots$. Here

$$b_{11}^{(k)} = (-1)^k, \quad b_{12}^{(k)} = 0, \quad b_{21}^{(k)} = e^k, \quad b_{22}^{(k)} = \frac{2k}{1+k}.$$

Evidently

$$\lim_{k\to\infty} b_{12}^{(k)} = \lim_{k\to\infty} 0 = 0, \quad \lim_{k\to\infty} b_{22}^{(k)} = \lim_{k\to\infty} \frac{2k}{1+k} = 2.$$

But $(-1)^k$ oscillates from -1 to 1 and hence there is no limit as $k \to \infty$. Also $e^k \to \infty$ when $k \to \infty$. Hence the sequence $B_1, B_2, \ldots$ is divergent.

(i) For any sequence $A_1, A_2, \ldots$ where $A_k = (a_{ij}^{(k)})$ we say that the sequence is divergent if for at least one element in A_k either the limit does not exist or the limit is $\pm\infty$.

The following properties are evident from the definition itself.

(ii) Let $A_1, A_2, \ldots$ and $B_1, B_2, \ldots$ be convergent sequences of matrices where $A_k \to A$ and $B_k \to B$ as $k \to \infty$. Then

$$A_k + B_k \to A + B, \quad A_k B_k \to AB,$$
$$Q A_k Q^{-1} \to QAQ^{-1}, \quad \text{diag}(A_k, B_k) \to \text{diag}(A, B),$$
$$\alpha_k A_k \to \alpha A$$

when $\alpha_k \to \alpha$, where $\alpha_1, \alpha_2, \ldots$ is a sequence of scalars.

By combining with the ideas of matrix polynomials from Section 6.1 we can establish the following properties: Since we have only considered Lagrange interpolating polynomials in Section 6.1 we will state the results when the eigenvalues of the $n \times n$ matrix A are distinct. But analogous results are available when some of the eigenvalues are repeated also.

(iii) Let the scalar functions $f_1(\lambda), f_2(\lambda), \ldots$ be defined on the spectrum of an $n \times n$ matrix A and let the sequence $A_1, A_2, \ldots$ be defined as $A_k = f_k(A)$, $k = 1, 2, \ldots$. Then the sequence $A_1, A_2, \ldots$ converges, for $k \to \infty$, if and only if the scalar sequences $\{f_1(\lambda_1), f_2(\lambda_1), \ldots\}, \{f_1(\lambda_2), f_2(\lambda_2), \ldots\}, \ldots, \{f_1(\lambda_n), f_2(\lambda_n), \ldots\}$ converge, as $k \to \infty$, where $\lambda_1, \ldots, \lambda_n$ are the eigenvalues of A.

Example 6.2.2. For the matrix A show that

$$e^{tA} = \begin{pmatrix} \cos t & \sin t \\ -\sin t & \cos t \end{pmatrix}, \quad \text{where } A = \begin{pmatrix} 0 & 1 \\ -1 & 0 \end{pmatrix}.$$

Solution 6.2.2. The eigenvalues of A are $\pm i$, $i = \sqrt{-1}$. Take $p(\lambda) = e^{\lambda t}$ and apply (6.1.5) of Section 6.1. Then

$$e^{tA} = e^{it} \frac{(A + iI)}{2i} + e^{-it} \frac{(A - iI)}{-2i}$$

$$= \frac{e^{it}}{2i} \begin{pmatrix} i & 1 \\ -1 & i \end{pmatrix} - \frac{e^{-it}}{2i} \begin{pmatrix} -i & 1 \\ -1 & -i \end{pmatrix}$$

$$= \begin{pmatrix} \frac{e^{it}+e^{-it}}{2} & \frac{e^{it}-e^{-it}}{2i} \\ \frac{-e^{it}+e^{-it}}{2i} & \frac{e^{it}+e^{-it}}{2} \end{pmatrix} = \begin{pmatrix} \cos t & \sin t \\ -\sin t & \cos t \end{pmatrix}.$$

6.2.2 Matrix series

A matrix series is obtained by adding up the matrices in a matrix sequence. For example if $A_0, A_1, A_2, \ldots$ is a matrix sequence then the corresponding matrix series is given by

$$f(A) = \sum_{k=0}^{\infty} A_k. \tag{6.2.1}$$

If the matrix series is a power series then we will be considering powers of matrices and hence in this case the series will be defined only for $n \times n$ matrices. For an $n \times n$ matrix A consider the series

$$g(A) = a_0 I + a_1 A + \cdots + a_k A^k + \cdots = \sum_{k=0}^{\infty} a_k A^k \tag{6.2.2}$$

where $a_0, a_1, \ldots$ are scalars. This is a matrix power series. As in the case of scalar series, convergence of a matrix series will be defined in terms of the convergence of the sequence of partial sums.

Definition 6.2.2 (Convergence of a matrix series). Let $f(A)$ be a matrix series as in (6.2.1). Consider the partial sums $S_0, S_1, \ldots$ where

$$S_k = A_0 + A_1 + \cdots + A_k.$$

If the sequence $S_0, S_1, \ldots$ is convergent then we say that the series in (6.2.1) is convergent. [If it is a power series as in (6.2.2) then $A_k = a_k A^k$ and then the above definition applies.]

Example 6.2.3. Check the convergence of the series $f_1(A)$ and $f_2(B)$ where

$$f_1(A) = \sum_{k=0}^{\infty} A_k, \quad A_k = \begin{bmatrix} y^k & 2^{-k} \\ \frac{x^k}{k!} & (-1)^k \frac{\theta^{2k}}{(2k)!} \end{bmatrix}$$

and

$$f_2(B) = \sum_{k=0}^{\infty} B_k, \quad B_k = \begin{bmatrix} \sin k\pi & \cos \frac{k\pi}{2} \end{bmatrix}.$$

Solution 6.2.3. The sum of the first $m + 1$ terms in $f_1(A)$ is given by

$$S_m = \sum_{k=0}^{m} A_k = \begin{bmatrix} \sum_{k=0}^{m} y^k & \sum_{k=0}^{m} 2^{-k} \\ \sum_{k=0}^{m} \frac{x^k}{k!} & \sum_{k=0}^{m} \frac{(-1)^k \theta^{2k}}{(2k)!} \end{bmatrix}.$$

Convergence of the series in $f_1(A)$ depends upon the convergence of the individual elements in S_m as $m \to \infty$. Note that

$$\sum_{k=0}^{\infty} y^k = 1 + y + y^2 + \cdots$$

$$= (1 - y)^{-1} \quad \text{if } |y| < 1 \text{ and } +\infty \text{ if } y \geq 1;$$

$$\sum_{k=0}^{\infty} 2^{-k} = \frac{1}{1 - \frac{1}{2}} = 2;$$

$$\sum_{k=0}^{\infty} \frac{x^k}{k!} = 1 + \frac{x}{1!} + \frac{x^2}{2!} + \cdots = e^x;$$

$$\sum_{k=0}^{\infty} \frac{(-1)^k \theta^{2k}}{(2k)!} = 1 - \frac{\theta^2}{2!} + \frac{\theta^4}{4!} - \cdots = \cos \theta.$$

Hence the series in $f_1(A)$ is convergent for $|y| < 1$ and diverges if $y \geq 1$. Now, consider $f_2(B)$. The partial sums are, for $m = 0, 1, \ldots$,

$$S_m = \sum_{k=0}^{m} B_k = \begin{bmatrix} \sum_{k=0}^{m} \sin k\pi, & \sum_{k=0}^{m} \cos \frac{k\pi}{2} \end{bmatrix}.$$

But $\sum_{k=0}^{m} \sin k\pi = 0$ for all m whereas $\sum_{k=0}^{m} \cos \frac{k\pi}{2}$ oscillates between 0 and 1 and hence the sequence of partial sums for this series is not convergent. Thus the series in $f_2(B)$ is not convergent.

Example 6.2.4. Check for the convergence of the following series in the $n \times n$ matrix A:

$$f(A) = I + A + A^2 + \cdots .$$

Solution 6.2.4. Let $\lambda_1, \lambda_2, \ldots, \lambda_n$ be the eigenvalues of A. Let us consider the case when the eigenvalues of A are distinct. Then there exists a nonsingular matrix Q such that

$$Q^{-1} A Q = D = \mathrm{diag}(\lambda_1, \ldots, \lambda_n)$$

and

$$Q^{-1} A^m Q = D^m = \mathrm{diag}(\lambda_1^m, \ldots, \lambda_n^m), \quad m = 1, 2, \ldots .$$

Then

$$Q^{-1} f(A) Q = I + D + D^2 + \cdots .$$

The j-th diagonal element on the right is then

$$1 + \lambda_j + \lambda_j^2 + \cdots = (1 - \lambda_j)^{-1} \quad \text{if } |\lambda_j| < 1, \, j = 1, \ldots, n$$

which are the eigenvalues of $(I - A)^{-1}$. Then if $|\lambda_j| < 1$ for $j = 1, 2, \ldots, n$ the series is convergent and the sum is $(I - A)^{-1}$ or

$$I + A + A^2 + \cdots = (I - A)^{-1} \quad \text{for } |\lambda_j| < 1, \, j = 1, \ldots, n.$$

We can also derive the result from (6.1.5) of Section 6.1. The result also holds good even if some eigenvalues are repeated. We can state the exponential and trigonometric series as follows: For any $n \times n$ matrix A,

$$\sin A = \sum_{k=0}^{\infty} \frac{(-1)^k A^{2k+1}}{(2k+1)!}, \quad \cos A = \sum_{k=0}^{\infty} \frac{(-1)^k A^{2k}}{(2k)!},$$

$$\sinh A = \sum_{k=0}^{\infty} \frac{A^{2k+1}}{(2k+1)!}, \quad \cosh A = \sum_{k=0}^{\infty} \frac{A^{2k}}{(2k)!},$$

$$e^A = \sum_{k=0}^{\infty} \frac{A^k}{k!} \tag{6.2.3}$$

and further, when the eigenvalues $\lambda_1, \ldots, \lambda_n$ of A are such that $|\lambda_j| < 1, \, j = 1, \ldots, n$ then the binomial and logarithmic series are given by the following:

$$(I - A)^{-1} = \sum_{k=0}^{\infty} A^k, \quad \ln(I + A) = \sum_{k=1}^{\infty} \frac{(-1)^{k-1} A^k}{k}. \tag{6.2.4}$$

6.2.3 Matrix hypergeometric series

A general hypergeometric series $_pF_q(\cdot)$ in a real scalar variable x is defined as follows:

$$_pF_q(a_1,\ldots,a_p;b_1,\ldots,b_q;x) = \sum_{r=0}^{\infty} \frac{(a_1)_r \cdots (a_p)_r}{(b_1)_r \cdots (b_q)_r} \frac{x^r}{r!} \tag{6.2.5}$$

where, for example,

$$(a)_m = a(a+1)\cdots(a+m-1), \quad (a)_0 = 1, \quad a \neq 0.$$

For example,

$$_0F_0(\ ;\ ;x) = e^x, \quad _1F_0(a;\ ;x) = (1-x)^{-a} \quad \text{for } |x| < 1.$$

In (6.2.5) there are p upper parameters $a_1,\ldots,a_p$ and q lower parameters $b_1,\ldots,b_q$. The series in (6.2.5) is convergent for all x if $q \geq p$, convergent for $|x| < 1$ if $p = q+1$, divergent if $p > q+1$ and the convergence conditions for $x = 1$ and $x = -1$ can also be worked out. A matrix series in an $n \times n$ matrix A, corresponding to the right side in (6.2.5) is obtained by replacing x by A. Thus we may define a hypergeometric series in an $n \times n$ matrix A as follows:

$$_pF_q(a_1,\ldots,a_p;b_1,\ldots,b_q;A) = \sum_{r=0}^{\infty} \frac{(a_1)_r \cdots (a_p)_r}{(b_1)_r \cdots (b_q)_r} \frac{A^r}{r!} \tag{6.2.6}$$

where $a_1,\ldots,a_p$, $b_1,\ldots,b_q$ are scalars. The series on the right in (6.2.6) is convergent for all A if $q \geq p$, convergent for $p = q+1$ when the eigenvalues of A are all less than 1 in absolute value, and divergent when $p > q+1$.

Example 6.2.5. If possible, sum up the series

$$I + 3A + \frac{1}{2}\{(3)(4)A^2 + (4)(5)A^3 + \cdots\}$$

where

$$A = \begin{bmatrix} \frac{1}{2} & 0 & 0 \\ 1 & -\frac{1}{3} & 0 \\ 2 & 3 & -\frac{1}{2} \end{bmatrix}.$$

Solution 6.2.5. Consider the scalar series

$$1 + 3x + \frac{1}{2}[(3)(4)x^2 + (4)(5)x^3 + \cdots]$$

$$= 1 + 3x + (3)(4)\frac{x^2}{2!} + (3)(4)(5)\frac{x^3}{3!} + \cdots$$

$$= (1-x)^{-3} \quad \text{for } |x| < 1.$$

In our matrix A, the eigenvalues are $\lambda_1 = \frac{1}{2}$, $\lambda_2 = -\frac{1}{3}$, $\lambda_3 = -\frac{1}{2}$ and therefore $|\lambda_j| < 1$, $j = 1,2,3$. Hence the series can be summed up into a $_1F_0$ type hypergeometric series or a binomial series and the sum is then

$$(I - A)^{-3} = \begin{bmatrix} \frac{1}{2} & 0 & 0 \\ 1 & \frac{4}{3} & 0 \\ 2 & 3 & \frac{3}{2} \end{bmatrix}^{-3}.$$

But

$$(I - A)^{-1} = \begin{bmatrix} 2 & 0 & 0 \\ -\frac{3}{2} & \frac{3}{4} & 0 \\ \frac{1}{3} & -\frac{3}{2} & \frac{2}{3} \end{bmatrix}$$

and

$$(I - A)^{-3} = [(I - A)^{-1}]^3 = \begin{bmatrix} 2 & 0 & 0 \\ -\frac{3}{2} & \frac{3}{4} & 0 \\ \frac{1}{3} & -\frac{3}{2} & \frac{2}{3} \end{bmatrix}^3$$

$$= \begin{bmatrix} 8 & 0 & 0 \\ -\frac{291}{32} & \frac{27}{64} & 0 \\ \frac{4153}{432} & -\frac{217}{96} & \frac{8}{27} \end{bmatrix}.$$

6.2.4 The norm of a matrix

For a 1×1 vector or a scalar quantity α the absolute value, $|\alpha|$, is a measure of its magnitude. For an $n \times 1$ vector X, $X' = (x_1, \ldots, x_n)$,

$$\|X\| = \{|x_1|^2 + \cdots + |x_n|^2\}^{\frac{1}{2}}, \tag{6.2.7}$$

where $|x_j|$ denotes the absolute value of x_j, $j = 1, \ldots, n$, and this can be taken as a measure of its magnitude. Equation (6.2.7) is its Euclidean length also. This Euclidean length satisfies some interesting properties:

(a) $\|X\| \geq 0$ for all X and $\|X\| = 0$ if and only if $X = O$ (null);

(b) $\|\alpha X\| = |\alpha| \|X\|$ where α is a scalar quantity;

(c) $\|X + Y\| \leq \|X\| + \|Y\|$, the triangular inequality. (6.2.8)

If (a), (b), (c) are taken as postulates or axioms to define a *norm* of the vector X, denoted by $\|X\|$, then one can see that, not only the Euclidean length but also other items satisfy (a), (b), (c).

Definition 6.2.3 (Norm of a vector and distance between vectors). For X and $n \times 1$ vector, or an element in a general vector subspace S where a norm can be defined, a measure satisfying (a), (b), (c) above will be called a *norm* of X and it will be denoted by $\|X\|$. Note that X replaced by $X - Y$ and satisfying (a), (b), (c) is called a *distance* between X and Y.

It is not difficult to show that the following measures are also *norms* of the vector X:

$$\|X\|_1 = \sum_{j=1}^{n} |x_j|;$$

$$\|X\|_2 = \left[\sum_{j=1}^{n} |x_j|^2 \right]^{\frac{1}{2}} = (X^*X)^{\frac{1}{2}} \quad \text{(the Euclidean norm)}$$

where X^* denotes the complex conjugate transpose of X

$$\|X\|_p = \left[\sum_{j=1}^{n} |x_j|^p \right]^{\frac{1}{p}}, \quad p \geq 1 \quad \text{(the Hölder norms)}$$

$$\|X\|_\infty = \max_{1 \leq j \leq n} |x_j| \quad \text{(the infinite norm).} \tag{6.2.9}$$

Example 6.2.6. Show that $\|X\|_1$ satisfies the conditions (a), (b), (c) in (6.2.8).

Solution 6.2.6. $|x_j|$ being the absolute value of x_j cannot be zero unless x_j itself is zero. If $x_j \neq 0$ then $|x_j| > 0$ by definition whether x_j is real or complex. Thus condition (a) is obviously satisfied. Note that for any two scalars α and x_j, $|\alpha x_j| = |\alpha| |x_j|$. Hence (b) is satisfied. Also for any two scalars x_j and y_j the triangular inequality holds. Thus $\|X\|_1$ satisfies (a), (b), (c) of (6.2.8).

The following properties are immediate from the definition itself:
(a) $|\|X\| - \|Y\|| \leq \|X + Y\| \leq \|X\| + \|Y\|$.
(b) $\|-X\| = \|X\|$.
(c) If $\|X\|$ is a norm of X then $k\|X\|$, $k > 0$ is also a norm of X.
(d) $|\|X\| - \|Y\|| \leq \|X - Y\|$.
(e) $\|U\|_2 = \|X\|_2$ where $U = AX$, A is a unitary matrix (orthonormal if real).
(f) $\|X\|_1 \geq \|X\|_2 \geq \cdots \geq \|X\|_\infty$.

Now let us see how we can define a *norm* of a matrix as a single number which should have the desirable properties (a), (b), (c) of (6.2.8). But there is an added difficulty here. If we consider two matrices, an $n \times n$ matrix A and an $n \times 1$ matrix X, then AX is again an $n \times 1$ matrix which is also an n-vector. Hence any definition that we take for the *norm* of a matrix must be compatible with matrix multiplication. Therefore an additional postulate is required.

Definition 6.2.4 (A *norm of a matrix A*)**.** A single number, denoted by $\|A\|$, is called a *norm* of the matrix A if it satisfies the following four postulates:
(a) $\|A\| \geq 0$ and $\|A\| = 0$ if and only if A is a null matrix.
(b) $\|cA\| = |c| \|A\|$ when c is a scalar.
(c) $\|A + B\| \leq \|A\| + \|B\|$ whenever $A + B$ is defined.
(d) $\|AB\| \leq \|A\| \|B\|$ whenever AB is defined.

It is not difficult to see that the following quantities qualify to be the *norms* of the matrix $A = (a_{ij})$:

$$\|A\|_{(p)} = \left(\sum_{i,j=1}^{n} |a_{ij}|^p \right)^{\frac{1}{p}}, \quad 1 \le p \le 2 \tag{6.2.10}$$

(Hölder norm, not a norm for $p > 2$),

$$\|A\|_2 = \left(\sum_{i,j=1}^{n} |a_{ij}|^2 \right)^{\frac{1}{2}} \quad \text{(Euclidean norm)}, \tag{6.2.11}$$

$$\|A\|_3 = n \max_{i,j} |a_{ij}|, \tag{6.2.12}$$

$$\|A\|_4 = \max_i \sum_j |a_{ij}|, \tag{6.2.13}$$

$$\|A\|_5 = \max_j \sum_i |a_{ij}|, \tag{6.2.14}$$

$$\|A\|_6 = s_1, \tag{6.2.15}$$

where s_1 is the largest singular value of A;

$$\|A\|_7 = \sup_{X \ne O} \frac{\|AX\|}{\|X\|} \tag{6.2.16}$$

where $\|AX\|$ and $\|X\|$ are vector norms, the same norm;

$$\|A\|_8 = \max_{X, \|X\|=1} \|AX\| \tag{6.2.17}$$

same vector norm is taken in each case. As a numerical example let us consider the following matrix:

$$A = \begin{bmatrix} 1+i & 0 \\ 1 & -1 \end{bmatrix}.$$

Then

$$\|A\|_1 = |(1+i)| + |(0)| + |(1)| + |(-1)| = \sqrt{2} + 0 + 1 + 1 = 2 + \sqrt{2};$$

$$\|A\|_2 = [2 + 0 + 1 + 1]^{\frac{1}{2}} = 2;$$

$$\|A\|_3 = 2\max[(\sqrt{2}, 0, 1, 1)] = 2\sqrt{2};$$

$$\|A\|_4 = \max[|(1+i)| + |(-1)|, |(0)| + |(-1)|] = 1 + \sqrt{2};$$

$$\|A\|_5 = \max[|(1+i)| + |(0)|, |(1)| + |(-1)|] = 2.$$

For computing $\|A\|_6$ we need the eigenvalues of A^*A:

$$A^*A = \begin{bmatrix} 1-i & 1 \\ 0 & -1 \end{bmatrix} \begin{bmatrix} 1+i & 0 \\ 1 & -1 \end{bmatrix} = \begin{bmatrix} 3 & -1 \\ -1 & 1 \end{bmatrix}.$$

The eigenvalues of A^*A are $2 \pm \sqrt{2}$ and then the largest singular value of A is

$$[(2+\sqrt{2})]^{\frac{1}{2}} = \|A\|_6.$$

Note that there are several possible values for $\|A\|_7$ and $\|A\|_8$ depending upon which vector norm is taken. For example, if we take the Euclidean norm and consider $\|A\|_8$ then it is a matter of maximizing $[Y^*Y]^{\frac{1}{2}}$ subject to the condition $X^*X = 1$ where $Y = AX$. But $Y^*Y = X^*A^*AX$. The problem reduces to the following:

Maximize X^*A^*AX subject to the condition $X^*X = 1$.

This is already done in Section 5.5 and the answer is the largest eigenvalue of A^*A and hence, when this particular vector norm is used,

$$\|A\|_8 = s_1 = \text{largest singular value of } A.$$

Note that for a vector norm $\|X\|$, $k\|X\|$ is also a vector norm when $k > 0$. This property need not hold for a matrix norm $\|A\|$ due to condition (d) of the definition.

Example 6.2.7. For an $n \times n$ matrix $A = (a_{ij})$ let $\alpha = \max_{i,j} |a_{ij}|$, that is, the largest of the absolute values of the elements. Is this a norm of A?

Solution 6.2.7. Obviously conditions (a), (b), (c) of Definition 6.2.4 are satisfied. Let us check condition (d). Let $B = (b_{ij})$ and $AB = C = (c_{ij})$. Then

$$c_{ij} = \sum_{k=1}^{n} a_{ik} b_{kj} \implies$$

$$\max_{i,j} |c_{ij}| = \max_{i,j} \left| \sum_{k=1}^{n} a_{ik} b_{kj} \right|.$$

Suppose that the elements are all real and positive and that the largest ones in A and B are $a_{11} = a$ and $b_{11} = b$. Then

$$\max_{i,j} |a_{ij}| = a, \quad \max_{i,j} |b_{ij}| = b, \quad \max_{i,j} |a_{ij}| \left[\max_{i,j} |b_{ij}| \right] = ab$$

whereas

$$\max_{i,j} \left| \sum_{k=1}^{n} a_{ik} b_{jk} \right| = ab + \delta, \quad \delta \geq 0.$$

Hence condition (d) is evidently violated. Thus α cannot be a norm of the matrix A. It is easy to note that $\beta = n\alpha$ is a norm of A, or

$$\beta = n\alpha = n \max_{i,j} |a_{ij}| = \|A\|_3. \tag{6.2.18}$$

Example 6.2.8. Let $\mu_A = \max_i |\lambda_i|$ where $\lambda_1, \ldots, \lambda_n$ be the eigenvalues of an $n \times n$ matrix A. Evidently μ is not a norm of A since condition (a) of Definition 6.2.4 is not satisfied by μ. [Take a non-null triangular matrix with the diagonal elements zeros. Then all eigenvalues are zeros.] Show that for any matrix norm $\|A\|$,

$$\|A\| \geq \mu_A. \tag{6.2.19}$$

This μ_A is called the *spectral radius of the matrix A*.

Solution 6.2.8. Let λ_1 be the eigenvalue of A such that $\mu_A = \lambda_1$. Then, by definition, there exists a non-null vector X such that

$$AX_1 = \lambda_1 X_1.$$

Consider the $n \times n$ matrix

$$B = (X_1, O, \dots, O).$$

Then

$$AB = (AX_1, O, \dots, O) = (\lambda_1 X_1, O, \dots, O) = \lambda_1 B.$$

From conditions (a) and (d) of Definition 6.2.4

$$|\lambda_1| \, \|B\| \leq \|A\| \, \|B\| \implies \|A\| \geq |\lambda_1|$$

since $\|B\| \neq O$ due to the fact that X_1 is non-null. This establishes the result. The result in (6.2.19) is a very important result which establishes a lower bound for norms of a matrix, whatever be the norm of a matrix.

6.2.5 Compatible norms

For any $n \times n$ matrix A and $n \times 1$ vector X if we take any matrix norm $\|A\|$ and any vector norm $\|X\|$ then condition (d) of the definition, namely,

$$\|AX\| \leq \|A\| \, \|X\| \tag{6.2.20}$$

need not be satisfied.

Definition 6.2.5. For any matrix A and any vector X, where AX is defined, if (6.2.20) is satisfied for a particular norm $\|A\|$ of A and $\|X\|$ of X then $\|A\|$ and $\|X\|$ are called *compatible norms*.

It is not difficult to show that the following are compatible norms:

Matrix norm	Vector norm
$\|A\|_4$ of (6.2.13)	$\|X\|_\infty$ of (6.2.9)
$\|A\|_5$ of (6.2.14)	$\|X\|_1$ of (6.2.9)
$\|A\|_6$ of (6.2.15)	$\|X\|_2$ of (6.2.9)
$\|A\|_7$ with any vector norm $\|X\|_v$	$\|X\|_v$
$\|A\|_8$ with any vector norm $\|X\|$	$\|X\|$

Example 6.2.9. Show that $\|A\|_4$ of (6.2.13) and $\|X\|_\infty$ of (6.2.9) are compatible norms.

Solution 6.2.9. Let X be an $n \times 1$ vector with $\|X\|_\infty = 1$. Consider the vector norm

$$\|AX\|_\infty = \max_i \left| \sum_{j=1}^n a_{ij} x_j \right| \leq \max_i \sum_{j=1}^n |a_{ij}| \, |x_j| \leq \|X\|_\infty \|A\|_4$$

which establishes the compatibility.

6.2.6 Matrix power series and rate of convergence

Let A be an $n \times n$ matrix and consider the power series

$$f(A) = I + A + A^2 + \cdots . \tag{6.2.21}$$

We have already seen that the power series in (6.2.21) is convergent when all the eigenvalues of A are less than 1 in absolute value, that is, $0 < |\lambda_j| < 1$, $j = 1, \dots, n$ where the λ_j's are the eigenvalues of A. If $\|A\|$ denotes a norm of A then evidently

$$\|A^k\| = \|AA \cdots A\| \leq \|A\|^k .$$

Then from (6.2.21) we have

$$\|I + A + A^2 + \cdots\| \leq 1 + \|A\| + \|A\|^2 + \cdots$$

$$= \frac{1}{1 - \|A\|} \quad \text{if } \|A\| < 1.$$

Therefore if the power series in (6.2.21) is approximated by taking the first k terms, that is,

$$f(A) \approx I + A + \cdots + A^{k-1} \tag{6.2.22}$$

then the error in this approximation is given by

$$A^k + A^{k+1} + \cdots = A^k [I + A + A^2 + \cdots] \Rightarrow$$

$$\|A^k + A^{k+1} + \cdots\| \leq \frac{\|A\|^k}{1 - \|A\|} \quad \text{if } \|A\| < 1. \tag{6.2.23}$$

Thus a measure of an upper bound for the error in the approximation in (6.2.22) is given by (6.2.23).

6.2.7 An application in statistics

In the field of design of experiments and analysis of variance, connected with two-way layouts with multiple observations per cell, the analysis of the data becomes quite complicated when the cell frequencies are unequal. Such a situation can arise, for example, in a simple randomized block experiment with replicates (the experiment is repeated a number of times under identical conditions). If some of the observations are missing in some of the replicates then in the final two-way layout (blocks versus treatments) the cell frequencies will be unequal. In such a situation, in order to estimate the treatment effects or block effects (main effects) one has to solve a singular system of a matrix equation of the following type: (This arises from the least square analysis.)

$$(I - A)\hat{\alpha} = Q \tag{6.2.24}$$

where $\alpha' = (\alpha_1, \ldots, \alpha_p)$ are the block effects to be estimated, $\hat{\alpha}$ denotes the estimated value, A is a $p \times p$ matrix

$$A = (a_{rs}), \quad a_{rs} = \sum_{j=1}^{q} \frac{(n_{rj} n_{sj})}{n_{r.} n_{.j}},$$
$$n_{r.} = \sum_k n_{rk}, \quad n_{.j} = \sum_k n_{kj},$$

and Q is a known column vector. The matrix A is the *incidence matrix* of this design. From the design itself α_j's satisfy the condition

$$\alpha_1 + \alpha_2 + \cdots + \alpha_p = 0. \tag{6.2.25}$$

Observe that A is a singular matrix (the sum of the elements in each row is 1). Obviously we cannot write and expand

$$\hat{\alpha} = (I - A)^{-1}Q = [I + A + A^2 + \cdots]Q$$

due to the singularity of A. Let $k_1, \ldots, k_p$ be the medians of the elements in the first, second, ..., p-th rows of A and consider a matrix $B = (b_{ij})$, $b_{ij} = (a_{ij} - k_i)$ for all i and j. Evidently $(I - B)$ is nonsingular. Consider

$$(I - B)\hat{\alpha} = (I - A - K)\hat{\alpha} = (I - A)\hat{\alpha} + K\hat{\alpha}$$

where K is a matrix in which all the elements in the i-th row are equal to k_i, $i = 1, \ldots, p$. Then with (6.2.25) we have $K\alpha = O$ and hence

$$(I - A)\hat{\alpha} = (I - B)\hat{\alpha} = Q \implies$$
$$\hat{\alpha} = (I - B)^{-1}Q = (I + B + B^2 + \cdots)Q.$$

Take the norm $\|B\|_4$ of (6.2.13). That is,

$$\|B\|_4 = \max_i \sum_{j=1}^{p} |b_{ij} - k_i|.$$

Since the mean deviation is least when the deviations are taken from the median $\|B\|_4$ is the least possible for the incidence matrix A so that the convergence of the series

$I + B + B^2 + \cdots$ is made the fastest possible. In fact, for all practical purposes of testing statistical hypotheses on α_j's a good approximation is available by taking

$$\hat{\alpha} \approx (I + B)Q$$

where inversion or taking powers of B is not necessary. For an application of the above procedure to a specific problem in testing of statistical hypothesis see [1].

Exercises 6.2

6.2.1. If $p(\lambda)$ is a polynomial defined on the spectrum of an $n \times n$ matrix A then show that $p(A') = [p(A)]'$.

6.2.2. For any $n \times n$ matrix A show that there exists a skew symmetric matrix B such that $A = e^B$ if and only if A is a real orthogonal matrix with its determinant 1.

6.2.3. For the matrix $A = \begin{bmatrix} \frac{1}{2} & \frac{1}{2} \\ \frac{1}{2} & -\frac{1}{2} \end{bmatrix}$ sum up the following matrix series, if possible:

$$I + A + A^2 + \cdots.$$

6.2.4. For the same matrix in Exercise 6.2.3 sum up the series

$$I + 2A + 3A^2 + 4A^3 + \cdots.$$

6.2.5. For the same matrix in Exercise 6.2.3 sum up the series

$$I + \frac{1}{2}A + \frac{3}{4}\frac{A^2}{2!} + \frac{(3)(5)}{8}\frac{A^3}{3!} + \cdots.$$

6.2.6. Show that the norms $\|X\|_p$ and $\|X\|_\infty$ in (6.2.9) satisfy all the conditions in (6.2.8).

6.2.7. Prove that the norm defined in (6.2.17) is a matrix norm, and from there prove that (6.2.16) is also a matrix norm.

6.2.8. For any $n \times n$ matrix A consider the Euclidean matrix norm $\|A\|_2$ of (6.2.11). Let $\lambda_1, \ldots, \lambda_n$ be the eigenvalues of A and let $\Re(\lambda_j) = $ real part of λ_j and $\Im(\lambda_j) = $ imaginary part of λ_j. Then show that

$$\sum_{i=1}^{n} |\lambda_i|^2 \le \|A\|^2$$

$$\sum_{i=1}^{n} |\Re(\lambda_i)|^2 \le \|B\|^2, \quad B = \frac{1}{2}(A + A^*)$$

$$\sum_{i=1}^{n} |\Im(\lambda_i)|^2 \le \|C\|^2, \quad C = \frac{1}{2}(A - A^*)$$

and that the equality in any one of these implies equality in all the three above and equality occurs if and only if A is a normal matrix.

6.2.9. For any $n \times n$ matrix $A = (a_{ij})$ let λ be any eigenvalue of A. Then show that

$$|\lambda| \le n\rho, \quad |\Re(\lambda)| \le n\sigma, \quad |\Im(\lambda)| \le n\gamma$$

where $\Re(\cdot)$ and $\Im(\cdot)$ denote the real part and the imaginary part of $(\cdot)$ respectively, and

$$\rho = \max_{i,j} |a_{ij}|,$$

$$\sigma = \max_{i,j} |b_{ij}|, \quad B = (b_{ij}) = \frac{1}{2}(A + A^*),$$

$$\gamma = \max_{i,j} |c_{ij}|, \quad C = (c_{ij}) = \frac{1}{2}(A - A^*).$$

6.2.10. For any $n \times n$ matrix $A = (a_{ij})$ and for any eigenvalue λ of A show that

$$|\Im(\lambda)| \le \alpha \sqrt{n(n-1)/2}, \quad \alpha = \frac{1}{2} \max_{i,j} |a_{ij} - a_{ji}|.$$

6.2.11. For any $n \times n$ matrix A let

$$B = \begin{bmatrix} I & A \\ A^* & I \end{bmatrix}.$$

Show that B is positive definite if and only if $\|A\|_6 < 1$ where $\|A\|_6$ is the norm defined in (6.2.15).

6.2.12. If B is positive definite and A is positive semi-definite then show that the eigenvalues λ_j's of $(A + B)^{-1}A$ are such that $0 \le \lambda_j \le 1$.

6.2.13. For an arbitrary $n \times n$ matrix A let $B = \begin{bmatrix} O & iA \\ -iA^* & O \end{bmatrix}$, $i = \sqrt{-1}$, then show that $\|B\|_6 = \|A\|_6$, see equation (6.2.15) for the norms.

6.2.14. For an arbitrary $n \times n$ matrix A show that $|A| \le \min(\|A\|_4^n, \|A\|_5^n)$, see equations (6.2.13) and (6.2.14) for the norms.

6.3 Singular value decomposition of a matrix

For the sake of readers who are interested in further results on matrices, a few more technical terms will be listed here. Recalling our standard notations, let A' be the transpose and A^* the conjugate transpose of a matrix A. Two matrices A and B are said to be *similar* if there exists a nonsingular matrix Q such that

$$A = QBQ^{-1}.$$

If Q is an $n \times n$ orthonormal matrix then $QQ' = I$, $Q'Q = I$ thereby $Q^{-1} = Q'$. If Q is unitary, that is, $QQ^* = I$, $Q^*Q = I$ then $Q^{-1} = Q^*$. If A and B are such that

$$A = UBU^*$$

for a unitary matrix U then A and B are said to be *unitarily similar*. If a square matrix is unitarily similar to a diagonal matrix, that is,

$$A = UDU^*$$

where U is unitary and D is diagonal, then A is called a *normal matrix*. If

$$A = UDU'$$

where U is an orthonormal matrix, $UU' = I$, $U'U = I$, and D is a diagonal matrix then A is called an *orthogonally similar matrix*. If there exists a nonsingular matrix P such that

$$A = PBP^*$$

then A and B are said to be *congruent*. Then if P is unitary then A and B are unitarily similar also. It is not difficult to establish the following results:

(i) An $n \times n$ matrix A is normal if and only if A^* is normal. If A is normal then A^p is also normal for any positive integer p or for any integer p if $|A| \neq 0$.
(ii) Any square matrix is unitarily similar to an upper triangular matrix.
(iii) An $n \times n$ matrix A is normal if and only if $AA^* = A^*A$.
(iv) A normal matrix A is Hermitian if and only if its spectrum lies on the real line (eigenvalues are real).
(v) A matrix A is normal if and only if its real part and imaginary part commute, that is, when $A = A_1 + iA_2$, $i = \sqrt{-1}$, A_1, A_2 real matrices, then $A_1A_2 = A_2A_1$.
(vi) A real symmetric matrix A is positive definite (or positive semi-definite) if and only if it has a positive definite (or positive semi-definite) square root B, that is, $A = B^2$, and further, $\text{rank}(A) = \text{rank}(B)$.

Definition 6.3.1 (Singular values). Consider an arbitrary $m \times n$ rectangular matrix A. Then A^*A is $n \times n$ and A^*A is nonnegative definite (positive definite or positive semi-definite). Then there exists a nonnegative square root B such that $A^*A = B^2$. The eigenvalues $s_1, \ldots, s_n$ of $B = (A^*A)^{\frac{1}{2}}$ are called the *singular values of the rectangular matrix A*. Thus the singular values are the eigenvalues of $(A^*A)^{\frac{1}{2}}$ thereby they are nonnegative real numbers.

Example 6.3.1. Evaluate the singular values of the matrix

$$A = \begin{bmatrix} 1 & 1 & 1 \\ 1 & 0 & 0 \end{bmatrix}.$$

Solution 6.3.1. Since A is real, $A^*A = A'A$. That is,

$$A'A = \begin{bmatrix} 1 & 1 \\ 1 & 0 \\ 1 & 0 \end{bmatrix} \begin{bmatrix} 1 & 1 & 1 \\ 1 & 0 & 0 \end{bmatrix} = \begin{bmatrix} 2 & 1 & 1 \\ 1 & 1 & 1 \\ 1 & 1 & 1 \end{bmatrix}.$$

The characteristic equation $|A'A - \lambda I| = 0$ gives

$$\lambda(\lambda^2 - 4\lambda + 2) = 0.$$

The solutions are $\lambda_1 = 2 + \sqrt{2}$, $\lambda_2 = 2 - \sqrt{2}$, $\lambda_3 = 0$. Hence the singular values of A are

$$s_1 = (2 + \sqrt{2})^{\frac{1}{2}}, \quad s_2 = (2 - \sqrt{2})^{\frac{1}{2}}, \quad s_3 = 0.$$

Let us see the eigenvalues of AA^*.

$$AA^* = AA' = \begin{bmatrix} 1 & 1 & 1 \\ 1 & 0 & 0 \end{bmatrix} \begin{bmatrix} 1 & 1 \\ 1 & 0 \\ 1 & 0 \end{bmatrix} = \begin{bmatrix} 3 & 1 \\ 1 & 1 \end{bmatrix}.$$

The eigenvalues of AA' are $\lambda_1 = 2 + \sqrt{2}$, and $\lambda_2 = 2 - \sqrt{2}$. Thus the nonzero eigenvalues of AA^* and A^*A coincide. This, in fact, is a general result.

(vii) For any rectangular $m \times n$ matrix A the nonzero eigenvalues of A^*A and AA^* coincide.

As a practical procedure, consider the eigenvalues of AA^* if A is $m \times n$ with $m \leq n$ or the eigenvalues of A^*A if $n \leq m$ so that the square roots of these nonzero eigenvalues provide all the nonzero singular values of A. If there are r such nonzero singular values then the remaining singular values are zeros and there are $n - r$ such zeros if the matrix A is $m \times n$.

6.3.1 A singular value decomposition

A very interesting representation of an arbitrary $m \times n$ matrix A is a decomposition in terms of the singular values of A. There exist an $m \times m$ unitary matrix U, $U^*U = I$, $UU^* = I$ (orthonormal if real), and an $n \times n$ unitary matrix V, $V^*V = I$, $VV^* = I$ (orthonormal if real), such that

$$A = UDV^* \tag{6.3.1}$$

with D an $m \times n$ matrix having $s_1, \dots, s_r$ at the leading diagonal positions and zeros elsewhere, where $s_1, \dots, s_r$ are the nonzero singular values of A with r denoting the rank of A. The representation in (6.3.1) is known as the *singular value decomposition of A*.

It is not difficult to prove the result in (6.3.1). Let A be an $m \times n$ matrix. Since A^*A and AA^* are both real and Hermitian symmetric we can always construct an orthonormal system of eigenvectors for each. Let $X_1, \dots, X_n$ and $Y_1, \dots, Y_m$ be orthonormal systems of eigenvectors of A^*A and AA^* respectively. Let λ_i be a nonzero eigenvalue of A^*A corresponding to X_i. Then

$$A^* A X_i = \lambda_i X_i \;\Rightarrow\; X_i^* A^* A = \lambda_i X_i^*$$
$$\Rightarrow\; X_i^* A^* A X_i = \lambda_i,$$

where $\lambda_i > 0$ since $A^* A$ is at least Hermitian positive semi-definite. But

$$X_i^* A^* A X_i = (A X_i)^* (A X_i) = \|A X_i\|^2 \;\Rightarrow\; \|A X_i\| = \sqrt{\lambda_i}$$

where $\|A X_i\|$ is the Euclidean length of the vector $A X_i$. Let

$$Y_i = \frac{1}{\|A X_i\|} A X_i \;\Rightarrow\; A A^* Y_i = \frac{1}{\|A X_i\|} (A A^*) A X_i$$
$$= \frac{A(A^* A) X_i}{\|A X_i\|} = \lambda_i \frac{A X_i}{\|A X_i\|} = \lambda_i Y_i.$$

Therefore $(A A^*) Y_i = \lambda_i Y_i$. Thus Y_i is an eigenvector of $A A^*$ corresponding to the same eigenvalue λ_i, and from the starting point above,

$$A X_i = \|A X_i\| Y_i = \sqrt{\lambda_i} Y_i = s_i Y_i \tag{6.3.2}$$

where s_i is the i-th singular value of A. Now, let

$$U = (Y_1, \dots, Y_m) \quad \text{and} \quad V = (X_1, \dots, X_n).$$

Then from (6.3.2)

$$AV = (s_1 Y_1, s_2 Y_2, \dots, s_r Y_r, 0, \dots, 0) = UD \tag{6.3.3}$$

where

$$U = (Y_1, \dots, Y_m)$$

and D is an $m \times n$ matrix with the leading diagonal positions having $s_1, \dots, s_r$ with r being the rank of A. Postmultiply (6.3.3) with V^* to obtain

$$A = UDV^*$$

and the result is established.

Example 6.3.2. Obtain the singular value decomposition of the matrix

$$A = \begin{bmatrix} 1 & 0 & -1 \\ 0 & 1 & 1 \end{bmatrix}.$$

Solution 6.3.2. Since A is 2×3 we consider the matrix $AA^* = AA'$:

$$AA' = \begin{bmatrix} 1 & 0 & -1 \\ 0 & 1 & 1 \end{bmatrix} \begin{bmatrix} 1 & 0 \\ 0 & 1 \\ -1 & 1 \end{bmatrix} = \begin{bmatrix} 2 & -1 \\ -1 & 2 \end{bmatrix}.$$

The eigenvalues of AA^* are evidently $\lambda_1 = 3$, $\lambda_2 = 1$ and hence the nonzero singular values of A are $s_1 = \sqrt{3}$, $s_2 = 1$.

$$A^*A = A'A = \begin{bmatrix} 1 & 0 \\ 0 & 1 \\ -1 & 1 \end{bmatrix} \begin{bmatrix} 1 & 0 & -1 \\ 0 & 1 & 1 \end{bmatrix}$$

$$= \begin{bmatrix} 1 & 0 & -1 \\ 0 & 1 & 1 \\ -1 & 1 & 2 \end{bmatrix}.$$

The singular values of A are therefore $s_1 = \sqrt{3}$, $s_2 = 1$, $s_3 = 0$. Let us compute the eigenvectors of $A^*A = A'A$:

$$(A'A - \lambda_1 I)Z_1 = O, \quad \lambda_1 = 3, \ \Rightarrow Z_1' = (1, -1, -2).$$

$$X_1 = \frac{Z_1}{\|Z_1\|} = \frac{1}{\sqrt{6}} \begin{pmatrix} 1 \\ -1 \\ -2 \end{pmatrix}$$

Corresponding to $\lambda_2 = 1$ and $\lambda_3 = 0$ we have the normalized eigenvectors of $A'A$ given by

$$X_2 = \frac{1}{\sqrt{2}} \begin{pmatrix} 1 \\ 1 \\ 0 \end{pmatrix}, \quad X_3 = \frac{1}{\sqrt{3}} \begin{pmatrix} 1 \\ -1 \\ 1 \end{pmatrix}.$$

These are the normalized eigenvectors of A^*A and forming an orthonormal system. Then

$$V = (X_1, X_2, X_3) = \begin{bmatrix} \frac{1}{\sqrt{6}} & \frac{1}{\sqrt{2}} & \frac{1}{\sqrt{3}} \\ -\frac{1}{\sqrt{6}} & \frac{1}{\sqrt{2}} & -\frac{1}{\sqrt{3}} \\ -\frac{2}{\sqrt{6}} & 0 & \frac{1}{\sqrt{3}} \end{bmatrix}.$$

Consider AX_1, AX_2, AX_3:

$$AX_1 = \frac{1}{\sqrt{6}} \begin{pmatrix} 1 & 0 & -1 \\ 0 & 1 & 1 \end{pmatrix} \begin{pmatrix} 1 \\ -1 \\ -2 \end{pmatrix} = \frac{3}{\sqrt{6}} \begin{pmatrix} 1 \\ -1 \end{pmatrix},$$

$$AX_2 = \frac{1}{\sqrt{2}} \begin{pmatrix} 1 \\ 1 \end{pmatrix},$$

$$AX_3 = O.$$

Therefore

$$Y_1 = \frac{AX_1}{\|AX_1\|} = \frac{1}{\sqrt{2}} \begin{pmatrix} 1 \\ -1 \end{pmatrix}, \quad Y_2 = \frac{AX_2}{\|AX_2\|} = \frac{1}{\sqrt{2}} \begin{pmatrix} 1 \\ 1 \end{pmatrix}$$

and hence

$$U = \begin{pmatrix} \frac{1}{\sqrt{2}} & \frac{1}{\sqrt{2}} \\ -\frac{1}{\sqrt{2}} & \frac{1}{\sqrt{2}} \end{pmatrix}$$

and

$$A = UDV^*$$

$$= \begin{bmatrix} \frac{1}{\sqrt{2}} & \frac{1}{\sqrt{2}} \\ -\frac{1}{\sqrt{2}} & \frac{1}{\sqrt{2}} \end{bmatrix} \begin{bmatrix} \sqrt{3} & 0 & 0 \\ 0 & 1 & 0 \end{bmatrix} \begin{bmatrix} \frac{1}{\sqrt{6}} & -\frac{1}{\sqrt{6}} & -\frac{2}{\sqrt{6}} \\ \frac{1}{\sqrt{2}} & \frac{1}{\sqrt{2}} & 0 \\ \frac{1}{\sqrt{3}} & -\frac{1}{\sqrt{3}} & \frac{1}{\sqrt{3}} \end{bmatrix}.$$

Note that equation (6.3.1) provides a way of defining *unitary equivalence* of rectangular matrices. Two $m \times n$ matrices A and B are said to be *unitarily equivalent* if there exist unitary matrices U and V such that

$$A = UBV^* \qquad\qquad (6.3.4)$$

and the matrix D in (6.3.1) is called the *canonical form* of the rectangular matrix A. It is not difficult to establish the following result:

(viii) Two $m \times n$ matrices are unitarily equivalent if and only if they have the same singular values.

6.3.2 Canonical form of a bilinear form

One interesting application of (6.3.1) is the reduction of a bilinear form in real or complex vectors. Let X be an $m \times 1$ and Y be an $n \times 1$ vectors and A an $m \times n$ matrix. Consider the bilinear form

$$\alpha = X^* A Y \qquad\qquad (6.3.5)$$

where α is linear in X as well as in Y, $A = (a_{ij})$ is free of X and Y. Consider the singular value decomposition of A as given in (6.3.1). Then

$$\alpha = X^* U D V^* Y, \quad D = \begin{bmatrix} S & O \\ O & O \end{bmatrix}$$

where $S = \mathrm{diag}(s_1, s_2, \ldots, s_r)$, with $s_1, \ldots, s_r$ being the nonzero singular values of A. r indicating the rank of A. Consider the unitary transformations (orthogonal transformations when U and V are orthonormal)

$$X^* U = T^* = \begin{pmatrix} t_1^* \\ \vdots \\ t_m^* \end{pmatrix} \quad \text{and} \quad V^* Y = W = \begin{pmatrix} w_1 \\ \vdots \\ w_n \end{pmatrix}.$$

Then

$$\alpha = T^* D W = s_1 t_1^* w_1 + \cdots + s_r t_r^* w_r. \tag{6.3.6}$$

This form in (6.3.6) is the canonical form of the bilinear form α. Several applications of bilinear forms may be found in [8].

Exercises 6.3

6.3.1. Construct a 3×3 nonsymmetric matrix A with positive eigenvalues for which there exists a non-null 3×1 vector X such that (1) $X'AX = 0$, (2) $X'AX < 0$, thereby showing that definiteness of a matrix cannot be associated with nonsymmetric or non-Hermitian matrices.

6.3.2. For any rectangular matrix A show that AA^* and A^*A are nonnegative definite (positive definite or positive semi-definite), where A^* denotes the conjugate transpose of A.

6.3.3. If A is a positive definite matrix then show that there exists a unique lower triangular matrix T with positive diagonal elements such that $A = TT^*$. [This is known as the *Cholesky factorization of A*.]

6.3.4. Show that any $n \times n$ Hermitian matrix A is congruent to the matrix

$$D = \begin{bmatrix} I_r & O & O \\ O & -I_{s-r} & O \\ O & O & O_{n-s} \end{bmatrix}$$

where s is the number of nonzero eigenvalues and r is the number of positive eigenvalues of A.

6.3.5. Show that a Hermitian matrix A is positive definite if and only if it is congruent to the identity matrix.

6.3.6. Show that two $n \times n$ Hermitian matrices A and B are congruent if and only if $\text{rank}(A) = \text{rank}(B)$ and the number of positive eigenvalues of both matrices is the same.

6.3.7. Compute the singular values of the following matrices:

$$A = \begin{bmatrix} 1 & 0 & -1 \\ 1 & 1 & 2 \end{bmatrix}, \quad B = \begin{bmatrix} 1 & -1 & 1 \\ 1 & 1 & 1 \\ 1 & 0 & -1 \end{bmatrix}, \quad C = [1, 1, -1].$$

6.3.8. Show that the singular values of square matrices are invariant under unitary transformations.

6.3.9. Obtain the singular value decompositions of the matrices in Exercise 6.3.7.

6.3.10. Show that two $m \times n$ matrices A and B are unitarily equivalent if and only if the matrices A^*A and B^*B are similar.

6.3.11. For any $n \times n$ matrix A show that there exists a unitary matrix U and an upper triangular matrix T whose diagonal elements are the eigenvalues of A, such that $U^*AU = T$.

6.3.12. For $n \times n$ matrices A and B where A is positive definite and B is positive semi-definite show that there exists a nonsingular matrix Q such that

$$A = QQ' \quad \text{and} \quad B = QDQ'$$

where D is a diagonal matrix.

6.3.13. Let A be positive definite and B positive semi-definite, where $A + B$ is defined. Then show that $|A + B| \geq |A|$ and equality if and only if $B = O$.

6.3.14. Let A and B be positive definite matrices then show that $A - B$ is positive definite if and only if $B^{-1} - A^{-1}$ is positive definite.

6.3.15. If A and B are positive definite and $A - B$ is positive semi-definite then show that $|A| \geq |B|$ with equality if $A = B$.

6.3.16. If A is positive definite and $I - A$ is positive semi-definite with $|A| = 1$ then show that $A = I$.

6.3.17. If A is positive definite then show that $A + A^{-1} - 2I$ is positive semi-definite.

6.3.18. Show that $(I - AB)^{-1} = I + A(I - BA)^{-1}A$ whenever the inverses exist.

6.3.19. Show that

$$(\alpha I - A)^{-1} - (\beta I - A)^{-1} = (\beta - \alpha)(\beta I - A)^{-1}(\alpha I - A)^{-1}$$

whenever the inverses exist, where α and β are scalars.

6.3.20. Let A be a positive definite matrix and α a positive scalar then show that $|B| \leq \alpha|A|$ where

$$B = \begin{bmatrix} A & b \\ b' & \alpha \end{bmatrix}, \quad b \text{ a vector.}$$

References

[1] A. M. Mathai, An approximate method of analysis for a two-way layout, Biometrics 21 (1965), 376–385.
[2] A. M. Mathai, A Handbook of Generalized Special Functions for Statistical and Physical Sciences, Oxford University Press, Oxford, 1993.
[3] A. M. Mathai, Jacobians of Matrix Transformations and Functions of Matrix Argument, World Scientic Publishing, New York, 1997.
[4] A. M. Mathai, An Introduction to Geometrical Probability: Distributional Aspects with Applications, Gordon and Breach, Newark, 1999.
[5] A. M. Mathai and T. A. Davis, Constructing the sunflower head, Mathematical Biosciences 20 (1974), 117–133.
[6] A. M. Mathai and R. S. Katiyar, A new algorithm for nonlinear least squares, Researches in Mathematical Statistics 207 (10) (1993), 143–157 (in Russian). English translation by the American Mathematical Society.
[7] A. M. Mathai and S. B. Provost, Quadratic Forms in Random Variables: Theory and Applications, Marcel Dekker, New York, 1992.
[8] A. M. Mathai, S. B. Provost, and T. Hayakawa, Bilinear Forms and Zonal Polynomials, Lecture Notes in Statistics, Springer-Verlag, New York, 1995.
[9] C. R. Rao and M. B. Rao, Matrix Algebra and Its Applications to Statistics and Econometrics, World Scientic Publishing, New York, 1998.

Index